The Algebra of Geometry

Cartesian, Areal and Projective Co-ordinates

Christopher J Bradley

ISBN 978-1-906338-00-8

UKMT
School of Mathematics Satellite
University of Leeds
Leeds
LS2 9JT
United Kingdom
www.ukmt.org.uk

Cover design by Andrew R Evans.

Introduction

This is a book about points, lines, triangles and conics situated in a plane. The circle, being a special conic in the Euclidean plane, is given due prominence. In fact Chapter 6 is entirely devoted to the circle and to the properties of cyclic quadrilaterals. Before that I attempt to describe comprehensively the standard properties of triangles, quadrangles, hexagons, circles and conics. In the later chapters, the amount of material competing for possible inclusion is so immense that choices have had to be made.

I decided to give no systematic account of triangle centres, which is a modern and competitive industry, now involving over three thousand points. I introduce a limited number of such centres, but only as they occur naturally in topics under discussion. A second choice is not to discuss in any detail the underlying groups of transformations that distinguish Euclidean, affine and projective geometry. These groups are of course mentioned, but as they have been the centre of attention in so many books during the twentieth century, it seems to me that one can be sure that readers are now well acquainted with the theoretical foundations of the subject, or, if not, that they have an abundance of splendid material to which they can refer. A third choice is to include nothing about cubic curves or curves of even higher degree. Some of these curves do, however, make an appearance in the examples and exercises.

What I have chosen to include has been dictated to a large extent by two major considerations. These are, first and foremost, what the book is attempting to provide for readers. In short that is an account of the use of co-ordinates in Euclidean, affine and projective geometry, with the aim of enabling others to use such techniques efficiently. A second aim was to incorporate some of my main interests over the last fifteen years.

Since I started to study geometry I have not become aware of any books written in English in recent years specialising in co-ordinate geometry. Sometimes one gets the impression that mathematicians regard those who practise the arts of synthetic geometry with greater esteem than those who profess skill with co-ordinate methods. To be fair, methods of studying geometry have swung in and out of fashion. So I prefer the point of view expressed by Maxwell, who wrote in [25], 'In a good geometrical technique the methods of Pure Geometry and Analytical Geometry should go hand in hand, each helping the other forward, and weakness in either will be in danger of leading to weakness in both'. However, I do not think I need to defend my point of view. It is a fact that as far as I am aware, there are no good modern sources written with the specific intention of describing co-ordinate methods and this has been most persuasive in framing the decision of what I should write about. The book covers the use of Cartesian co-ordinates, areal co-ordinates and homogeneous projective co-ordinates and describes the main properties and applications, as appropriate, of each of these disciplines. A further decision had to be made. The temptation is always to write an introductory text, but elementary geometrical texts are in fact available that deal with both pure and analytic methods, such as the Open University Course by Brannan, Esplen and Gray [12]. So, quite deliberately, I have written a book that starts by quoting a large number of introductory results in its first two chapters, with the intention of leaving room in the text for more advanced applications, illustrating to greater advantage the power of co-ordinate methods. Some of these preliminary results I first heard of as a boy at Rossall School, Fleetwood, and I have been reminded of some of them by looking again at Carr's Synopsis [13].

However, the book is not exclusively about co-ordinates. It would be preposterous to neglect synthetic methods, particularly in dealing with topics such as circles. It would also seem absurd to exclude complex variable methods and polar reciprocation, if readers are to be given a reasonably comprehensive account of the properties of triangles, circles and conics. Chapters 14 and 15 respectively provide some account of these topics.

The secondary reasons dictating the choice of contents are concerned with what I have been doing these last fifteen years besides teaching at Clifton College, Bristol. Thus certain of the chapters in the book contain an account of my own work, particularly research in collaboration with Dr Geoff Smith of the University of Bath. An attempt has been made to write an alternative and more co-ordinate based account of this work and my thanks to him for his support and his patience with some of my more involved calculations. My other contribution has been in the training of UK teams for the International Mathematical Olympiad and it has been an immense privilege to be involved with so many brilliant students. Though co-ordinate methods are regarded as non-essential by the IMO authorities and indeed great effort is made in the composition of the examinations to ensure that geometrical problems are not easily treated by co-ordinate methods, good pupils are not to be denied knowledge that may be useful to them. Thus the emphasis in the text on problem solving derives from IMO activities. The text, in fact, includes dozens of worked examples and over eight hundred problems, many of them original. Some of them have been composed by Dr David Monk [28], formerly of the University of Edinburgh, with whom I have maintained an active correspondence over the years, and who has been a great influence. I thank him for allowing me to publish about twenty of his original problems and for his comments on the material of several chapters. About thirty problems in the book are taken from Wolstenholme's famous book of problems [37], first published in the middle of the 19th century. These are labelled in the text with the initials JW. One problem was devised by Dr Gerry Leversha of St Paul's School, London and two by Prof Ben Green of the University of Cambridge. These are labelled in the text by their initals GL and BG respectively.

The book contains a certain amount of original material, not previously published. I mention, in this context, part of Chapter 8 and the whole of Chapter 12, though one of the main theorems in Chapter 12 is due to Vin de Silva [16]. Until very recently Chapter 17 fell into this category and is perhaps the chapter that has given me the greatest pleasure to devise. This is partly because of the way it has enabled me to draw generalisations of Euclidean results in the projective plane. The book is illustrated with figures drawn using the geometrical software CABRI. But drawing diagrams in books is not the main use of geometrical software. Many of the recent developments in geometry have been made possible by the use of such software, with its highly accurate numerical background, allowing vast numbers of complicated constructions and numerical tests to be carried out in a single day. Such software allows conclusive checks to be carried out, for example, on whether three points are collinear or three lines are concurrent or parallel, and this greatly facilitates the framing of hypotheses. In fact geometry software is one of the main reasons for the revival of interest in the subject over the past ten years.

I confess that the book has not been written with any undergraduate university course in mind. In fact it is not written for a particular student audience. It is written for anyone interested in geometry, of whatever age and wherever they may be studying, teaching or working. I am delighted therefore to have the backing of the present publishers, and thank them for the care that has been taken with the final manuscript and the final product.

Christopher Bradley, Bristol, UK, May 2007.

Publisher's remarks

In the 19th century the algebra of the plane was common currency among mathematicians. A hundred years later the main fronts of mathematical research have moved elsewhere. However, the recent ready availability of computer graphics packages make it possible for a skilled enthusiast to discover new empirical truths about geometrical configurations with great rapidity and confidence. However, such conjectures must be proved, else we have no idea why they are true.

Complicated geometrical configurations are often not amenable to synthetic proofs (the combining of previously known results in a cunning way). Sometimes one must calculate, and there is nothing wrong with that. If you are going to calculate, then of course you must calculate well. Christopher Bradley's dexterity with the algebra of the plane is marvellous to behold.

$\mathtt{HP}^{\mathtt{n}}$, 2007.

Contents

Chapter 1

Rectangular Cartesian Co-ordinates

1.1 Introduction

This chapter is concerned with Euclidean geometry, in which the concepts of distance, angle, congruence and similarity play a major role. This is because Euclidean geometry is concerned with those properties that are invariant under the action of the Euclidean group $\mathbb{E}(2, \mathbb{R})$. The elements of this group are the isometries and similarities. The former are the distance preserving transformations: the rotations, reflections, translations and combinations of them (such as glide reflections) which must map any triangle to a congruent triangle. The latter are the shape preserving transformations which enlarge (or reduce) every line segment by a constant factor and which map any triangle to a similar triangle. Isometries are, of course, similarities, when the constant factor is 1.

A proper conic is one that does not degenerate into a pair of straight lines, or degenerate in some other way or is empty over the real field. Any proper conic can be transformed by an isometry into an ellipse with equation $\frac{x^2}{a^2} + \frac{y^2}{b^2} = 1$, a hyperbola with equation $\frac{x^2}{a^2} - \frac{y^2}{b^2} = 1$ or a parabola with equation $y^2 = 4ax$. An ellipse has an eccentricity e, where $0 < e < 1$; a circle has eccentricity $e = 0$; a parabola has eccentricity $e = 1$; a hyperbola has eccentricity $e > 1$. Here b is expressible as a simple function of a and e. This means that any conic belongs to one of these congruence classes and the constants a and e determine which congruence class. When similarities are introduced the factor $1/a$ can be used as the constant of the similarity, so that any two conics with the same value of e are similar. The detailed mathematical definition of the eccentricity of a conic is given in Section 1.5.3. It follows that all circles are similar, which is obvious, and that all parabolas are similar, which may be less obvious.

It is not our intention to give an account of the properties of the elements of the Euclidean group, as this is covered splendidly by Silvester [31]. We are concerned at first with the basic properties of Euclidean geometry, most of which are stated but not proved, and also with the solution of problems of varying degrees of difficulty. As the book progresses more advanced topics are treated, and more advanced problems are posed. We assume the reader is acquainted with the ideas of congruence and similarity, as well as having an elementary knowledge of Cartesian co-ordinate geometry.

1.2 Basic formulas

We assume knowledge of basic formulas that hold when using rectangular Cartesian co-ordinates. In what follows the points P_1, P_2 and P_3 have co-ordinates $P_1(x_1, y_1)$, $P_2(x_2, y_2)$ and $P_3(x_3, y_3)$. These points are generic, so they are arbitrary but, unless otherwise stated, they are not collinear (and so they are distinct). The point $O(0, 0)$ is the origin of co-ordinates. We state without proof the following results, where if a formula is not known it should be revised or treated as an exercise.

1.2.1 Distance

$$P_1P_2^2 = (x_2 - x_1)^2 + (y_2 - y_1)^2.$$

1.2.2 Section Theorem

The point P that divides P_1P_2 in the ratio $u : v$, so that $\frac{P_1P}{PP_2} = \frac{u}{v}$ has co-ordinates

$$\frac{P((ux_2 + vx_1), (uy_2 + vy_1))}{u + v}.$$

This is a useful shorthand for

$$P\left(\frac{ux_2 + vx_1}{u + v}, \frac{uy_2 + vy_1}{u + v}\right).$$

Here line segments are signed. When $u, v > 0$, P lies between P_1 and P_2. When $u = v$, then P is the midpoint of P_1P_2. When $u = 1, v = 0$, then P coincides with P_2 and when $u = 0, v = 1$, then P coincides with P_1. When $u < 0, v > 0$, then the ordering of the points on P_1P_2 is PP_1P_2 and when $u > 0, v < 0$ the ordering is P_1P_2P. It is always possible to insist that $u + v = 1$ and then (u, v) may be thought of as the normalized line co-ordinates of points on P_1P_2.

1.2.3 Equation of a line

The co-ordinates (x, y) of all points $P(x, y)$ on the line P_1P_2 satisfy

$$(x_2 - x_1)y - (y_2 - y_1)x = (x_2y_1 - x_1y_2).$$

1.2.4 Gradient and point of intersection

The gradient or slope of the line P_1P_2 is $m = \frac{y_2-y_1}{x_2-x_1}$ provided that $x_1 \neq x_2$. Then the equation of the line P_1P_2 may be written in the form $(y - y_1) = m(x - x_1)$ or $y = mx + c$, where c is the y-intercept (the value of y when $x = 0$). When $x_1 = x_2$, then the line P_1P_2 is parallel to the y-axis and has equation $x = x_1$. The point of intersection of two lines $L_k : y - (m_kx + c_k) = 0,\ k = 1, 2 (m_1 \neq m_2)$ is given by $x = (c_1 - c_2)/(m_2 - m_1)$, $y = (c_1m_2 - c_2m_1)/(m_2 - m_1)$. For all values of s, t the line with equation $sL_1 + tL_2 = 0$ passes through this point.

1.2.5 Condition for a point to lie on a line

P_3 lies on P_1P_2 if, and only if,

$$(x_1y_2 + x_2y_3 + x_3y_1) - (x_1y_3 + x_2y_1 + x_3y_2) = 0.$$

1.2.6 Perpendicular distance on to a line

All lines may be written in the form $ax + by = c$ for some constants a, b and c. The perpendicular distance p from O to this line is given by $p = |c|/\sqrt{a^2 + b^2}$. The perpendicular distance of P_3 on to this line is given by $|c - ax_3 - by_3|/\sqrt{a^2 + b^2}$.

1.2.7 Parallel lines

Distinct lines $a_1x + b_1y = c_1$ and $a_2x + b_2y = c_2$ meet at the point with co-ordinates (x, y) found by solving the two equations as a pair of simultaneous equations. Since such a solution exists if, and only if, $a_1b_2 \neq a_2b_1$, the two lines are parallel if, and only if, $a_1 : b_1 = a_2 : b_2$ (even if the two lines are the same). When $a_1 = a_2 = a$ and $b_1 = b_2 = b$ then the distance between the two lines is $|c_1 - c_2|/(a^2 + b^2)$.

1.2.8 Perpendicular lines

The two lines with equations $a_1x+b_1y = c_1$ and $a_2x+b_2y = c_2$ are perpendicular if, and only if, $a_1a_2 + b_1b_2 = 0$. Trivially, the lines with equations $y = c$ and $x = d$ are at right angles. Also the lines with equations $y = m_1x+c_1$ and $y = m_2x + c_2$ are perpendicular if, and only if, $m_1m_2 = -1$. It follows, for example, that the line perpendicular to the line with equation $3x + 4y = 5$ passing through the point with co-ordinates $(3, 2)$ is $4x - 3y = 6$. The prescription 'exchange the coefficients of x and y and change the sign of one of them' is extremely useful for remembering how to write down the equation of a line perpendicular to a given line.

1.2.9 Area

The area of triangle OP_1P_2 is given by

$$[OP_1P_2] = \frac{1}{2}|x_1y_2 - x_2y_1|.$$

The area of the triangle $P_1P_2P_3$ is given by

$$[P_1P_2P_3] = \frac{1}{2}|x_1y_2 + x_2y_3 + x_3y_1 - x_1y_3 - x_2y_1 - x_3y_2|.$$

These formulas may be expressed neatly as the moduli of 2×2 and 3×3 determinants respectively. The area of the triangle formed by the three lines with equations $y = m_kx + c_k$, $k = 1, 2, 3$ is given by

$$\Delta = \frac{1}{2}\left(\frac{(c_1 - c_2)^2}{m_1 - m_2} + \frac{(c_2 - c_3)^2}{m_2 - m_3} + \frac{(c_3 - c_1)^2}{m_3 - m_1}\right).$$

1.2.10 Angles

The angle θ between the line $y = mx + c$ and the x-axis is given by $\tan\theta = m$. The angle ϕ between the lines with equations $y = m_1x + c_1$ and $y = m_2x + c_2$ is given by $\tan\phi = \frac{m_2 - m_1}{1 + m_1m_2}$, $(m_1m_2 \neq -1)$ and is $90°$ when $m_1m_2 = -1$. The angle ψ between the lines with equations $a_1x + b_1y = c_1$ and $a_2x + b_2y = c_2$ is given by $\cos\psi = \frac{a_1a_2 + b_1b_2}{\sqrt{a_1^2 + b_1^2}\sqrt{a_2^2 + b_2^2}}$. The angle χ between the vectors $\mathbf{OP_1}$ and $\mathbf{OP_2}$ is given by

$$\cos\chi = \frac{x_1x_2 + y_1y_2}{\sqrt{x_1^2 + y_1^2}\sqrt{x_2^2 + y_2^2}}.$$

1.2.11 Change of origin and rotation of axes

To change the origin to the point (h, k) use the change of variables $x' = x + h$, $y' = y + k$. To transfer to rectangular axes inclined at an angle θ to the original axes use the change of variables $x' = x\cos\theta - y\sin\theta$, $y' = x\sin\theta + y\cos\theta$. When the reflection $x' = -x$, $y' = y$ is added to these, they generate the group of isometries mentioned in Section 1.1.

1.2.12 Two straight lines through the origin

The equation $ax^2 + 2hxy + by^2 = 0$ represents two straight lines through the origin, provided that $h^2 > ab$. If $h^2 = ab$, then the two lines are coincident and if $h^2 < ab$, then they are imaginary, and since we deal only with the real plane, they do not exist. The angle ϕ between these lines is given by $\tan\phi = \frac{\sqrt{h^2 - ab}}{a + b}$, so they are perpendicular if $a + b = 0$. The equation of the angle bisectors of these two lines is $hx^2 - (a - b)xy - hy^2 = 0$.

Example 1.2.1

Consider two fixed rays meeting at an acute angle at O, with a point X lying between them, as in Fig. 1.1. There are two problems:

1. Find the line AXB such that the area $[OAB]$ of triangle OAB is minimized;
2. Find the line CXD such that the length of CXD is minimized.

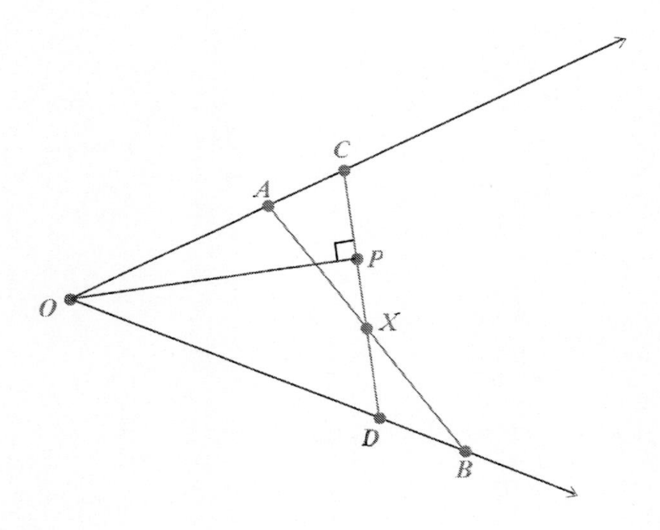

Figure 1.1: Example 1.2.1

Suppose the lower of the two lines has equation $y = 0$, with O the origin. Let the upper line have equation $y = mx$, $m > 0$. Note that it is always open to us, in any problem, to choose a suitable origin and a suitable direction for the axes. The success of a solution using rectangular Cartesian co-ordinates very often depends on making good choices. It is also open to us to impose a scale on a given configuration, if one is not already supplied by the problem itself. So here we may suppose that X has co-ordinates $(1, n)$, where, since X lies between the two lines, we have $0 < n < m$.

We choose a point $K(k, 0)$ on the lower line, where k is a variable whose value we adjust to solve each of the two problems. First we find the equation of KX and determine the point J where this line meets the upper line.

The equation of KX is $y/n = (x - k)/(1 - k)$, that is $nx + (k - 1)y = nk$. This meets the line with equation $y = mx$ where $nx + mx(k - 1) = nk$, so J has co-ordinates

$$J(nk/(n - m + mk), mnk/(n - m + mk)).$$

For the solution of Problem 1, since the angle between the two lines is fixed, we wish to choose k to minimize $(OK)(OJ)$, for which it is sufficient to choose k to minimize $f(k) = k^2/(n - m + mk)$. $df/dk = 0$ gives $k = 2(m - n)/m$. Geometrical considerations show this to be a minimum. It follows that the x-co-ordinates of A and B are $2n/m$ and $2(m - n)/m$ respectively. The midpoint of AB has x-co-ordinate $\frac{1}{2}(2n/m + 2(m - n)/m) = 1$, which is the x-co-ordinate of X. Problem 1 is therefore solved by choosing X to be the midpoint of AB.

Problem 2 is more difficult, so, to understand the plan of campaign, we first state the solution. CXD is minimized in length when the point P, which is the foot of the perpendicular from O on to CXD is such that $CP = XD$. In fact we cannot find the value of k explicitly, but we show that it satisfies a cubic equation, which is also satisfied when $CP = XD$.

Using the co-ordinates of J and K from above we find

$$JK^2 = \frac{m^2k^2((1-k)^2 + n^2)}{(n - m + mk)^2}.$$

Differentiating and putting the result equal to zero, we find that k is a solution of the cubic equation

$$mk^3 - (3m - 2n)k^2 + 3(m - n)k - (m - n)(1 + n^2) = 0. \qquad (1.1)$$

The one positive real root of this equation provides the solution to the problem.

Now the perpendicular from O onto JK has equation $(1 - k)x + ny = 0$, and this meets JK at the point P whose x-co-ordinate is $\frac{n^2k}{(1-k)^2+n^2}$. The condition $CP = XD$ implies $nk/(n-m+mk) - n^2k/((1-k)^2+n^2) = (1-k)$, which on simplification leads to the same cubic Equation (1.1). The line CXD is called Philo's line.

Exercise 1.2.2

1. Points A, B, C, O, H have co-ordinates

$$A(0,0), B(2,2v), C(2,2w), H(2(1+vw),0), O(1-vw, v+w)$$

respectively, where $1 + vw \neq 0$. Prove that $OA = OB = OC$ and that $AH \perp BC, BH \perp CA, CH \perp AB$. The line through H parallel to CA meets BA at X and the line through H parallel to BA meets CA at Y. Prove that $OX = OY$. (Note: This way of assigning co-ordinates to the vertices, to the circumcentre O and to the orthocentre H of a triangle, can be useful, particularly when A is a special point in the configuration. If $1 + vw = 0$, then $\angle BAC = 90°$ and H coincides with A.)

2. Let ABC be a triangle with $\angle BAC = 90°$. Let D be the foot of the perpendicular from A to BC and M the midpoint of DC. Let AB be extended to E so that $AB = BE$. Prove that ED is perpendicular to AM.

3. The point P lies inside triangle ABC. BP is extended to a point Q such that AQ is parallel to PC. A line through Q parallel to CB meets AC at R. Prove that RP is parallel to AB. *Hint: Choose B to be the origin and C to be the point with co-ordinates* $(c, 0)$.

4. $ABCD$ is a convex quadrilateral. Prove that $ABCD$ is a trapezium with parallel sides AB and CD if, and only if, $BC^2 + AD^2 = BD^2 + AC^2 - 2(CD)(AB)$.

5. A point P, not on either diagonal, is taken within a rectangle $ABCD$. Prove that the value of $(\cot \angle APC)[APC] - (\cot \angle BPD)[BPD]$ is independent of the position of P.

6. Through the midpoint of side BC of triangle ABC a line is drawn parallel to the internal bisector of $\angle BAC$, meeting the lines AB and AC at X and Y respectively. The line XY is extended to Z so that $XY = YZ$. BY and CZ meet at D. Prove that the internal bisector of $\angle BDC$ is parallel to XY. *Hint: Take A as origin and the internal and external bisectors of $\angle BAC$ to be the x- and y-axes. The following notation may prove useful: $b = AC$, $c = AB$, $\theta = \frac{1}{2}\angle BAC$.*

7. Using the same notation as in Exercise 1.2.2.1, let G be the point with co-ordinates $G(4/3, (2/3)(v+w))$. Prove that O, G, H are collinear and that $OG = (1/3)OH$. (Note: The line OGH is called the *Euler line* of triangle ABC.)

8. Using the same notation as in Exercises 1.2.2.1 and 1.2.2.7, let T be the midpoint of OH. Show that T is equidistant from the midpoints of the sides BC, CA, AB, from the feet of the altitudes and from the

midpoints of AH, BH, CH. (Note: T is the centre of the *nine-point circle*.)

9. Let ABC be an equilateral triangle and P be a variable point in the plane of ABC. Find the locus of P such that $BP^2 + CP^2 = 2AP^2$. More generally, find the locus of P if $pBP^2 + qCP^2 + rAP^2 = 0$

   (i) when $p + q + r = 0$ and

   (ii) when $pq + qr + rp = 0$.

1.3 Formulas for circles

In this section either we give an outline derivation of a formula or we state it without proof.

1.3.1 Equation of a circle

The equation of a circle centre P_1 and radius r is

$$(x - x_1)^2 + (y - y_1)^2 = r^2. \tag{1.2}$$

Equation (1.2) expresses the fact that the distance of $P(x, y)$ from $P_1(x_1, y_1)$ is equal to the radius r of the circle, which is the usual defining property of a circle.

It follows immediately that the circle S with equation

$$x^2 + y^2 + 2gx + 2fy + c = 0, \ \ c < g^2 + f^2, \tag{1.3}$$

has centre $C(-g, -f)$ and radius $r = \sqrt{g^2 + f^2 - c}$. The equation of the circle passing through non-collinear points (x_k, y_k, z_k), $k = 2, 3, 4$ is the 4×4 determinant with first row $(x^2 + y^2, x, y, 1)$ and with second, third and fourth rows $(x_k^2 + y_k^2, x_k, y_k, 1), k = 2, 3, 4$ respectively.

1.3.2 Angle in a semi-circle is a right angle

Consider the circle with equation

$$(x - x_1)(x - x_2) + (y - y_1)(y - y_2) = 0. \tag{1.4}$$

Clearly it passes through P_1 and P_2 and from Section 1.3.1 its centre is at the point with co-ordinates $(\frac{1}{2}(x_1 + x_2), \frac{1}{2}(y_1 + y_2))$, which is the midpoint of the line P_1P_2. It is therefore the equation of the circle having P_1P_2 as diameter. From Equation (1.4), it follows that if $P(x, y)$ lies on the circle, then the lines P_1P and P_2P are at right angles.

1.3.3 Tangents

The equation of the tangent at P_1 to the circle with Equation (1.3) is

$$x_1x + y_1y + g(x + x_1) + f(y + y_1) + c = 0,$$

where (x_1, y_1) satisfies Equation (1.3). This may be proved using calculus, or alternatively by using the property that the tangent is at right angles to the radius vector. The equation of the tangent at $(r\cos\theta, r\sin\theta)$ to the circle S_0 with equation $x^2 + y^2 = r^2$ is $x\cos\theta + y\sin\theta = r$.

1.3.4 Power of a point

The *power* of the point P_1 with respect to the circle S with Equation (1.3) is given by $x_1^2 + y_1^2 + 2gx_1 + 2fy_1 + c$. It is easy to show that this is equal to $P_1C^2 - r^2$, where C is the centre of S and r is its radius. If P_1 is external to the circle then, by Pythagoras's theorem, the power of P_1 is the square of the length of the tangent from P_1 to S, if P_1 lies on S, then its power is zero, and if P_1 lies internal to S, then its power is negative. If S is replaced by S_0, then the power of P_1 is $x_1^2 + y_1^2 - r^2$.

1.3.5 Parameters

For the circle S_0 there are two commonly used sets of parameters. These are

1. $x = r\cos\theta, y = r\sin\theta$, $(0 \le \theta < 2\pi$ or $-\pi < \theta \le \pi)$; and

2. $$x = \frac{r(1-t^2)}{1+t^2}, y = \frac{2rt}{1+t^2}, (-\infty < t \le \infty).$$

These are related by the equation $t = \tan\frac{1}{2}\theta$. ($t = \infty$ is taken to refer to the point with co-ordinates $(-1, 0)$.) The corresponding equations of tangents are $x\cos\theta + y\sin\theta = r$ and $(1-t^2)x + 2ty = (1+t^2)r$. The equations of chords joining 'θ' and ϕ and 't' and 's' in the two systems respectively are

$$x\cos\frac{\theta+\phi}{2} + y\sin\frac{\theta+\phi}{2} = r\cos\frac{\theta-\phi}{2} \tag{1.5}$$

and

$$(1-st)x + (s+t)y = (1+st)r. \tag{1.6}$$

The complicated form of Equation (1.5) makes the use of the second parameter system preferable in most problems dealing with tangents and chords.

1.3.6 Orthogonal circles

The condition for two circles S_1 and S_2 with equations

$$S_1: \quad x^2 + y^2 + 2g_1x + 2f_1y + c_1 = 0, \tag{1.7}$$

$$S_2: \quad x^2 + y^2 + 2g_2x + 2f_2y + c_2 = 0, \tag{1.8}$$

to be *orthogonal* (that is, their tangents at both points of intersection are at right angles) is that the distance between their centres is equal to $\sqrt{r_1^2 + r_2^2}$, where r_1, r_2 are the radii of the two circles. This condition simplifies algebraically to

$$2(g_1g_2 + f_1f_2) = (c_1 + c_2). \tag{1.9}$$

By solving three such equations it is possible to find the equation of a circle that is orthogonal to three given circles. More generally the two circles with equations $(x-a)^2 + (y-b)^2 = R^2$ and $(x-h)^2 + (y-k)^2 = r^2$ meet at an angle θ given by

$$(R^2 - 2Rr\cos\theta + r^2) = (h-a)^2 + (k-b)^2.$$

1.3.7 Conjugate points and inverse points

P_1 and P_2 are said to be *conjugate* with respect to the circle S_0 if, and only if,

$$P_1P_2^2 = OP_1^2 + OP_2^2 - 2r^2. \tag{1.10}$$

This condition simplifies to

$$x_1x_2 + y_1y_2 = r^2. \tag{1.11}$$

For the circle S the condition is

$$x_1x_2 + y_1y_2 + g(x_1 + x_2) + f(y_1 + y_2) + c = 0. \tag{1.12}$$

When O, P_1, P_2 are collinear Equation (1.10) becomes $(OP_1)(OP_2) = r^2$ and P_1, P_2 are said to be *inverse points.*

1.3.8 Polar

The locus of points conjugate to P_1 with respect to S_0 has equation

$$x_1x + y_1y = r^2. \tag{1.13}$$

This line is called the *polar* of P_1 with respect to S_0. Likewise the polar of P_1 with respect to S is

$$x_1x + y_1y + g(x + x_1) + f(y + y_1) + c = 0. \tag{1.14}$$

P_1 is called the *pole* of these lines. It is easy to check that the line from the centre of the circle to P_1 is perpendicular to the polar of P_1. It is also clear from the symmetry of Equations (1.11) and (1.12) that a fundamental theorem holds, namely that if P_2 lies on the polar of P_1, then P_1 lies on the polar of P_2. Observe that Equations (1.13) and (1.14) become the equations of the tangents when P_1 lies on the circle concerned. When P_1 is external to the circle, its polar is the chord of contact of the tangents from P_1 to the circle. When P_1 is internal to the circle, its polar consists of those points whose polars pass through P_1.

1.3.9 Radical axis

The *radical axis* of the two circles S_1 and S_2 given by Equations (1.7) and (1.8) is the locus of points P whose powers with respect to S_1 and S_2 are equal. From Section 1.3.4 it follows that the radical axis is a line with equation

$$2(g_1 - g_2)x + 2(f_1 - f_2)y + (c_1 - c_2) = 0.$$

It is perpendicular to the line of centres of the two circles and when S_1 and S_2 intersect it is the common chord and its extension. The radical axes of three circles, taken in pairs and whose centres are not collinear, meet in a point called their *radical centre.*

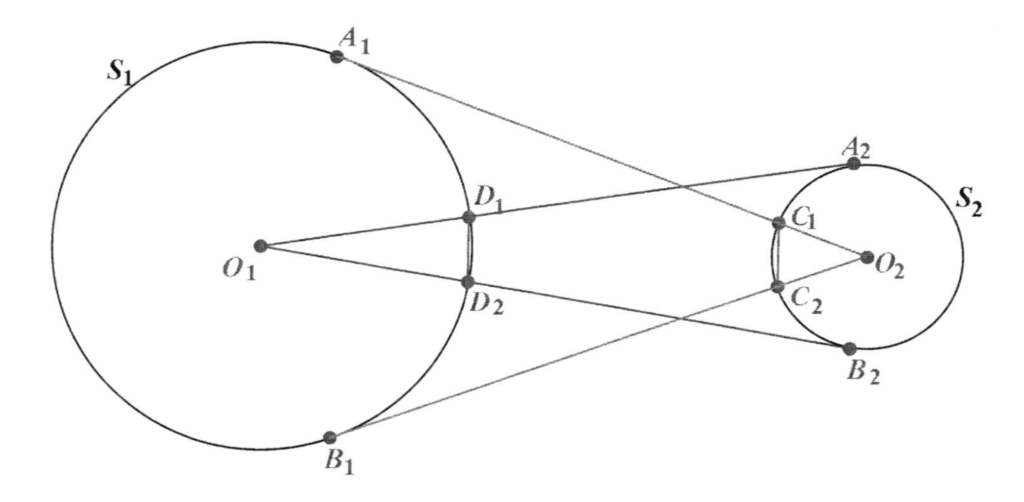

Figure 1.2: The eyeball theorem

Example 1.3.1

Let S_1, S_2 be two circles centres O_1, O_2 and radii r, s respectively. Suppose the tangents from O_2 to S_1 meet S_1 at A_1 and B_1 and S_2 at C_1 and C_2. Suppose the tangents from O_1 to S_2 meet S_2 at A_2 and B_2 and meet S_1 at D_1 and D_2. Then $C_1C_2 = D_1D_2$. See Fig. 1.2.

Let S_1 have equation $(x-d)^2+y^2 = r^2$ and S_2 have equation $(x-c)^2+y^2 = s^2$, where the centres are at $(d, 0)$ and $(c, 0)$ respectively, and without loss of generality we may take $c > d$.

For tangents to exist the distance between the centres must be greater than the radius of either circle, that is $(c - d) > r, s$. The plan we adopt is to calculate C_1C_2, and noting that the expression for this is symmetrical in r and s, we deduce that $C_1C_2 = D_1D_2$.

The tangent to S_1 at $A_1(x_1, y_1)$ is $xx_1 + yy_1 - d(x + x_1) + d^2 - r^2 = 0$. If this passes through $O_2(c, 0)$, then $cx_1 - d(c + x_1) + d^2 - r^2 = 0$ and $x_1 = (r^2 - d^2 + cd)/(c - d)$. Since (x_1, y_1) lies on S_1 this means

$$y_1^2 = \frac{r^2((c-d)^2 - r^2)}{(c-d)^2}.$$

The distance formula now gives $A_1O_2^2 = (c - d)^2 - r^2$. It follows that $O_2C_1/A_1O_2 = s/\sqrt{((c-d)^2 - r^2)}$ and hence, by similar triangles, $C_1C_2 = 2rs/(c - d) = D_1D_2$.

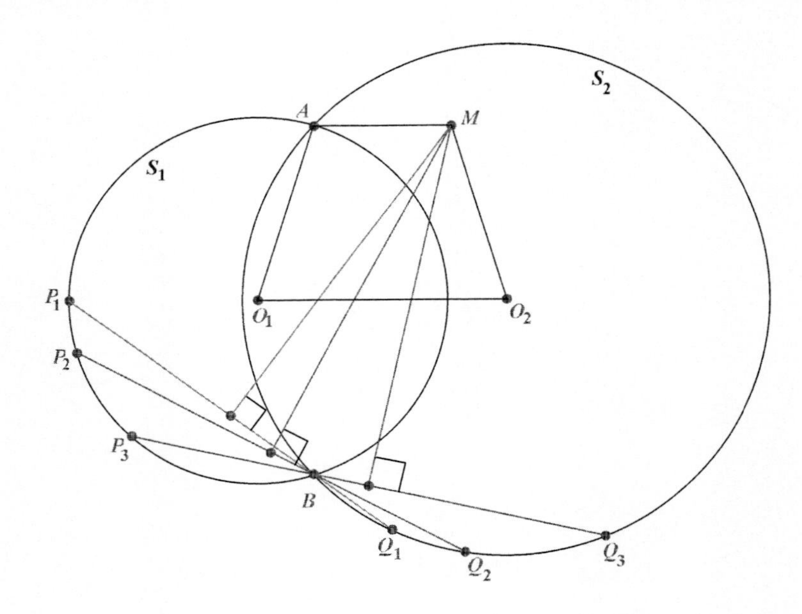

Figure 1.3: Example 1.3.2

Example 1.3.2

Another striking result, involving the relationship between two intersecting circles, appeared in the 36th Spanish Mathematical Olympiad and is as follows:

Let S_1 and S_2 be two circles of different radii, centres O_1 and O_2 respectively, intersecting at A and B. Let PQ be a line through B meeting S_1 at P and S_2 at Q. Then the perpendicular bisector of PQ passes through a fixed point M. Furthermore AMO_2O_1 is an isosceles trapezium. See Fig. 1.3. (Note that when the circles have equal radii it is trivial to prove that M coincides with A.)

Let the equations of S_1 and S_2 be $x^2 + y^2 + 2ax = c^2$ and $x^2 + y^2 - 2bx = c^2$, so that $A(0, c)$ and $B(0, -c)$. The centres lie on the x-axis and have co-ordinates $O_1(-a, 0)$ and $O_2(b, 0)$, where we suppose, without loss of generality, that $b > a$ and the common chord has equation $x = 0$. We show that M has co-ordinates $(b-a, c)$ so that AMO_2O_1 is an isosceles trapezium.

The equation of any line through B has the form $y = mx - c$ (where m is a variable) and then it is easily calculated that the x-co-ordinates of P and

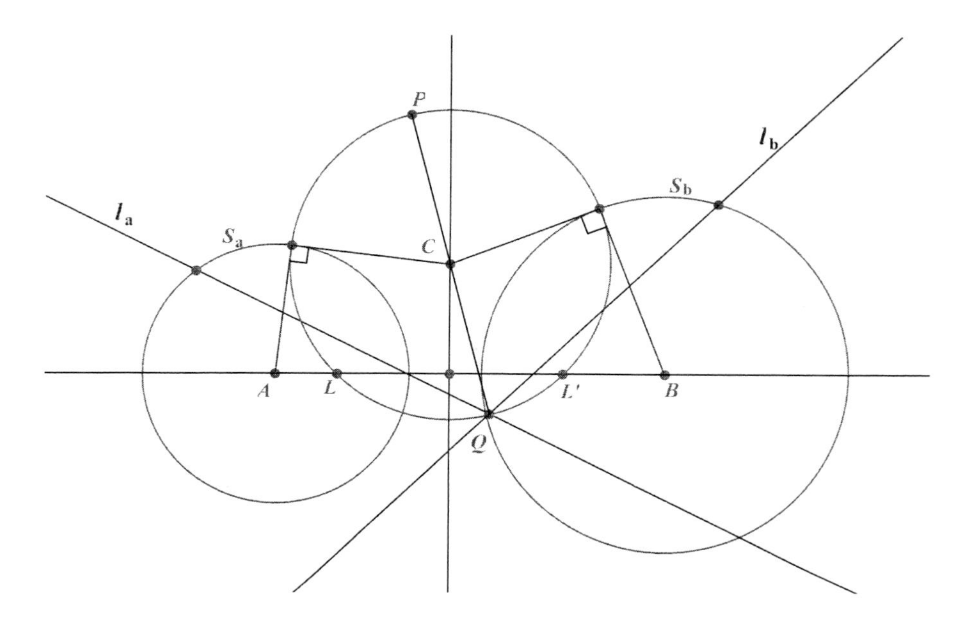

Figure 1.4: Example 1.3.3

Q are respectively

$$x_P = \frac{2(cm-a)}{1+m^2} \text{ and } x_Q = \frac{2(cm+b)}{1+m^2}.$$

The line perpendicular to PQ through $(b-a,c)$ has equation $my+x = mc+(b-a)$ and this meets PQ, with equation $y = mx-c$, at a point whose x-co-ordinate is $\frac{2cm+b-a}{1+m^2}$. As this is equal to $\frac{1}{2}(x_P+x_Q)$ we conclude that the perpendicular bisector passes through $M(b-a,c)$.

Example 1.3.3

Two non-intersecting circles S_a and S_b have centres A and B respectively. P is a variable point. The polars l_a, l_b of P with respect to the two circles meet at Q. Prove that the circle on PQ as diameter passes through two fixed points L and L' and is orthogonal to S_a and S_b. This result is illustrated in Fig. 1.4.

Let the equations of S_a and S_b respectively be

$$x^2+y^2-2ax+1=0$$

and

$$x^2 + y^2 - 2bx + 1 = 0.$$

These are non-intersecting circles with radii $\sqrt{a^2 - 1}$ and $\sqrt{b^2 - 1}$ so we require $|a|, |b| > 1$, $a \neq b$. In the figure $a < -1$ and $b > 1$, but the result still holds if both a and b are positive (or negative) and one circle encloses the other. Let the variable point P have co-ordinates $P(x_0, y_0)$. The equation of the polar of P with respect to S_a is the line l_a with equation $x_0x + y_0y - a(x + x_0) + 1 = 0$ and the equation of the polar of P with respect to S_b is the line l_b with equation $x_0x + y_0y - b(x + x_0) + 1 = 0$. These lines meet at Q with co-ordinates $Q(-x_0, (x_0^2 - 1)/y_0)$. The equation of the circle on PQ as diameter is

$$(x - x_0)(x + x_0) + (y - y_0)(y - (x_0^2 - 1)/y_0) = 0,$$

which reduces to

$$x^2 + y^2 - (y/y_0)(x_0^2 + y_0^2 - 1) - 1 = 0.$$

This always passes through the fixed points $L(-1, 0)$ and $L'(1, 0)$ and from Equation (1.9) is orthogonal to S_a and S_b.

This example illustrates theory that we have not so far mentioned. Any two non-intersecting circles are members of a coaxal system, and their equations are chosen above to be in a canonical form $x^2 + y^2 + 2gx + 1 = 0$, $|g| \geq 1$, in which the limiting points L and L' (which are degenerate members of the system of zero radius, corresponding to $|g| = 1$) have co-ordinates $(-1, 0)$ and $(1, 0)$. The circles on PQ as diameter turn out to be members of the complementary coaxal system of circles, which are intersecting circles, all of which pass through L and L'. If you take any member of a non-intersecting coaxal system and any member of the complementary intersecting coaxal system, then it is a general property of coaxal systems that these circles are orthogonal. See Fig. 1.5. There is a general theorem that states that the polars of a given point P with respect to the members of a coaxal system are concurrent. In the example the point of concurrence is the point Q. Note that its co-ordinates do not depend on a or b, which is sufficient proof of the general theorem. The midpoint C of PQ lies on the common radical axis of the non-intersecting system, of which S_a and S_b are two examples. In the above example this radical axis is the line with equation $x = 0$. The circle, centre C, passing through P and Q is, as proved above, orthogonal to S_a

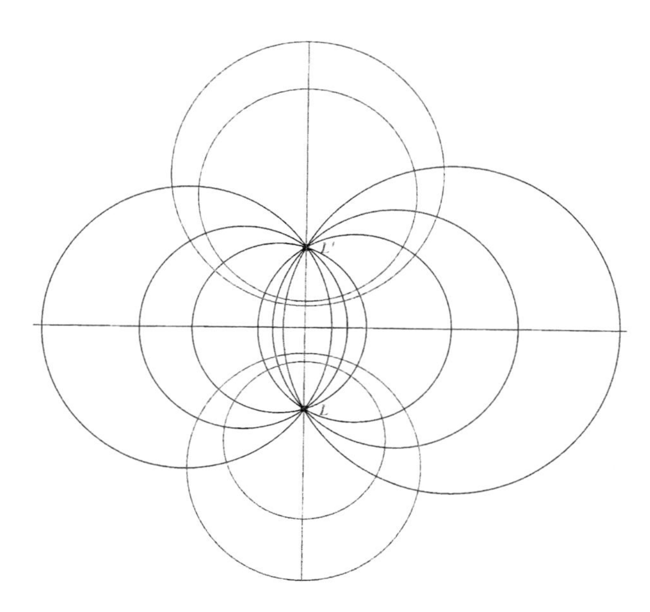

Figure 1.5: Systems of coaxal circles

and S_b and is therefore a member of the complementary coaxal system, and hence passes through the fixed points L and L'. The above solution contains implicitly all these general results, which makes it a particularly instructive example.

Exercise 1.3.4

1. The diameter AB of a circle is extended to C so that $BC = AB$. The tangent at A and a line parallel to it through C are drawn and any point P being taken on the latter, the two tangents from P are drawn, forming a triangle with the tangent at A. Prove that, as P varies, this triangle has a fixed centroid.

2. AB is a diameter of a circle and a chord through A meets the tangent at B at a point P. On the extension of AP a point Q is chosen and QR, QS are the tangents from Q to the circle. The extensions of AR, AS meet the tangent at B in X, Y respectively. Prove that $PX = PY$.

3. A tangent is drawn at a fixed point A on a circle. P is a variable point on a line parallel to the tangent, on the other side of it from the circle.

The tangents from P to the circle meet the tangent at A at Q and R. Prove that $(QA)(AR)$ is constant.

4. The tangents at A and B to circle ABC meet at P. Prove that the perpendicular bisector of BC and the line through P parallel to BC meet at a point on AC. *Hint: Use parameters and choose BC to be parallel to the x-axis.*

5. C and C^* are circles in a plane, which meet at two distinct points. Consider points P with the following property: $(*)$ If T is a point of contact of a tangent from P to C and if circle S_P with centre P and radius PT meets C^* at Q, Q^*, then QQ^* passes through P. Show that points P with property $(*)$ exist and that they all lie on a circle K. *Hint: Take C and C^* to have equations $x^2 + y^2 - 2ax - 1 = 0$ and $x^2 + y^2 - 2bx - 1 = 0$ respectively.*

6. P, Q are two points subtending an angle of $90°$ at a point T on a circle. Prove that if the tangent at T bisects PQ, then P, Q are conjugate points with respect to the circle.

7. A point O lies outside a fixed circle C_1. P_1 is a variable point on C_1. P_2 is a point on OP_1 such that OP_2/OP_1 is constant. Prove that the locus of P_2 is a circle.

8. K is a fixed point inside a circle. Find the locus of the midpoints of chords PQ such that $\angle PKQ = 90°$.

9. Four points A, P, Q, B lie in that order in a straight line, and the distances AQ, PB and AB are $2a$, $2b$ and $2c$ respectively. Circles are drawn with diameters AQ, PB and AB. Prove that the radius of the circle that touches the three circles is

$$\frac{c(c-a)(c-b)}{c^2 - ab}.$$

10. In a circle AB is a fixed chord, which is not a diameter, and PQ is a variable diameter. Find the locus of R, the intersection of AP and BQ.

11. Circles C_k, $k = 1$ to n, are given with equations $x^2 + y^2 + 2g_kx + 2f_ky = 0$. Find the condition that the n circles should touch at the origin.

12. ABC is a triangle in which $\angle BCA = 90°$. M is the midpoint of AB. MC is extended to N so that C is the midpoint of MN. K is any point on the circle S, centre N radius NC. Points D, E are chosen so that the figures $KCAD, KCBE$ are both parallelograms. Prove that DEC is a triangle in which $\angle DCE = 90°$. Describe the locus of D as K moves round the circle S.

13. ABC is a triangle right-angled at C. X lies on the extension of BC and Y lies on the extension of AC and XY is not parallel to AB. Circles ACX, BCY meet again at K. L, M are the centres of these circles. The line through K parallel to LM meets AB at Z. Prove that X, Y, Z are collinear.

14. Three circles of equal radius have a point in common. Prove that the other three points of intersection lie on a circle of the same radius.

15. PQ is a variable diameter of a fixed circle. A is any fixed point. AP, AQ cut the circle again at R, S. Prove that RS passes through a fixed point.

16. Let P, Q be conjugate points with respect to a circle S. Prove that the circle on PQ as diameter is orthogonal to S.

17. A variable chord PQ of a fixed circle passes through a fixed point. Prove that the tangents at P, Q intersect on a fixed line.

18. Let S_1 and S_2 be two orthogonal circles. Prove that the extremities of any diameter of S_1 are conjugate with respect to S_2.

19. Prove that the locus of a point, whose polars with respect to three given circles in general position are concurrent, is the circle orthogonal to each of the given circles.

1.4 The cyclic quadrangle

In this section we review the basic properties of an important configuration, that of the cyclic quadrangle. A later chapter gives a more thorough account of the properties of quadrangles. The configuration is shown in Fig. 1.6. $ABCD$ is a quadrangle inscribed in a circle centre O, in which neither pair of opposite sides is parallel. The diagonals AC and BD meet at E, the

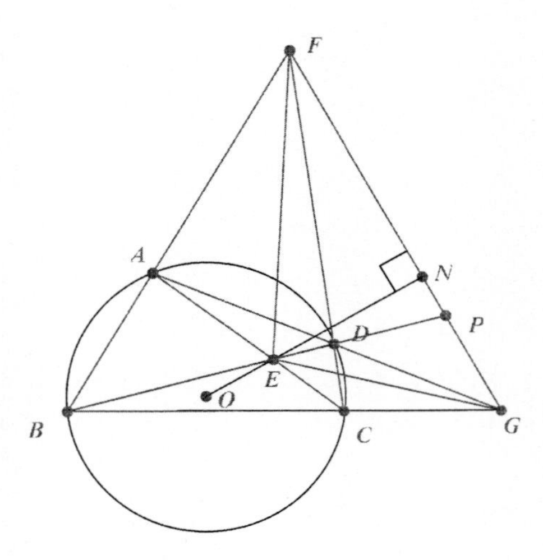

Figure 1.6: A cyclic quadrangle

diagonals BA and CD meet at F and the diagonals AD and BC meet at G. The triangle EFG is called the *diagonal point triangle*. In this configuration the following properties hold:

1.4.1 Self-conjugate triangle

EFG is a self-conjugate triangle with respect to circle $ABCD$. That is to say, FG is the polar of E, GE is the polar of F and EF is the polar of G.

1.4.2 Its orthocentre

The orthocentre of triangle EFG (the point where the altitudes of triangle EFG meet) is O, the centre of circle $ABCD$. This result 1.4.2 is trivial once the result 1.4.1 has been proved, since if l is the polar of a point P with respect to a circle centre O, then OP is perpendicular to l. Hence, for a self-conjugate triangle EFG, OE is perpendicular to FG, OF is perpendicular to GE and OG is perpendicular to EF.

To prove result 1.4.1, take the circle to have centre the origin and radius 1 and use the parameter 't' for points on the circle. Let A, B, C, D have parameters $t = 0, u, v, w$ respectively. See Section 1.3.5. The equation of AC is $x + vy = 1$ and that of BD is $(1 - wu)x + (w + u)y = (1 + wu)$. These two lines meet at E with co-ordinates $E((u-v+w-uvw), 2wu)/(uvw+u-v+w)$. The co-ordinates of G and F follow from those of E by cyclic change of u, v, w. Now from Section 1.3.8 the polar of E with respect to circle $ABCD$ has equation

$$(u-v+w-uvw)x+2wuy=(uvw+u-v+w).$$

It may now be verified by direct substitution that F and G lie on this line. Similarly, by cyclic change of u, v, w, E and F lie on the polar of G and G and E lie on the polar of F.

Exercise 1.4.1

1. Let ABC be a triangle in which there exists a circle such that BC is the polar of A, CA is the polar of B and AB is the polar of C. Prove that triangle ABC is obtuse and that the circle has centre at the orthocentre H and radius ρ, given by

$$\rho^2 = -4R^2 \cos A \cos B \cos C.$$

   Here R is the radius of the circumcircle of triangle ABC. *Hint: You will need to use Section 1.3.7 and some trigonometry to work out the lengths of HA and HD, where D is the foot of the altitude from A.*

2. Prove that in Fig. 1.6 the circle on FG as diameter is orthogonal to the circle $ABCD$.

3. In Fig. 1.6, suppose that BD meets FG at P. Prove that the circle on BD as diameter is orthogonal to the circle PEF.

4. Give a straight edge only construction of how to draw the tangents to a circle from an external point.

5. Prove that the circumcircle of the triangle formed by the three diagonals of a cyclic quadrangle is orthogonal to each of the three circles whose diameters are the three diagonals.

1.5 Formulas for conics

We do not propose to provide an elementary course on conics, as our interest lies in their more advanced properties in relation to lines, triangles, quadrangles and hexagons. However, in order to deal with such topics, it is necessary to have a firm knowledge of basic concepts and formulas. This section is designed with that in mind. Formulas involving the equations of parabolas, ellipses and hyperbolas and the parameter systems associated with these curves are stated without proof.

1.5.1 General equation of a conic

We define a conic S to be a curve having an equation of the second degree, so its general equation is of the form

$$S(x, y) \equiv ax^2 + 2hxy + by^2 + 2gx + 2fy + c = 0, \tag{1.15}$$

where not all of a, b, h are zero. If $S(x, y)$ factorises into two linear factors over the real field, then it degenerates into two straight lines. It can be shown that this occurs if, and only if,

$$\triangle \equiv abc + 2fgh - af^2 - bg^2 - ch^2 = 0.$$

Otherwise, if $ab - h^2 > 0$, the conic is an ellipse; if $ab - h^2 = 0$, the conic is a parabola; and if $ab - h^2 < 0$, the conic is a hyperbola. If $f = g = 0$ and $|ab - h^2| \neq 0$, the ellipse or hyperbola concerned has its centre at the origin (the centre having the property that it is the midpoint of any chord through it). It is customary to denote the minors of $\triangle$, when written as a determinant of the coefficients of the quadratic form 1.15 by $A = bc - f^2$, $B = ca - g^2$, $C = ab - h^2$, $F = gh - af$, $G = hf - bg$, $H = fg - ch$. When the conic is a central conic then the co-ordinates of its centre are $x = G/C$ and $y = F/C$. As stated above when $C = 0$, $\triangle \neq 0$ the conic is a parabola, for which the centre may be regarded as having receded to infinity. The angle θ made by a principal axis of the conic with the x-axis is given by $\tan 2\theta = 2h/(a - b)$. A Euclidean transformation of co-ordinates leaves $a + b$ and $C = ab - h^2$ invariant. When the conic passes through the origin O the equation of the tangent at O is $gx + fy = 0$. If the conic does not pass through O, then the equation of the pair of tangents passing through O is $Bx^2 - 2Hxy + Ay^2 = 0$.

1.5.2 Canonical forms of the equations of the ellipse, parabola and hyperbola

By a canonical form we mean the standard equation of the curve when the origin and x- and y-axes are chosen suitably. The suitable choice is partly conventional. For example the axis of a parabola is normally chosen to be the x-axis, rather than the y-axis.

Ellipse: the canonical form is $x^2/a^2 + y^2/b^2 = 1$, $(a > b)$. See Fig. 1.7. The origin is at the centre, the major axis is of length $2a$ and is in the direction of the x-axis, the minor axis is of length $2b$ and is in the direction of the y-axis. The circle with equation $x^2 + y^2 = a^2$, which touches the ellipse at the ends of the major axis is called the *director circle.*

Using the notation of Section 1.2.2, let the line P_1P_2 cut the ellipse at the point P with line co-ordinates (u, v). Then the equation

$$u^2\left(\frac{x_2^2}{a^2}+\frac{y_2^2}{b^2}-1\right)+2uv\left(\frac{x_1x_2}{a^2}+\frac{y_1y_2}{b^2}-1\right)$$

$$+v^2\left(\frac{x_1^2}{a^2}+\frac{y_1^2}{b^2}-1\right)=0 \tag{1.16}$$

provides the two values of the ratio u/v where the line meets the ellipse. This means, by putting in the condition for equal roots in Equation (1.16), that the pair of tangents from (x_1, y_1) to the ellipse has equation

$$\left(\frac{x_1^2}{a^2}+\frac{y_1^2}{b^2}-1\right)\left(\frac{x^2}{a^2}+\frac{y^2}{b^2}-1\right)=\left(\frac{x_1x}{a^2}+\frac{y_1y}{b^2}-1\right)^2.$$

If the line with equation $px + qy + r = 0$ meets the ellipse in two points, then the length of the intercepted chord is

$$\frac{2ab\sqrt{(p^2+q^2)(p^2a^2+q^2b^2-r^2)}}{p^2a^2+q^2b^2}.$$

The condition that the line touches the ellipse is therefore

$$p^2a^2+q^2b^2=r^2.$$

Parabola: the canonical form is $y^2 = 4ax$. See Fig. 1.8. Here a is a positive constant that distinguishes between different parabolas having the same axis

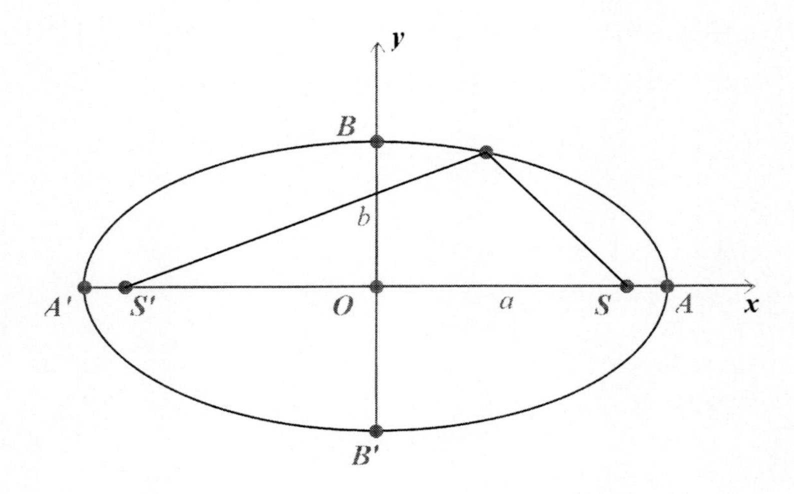

Figure 1.7: The canonical form of the ellipse

of symmetry, which is taken to be the x-axis. The parabola has no centre though, in a sense that will be described later, it may be thought of as being at infinity.

Using the notation of Section 1.2.2, let the line P_1P_2 cut the parabola at the point P with line co-ordinates (u, v). Then the equation

$$u^2(y_2^2 - 4ax_2) + 2uv(y_1y_2 - 2a(x_1 + x_2)) + v^2(y_1^2 - 4ax_1) = 0 \qquad (1.17)$$

provides the two values of the ratio u/v where the line meets the parabola. This means, by putting in the condition for equal roots in Equation (1.17), that the pair of tangents from (x_1, y_1) to the parabola has equation

$$(y^2 - 4ax)(y_1^2 - 4ax_1) = (yy_1 - 2a(x + x_1))^2.$$

If the line with equation $px + qy + r = 0$ meets the parabola in two points, then the length of the intercepted chord is

$$\frac{4\sqrt{(q^2a^2 - pra)(p^2 + q^2)}}{p^2}.$$

The condition that the line should touch the parabola is therefore $q^2a = pr$.
Hyperbola: the canonical form is $x^2/a^2 - y^2/b^2 = 1$. See Fig. 1.9. The axes of symmetry are the x- and y-axes and the equations of the asymptotes are $x/a = \pm y/b$.

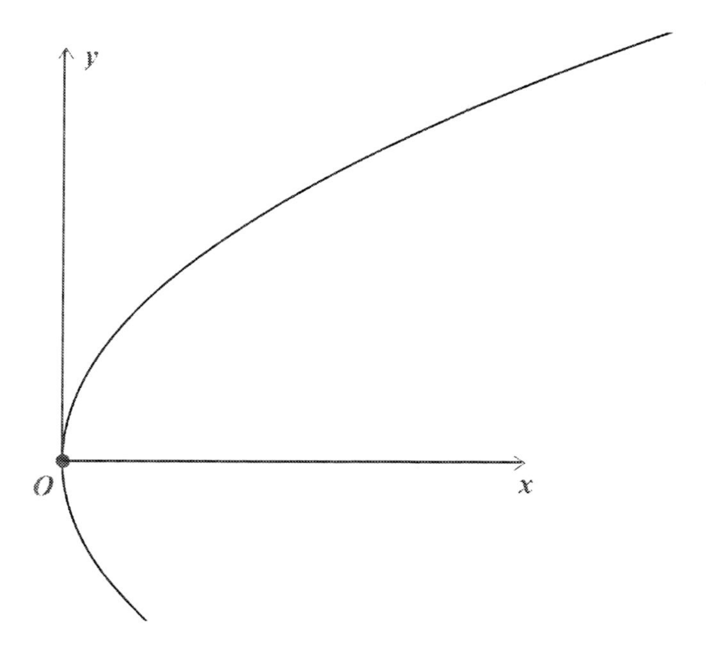

Figure 1.8: The canonical form of the parabola

Using the notation of Section 1.2.2, let the line P_1P_2 cut the hyperbola at the point P with line co-ordinates (u, v). Then the equation

$$u^2\left(\frac{x_2^2}{a^2} - \frac{y_2^2}{b^2} - 1\right) + 2uv\left(\frac{x_1x_2}{a^2} - \frac{y_1y_2}{b^2} - 1\right)$$
$$+v^2\left(\frac{x_1^2}{a^2} - \frac{y_1^2}{b^2} - 1\right) = 0 \qquad (1.18)$$

provides the two values of the ratio u/v where the line meets the hyperbola. This means, by putting in the condition for equal roots in Equation (1.18), that the pair of tangents from (x_1, y_1) to the hyperbola has equation

$$\left(\frac{x_1^2}{a^2} - \frac{y_1^2}{b^2} - 1\right)\left(\frac{x^2}{a^2} - \frac{y^2}{b^2} - 1\right) = \left(\frac{x_1x}{a^2} - \frac{y_1y}{b^2} - 1\right)^2.$$

If the line with equation $px + qy + r = 0$ meets the hyperbola in two points, then the length of the intercepted chord is

$$\frac{2ab\sqrt{(p^2+q^2)(p^2a^2 - q^2b^2 - r^2)}}{(p^2a^2 - q^2b^2)}, \quad pa \neq qb.$$

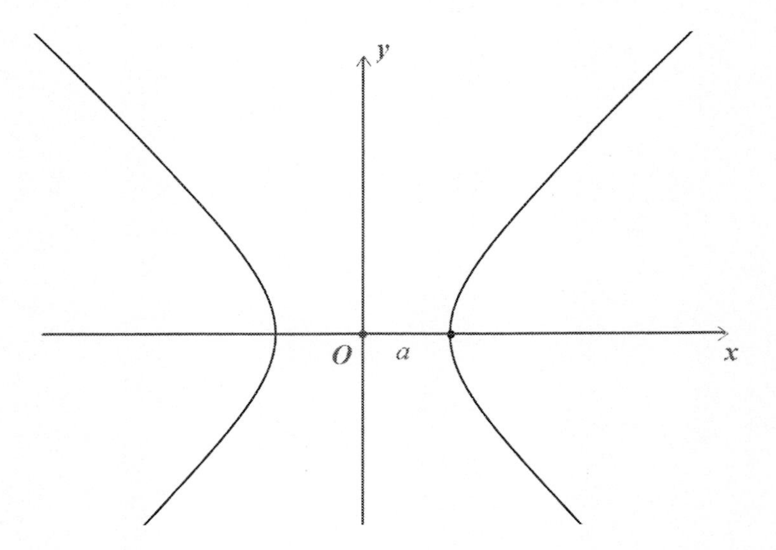

Figure 1.9: The canonical form of the hyperbola

The condition that the line touches the hyperbola is therefore $p^2a^2 - q^2b^2 = r^2$.

1.5.3 Eccentricity and foci

If we take the equation of an ellipse in canonical form, then its eccentricity e, $0 \leq e < 1$, is defined by the equation $b^2 = a^2(1 - e^2)$. It follows that a circle is an ellipse in which $e = 0$. The *foci* are defined to be the points S and S' with co-ordinates $S(ae, 0)$ and $S'(-ae, 0)$. It is a key property of an ellipse that, if P lies on the ellipse and N is the foot of the perpendicular from P on to the directrix with equation $x = a/e$, then $SP = ePN$. There is, of course, a second directrix $x = -a/e$ and it is then also the case that $S'P = ePN'$, where N' is the foot of the perpendicular from P on to the second directrix. Consequently it is true for all points P on the ellipse that $SP+S'P = e(PN+PN') = 2a$. This property may be taken as an alternative defining property of an ellipse and leads to a construction for an ellipse in which pegs are placed at the foci and a string is attached of length greater than the distance between the pegs. A variable point on the string is then selected and moved in a plane so that the string remains taut. The locus of such points when the string is taut is an ellipse. This is useful in the construction of ponds and garden beds.

If we take the equation of a parabola in the canonical form $y^2 = 4ax$, $a > 0$, then the focus is the point S with co-ordinates $(a, 0)$. The directrix is the line with equation $x = -a$. It is a key property of the parabola that, if P lies on the parabola and N is the foot of the perpendicular from P on to the directrix, then $SP = PN$.

If we take the equation of a hyperbola in canonical form, then the eccentricity e, $e > 1$, is defined by the equation $b^2 = a^2(e^2 - 1)$. Hence, if $e = \sqrt{2}$, then $b = a$ and the asymptotes are at right angles. The hyperbola is then called a *rectangular hyperbola*. An alternative canonical form for the rectangular hyperbola is $xy = c^2$, $c > 0$. This places the asymptotes along the x- and y-axes. In the original canonical form the foci S and S' have co-ordinates $S(ae, 0)$ and $S'(-ae, 0)$. It is a key property of a hyperbola that if P lies on the hyperbola and N is the foot of the perpendicular from S on to the directrix with equation $x = a/e$, then $SP = ePN$. The second directrix has equation $x = -a/e$. As with the ellipse there is a second relationship $S'P = ePN'$, where N' is the foot of the perpendicular on to the second directrix. It now follows by similar working as for the ellipse that $|SP - S'P| = 2a$.

1.5.4 The parametric representation of the parabola

It is conventional to use the parameter t, in terms of which $x = at^2$, $y = 2at$, $-\infty < t < \infty$. The origin corresponds to $t = 0$. The equation of the tangent at 't' is

$$ty - x = at^2$$

and the equation of the normal at 't' is

$$y + tx = at(2 + t^2).$$

The equation of the chord joining 't' and 's' is

$$(t + s)y - 2x = 2ats.$$

The co-ordinates of the point of intersection of the tangents at 't' and 's' are $x = ats, y = a(t + s)$. Sometimes it is convenient to use the parameter $m = 1/t$, since this is actually the value of the gradient of a tangent.

1.5.5 The parametric representation of an ellipse

It is conventional to use the parameter θ, in terms of which $x = a\cos\theta,\ y = b\sin\theta,\ 0 \leq \theta < 2\pi$. The angle θ here is called the eccentric angle of the point (x, y). However convention is at odds with practicality, as the equations of tangent, normal and chord are very cumbersome using the parameter 'θ'. In my experience it is preferable, unless the problem is specifically about the eccentric angle, to use a parameter 't', $-\infty < t \leq \infty$, in terms of which $x = a(1-t^2)/(1+t^2)$, $y = 2bt/(1+t^2)$. The equation of the tangent at t is

$$(1-t^2)x/a + 2ty/b = (1+t^2)$$

and the equation of the normal at 't' is

$$(1+t^2)(2tax - (1-t^2)by) = 2(a^2-b^2)t(1-t^2).$$

The equation of the chord joining 't' and 's' is

$$(1-ts)\frac{x}{a} + (t+s)\frac{y}{b} = 1 + ts. \tag{1.19}$$

The co-ordinates of the point of intersection of the tangents at 't' and 's' are

$$x = \frac{a(1-ts)}{1+ts}, y = \frac{b(t+s)}{1+ts}.$$

This is also the pole of the chord with Equation (1.19).

1.5.6 The parametric representation of the rectangular hyperbola

If we take the canonical form to be $xy = c^2$, then the conventional parameter is 't', $-\infty < t < \infty$, in terms of which $x = ct$, $y = c/t$. The equation of the tangent at 't' is $t^2y + x = 2ct$ and the equation of the normal at 't' is $ty - t^3x = c(1-t^4)$. The equation of the chord joining t and s is

$$tsy + x = c(t+s).$$

The tangents at 't' and 's' meet at the point with co-ordinates

$$x = \frac{2cts}{t+s}, y = \frac{2c}{t+s}.$$

1.5.7 The parametric representation of the hyperbola

It is conventional to use the parameter θ, in terms of which $x = a\sec\theta$, $y = b\tan\theta$, $0 < \theta < 2\pi, \theta \neq \pi/2$ or $3\pi/2$. However, once again convention is at odds with practicality, and it is much better to use a parameter 't', $-\infty < t \leq \infty, t \neq \pm 1$, in terms of which

$$x = \frac{a(1+t^2)}{1-t^2}, y = \frac{2bt}{1-t^2}.$$

The equation of the tangent at t is

$$(1+t^2)\frac{x}{a} - 2t\frac{y}{b} = 1 - t^2$$

and the equation of the normal at t is

$$(1-t^2)(2atx + (1+t^2)by) = 2(a^2+b^2)t(1+t^2).$$

The equation of the chord joining 't' and 's' is

$$(1+ts)\frac{x}{a} - (t+s)\frac{y}{b} = (1-ts).$$

The tangents at 't' and 's' meet at the point with co-ordinates

$$x = \frac{a(1+ts)}{1-ts}, y = \frac{b(t+s)}{1-ts}.$$

1.5.8 Polars

The concept of pole and polar carries over from circles to conics. From an external point P to a conic it is always possible to draw two tangents. Then the polar of P is the chord of contact. If the point P lies on the conic then the polar coincides with the tangent at P. If the point P is interior to the conic, then its polar consists of all those points whose polars pass through P. It is a standard exercise to find the equation of the polar of (x_0, y_0) with respect to the conic $S(x,y) = 0$, given by Equation (1.15). It is

$$ax_0x + h(y_0x + x_0y) + by_0y + g(x + x_0) + f(y + y_0) + c = 0. \qquad (1.20)$$

The points of intersection of the tangents from an external point (x_0, y_0) to the conic are given by the intersections of Equations (1.15) and (1.20).

1.5.9 Confocal conics

Two ellipses with equations in canonical from are confocal if

$$a^2 - b^2 = a'^2 - b'^2.$$

Two hyperbolas with equations in canonical form are confocal if $a^2 + b^2 = a'^2 + b'^2$. An ellipse $x^2/a^2 + y^2/b^2 = 1$ is confocal with the hyperbola $x^2/a'^2 - y^2/b'^2 = 1$ if $a^2 - b^2 = a'^2 + b'^2$. If an ellipse and a hyperbola are confocal, then they intersect at right angles.

Example 1.5.1

R is a point from which a pair of tangents is drawn to a parabola. Prove that the tangents are at right angles if, and only if, R lies on the directrix.

Suppose the tangents at P and Q have equations $ty - x = at^2$ and $sy - x = as^2$. These meet at $R(ast, a(s+t))$, which lies on the directrix with equation $x = -a$ if, and only if, $st = -1$. However, the gradients of the two tangents are $1/t$ and $1/s$, so the condition is equivalent to the tangents being at right angles. This is illustrated in Fig. 1.10.

Example 1.5.2

Describe a straight edge and compass construction for drawing a tangent at a given point P on a parabola.

This may be done by constructing the Frégier point F and then the tangent at P is the line through P perpendicular to PF. The construction of F is as follows. Draw two chords PA and PB at right angles and join AB. Then it turns out that AB always passes through a fixed point F, depending only on P, so by repeating the process and forming two chords AB and $A'B'$, F is their point of intersection. The construction works, not only for parabolas, but also for all non-degenerate conics.

Let the parabola have equation $y^2 = 4ax$, and suppose P is the point $(at^2, 2at)$, where t is the parameter of P. Two chords at right angles may be written as $y - mx = 2at - amt^2$ and $my + x = 2amt + at^2$, where m is a variable. These meet the parabola again at points A and B with parameters $s_1 = 2/m - t$ and $s_2 = -2m - t$ respectively. The equation of the chord AB is $(s_1 + s_2)y - 2x = 2as_1s_2$, that is

$$(y + 2at)(2/m - 2m) - 2ty - 2x + 8a - 2at^2 = 0.$$

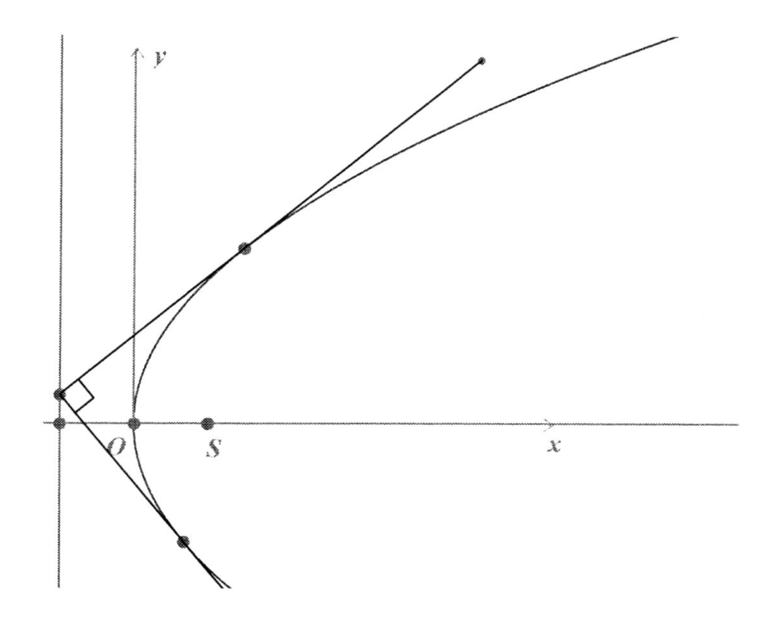

Figure 1.10: A key property of a parabola

For all values of m, this passes through the point given by the equations $y + 2at = 0$ and $ty + x - 4a + at^2 = 0$. Solving these gives the co-ordinates of F as $F(4a + at^2, -2at)$. Now FP has gradient $-4at/4a = -t$. Thus the line perpendicular to FP has gradient $1/t$. Since this is equal to the value of $(dy/dt)/(dx/dt)$ at P, it is evident that the line through P perpendicular to PF is the tangent at P. Note that as P varies the locus of the Frégier point is the curve with equation $y^2 = 4a(x - a)$, which is similar to the original parabola, but with vertex at the focus $S(a, 0)$. The construction is illustrated in Fig. 1.11.

In Fig. 1.12 and Fig. 1.13 we show diagrams of the same construction for an ellipse and a hyperbola respectively.

Example 1.5.3

Prove that, if the vertices A, B, C of a triangle lie on a rectangular hyperbola S, then S passes through the orthocentre H of triangle ABC.

Let the rectangular hyperbola have equation $xy = c^2$ and suppose that A, B, C have parameters u, v, w so that $A(cu, c/u), B(cv, c/v), C(cw, c/w)$. The equation of BC is $vwy + x = c(v + w)$. The perpendicular to this line

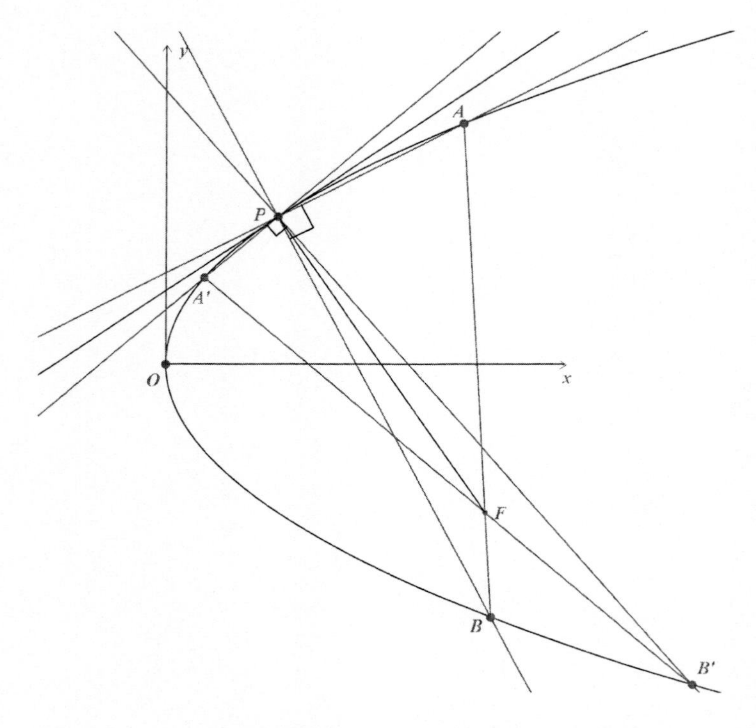

Figure 1.11: Drawing a tangent to a parabola

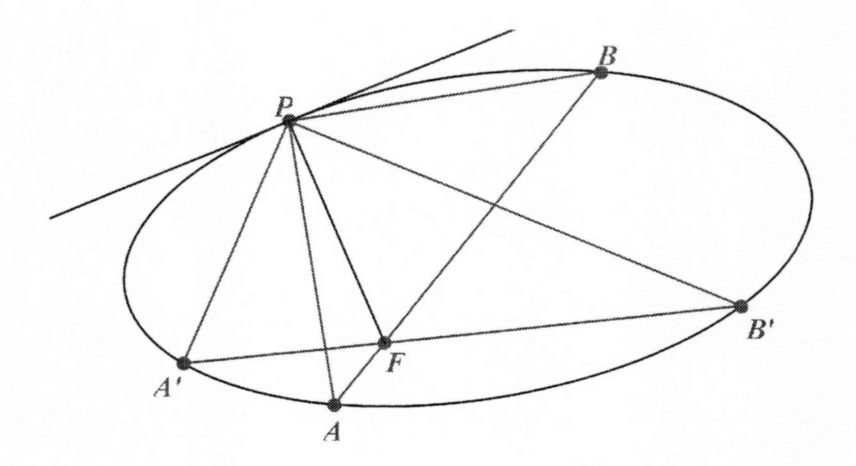

Figure 1.12: Drawing a tangent to an ellipse

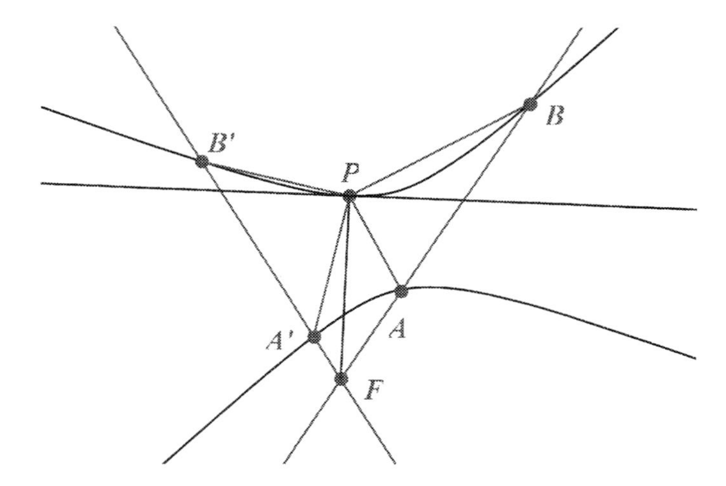

Figure 1.13: Drawing a tangent to a hyperbola

through A has equation $vwx - y = cuvw - c/u$. This line passes through the point H with co-ordinates $(-c/uvw, -cuvw)$. By symmetry H also lies on the altitudes through B and C. Furthermore H lies on the rectangular hyperbola with parameter $-1/uvw$. See Fig. 1.14.

Example 1.5.4

Conjugate diameters of an ellipse have the property that one bisects all chords parallel to the other. Prove that if the ellipse has equation $x^2/a^2 + y^2/b^2 = 1$, then $y = mx$ and $y = nx$ are conjugate if, and only if, $mn = -b^2/a^2$.

The condition is symmetric on interchange of m and n, which is why the word conjugate is appropriate. The chord joining 't' and 's' has gradient

$$n = -\frac{(1 - ts)b}{a(t + s)}.$$

Now the midpoint of this chord has co-ordinates

$$\left(\frac{(1 - ts)(1 + ts)a}{(1 + s^2)(1 + t^2)}, \frac{(t + s)(1 + ts)b}{(1 + s^2)(1 + t^2)}\right).$$

If the diameter $y = mx$ passes through this point then

$$m = \frac{b(t + s)}{a(1 - ts)},$$

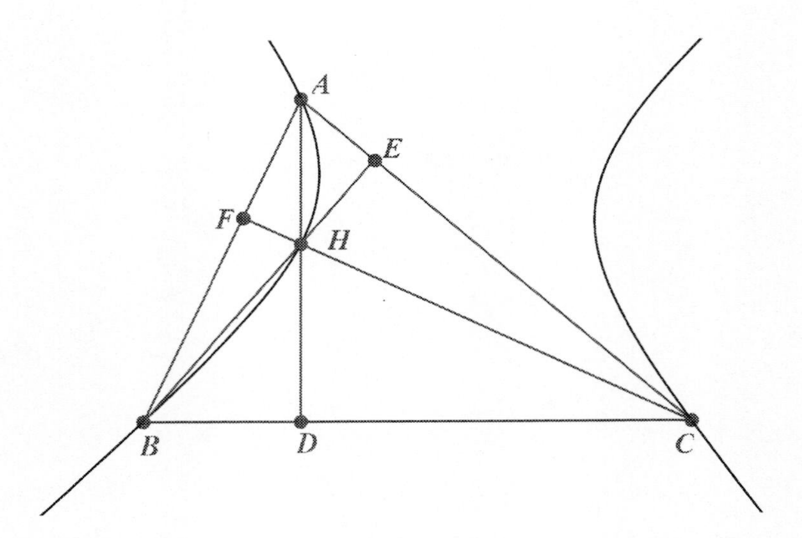

Figure 1.14: A key property of a rectangular hyperbola

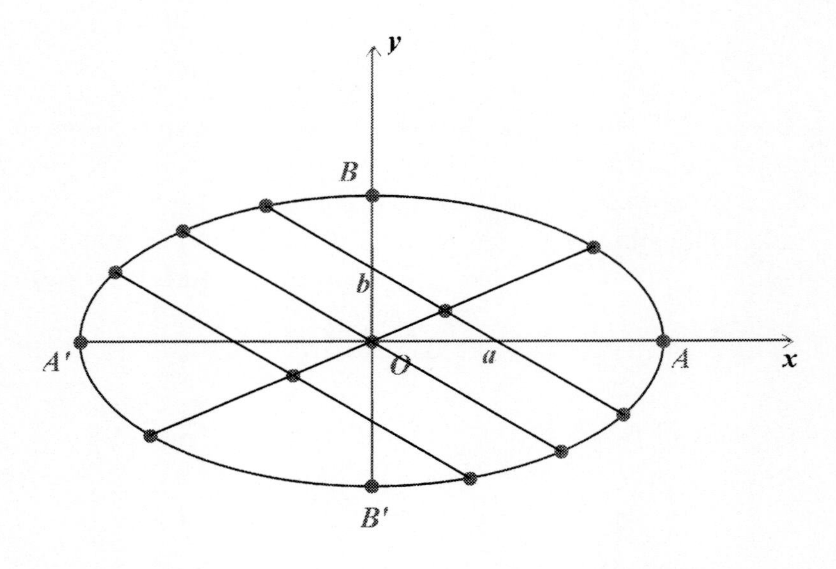

Figure 1.15: Conjugate diameters of an ellipse

and $mn = -b^2/a^2$. The condition is clearly both necessary and sufficient. See Fig. 1.15.

Example 1.5.5

An ellipse has foci at S and S' and P is any point on the ellipse. Prove that the normal at P makes equal angles with PS and PS'.

If the normal at P meets the major axis at N, then it is sufficient, by the internal angle bisector theorem (see Example 2.2.2), to establish that $PS/PS' = NS/NS'$. Now the normal at P meets the major axis at the point N with x-co-ordinate $ae^2(1-t^2)/(1+t^2)$. It follows that $NS = ae(1-e(1-t^2)/(1+t^2))$, with a similar expression for NS'. Hence

$$\frac{NS}{NS'} = \frac{(1-e)+(1+e)t^2}{(1+e)+(1-e)t^2}.$$

Now

$$PS^2 = a^2(1-t^2)^2/(1+t^2)^2 + a^2e^2 - 2a^2e(1-t^2)/(1+t^2) + 4b^2t^2/(1+t^2)^2$$

and after some algebra, using $b^2 = a^2(1-e^2)$, this is equal to

$$\begin{aligned}&(a^2/(1+t^2))((1+t^2)^2 - 2e(1-t^2)(1+t^2) + e^2(1-t^2)^2)\\ &= (a^2/(1+t^2)^2)((1-e)+(1+e)t^2)^2\end{aligned}$$

with a similar expression for PS'. It follows that $PS/PS' = NS/NS'$.

Exercise 1.5.6

1. Let S be a general conic through the origin with equation $ax^2+2hxy+by^2 = 2x$. Let $x^2+2kxy-y^2 = 0$ be the equation of a pair of straight lines through the origin that are at right angles to one another. Find the equation of the corresponding chord subtending the right angle at the origin, and show that this chord passes through a fixed point (the Frégier point) on the normal. (JW)
(Note: This problem generalizes Example 1.4.3, showing the construction of a tangent to be valid for any point on any non-degenerate conic.)

2. Prove that if two straight lines through the origin have equation $ax^2+2hxy+by^2 = 0$, then the pair of straight lines at right angles respectively to the given lines have equation $bx^2-2hxy+ay^2 = 0$. (JW)

3. Given the parabola with equation $y^2 = 4ax$, show that the locus of the point of intersection of normals to the parabola that are at right angles to one another has equation $y^2 = a(x - 3a)$. (JW)
Hint: There are 3 normals through a given point. Let these normals have parameters t_1, t_2 *and* t_3 *and put* $t_2t_3 = -1$.

4. A triangle is inscribed in a parabola and a similar and similarly situated triangle circumscribes it. Prove that the sides of the former triangle are four times the corresponding sides of the latter. (JW)

5. A chord of the parabola $y^2 = 4ax$ is drawn through the point with co-ordinates $(-2a, 0)$. Prove that the normals at its extremities intersect at a point on the parabola. (JW)

6. Consider triangle ABC. The point X lies on AB and is equidistant from A and BC. The point Y lies on AC and is also equidistant from A and BC. Prove that XY, BC and the external bisector of $\angle BAC$ are concurrent. *Hint: Let* A *be the focus and* BC *the directrix of the parabola with equation* $y^2 = 4ax$.

7. Normals are drawn at the extremities of a variable chord which passes through a fixed point on the axis of a parabola. Prove that their point of intersection lies on a fixed parabola. (JW)

8. The normals to the parabola $y^2 = 4ax$ at A, B, C meet at the point $(5a, 2a)$. Prove that the orthocentre of triangle ABC lies on the directrix.

9. A chord is drawn through the focus S of a parabola. The circle on this chord as diameter meets the parabola again at P and Q. Prove that the circle PSQ touches the parabola. (JW)

10. A parabola is drawn touching the sides AB and AC of triangle ABC at B and C, and passing through its orthocentre H. Prove that H is the vertex of the parabola. (JW)

11. The line segment QR is a chord of an ellipse, centre O, foci S, S'. The chord is normal to the ellipse at Q and P is the pole of QR. Let M, N be the feet of the perpendiculars from O, P respectively on to QR. Prove that $(OM)(PN) = (PS)(PS')$. (JW)

12. An ellipse is given whose axes are in the ratio $1+\sqrt{2}:1$. Prove that a circle whose diameter joins the ends of two conjugate diameters touches the ellipse. (JW)

13. The normals to an ellipse at P, Q, R, S meet in a point. The circles QRS, RSP, SPQ, PQR meet the ellipse again at P', Q', R' and S'. Prove that the normals at P', Q', R', S' meet in a point. (JW)

14. A variable chord PQ of an ellipse passes through the focus S. Normals are drawn at P and Q to meet at the point R. Prove that R lies on an ellipse. (Adapted from JW)

15. The line segment QR is a chord of the ellipse $x^2/a^2 + y^2/b^2 = 1$ and is normal to the ellipse at Q. The point P is the pole of QR. Calculate (i) the perpendicular distance from the centre of the ellipse on to the tangent at Q; (ii) QR; (iii) PQ. (JW)

16. Two points H, H' are conjugate with respect to an ellipse and P is any point on that ellipse. Suppose that PH, PH' meet the ellipse in Q, Q' respectively. Prove that QQ' passes through the pole of HH'. (JW)

17. Suppose that ABC is a variable triangle inscribed in an ellipse in which $\angle BAC = 90°$. Find the locus of the midpoint of BC.

18. Prove that the equation of the chord of the hyperbola $xy = c^2$ whose extremities are the points (x_1, y_1) and (x_2, y_2) is

$$\frac{x}{x_1 + x_2} + \frac{y}{y_1 + y_2} = 1.$$

(JW)

19. Prove that for all values of k, the rectangular hyperbola $xy = c^2$ is cut orthogonally by all ellipses with equations either $x^2 - xy + y^2 + k(x - y) - 3a^2 = 0$ or $x^2 + xy + y^2 + k(x + y) + 3a^2 = 0$. (JW)

20. Prove that the asymptotes of the hyperbola $x^2/a^2 - y^2/b^2 = 1$ are conjugate diameters of the ellipse $x^2/a^2 + y^2/b^2 = 1$. (JW)

21. Three tangents are drawn to the rectangular hyperbola $xy = c^2$ at the points $(x_1, y_1), (x_2, y_2), (x_3, y_3)$ and form a triangle whose circumcircle

passes through the centre of the hyperbola. Prove that

$$\frac{x_1 + x_2 + x_3}{x_1 x_2 x_3} + \frac{y_1 + y_2 + y_3}{y_1 y_2 y_3} = 0.$$

(JW)

22. A triangle is inscribed in the hyperbola $xy = c^2$ so that its centroid is the point (c, c). Prove that its sides touch the ellipse with equation $(3x + 3y - 8c)^2 = 4xy$. (JW)

23. Prove that every conic passing through the vertices A, B, C of a triangle and its orthocentre H is a rectangular hyperbola. *Hint: If D is the foot of the altitude from A on to BC, then the line pair AD, BC is a degenerate rectangular hyperbola.*

1.6 The radical centre of three circles

Example 1.6.1

Four fixed points lie on a circle, and two other variable circles are drawn touching each other, one passing through two fixed points of the four and the other through the other two fixed points. Prove that the point of contact of the variable circles lies on a circle.

The power of a point P with respect to a circle, centre O, radius R is given by $OP^2 - R^2$. The locus of a point P such that it has equal power with respect to two given circles is a straight line called the radical axis of the two circles. When the two circles intersect then the radical axis is their common chord and its extension. This is because, when the two circles intersect at A and B, then the power of a point P lying on AB is easily seen to be equal to $(PA)(PB)$, being the square of the length of the tangent from P to either circle. Suppose now that we have three circles S_1, S_2 and S_3 and suppose the radical axis of S_1 and S_2 is l_{12}, and that the radical axis of S_2 and S_3 is l_{23}. Suppose that l_{12} and l_{23} are not coincident or parallel and meet at P. Then P has the same power with respect to S_1 and S_2, since it lies on l_{12}. Similarly it has the same power with respect to S_2 and S_3. Hence it has equal power with respect to S_3 and S_1 and so lies on l_{31}, the radical axis of S_3 and S_1. The point P where the three radical axes meet is called the radical centre of S_1, S_2 and S_3.

Now let us move to the problem stated above. Let A, B be two of the fixed points and C, D be the other two. Suppose that the circles TAB and TCD touch at T. Let AB and CD meet at P. Now consider the three circles $ABCD$, TAB and TCD. They have radical centre P. Now $PT^2 = (PA)(PB) = (PC)(PD) = c^2$, where c is a constant. Hence T lies on the circle centre P and radius c. In the event that $AB \parallel CD$ then AB and CD do not meet at a finite point so the analysis above will not work, but then the point T clearly lies on the common perpendicular bisector of AB and CD, a straight line which for this purpose we must think of as a circle of infinite radius.

Exercise 1.6.2

1. A, B, C are the points of contact of three circles, each of which touches the other two. Prove that the tangents at A, B, C are concurrent.

2. A variable circle passes through two fixed points A, B and cuts a fixed circle at P, Q. Prove that PQ passes through a fixed point.

3. A common tangent to two non-intersecting circles C, D touches them at P, R respectively. The points K and L are internal to C and D respectively, and are the limiting points of the coaxal system determined by C and D. Suppose that PL meets C again at Q and RL meets D again at S. Prove that QS is a common tangent to C and D.

4. Let $ABCD$ be a cyclic quadrilateral. Let S be a variable circle having A and B as limiting points. and Σ a variable circle having C and D as limiting points. Prove that the radical axis of S and Σ passes through a fixed point.

Chapter 2

Areal co-ordinates

2.1 Introduction

A more detailed account of areal co-ordinates and their properties has been given in a previous text, Bradley [1]. We now give a summary of that account, and we will illustrate the theory by including more examples of their use. This is important as the content of several chapters in the present book depend on understanding how to use areal co-ordinates advantageously.

In most geometry books areal co-ordinates are mentioned only briefly, usually in the context of the theorems of Ceva and Menelaus, where the connection with areas is most obvious.

We give a treatment in which areal co-ordinates are developed as tools of Euclidean geometry, with a distance function duly established. This enables circles and conics to be classified in areal terms, just as they are in Chapter 1 for rectangular Cartesian co-ordinates. Indeed the use of areal co-ordinates for dealing with problems involving the relationship of a triangle with its associated circles and conics is one of the main features of this book. The fact that an obvious 'line at infinity' exists when using areal co-ordinates allows us to identify parallel lines as those that meet on that line. By refusing to allow the line at infinity to have any greater significance than any other line, we can obtain somewhat artificial entry to the geometry of the projective plane. It must be emphasised, however, that areals (as we shall now call them) are not suitable for use in the projective plane. Whereas in analytic projective geometry a triangle of reference certainly has to be chosen, it is, however, a triangle of no particular shape or size. No notion of its side lengths (or indeed of any lengths) is admitted in projective geometry, whereas with areals, knowledge of the side lengths of the triangle of reference is crucial for their role in solving problems in the Euclidean plane.

We are not concerned in this book with an account of the underlying groups of transformations that are employed when dealing with Euclidean, affine and projective geometry, as these have been given much prominence in the literature for well over a century. For these topics we refer the reader to the books by Coxeter [14], Pedoe [29] and more recently by Silvester [31].

It is true that one can start with a projective plane, and specify a line in it, which may be called 'the line at infinity', and then designate as affine geometry the study of those properties invariant under the projective transformations which map this distinguished line to itself (and then we may as well call these transformations 'affine maps'). Furthermore, if two points on the line at infinity are selected as the 'circular points at infinity', then Euclidean geometry may be regarded as the study of those properties invariant under the affine maps which preserve or exchange these circular points at infinity. This is all very satisfactory from an intellectual point of view, but Euclidean geometry is studied before projective geometry by almost everyone, so an alternative development, in which the Euclidean plane is viewed as a subset of the projective plane, is more common, even though the method of doing this appears somewhat artificial. It is this latter approach that we shall adopt in Chapter 3.

2.2 Definition and basic properties of areals

2.2.1 Line co-ordinates

If PQ is a line, then we may designate as the line co-ordinates of a point R on PQ an ordered pair of numbers (l, m), where $l+m = 1$, with the following properties. If $l = 1$ and $m = 0$, then R coincides with P. If $l = 0$ and $m = 1$, then R coincides with Q. If l and m are both positive, then R lies between P and Q and divides PQ *internally* in the ratio $m : l$, so that $PR/RQ = m/l$. If l is negative and m is positive then the order of the points on the line is PQR and if l is positive and m is negative then the order of the points on the line is RPQ. In either of the latter cases, provided a convention is used that segments have a sign associated with them, so that those lengths in the direction from P to Q are positive and those from Q to P are negative, we may say that R divides PQ *externally* in the ratio $m : l$ and once again $PR/RQ = m/l$.

If R and S are two points on the line PQ with line co-ordinates $R(l, m)$ and $S(r, s)$, where $l + m = 1$ and $r + s = 1$, then the displacement $\mathbf{RS} = (r - l, s - m) = (u, v)$, where $u + v = 0$. Moreover, if $PQ = c$, then

$$RS^2 = -c^2uv, \tag{2.1}$$

thereby providing a formula for the distance from R to S in terms of the distance from P to Q. Equation (2.1) follows from the fact that $\mathbf{RS} = v\mathbf{PQ}$, so that $RS^2 = c^2v^2 = -c^2uv$.

Note that it is possible to choose an origin O such that $\mathbf{OR} = l\mathbf{OP} + m\mathbf{OQ}$, but that if the origin is translated the co-ordinates (l, m) remain the same.

2.2.2 Areal co-ordinates

If O is any origin, whether or not in the plane Π of a triangle ABC, then for any point $P \in \Pi$ there exist constants l, m, n such that $l + m + n = 1$ and

$$\mathbf{OP} = l\mathbf{OA} + m\mathbf{OB} + n\mathbf{OC}.$$

Furthermore l, m, n are unique and independent of the origin O. We may therefore use the ordered triple (l, m, n) as the co-ordinates of P. ABC is called the triangle of reference.

2.2.3 The reason

The reason that these co-ordinates are called areal co-ordinates is that, for any point P it may be shown by standard vector methods that

$$l = \frac{[PBC]}{[ABC]}, m = \frac{[PCA]}{[ABC]}, n = \frac{[PAB]}{[ABC]},$$

where $[XYZ]$ denotes the area of triangle XYZ and the co-ordinates are signed so, for example, if P lies on the same side of BC as A then $l > 0$, if P lies on BC, then $l = 0$ and if P lies on the other side of BC to A then $l < 0$, and in all cases $l + m + n = 1$. This is because, wherever P is situated, we have

$$[PBC] + [PCA] + [PAB] = [ABC].$$

Fig. 2.1 shows the signs of l, m, n in the seven regions created by triangle ABC, where, for example, $- - +$ means $l < 0, m < 0, n > 0$.

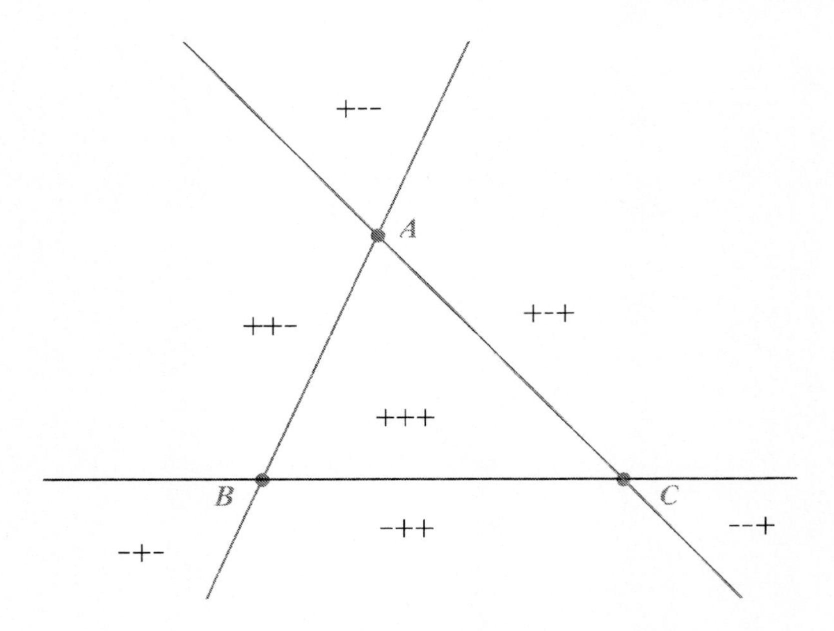

Figure 2.1: Sign convention for areal co-ordinates

2.2.4 Formula for the area of a triangle

If the points P, Q, R have areals (d, e, f), (g, h, k), (l, m, n) respectively then $[PQR]/[ABC]$ is equal to the 3×3 determinant whose rows (or columns) are equal respectively to these triples of real numbers. The sign of $[PQR]$ may be suppressed by taking the modulus of the determinant.

2.2.5 The equation of a line

If Q and R are fixed points, then the locus of the point P such that

$$[PQR]/[ABC] = 0$$

is the line QR, which therefore has the equation

$$x(hn - km) + y(kl - gn) + z(gm - hl) = 0. \tag{2.2}$$

As one would expect, Equation 2.2 is homogeneous of degree 1 in x, y, z and it may be verified that it contains the points Q and R.

2.2.6 Normalized and unnormalized co-ordinates

If $l + m + n = 1$, then the co-ordinates of the point $P(l, m, n)$ are said to be normalized. If this condition is relaxed so that $l + m + n = t \neq 0, 1$ then the co-ordinates are said to be *unnormalized.* They can be normalized by dividing all of them by t. The reason that unnormalized co-ordinates are often used is that the equation of QR in Equation (2.2) is the same whether the co-ordinates of Q and R are normalized or not. Normalized co-ordinates are often very cumbersome expressions, so it is useful to be able to drop the normalization. However it is important that co-ordinates are normalized when problems involve distance, as in calculating the co-ordinates of a midpoint or in calculating the distance between two points using the areal distance function; see Section 2.3.2.

2.2.7 The line at infinity

A point P with co-ordinates (x, y, z) such that $x + y + z = 0$ is such that its co-ordinates cannot be normalized. It is said to lie on the *line at infinity,* the equation of which is therefore $x + y + z = 0$. The equation $x + y + z = 0$ is impossible for any finite point and so is the concept of a line at infinite distance from the triangle of reference. However, the two ideas are consistent.

2.2.8 Parallel lines

The lines parallel to the line with equation $px + qy + rz = 0$ have equations $(p+s)x + (q+s)y + (r+s)z = 0$ for variable s. This is because any two such distinct lines do not meet at any point $P(x, y, z)$ whose co-ordinates can be normalized. Or you can say they meet at a point $(q - r, r - p, p - q)$ with $x + y + z = 0$, that is, at no finite point. Thus they meet on the line at infinity and are therefore parallel. Equivalently the two lines with equations $p_1x + q_1y + r_1z = 0$ and $p_2x + q_2y + r_2z = 0$ are parallel if, and only if, the determinant with rows (or columns) (p_1, q_1, r_1), (p_2, q_2, r_2), $(1, 1, 1)$ vanishes. As an example, the equation of BC is $x = 0$, so lines parallel to BC have equations of the form $(1 + s)x + sy + sz = 0$. The line that passes through $A(1, 0, 0)$ evidently has parameter $s = -1$ and so the equation of the line parallel to BC through A is $y + z = 0$.

2.2.9 Areals

We now discuss areals for key points of a triangle ABC, when it is chosen as the triangle of reference. L, M, N are the midpoints of the sides BC, CA, AB respectively, D, E, F are the feet of the altitudes through A, B, C respectively, G is the centroid, H is the orthocentre, O is the circumcentre, I is the incentre, and I_1, I_2, I_3 are the excentres opposite A, B, C respectively. $BC = a, CA = b, AB = c$.

Point	Normalized co-ordinates	Unnormalized co-ordinates
A	$(1,0,0)$	–
B	$(0,1,0)$	–
C	$(0,0,1)$	–
G	$(1/3,1/3,1/3)$	$(1,1,1)$
L	$(0,\frac{1}{2},\frac{1}{2})$	$(0,1,1)$
M	$(\frac{1}{2},0,\frac{1}{2})$	$(1,0,1)$
N	$(\frac{1}{2},\frac{1}{2},0)$	$(1,1,0)$
H	$(\cot B\cot C, \cot C\cot A, \cot A\cot B)$	$(\tan A, \tan B, \tan C)$
D	$(0, \sin B\cos C, \sin C\cos B)/\sin A$	$(0, \tan B, \tan C)$
E	$(\sin A\cos C, 0, \sin C\cos A)/\sin B$	$(\tan A, 0, \tan C)$
F	$(\sin A\cos B, \sin B\cos A, 0)/\sin C$	$(\tan A, \tan B, 0)$
O	$(\sin 2A, \sin 2B, \sin 2C)/4\sin A\sin B\sin C$	$(\sin 2A, \sin 2B, \sin 2C)$
I	$(a,b,c)/(a+b+c)$	(a,b,c)
I_1	$(-a,b,c)/(b+c-a)$	$(-a,b,c)$
I_2	$(a,-b,c)/(c+a-b)$	$(a,-b,c)$
I_3	$(a,b,-c)/(a+b-c)$	$(a,b,-c)$

Example 2.2.1

Let ABC be a triangle and AKD, BKE, CKF a set of three Cevians. That is, K is a point in the plane of ABC, not on its sides or their extensions nor on a line through a vertex parallel to the opposite side, and D, E, F are the intersections of AK, BK, CK with the lines BC, CA, AB respectively. (Note that K does not necessarily lie inside ABC, but D, E, F must exist at finite points so that, for example, K cannot lie on the line parallel to BC through A.) Let L, M, N be the midpoints of EF, FD, DE respectively. Then AL, BM, CN are concurrent. See Fig. 2.2.

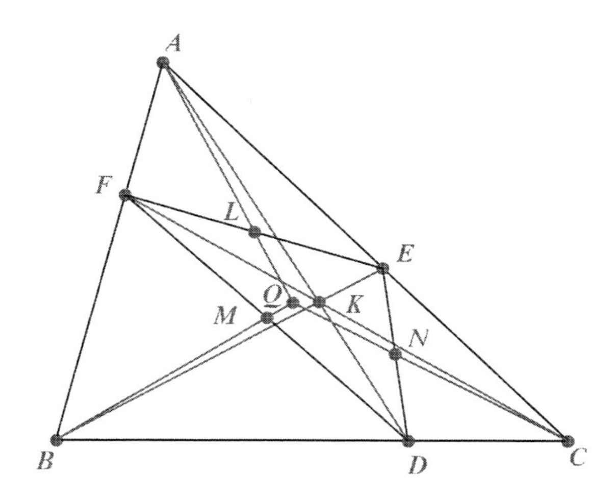

Figure 2.2: Example 2.2.1

Use areals and suppose those of K are (l, m, n). It follows that E and F have co-ordinates $E(l, 0, n)/(n+l)$ and $F(l, m, 0)/(l+m)$. L is the midpoint of EF and so has co-ordinates $\frac{1}{2}(l/(n+l)+l/(l+m), m/(l+m), n/(n+l))$. The equation of AL is therefore $yn(l+m) = zm(n+l)$. This evidently contains the point Q with unnormalized areals $Q(l(m+n), m(n+l), n(l+m))$ and, by symmetry, this also lies on BM and CN.

2.2.10 Menelaus's theorem

Let ABC be a triangle and L, M, N points with L on BC, M on CA, N on AB, then LMN is a straight line (a *transversal*) if, and only if,

$$\frac{BL}{LC} \times \frac{CM}{MA} \times \frac{AN}{NB} = -1, \tag{2.3}$$

where the sign convention explained in Section 2.2.1 is used. See Fig. 2.3.

Suppose first that LMN is a transversal with equation $px + qy + rz = 0$, then L, the point where the transversal meets BC, $x = 0$, has areals $L(0, r, -q)$ and $(BL/LC) = -q/r$. Similarly $(CM/MA) = -r/p$ and $(AN/NB) = -p/q$ and hence Equation (2.3) holds. (Note that no two of p, q, r are equal, otherwise the line would be parallel to one of the sides.)

Conversely, suppose that Equation (2.3) holds. Join LM and extend to meet AB at N'. Then LMN' is a transversal and Equation (2.3) holds with

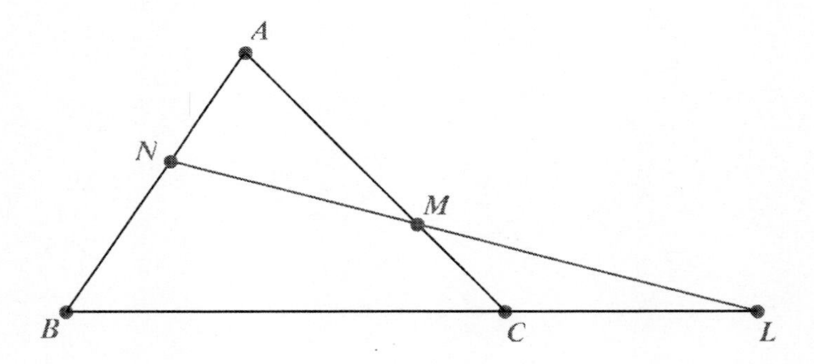

Figure 2.3: Menelaus's theorem

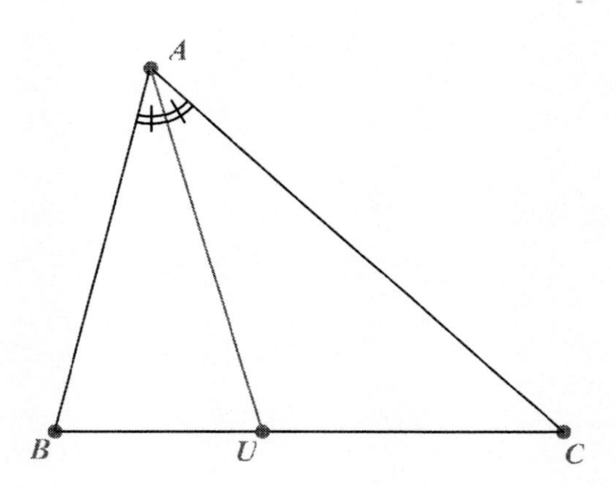

Figure 2.4: The internal bisector theorem

N' replacing N. It follows that $(AN/NB) = (AN'/N'B)$, and so N' coincides with N and LMN is a straight line.

Example 2.2.2 *(Internal bisector theorem)*

Let ABC be a triangle and AU the internal bisector of $\angle BAC$, with U on BC. Then $BU/UC = AB/AC$. This is illustrated in Fig. 2.4.

The line AU passes through the incentre $I(a, b, c)$ and therefore has equation $cy = bz$. This line meets BC, $x = 0$, at $U(0, b, c)$ and so $BU/UC = c/b = AB/AC$. The reader is invited to consider what change is required when AV is the external bisector of $\angle BAC$ and V lies on BC.

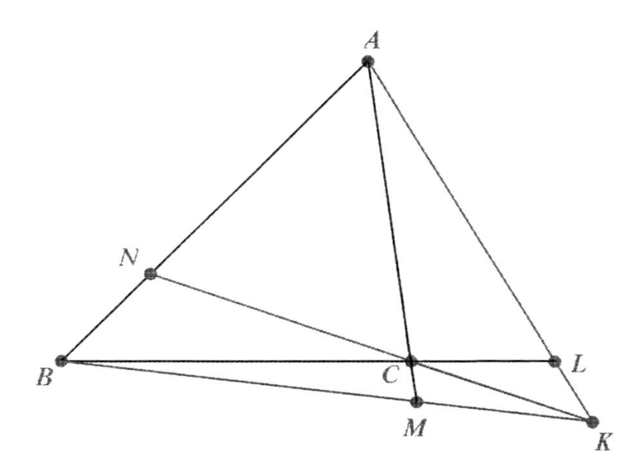

Figure 2.5: Ceva's Theorem

2.2.11 Ceva's theorem

Let ABC be a triangle and AKL, BKM, CKN a set of Cevians through a point K (as described in Example 2.2.1). Then

$$\frac{BL}{LC} \times \frac{CM}{MA} \times \frac{AN}{NB} = +1, \tag{2.4}$$

In Equation (2.4) the sign convention is used, which is why the plus sign is displayed. Conversely if L, M, N lie on BC, CA, AB respectively and Equation (2.4) holds, then AL, BM, CN are concurrent or parallel. See Fig. 2.5.

Suppose that K has areals (l, m, n), where, since K does not lie on any side and K does not lie on any parallel to a side through an opposite vertex, then $l, m, n, m+n, n+l, l+m \neq 0$. Then the equation of AK is $ny = mz$ and this line meets BC, $x = 0$, at $L(0, m, n)/(m+n)$. Then $BL/LC = n/m$. Similarly $CM/MA = l/n$ and $AN/NB = m/l$. Equation (2.4) follows.

Conversely, suppose Equation (2.4) holds. Then either AL, BM, CN are parallel or two of these lines meet. Suppose, without loss of generality, that AL, BM meet at K. Then draw CK and suppose it meets AB at N'. Then AL, BM, CN' are a set of Cevians and Equation (2.4) holds with N' replacing N. It follows that $AN/NB = AN'/N'B$ and hence N and N' coincide. Hence, if AL, BM, CN are not parallel, then they are concurrent. The parallel case can occur, as the reader should verify.

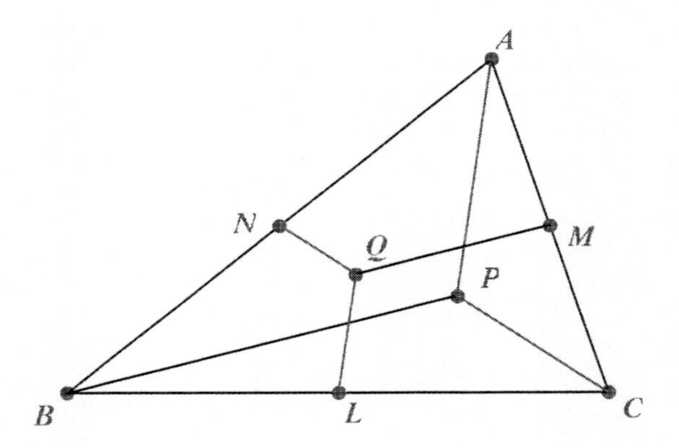

Figure 2.6: Example 2.2.3

Example 2.2.3

P, A, B, C are distinct points in a plane. Through the midpoints of BC, CA, AB lines are drawn parallel to PA, PB, PC respectively. Prove that these lines meet in a point. Identify the point of concurrency when P is the orthocentre of ABC.

Suppose P has areals (l, m, n) then the equation of AP is $ny = mz$ and a line parallel to AP has an equation of the form $sx+(s+n)y+(s-m)z = 0$. If it passes through the midpoint $L(0, 1, 1)$ of BC, then $s = \frac{1}{2}(m-n)$ and so the line parallel to AP through L has equation $(m-n)x+(m+n)y-(m+n)z = 0$. This passes through the point Q with areals $Q(m+n, n+l, l+m)$. By symmetry this point also lies on the line through M parallel to BP and on the line through N parallel to CP. See Fig. 2.6.

When P is the orthocentre, then Q has unnormalized co-ordinates

$$(\tan B + \tan C, \tan C + \tan A, \tan A + \tan B) \propto (\sin 2A, \sin 2B, \sin 2C)$$

and hence Q is the circumcentre.

When P is the circumcentre, then Q is the nine-point centre (the midpoint of OH). This is left as an exercise for the reader to complete.

Example 2.2.4

ABC is a triangle and K is a general point in the plane of ABC, not lying on its sides or the medians. The lines AK, BK, CK meet BC, CA, AB

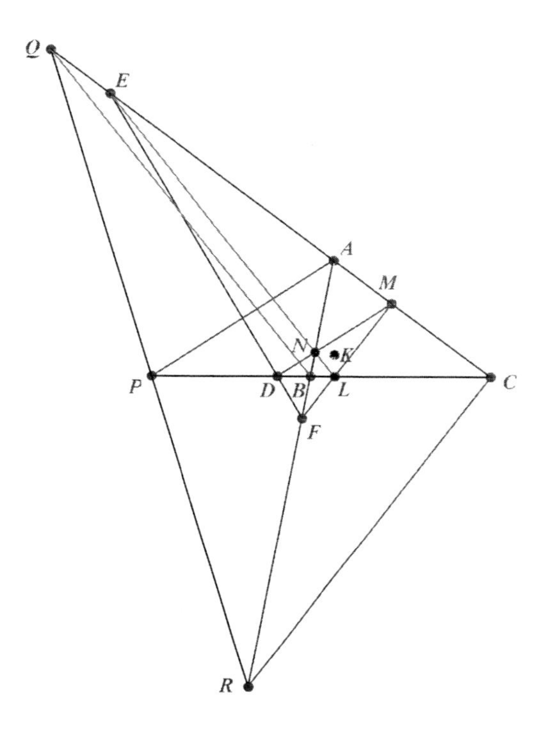

Figure 2.7: Example 2.2.4

respectively at L, M, N. The line MN meets BC at D, NL meets CA at E and LM meets AB at F. Prove that D, E, F are collinear. Also the line through A parallel to MN meets BC at P, the line through B parallel to NL meets CA at Q and the line through C parallel to LM meets AB at R. Prove that P, Q, R are collinear. See Fig. 2.7.

Let K have areals $K(l, m, n)$, where, since K does not lie on the sides or medians $l, m, n \neq 0$ and no two of l, m, n are equal. Then $L(0, m, n)$, $M(l, 0, n)$, $N(l, m, 0)$. The equation of MN is $-mnx + nly + lmz = 0$ and this meets BC, $x = 0$, at $D(0, -m, n)$. Similarly $E(l, 0, -n)$, $F(-l, m, 0)$ and D, E, F are collinear on the line with equation $x/l + y/m + z/n = 0$.

Now a line parallel to MN has an equation of the form $(s - mn)x + (s + nl)y + (s + lm)z = 0$ and if it passes through $A(1, 0, 0)$ then $s = mn$ and its equation is $n(l+m)y+m(n+l)z = 0$. This meets BC at $P(0, -m(n+l), n(l+$

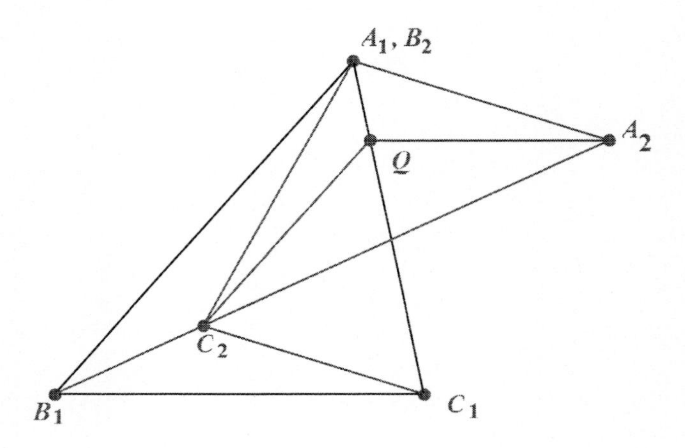

Figure 2.8: Example 2.2.5

m)) and similarly $Q(l(m+n), 0, -n(l+m))$, $R(-l(m+n), m(n+l), 0)$. The points P, Q, R are collinear on a line with equation

$$\frac{x}{l(m+n)} + \frac{y}{m(n+l)} + \frac{z}{n(l+m)} = 0.$$

Example 2.2.5

Two triangles T_1 and T_2 are so related that straight lines drawn through the vertices of T_1 parallel to the sides of T_2 meet in a point P. Prove that straight lines drawn through the vertices of T_2 parallel to the sides of T_1 also meet in a point.

Let the vertices of T_1 be A_1, B_1, C_1 and those of T_2 be A_2, B_2, C_2. Adjust T_2 to produce a similar triangle (carrying the same labels as its vertices, for convenience) in which B_2 coincides with A_1 and C_2 coincides with the point of concurrence P. Then A_2 lies at the point on B_1C_2 (extended) so that A_2A_1 is parallel to C_1C_2. Now draw a line through A_2 parallel to B_1C_1 to meet A_1C_1 at Q. The result now holds provided we can establish that QC_2 is parallel to A_1B_1. See Fig. 2.8.

We use areals with co-ordinates as follows:

$$A_1(1,0,0), B_1(0,1,0), C_1(0,0,1), B_2(1,0,0), C_2(p,q,r)$$

(with $p + q + r = 1$). Then $\mathbf{C_1C_2} = (p, q, r - 1)$. Now A_2 lies on B_1C_2 and is such that A_2A_1 is parallel to C_1C_2. It follows that constants s, t exist such that its co-ordinates may be expressed as $s(p, q, r) + (1 - s)(0, 1, 0)$ and also as $(1, 0, 0) - t(p, q, r - 1)$. Equating co-ordinates one finds $t = r/p$ and $s = (1 - r)/p$ and so $A_2(1 - r, -qr/p, -r(r - 1)/p)$. Now Q lies on the line parallel to B_1C_1 through A_2 and also lies on A_1C_1. Hence a constant u exists so that the co-ordinates of Q are $(1 - r, -qr/p, -r(r - 1)/p) + u(0, 1, -1)$. But the value of u must be chosen so that the second co-ordinate vanishes. Hence $u = qr/p$ and Q has co-ordinates $(1-r, 0, r)$. Now C_2 has co-ordinates (p, q, r), so $\mathbf{QC_2} = (p - 1 + r, q, 0) = (-q, q, 0)$, which is parallel to $\mathbf{A_1B_1} = (-1, 1, 0)$, as required.

Exercise 2.2.6

1. P and Q are variable points on the sides CA, AB of triangle ABC such that $CP/PA + BQ/QA = 1$. Show that the line PQ passes through a fixed point.

2. ABC is a triangle and LMN is a transversal with L, M, N lying on BC, CA, AB respectively. Let X, Y, Z be the midpoints of MN, NL, LM respectively. Let BY, CZ meet at P; let CZ, AX meet at Q; let AX, BY meet at R. Prove that AP, BQ, CR are parallel.

3. ABC is a triangle and D, E, F are points on BC, CA, AB respectively such that AD, BE, CF meet at a point U. P, Q, R lie on EF, FD, DE respectively and are such that DP, EQ, FR meet at a point V. Prove that AP, BQ, CR are concurrent or parallel.

4. ABC is an acute-angled triangle and P, Q, R are points on BC, CA, AB respectively such that AP, BQ, CR are concurrent at an internal point K. L, M, N are the midpoints of BC, CA, AB respectively and D, E, F are the midpoints of AP, BQ, CR. Prove that LD, ME, NF are concurrent.

5. ABC is a triangle and K is an internal point of ABC. AK, BK, CK meet the sides BC, CA, AB at L, M, N respectively. L^*, M^*, N^* are the midpoints of MN, NL, LM respectively. Prove that AL^*, BM^*, CN^* are concurrent. The line through A parallel to MN meets BC at L'. Points M', N' are similarly defined. Prove that L', M', N' are collinear.

6. In triangle ABC, I is the incentre and I_1 the centre of the escribed circle opposite A. II_1 meets BC at U. A straight line through U meets AB and AC at S and T respectively. IS, I_1T intersect at P, I_1S, IT intersect at Q. Prove that PAQ is a straight line perpendicular to II_1. *Hint: the internal and external bisectors of an angle are at right angles.*

7. ABC is a triangle and K an interior point. AK, BK, CK meet the sides BC, CA, AB respectively at L, M, N. The circle LMN meets the sides BC, CA, AB again at X, Y, Z. Prove that AX, BY, CZ are concurrent. *Hint: Use the intersecting chord theorem and Ceva's theorem.*

8. ABC is a triangle and P is a point in the plane of ABC, not lying on any of its sides. PA, PB, PC meet the sides BC, CA, AB respectively at D, E, F and it is found that $AD = BE = CF$. Prove that $AP + BP + CP = 2(PD + PE + PF)$.

9. Within a triangle ABC are taken two points U and V. AU, BU, CU meet the opposite sides at L, M, N respectively. The points of intersection of $VA, MN; VB, NL; VC, LM$ are denoted by D, E, F respectively. Prove that LD, ME, NF are concurrent at a point P, which remains the same if in the above construction U and V are interchanged. (JW)

10. ABC is a triangle. P is an internal point of ABC. APL, BPM, CPN is a set of Cevians. Locate the possible positions of P if the areas of the quadrilaterals $PLCM$, $PMAN$, $PNBL$ are equal.

11. Let I be the incentre of triangle ABC. The incircle touches BC, CA, AB at X, Y, Z. Prove that AX, BY, CZ are concurrent. If XI meets YZ at K, prove that AK bisects BC.

12. A triangle ABC is given with $BC = 6, CA = 4, AB = 3$. L lies on BC with $BL = 3.6$, $LC = 2.4$; M lies on the extension of CA with $CM = 8, AM = 4$; N lies on AB with $AN = 0.75$, $NB = 2.25$. Prove that L, M, N are collinear and show that LMN passes through (i) the centroid of ABC; (ii) the incentre of ABC; (iii) the centre of mass of ABC considered as a uniform wire framework (with no interior).

13. ABC is a triangle with incentre I. The incircle touches BC at X. Prove that I, the midpoint of BC and the midpoint of AX are collinear.

14. A point M is chosen on the side BC of triangle ABC. N and P are points on AC and AB respectively so that MN is parallel to AB and MP is parallel to AC. Q is the intersection of BN and PC, and R is the intersection of AQ and PN. Show that $PR/RN = BM/MC$. (GL)

15. ABC is a triangle and L, M, N are the midpoints of BC, CA, AB respectively. U is the point on CA or its extension such that $LC + CU = \frac{1}{2}(a + b + c)$, and V, W are similarly defined on AB, BC. Prove that LU, MV, NW are concurrent.

16. ABC is a triangle that is not equilateral. I is the incentre of ABC. Lines through I parallel to the medians meet the sides BC, CA, AB at L, M, N respectively. X, Y, Z are the points on BC, CA, AB where the incircle touches ABC. Prove that $LX + MY + NZ = 0$, where the line segments obey the usual sign convention.

17. ABC is a triangle and P is any point in the plane of ABC not on its sides. Lines through P parallel to the medians AL, BM, CN of ABC meet the sides BC, CA, AB respectively at X, Y, Z. Prove that the lines through A, B, C parallel to the medians XD, YE, ZF respectively of triangle XYZ are concurrent or parallel.

18. Medians AL, BM, CN are drawn in a triangle ABC. Points P, Q, R on AL, BM, CN are chosen with $AP = kAL$, $BQ = kBM$, $CR = kCN$, where $0 < k < 1$. Determine $[PQR]/[ABC]$.

19. ABC is a triangle and P is a point in the plane of ABC that does not lie on the sides or medians. APD, BPE, CPF are the Cevians through P. EF meets BC at L, FD meets CA at M, DE meets AB at N. X, Y, Z are the midpoints of DL, EM, FN. Prove that X, Y, Z are collinear.

20. P is a variable point inside triangle ABC and APL, BPM, CPN are the Cevians through P. Locate the position of P that minimises $(AP/AL)^2 + (BP/BM)^2 + (CP/CN)^2$.

21. ABC is a triangle which is not equilateral. O is the circumcentre and H is the orthocentre. K lies on OH and is such that O is the midpoint of HK. AK meets BC in X and Y, Z are the feet of the perpendiculars

from X on to the sides AC, AB respectively. Prove that AX, BY, CZ are concurrent or parallel.

22. In triangle ABC it is given that $\angle BAC = 90°$. G is the centroid and I the incentre. If IG is parallel to AC, determine the ratio AC/BC.

2.3 Distance, circles and conics

2.3.1 The line with equation $px + qy + rz = 0$

The signed perpendicular distances from A, B, C on to the line with equation $px + qy + rz = 0$ are proportional to p, q, r respectively. If the line meets BC, CA, AB at P, Q, R respectively, this result may be proved by working out $[AQR], [BRP], [CPQ]$. The perpendicular distance from A on to the line comes to $p[ABC]/|\mathbf{k}|$, where $\mathbf{k} = (q - r, r - p, p - q)$.

2.3.2 The distance function

If the displacement $\mathbf{PQ} = (u, v, w)$, where $u + v + w = 0$, then

$$PQ^2 = -a^2vw - b^2wu - c^2uv. \tag{2.5}$$

Equation (2.5) also holds for an infinitesimal displacement (du, dv, dw) of course, and the square of the arc length may be re-expressed as

$$ds^2 = bc\cos A du^2 + ca\cos B dv^2 + ab\cos C dw^2,$$

a form suitable for working out arc lengths using calculus.

2.3.3 The condition for displacements to be perpendicular

Suppose that $\mathbf{PQ} = (u, v, w)$ and $\mathbf{RS} = (f, g, h)$ are two displacements in the plane of ABC, so that $u+v+w = f+g+h = 0$, then the condition for these displacements to be at right angles is

$$a^2(gw + hv) + b^2(hu + fw) + c^2(fv + gu) = 0. \tag{2.6}$$

Note that a line with equation $px+qy+rz = 0$ contains a displacement along its length equal to $(q-r, r-p, p-q)$, so given the equations of two lines, the coefficients may be used, with the aid of Equation (2.6), to test whether they are perpendicular. For example, the line with equation $px + qy + rz = 0$ is perpendicular to BC, $x = 0$, if, and only if, $a^2(q+r-2p)+(b^2-c^2)(q-r) = 0$.

2.3.4 The equation of a conic

The property of a conic that it is met by a line in two points ensures that the general equation of a conic in areals is homogeneous of the second degree and has the form

$$\Phi(x, y, z) \equiv ux^2 + vy^2 + wz^2 + 2fyz + 2gzx + 2hxy = 0. \qquad (2.7)$$

If this quadratic form factorises over the real numbers into the product of two linear factors, then the conic degenerates into two straight lines. The necessary and sufficient condition for this to be the case is

$$uvw + 2fgh - uf^2 - vg^2 - wh^2 = 0.$$

2.3.5 Polar and tangent to a conic

The equation of the polar of the point with areals (X, Y, Z) with respect to the conic $\Phi(x, y, z) = 0$ is

$$uXx + vYy + wZz + f(Zy + Yz) + g(Xz + Zx) + h(Yx + Xy) = 0. \quad (2.8)$$

This is not a subtle result and may be established using standard calculus techniques. If (X, Y, Z) lies on the conic, then Equation (2.8) represents the equation of the tangent to the conic at that point.

2.3.6 The centre of a conic

Now the polar of the centre of a conic is the line at infinity. Hence, if (X, Y, Z) is the centre, then the coefficients of x, y, z in Equation (2.8) must be equal, in order that it should reduce to $x + y + z = 0$. This provides two equations that can be solved for the ratio $X : Y : Z$. The result is

$$X : Y : Z = vw - gv - hw - f^2 + fg + hf$$

$$: wu - hw - fu - g^2 + gh + fg : uv - fu - gv - h^2 + hf + gh. \qquad (2.9)$$

2.3.7 Ellipses, parabolas and hyperbolas

The conic $\Phi(x, y, z) = 0$ is a hyperbola, parabola or ellipse according as it meets the line at infinity $x+y+z=0$ at two real, coincident real or non-real values of x, y, z. This involves the function $F(u, v, w, f, g, h)$

$$\equiv f^2+g^2+h^2-vw-wu-uv+2fu+2gv+2hw-2gh-2hf-2fg \quad (2.10)$$

If $F > 0$ then the conic is a hyperbola. If $F = 0$ the conic is a parabola. If $F < 0$, then the conic is an ellipse. It can be seen that this tallies with the notion that if the conic is a parabola, then its centre (X, Y, Z) lies on the line at infinity. For, if (X, Y, Z) is the centre, then $X + Y + Z = 0$ is equivalent to $F = 0$.

2.3.8 Conics passing through A, B, C

A conic passing through the vertices of the triangle of reference must have an equation in which $u = v = w = 0$. It therefore has an equation of the form

$$fyz + gzx + hxy = 0.$$

The function F is now given by

$$F(f, g, h) = f^2 + g^2 + h^2 - 2gh - 2hf - 2fg$$

and the areals (X, Y, Z) of the centre are given by

$$X : Y : Z = f(g + h - f) : g(h + f - g) : h(f + g - h).$$

The general equation in areals of a rectangular hyperbola through A, B, C is

$$\sin^2 A(v \sin 2C - w \sin 2B)yz + \sin^2 B(w \sin 2A - u \sin 2C)zx$$

$$+ \sin^2 C(u \sin 2B - v \sin 2A)xy = 0.$$

See Problems 5 and 6 of Exercise 10.2.4.

2.3.9 The circumcircle of triangle ABC

When $f = a^2, g = b^2, h = c^2$, then

$$f(g + h - f) = a^2(b^2 + c^2 - a^2) = (2abc)(a \cos A)$$

and so the centre has areals $\propto (\sin 2A, \sin 2B, \sin 2C)$, which are the unnormalized co-ordinates of the centre of the circumcircle. It follows that the equation of the circumcircle of triangle ABC is

$$a^2yz + b^2zx + c^2xy = 0. \tag{2.11}$$

This meets the line at infinity at its (non-real) intersections with $x + y + z = 0$, and these two points are called the *circular points at infinity*. Their co-ordinates are rather complicated and are not written down, as they are not subsequently used.

2.3.10 The equation of a circle

Since all circles pass through the circular points at infinity, it follows that the equation of any circle is of the form

$$a^2yz + b^2zx + c^2xy - (x + y + z)(ux + vy + wz) = 0. \tag{2.12}$$

Changing the sign and comparing Equations (2.12) and (2.7) we see that

$$\frac{v + w - 2f}{a^2} = \frac{w + u - 2g}{b^2} = \frac{u + v - 2h}{c^2}. \tag{2.13}$$

Note that if the equation of a circle is expressed in the form of Equation (2.12), then it may established that u, v, w are the powers of A, B, C with respect to the circle. See equation $A10.4$ of Bradley [1]. If the equation of a circle with a given centre (l, m, n) and a given radius R is required, it may be obtained by equating to R^2 the square of the distance from (x, y, z) to (l, m, n). And then making the equation homogeneous in x, y, z by using the condition $x + y + z = 1$.

2.3.11 Other common circles in the geometry of the triangle

We refer to Bradley [1] for the derivation of these equations.

The incircle

This has equation

$$(s - a)^2x^2 + (s - b)^2y^2 + (s - c)^2z^2 - 2(s - b)(s - c)yz$$

$$-2(s-c)(s-a)zx - 2(s-a)(s-b)xy = 0.$$

Here $s = (a+b+c)/2$. It is easily checked that this satisfies the conditions for being a circle and that it touches the sides BC, CA, AB internally.

The excircle opposite A

This has equation

$$s^2x^2+(s-c)^2y^2+(s-b)^2z^2-2(s-b)(s-c)yz+2s(s-b)zx+2s(s-c)xy = 0.$$

It is easily checked that this satisfies the conditions for being a circle and that it touches BC internally and CA and AB externally.

The nine-point circle

This is the circle that passes through the midpoints of the sides $L(0, \frac{1}{2}, \frac{1}{2})$, $M(\frac{1}{2}, 0, \frac{1}{2})$, $N(\frac{1}{2}, \frac{1}{2}, 0)$ and therefore has the equation

$$(b^2+c^2-a^2)x^2+(c^2+a^2-b^2)y^2+(a^2+b^2-c^2)z^2-2a^2yz-2b^2zx-2c^2xy = 0. \tag{2.14}$$

This meets BC where $(y-z)(c\cos By - b\cos Cz) = 0$ and so passes through D, the foot of the altitude through A. Similarly it passes through E and F.

The polar circle

This is the circle with respect to which the triangle of reference is self-conjugate and it has equation

$$\cot Ax^2 + \cot By^2 + \cot Cz^2 = 0.$$

This circle exists if, and only if, one of A, B, C is obtuse. From Equation (2.9) its centre is the point with areals $(\cot B\cot C, \cot C\cot A, \cot B\cot C)$, that is, the orthocentre H.

2.3.12 Asymptotes of a hyperbola

Suppose the hyperbola has Equation (2.7), with $F > 0$. If the line $px+qy+rz = 0$ is an asymptote, then the hyperbola touches this line where it meets the line at infinity, that is, at the point with co-ordinates $(q-r, r-p, p-q)$. Using Equation (2.8) this tangent has equation

$$(u(q-r)+g(p-q)+h(r-p))x+(v(r-p)+h(q-r)$$
$$+f(p-q))y+(w(p-q)+f(r-p)+g(q-r))z=0$$

and the left hand side must be proportional to $(px+qy+rz)$. It follows that p, q, r must satisfy the eigenvalue equations

$$(g-h-k)p+(u-g)q+(h-u)r=0,$$
$$(f-v)p+(h-f-k)q+(v-h)r=0,$$
$$(w-f)p+(g-w)q+(f-g-k)r=0.$$

The eigenvalues k of these equations are $0, \pm\sqrt{F}$. The zero value has no geometrical significance. Also note that the eigenvalues are all zero for a parabola, and non-real for an ellipse, as one would expect. For a hyperbola the values $k=\pm\sqrt{F}$ provide two sets of values $p:q:r$, thereby determining the equations of the two asymptotes. The general expression for $p:q:r$ is too complicated to be of interest, so we give an example to illustrate the procedure. Consider the hyperbola with equation $x^2+2y^2-6z^2=0$. Here $u=1, v=2, w=-6$. $F=-(uv+vw+wu)=16$, so the eigenvalues are $k=\pm 4$. When $k=4$ we have $-4p+q-r=0$, $-2p-4q+2r=0$, $-6p+6q-4r=0$ and the corresponding asymptote has equation $-x+5y+9z=0$. When $k=-4$ we have $4p+q-r=0$, $-2p+4q+2r=0$ and $-6p+6q+4r=0$ and the corresponding asymptote has equation $x-y+3z=0$. If the asymptotes are at right angles the conic is a rectangular hyperbola. See Section 2.3.3 for the condition for lines to be at right angles.

2.3.13 The radical axis of two circles

If the equations of the two circles are put in the canonical form given by Equation (2.12), then all that needs to be done to find their radical axis is to subtract one equation from the other. The result after division by $(x+y+z)$ is an equation of the form $(u_1-u_2)x+(v_1-v_2)y+(w_1-w_2)z=0$. For example the circles with equation

$$a^2yz+b^2zx+c^2xy-ux(x+y+z)=0$$

for varying u form the intersecting coaxal system of circles passing through B and C.

Example 2.3.1

In triangle ABC, the centres of the escribed circles are denoted by I_1, I_2, I_3 and their radii by r_1, r_2, r_3. The radius of the circumcircle is R. Prove that

(i)

$$\frac{(I_1I_2)^2}{r_1+r_2} = \frac{(r_2+r_3)(r_3+r_1)(r_1+r_2)}{r_2r_3+r_3r_1+r_1r_2} = 4R;$$

(ii)

$$\frac{8r_1r_2r_3}{(I_1I_2)(I_2I_3)(I_3I_1)} = \sin A \sin B \sin C.$$

Using areal co-ordinates we have

$$\frac{I_1(-a,b,c)}{b+c-a} \text{ and } \frac{I_2=(a,-b,c)}{a-b+c}.$$

So the displacement

$$\mathbf{I_1I_2} = \frac{2c(a,-b,b-a)}{(b+c-a)(a-b+c)}.$$

Now, given a displacement (dx, dy, dz) in areal co-ordinates, with $dx + dy + dz = 0$, the distance formula is

$$(ds)^2 = -a^2(dy)(dz) - b^2(dz)(dx) - c^2(dx)(dy).$$

Using this, we obtain,

$$I_1I_2^2 = \frac{4c^2ab}{(b+c-a)(a-b+c)} = \frac{c^2ab}{(s-a)(s-b)},$$

where s is the semiperimeter. The formulas

$$[ABC] = \frac{abc}{4R} = r_1(s-a) = r_2(s-b) = r_3(s-c) = \sqrt{s(s-a)(s-b)(s-c)}$$

are used in what follows, without proof. We have

$$r_1 + r_2 = \frac{[ABC]}{s-a} + \frac{[ABC]}{s-b} = \frac{c[ABC]}{(s-a)(s-b)}.$$

Hence

$$\frac{(I_1I_2)^2}{r_1+r_2} = \frac{c^2ab}{(s-a)(s-b)} \times \frac{(s-a)(s-b)}{c[ABC]} = \frac{abc}{[ABC]} = 4R.$$

Now

$$(r_2+r_3)(r_3+r_1)(r_1+r_2) - \frac{abc[ABC]^3}{(s-a)^2(s-b)^2(s-c)^2}$$

and since

$$r_1r_2 = \frac{[ABC]^2}{(s-a)(s-b)}$$

we have

$$r_1r_2+r_2r_3+r_3r_1 = \frac{[ABC]^2s}{(s-a)(s-b)(s-c)},$$

so their quotient is

$$\frac{abc[ABC]}{s(s-a)(s-b)(s-c)} = 4R.$$

Part (ii) is almost immediate, the left hand side being equal to

$$\frac{8[ABC]^3}{a^2b^2c^2} = \frac{abc}{8R^3} = \sin A \sin B \sin C.$$

Example 2.3.2

Let ABC be a triangle and let K be a point in the plane of the triangle not on its sides and not on a line through a vertex parallel to an opposite side. Let AKL, BKM, CKN be the Cevians through K. Through A draw the lines parallel to BKM and CKN to meet BC in P and U. Let Q, V, R, W be similarly defined. Prove that P, U, Q, V, R, W lie on a conic. Show that when K is an internal point, there exists, in general, a triangle for which the conic is a circle. See Fig. 2.9.

Use areal co-ordinates and let $K(l, m, n)$. The Cevian BKM has equation $nx - lz = 0$. Parallels to BKM have equations of the form $(k+n)x + ky + (k-l)z = 0$ for some non-zero constant k. If such a line passes through A, then $k = -n$ and so P is defined by $ny + (n+l)z = 0$. Similarly U is defined by $(l+m)y + mz = 0$. P and U are therefore determined by the equation

$$nl(m+n)(l+m)y^2 + lm(n+l)(m+n)z^2 + l(m+n)(mn+(l+m)(n+l))yz = 0.$$

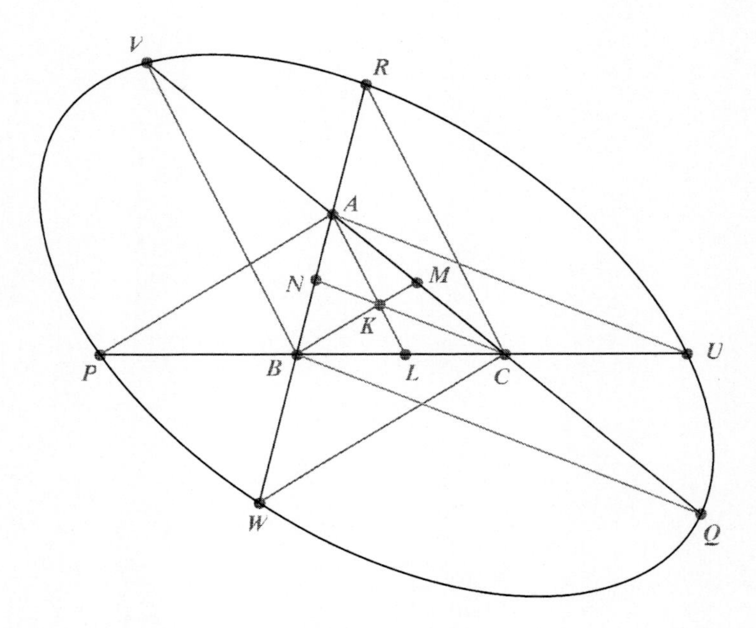

Figure 2.9: Example 2.3.2

By symmetry the six points therefore lie on the conic with equation

$$mn(l+m)(n+l)x^2 + nl(m+n)(l+m)y^2 + lm(n+l)(m+n)z^2$$

$$+l(m+n)(mn+(l+m)(n+l))yz + m(n+l)(nl+(m+n)(l+m))zx$$

$$+n(l+m)(lm+(n+l)(m+n))xy = 0.$$

The condition for this to be a circle is

$$l^3(m+n) = ta^2, m^3(n+l) = tb^2, n^3(l+m) = tc^2$$

for some constant t. With $l = 3/10, m = 1/3, n = 11/30, t = 1/1000$ we get the following lengths for a, b, c to 2 decimal places: $a = 4.35, b = 4.97, c = 5.59$.

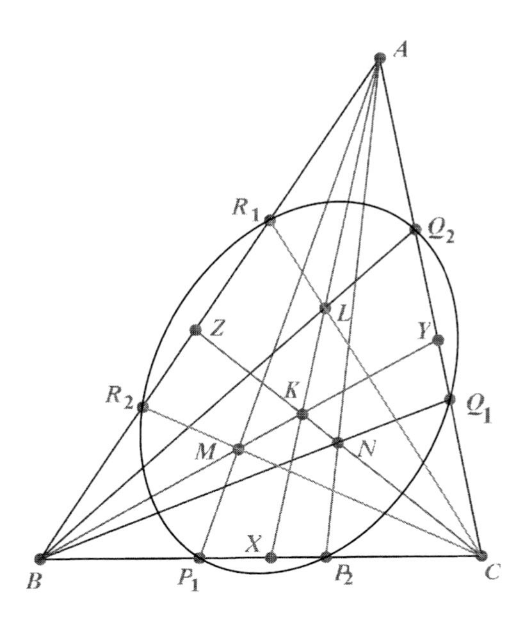

Figure 2.10: Example 2.3.3

Example 2.3.3

ABC is a triangle and AX, BY, CZ are three Cevians. L, M, N are the midpoints of AX, BY, CZ respectively. AM, AN meet BC at P_1, P_2 respectively; BN, BL meet CA at Q_1, Q_2 respectively; and CL, CM meet AB at R_1, R_2 respectively. Prove that $P_1, P_2, Q_1, Q_2, R_1, R_2$ lie on a conic. See Fig. 2.10.

Use areal co-ordinates. Let $K(l, m, n)$ be the point through which the Cevians pass. Then X has co-ordinates $(0, m/(m+n), n/(m+n))$ Since $A(1,0,0)$, the midpoint of AX has co-ordinates $L(\frac{1}{2}, \frac{1}{2}(m/(m+n)), \frac{1}{2}(n/(m+n)))$. The equation of BL is therefore $nx = (m+n)z$ and hence Q_2 has unnormalized co-ordinates $(m+n, 0, n)$. Likewise N has co-ordinates $N(\frac{1}{2}(l/(l+m)), \frac{1}{2}(m/(l+m)), \frac{1}{2})$ and the equation of BN is $(l+m)x = lz$. Hence Q_1 has unnormalized co-ordinates $(l, 0, l+m)$. The co-ordinates of Q_1 and Q_2 therefore satisfy $n(l+m)x^2 - (2ln+m)zx + l(m+n)z^2 = 0$, where we have used $l+m+n = 1$. It follows, by symmetry that $P_1, P_2, Q_1, Q_2, R_1, R_2$ all lie on the conic with equation

$$x^2/l(m+n) + y^2/m(n+l) + z^2/n(l+m) - (2ln+m)zx/ln(ln+m)$$
$$-(2ml+n)xy/ml(ml+n) - (2nm+l)yz/nm(nm+l) = 0.$$

Example 2.3.4

Show that the equation of the circle $I_1I_2I_3$ is

$$bcx^2 + cay^2 + abz^2 + (a+b+c)(ayz + bzx + cxy) = 0.$$

Find the areals of the centre of this circle.

Suppose the equation of the circle is given by Equation (2.7) with conditions (2.13). Now I_1 has areals $(-a, b, c)$ and hence

$$ua^2 + vb^2 + wc^2 + 2fbc - 2gca - 2hab = 0,$$

and similarly

$$ua^2 + vb^2 + wc^2 - 2fbc + 2gca - 2hab = 0,$$

and

$$ua^2 + vb^2 + wc^2 - 2fbc - 2gca + 2hab = 0.$$

It follows that $fbc = gca = hab$, so we may choose $2f = (a+b+c)a, 2g = (a+b+c)b$ and $2h = (a+b+c)c$ (the Equation (2.7) is linear and homogeneous in u, v, w, f, g, h so we may choose the multiple of a, b, c in f, g, h respectively to suit our convenience). This means we also have $ua^2+vb^2+wc^2 = (a+b+c)abc$ and conditions (2.13) become

$$\frac{v + w - (a+b+c)a}{a^2} = \frac{w + u - (a+b+c)b}{b^2} = \frac{u + v - (a+b+c)c}{c^2}.$$

It may now be checked that these three remaining equations are satisfied by $u = bc, v = ca, w = ab$. Using Equation (2.9) for the centre we find its unnormalized co-ordinates to be (X, Y, Z) where

$$X = a^3 - b^3 - c^3 + a^2b + a^2c - ab^2 - ac^2 - 2abc + bc^2 + cb^2,$$

and Y, Z may be obtained from X by cyclic permutation of a, b, c.

Exercise 2.3.5

1. The unnormalized areal co-ordinates of the Brocard points Ω and Ω' are $(1/b^2, 1/c^2, 1/a^2)$ and $(1/c^2, 1/a^2, 1/b^2)$ respectively. Prove that the square of the distance between them is given by

   $$\Omega\Omega'^2 = \frac{a^2b^2c^2(a^4 + b^4 + c^4 - d^4)}{d^8},$$

   where $d^4 = b^2c^2 + c^2a^2 + a^2b^2$.

2. Find, in terms of a, b, c an expression for IG^2, where I and G are the incentre and centroid of triangle ABC. Let ABC be a triangle with centroid G and excentres I_1, I_2, I_3. What can be said about ABC if $I_1G > I_2G > I_3G$?

3. Let ABC be a triangle and suppose that AL, BM, CN are Cevians. Prove that there is a conic touching BC at L, CA at M and AB at N.

4. A straight line through the circumcentre O of triangle ABC meets BC, CA, AB respectively at P, Q, R. Points X, Y, Z are taken so that PX, QY, RZ have O as midpoint. Prove that AX, BY, CZ meet at a point on the circumcircle.

5. Let ABC be a triangle and P a point in the plane of ABC not lying on the sides. Let APD, BPE, CPF be the Cevians through P. The lines through D parallel to BE, CF meet CA, AB respectively at M, N'. Points N, L' and L, M' are similarly defined on lines through E and F. Prove that L, L', M, M', N, N' lie on a conic.

6. Let ABC be a triangle and P a point in the plane of ABC not lying at a vertex of the triangle. The perpendicular bisector of AP meets AB at N and AC at M'. Points L, N' and M, L' are similarly defined on the perpendicular bisectors of BP, CP. Find the locus of P such that L, L', M, M', N, N' lie on a conic.

7. Let ABC be a triangle and P a point in the plane of ABC not at any of the vertices and not at the orthocentre. The perpendicular bisector of AP meets BC at L. Points M and N on CA and AB are similarly defined. Find the locus of P if L, M, N are collinear.

8. ABC is a triangle and AKL, BKM, CKN are three Cevians. D, E, F are the midpoints of AK, BK, CK respectively. AE, AF meet BC at P, Q; BF, BD meet CA at R, S and CD, CE meet AB at T, U respectively. Prove that P, Q, R, S, T, U lie on a conic. Is the result true if D, E, F are the midpoints of KL, KM, KN respectively?

9. A circle meets the sides of a triangle ABC at P, Q, R, P', Q', R'. Suppose that AP, BQ, CR meet at the point $x : y : z$. Show that AP', BQ', CR' meet at the point $x' : y' : z'$ where

$$\frac{xx'(y+z)(y'+z')}{a^2} = \frac{yy'(z+x)(z'+x')}{b^2} = \frac{zz'(x+y)(x'+y')}{c^2}.$$

10. L, M, N are the midpoints of sides BC, CA, AB of triangle ABC. P and Q are points on AB and BC respectively; R and S are points such that N is the midpoint of PR and L is the midpoint of QS. Prove that if PS and QR meet at right angles at T, then T lies on the circle LMN. *Hint: Take LMN to be the triangle of reference.*

11. The tangents to the circumcircle of triangle ABC at B and C meet at L. M and N are similarly defined. Prove that AL, BM, CN are concurrent at a point S, whose areals should be determined.

12. $PACB$ is a trapezium inscribed in a circle with PA parallel to BC. If $AB > AC$ prove that $a^2[PAB] = (c^2 - b^2)[ABC]$.

13. The point S in Problem 11 of this Exercise is called the symmedian point of triangle ABC and has areals (a^2, b^2, c^2). ABC is a triangle and L, M, N are the midpoints of BC, CA, AB respectively. G is the centroid and S the symmedian point. Let X, Y, Z be the poles of MN, NL, LM respectively with respect to the nine-point circle. Prove that XL, YM, ZN, SG are concurrent at a point V such that $VS = 3VG$.

14. ABC is a triangle that is not isosceles. L, M, N are the midpoints of BC, CA, AB respectively. Points X, Y, Z on AL, BM, CN are chosen such that $AL = LX, BM = MY, CN = NZ$. Tangents at X, Y, Z to the circle XYZ meet BC, CA, AB respectively at D, E, F. Prove that D, E, F are collinear and that the line DEF lies outside triangle ABC.

15. The tangents to the circumcircle of triangle ABC at A, B, C meet the opposite sides at L, M, N respectively. Prove that L, M, N are collinear and that

$$(AL)(BM)(CN) = (AM)(BN)(CL) = (AN)(BL)(CM).$$

16. The point A does not lie on the circle C. The lines joining A to the vertices P, Q, R of a variable equilateral triangle inscribed in C meet C again at U, V, W respectively. Prove that $AP/AU + AQ/AV + AR/AW$ is constant. *Hint : Instead, fix C and move A on a concentric circle.*

17. The triangle ABC with sides $a = 24, b = 11, c = 17$ has the property that the perpendicular bisector of BC, the internal angle bisector

of $\angle ABC$ and the median from C to AB are concurrent. Find other integer-sided triangles which possess this property and which also satisfy $3c = a + 2b + 5$.

18. ABC is a triangle. The perpendicular bisectors of BC, CA, AB are denoted by l, m, n respectively and the altitudes through A, B, C are denoted by l', m', n' respectively. Denote the intersections $m \wedge n', n \wedge m'$ by L, L' respectively and let M, M' and N, N' be defined similarly. Prove that L, L', M, M', N, N' lie on a hyperbola and find the areals of its centre. (It may be shown that the conic is actually a rectangular hyperbola.)

19. P is an internal point of triangle ABC. The line through P parallel to AB meets BC at L and CA at M'. The line through P parallel to BC meets CA at M and AB at N'. The line through P parallel to CA meets AB at N and BC at L'. Prove that L, L', M, M', N, N' lie on a conic. Show that this conic is a circle if, and only if, P lies at the symmedian point S and then the centre of the circle is the midpoint of OS. Prove further

    (i) $(BL/LC)(CM/MA)(AN/NB) \leq 1/8$;
    (ii) $[LMN] = [L'M'N']$;
    (iii) $[LMN] \leq \frac{1}{3}[ABC]$.

20. Prove that the equation of any parabola through the vertices of the triangle of reference may be written in the form $(q - r)^2 yz + (r - p)^2 zx + (p - q)^2 xy = 0$ for some constants p, q, r.

21. Show that the polar circle, if it exists, passes through the points of intersection of the circumcircle and the nine-point circle and that the same is true for the circle on GH as diameter (the *orthocentroidal circle*).

22. ABC is a triangle and L, M, N are points on BC, CA, AB respectively, but not at the vertices, Prove that the circles AMN, BNL, CMN meet at a point, called the *pivot point*. If L has areals $(0, l, (1 - l))$, M has areals $((1 - m), 0, m)$ and N has areals $(n, (1 - n), 0)$, find the areals of P in terms of l, m, n, a, b, c.

23. If in Problem 22, L, M, N are collinear, prove that the pivot point lies on the circumcircle of ABC, thereby proving that the circumcircles of the four triangles formed by four lines in general position concur at a point called the *Miquel point.*

24. Prove that the nine-point circle touches the incircle and the three ex-circles; and that their common tangents are $x/(b-c) + y/(c-a) + z/(a-b) = 0$ and the three lines obtained from this by changing a into $-a$, b into $-b, c$ into $-c$. (This is *Feuerbach's theorem.*)

25. Is it possible to draw a circle meeting BC at D, D', CA at E, E' and AB at F, F' such that AD, BE, CF are parallel and AD', BE', CF' are also parallel?

26. ABC is an equilateral triangle and P is any point in the plane of ABC other than the vertices. The perpendicular bisectors of AP, BP, CP meet BC, CA, AB respectively in L, M, N. Prove that L, M, N are collinear.

27. Identify the conic with equation $4yz + zx + xy = 0$.

28. Let ABC be an equilateral triangle and P any point on the circumference of the circumcircle (other than at the vertices of ABC). Let AP meet BC at W and let the perpendicular through P to AP meet BC at V. Finally let the perpendicular bisector of AP meet BC at U. Prove that $BU = VC$ and $BV/VC \div BW/WC = -1$ (where the sign convention for line segments on BC is used).

29. Find the conditions on f, g, h so that the conic $fyz + gzx + hxy = 0$ should have $x = 0$ as an axis of symmetry.

2.4 Trilinear co-ordinates

These are used by some authors in preference to areal co-ordinates. They are not used in this book, but for completeness a short account of them is included.

2.4.1 Definition

Let ABC be the triangle of reference and P a point in the plane Π of ABC. Then the trilinear co-ordinates of P are (d, e, f), where d, e, f are the signed perpendicular distances of P from the sides BC, CA, AB respectively. The signs are the same as with areal co-ordinates, so that, for example if P lies on the same side of BC as A, then d is positive and if P lies on the other side of BC from A, then d is negative. To distinguish trilinear from areal co-ordinates in what follows we use (x, y, z) for the current co-ordinates using areals and (X, Y, Z) for the current co-ordinates using trilinears. We use (d, e, f) for the trilinear co-ordinates of a fixed point P and (l, m, n) for its normalized areal co-ordinates.

2.4.2 The connection between areals and trilinears

Since $l = [PBC]/[ABC]$ and since $[PBC] = \frac{1}{2}ad$, we have $l = ad/(2\Delta)$, where $\Delta = [ABC]$, which is consistent with $l+m+n = 1$, since $ad+be+cf = 2\Delta$. Also $d = 2l\Delta/a$. If one drops the normalization of areals then one can write $(l, m, n) \propto (ad, be, cf)$ or $(d, e, f) = (l/a, m/b, n/c)$.

2.4.3 The equation of a line

The equation in trilinears of the line joining $P_1(d_1, e_1, f_1)$ and $P_2(d_2, e_2, f_2)$ is

$$(e_1 f_2 - f_1 e_2)X + (f_1 d_2 - d_1 f_2)Y + (d_1 e_2 - e_1 d_2)Z = 0.$$

2.4.4 The line at infinity

This corresponds to $x + y + z = 0$ in areals and so has equation

$$aX + bY + cZ = 0$$

in trilinears.

2.4.5 The point of intersection of two lines

The lines with equations $pX + qY + rZ = 0$ and $p'X + q'Y + r'Z = 0$ meet at $X/(qr' - rq') = Y/(rp' - pr') = Z/(pq' - qp')$, where $aX + bY + cZ = 2\Delta$.

2.4.6 Collinear points and concurrent lines

The condition that $P_3(d_3, e_3, f_3)$ lies on P_1P_2 is that the determinant with rows (or columns) the co-ordinates of P_1, P_2, P_3 should vanish. The condition that the line with equation $p''X + q''Y + r''Z = 0$ is concurrent with the lines in Section 2.4.5 is that the determinant with rows (or columns) (p, q, r), (p', q', r') (p'', q'', r'') should vanish.

2.4.7 Parallel lines

These are lines that meet on the line at infinity, so by Section 2.4.6, lines with equations $pX + qY + rZ = 0$ and $p'X + q'Y + r'Z = 0$ are parallel if, and only if, the determinant with rows (or columns) equal to (p, q, r), (p', q', r'), (a, b, c) vanish. Alternatively we may write down the equation of the lines parallel to $pX + qY + rZ = 0$ as $(p + as)X + (q + bs)Y + (r + cs)Z = 0$ for varying s. For example, the line BC has equation $X = 0$, so lines parallel to BC have equations of the form $(1 + as)X + bsY + csZ = 0$. For the line through A parallel to BC we have $s = -1/a$ and so it has equation $bY + cZ = 0$.

2.4.8 Co-ordinates of key points

These are given as ratios, and not as actual perpendicular distances, as these are easily deduced from the ratios by means of Section 2.4.2. A $1 : 0 : 0$; B $0 : 1 : 0$; C $0 : 0 : 1$; L $0 : 1/b : 1/c$; M $1/a : 0 : 1/c$; N $1/a : 1/b : 0$; G $1/a : 1/b : 1/c$; I $1 : 1 : 1$; I_1 $-1 : 1 : 1$; I_2 $1 : -1 : 1$; I_3 $1 : 1 : -1$; H $\sec A : \sec B : \sec C$; O $\cos A : \cos B : \cos C$; D $0 : \sec B : \sec C$; E $\sec A : 0 : \sec C$; F $\sec A : \sec B : 0$; T $\cos(B - C) : \cos(C - A) : \cos(A - B)$.

2.4.9 Distance between two points

The best thing to do is to convert into areal co-ordinates and use the areal distance formula. Alternatively it is possible to find the trilinear distance formula from the areal one. This is left to the reader as an exercise.

2.4.10 Condition for two displacements to be at right angles

Use Section 2.4.9 and Pythagoras's theorem.

2.4.11 The equation of a conic

As in areals this is a homogeneous equation of degree two in X, Y, Z and so may be written as

$$\Phi(X, Y, Z) \equiv uX^2 + vY^2 + wZ^2 + 2fYZ + 2gZX + 2hXY = 0. \tag{2.15}$$

For a conic through the vertices of the triangle of reference $u = v = w = 0$.

2.4.12 Tangent and polar

The polar of the point with co-ordinates (X', Y', Z') is

$$uX'X + vY'Y + wZ'Z + f(Y'Z + Z'Y) + g(Z'X + X'Z) + h(X'Y + Y'X) = 0. \tag{2.16}$$

If (X', Y', Z') lies on the conic, then this is the equation of the tangent at this point.

2.4.13 Centre of a conic

The polar of the centre (X', Y', Z') is the line at infinity $aX + bY + cZ = 0$, so comparing coefficients with Equation (2.16) we get

$$\begin{aligned} X' : Y' : Z' = {} & a(vw - f^2) + b(fg - hw) + c(hf - gv) \\ & : a(fg - hw) + b(wu - g^2) + c(gh - fu) \\ & : a(hf - gv) + b(gh - fu) + c(uv - h^2). \end{aligned} \tag{2.17}$$

2.4.14 The equation of the circumcircle

If we put $f = a, g = b, c = w$ in Equation (2.17) we get the centre $\cos A : \cos B : \cos C$, which is the centre of the circumcircle. The equation of the circumcircle is therefore

$$aYZ + bZX + cXY = 0.$$

2.4.15 The equations of other circles

These may be written in the canonical form

$$aYZ + bZX + cXY - (aX + bY + cZ)(UX + VY + WZ) = 0 \qquad (2.18)$$

since every circle passes through the circular points at infinity. Here Ubc, Vca, Wab are the powers of A, B, C with respect to the circle. Alternatively, a circle has Equation (2.15), where

$$b^2w + c^2v - 2fbc = c^2u + a^2w - 2gca = a^2v + b^2u - 2hab.$$

The incircle

This has equation

$$(b+c-a)^2a^2X^2 + (c+a-b)^2b^2Y^2 + (a+b-c)^2c^2Z^2$$
$$-2bc(c+a-b)(a+b-c)YZ - 2ca(a+b-c)(b+c-a)ZX$$
$$-2ab(b+c-a)(c+a-b)XY = 0.$$

The excircle opposite A

This has equation

$$(a+b+c)^2a^2X^2 + (c+a-b)^2b^2Y^2 + (a+b-c)^2c^2Z^2$$
$$-2ca(c+a-b)(a+b-c)YZ + 2ca(a+b+c)(c+a-b)ZX$$
$$+2ab(a+b+c)(a+b-c)XY = 0.$$

The nine-point circle

This has equation

$$(b^2+c^2-a^2)a^2X^2 + (c^2+a^2-b^2)b^2Y^2 + (a^2+b^2-c^2)c^2Z^2$$
$$-2a^2bcYZ - 2ab^2cZX - 2abc^2XY = 0.$$

An alternative form is

$$\sin 2AX^2 + \sin 2BY^2 + \sin 2CZ^2 - 2\sin AYZ - 2\sin BZX - 2\sin CXY = 0.$$

The polar circle

This has equation

$$\sin 2AX^2 + \sin 2BY^2 + \sin 2CZ^2 = 0.$$

2.4.16 The radical axis of two circles

If the equations of the two circles are in the canonical form described by Equation (2.18), one with U, V, W, the other with U', V', W' then the equation of the radical axis is

$$(U - U')X + (V - V')Y + (W - W')Z = 0.$$

2.4.17 Ellipses, parabolas and hyperbolas

The conic with Equation (2.15) is an ellipse, parabola or hyperbola according as it meets the line at infinity, with equation $aX + bY + cZ = 0$ in non-real, coincident real or two real values of x, y, z. This involves the function $F(u, v, w, f, g, h)$ given by

$$F = a^2f^2 + b^2g^2 + c^2h^2 - 2bcgh - 2cahf - 2abfg$$

$$+2bcuf + 2cavg + 2abwh - a^2vw - b^2wu - c^2uv.$$

If $F > 0$ then the conic is a hyperbola, if $F = 0$ it is a parabola and if $F < 0$ it is an ellipse. $F = 0$ is also the condition that the centre of the conic lies on the line at infinity, showing that this is consistent with the idea that a parabola touches the line at infinity.

Exercise 2.4.1

1. Show that in trilinears the equations of

   (i) the angle bisectors are $Y = Z$, $Z = X$, $X = Y$;

   (ii) the medians are $bY = cZ, cZ = aX, aX = bY$;

   (iii) the altitudes are $Y \cos B = Z \cos C, Z \cos C = X \cos A, X \cos A = Y \cos B$;

   (iv) the perpendicular bisector of BC is $X \sin(C - B) - Y \sin B + Z \sin C = 0$.

2. Show that in trilinears the equation of the Euler line (the line joining G and H) is

$$X \sin 2A \sin(B - C) + Y \sin 2B \sin(C - A) + Z \sin 2C \sin(A - B) = 0.$$

3. Show that the co-ordinates of the pole of the line with equation $pX + qY + rZ = 0$ with respect to the conic with Equation (2.16) are as

$$Up + Hq + Gr : Hp + Vq + Fr : Gp + Fq + Wr,$$

where $U = vw - f^2$ etc., and $F = gh - uf$ etc.

4. Writing $U_p = Up + Hq + Gr$, $U_q = Hp + Vq + Fr$, $U_r = Gp + Fq + Wr$ prove that the condition the lines with equations $pX + qY + rZ = 0$ and $p'X + q'Y + r'Z = 0$ are conjugate (pass through each other's poles) is $p'U_p + q'U_q + r'U_r = 0$ (or what is the same thing that $pU_{p'} + qU_{q'} + rU_{r'} = 0$).

5. Show that the co-ordinates of the centre of the conic (2.16) may be re-expressed in the form

$$Ua + Hb + Gc : Ha + Vb + Fc : Ga + Fb + Wc.$$

6. It may be shown that in trilinears the condition the conic with Equation (2.16) is a rectangular hyperbola is $u + v + w = 2f \cos A + 2g \cos B + 2h \cos C$. Use this to show that if a rectangular hyperbola passes through the vertices of the triangle of reference, then it must pass through the orthocentre.

Chapter 3

Projective co-ordinates

3.1 The embedding of the Euclidean plane in the projective plane

The initial step is to take any equation in Cartesian co-ordinates and to write $x = X/Z$ and $y = Y/Z$. In this way an equation in Cartesian co-ordinates becomes homogeneous in X, Y, Z. For example the line with equation $lx + my + n = 0$ becomes $lX + mY + nZ = 0$, the parabola $y^2 = 4ax$ becomes $Y^2 = 4aXZ$, the circle $x^2+y^2+2gx+2fy+c = 0$ becomes $X^2+Y^2+2gXZ+2fYZ+cZ^2 = 0$, the ellipse $x^2/a^2+y^2/b^2 = 1$ becomes $X^2/a^2+Y^2/b^2 = Z^2$ and the hyperbola $x^2/a^2 - y^2/b^2 = 1$ becomes $X^2/a^2 - Y^2/b^2 = Z^2$. The process may be reversed by setting $Z = 1, X = x, Y = y$.

Nothing much appears to have happened. But in fact this is not the case. Note that the points $(X, Y, Z) = (kx, ky, k)$, $k \neq 0$ represent, for varying k, the same point (x, y) in the Cartesian plane. What, in fact, has been created is a real 3-dimensional vector space V_3 with the origin $(0, 0, 0)$ deleted, in which the rays through the origin are placed in correspondence with the points of $\mathbb{E}^2$, the 2-dimensional real Euclidean space.

In this new space the tangents to the above conics at (X', Y', Z') have equations $Y'Y = 2a(Z'X + X'Z)$ for the parabola, $X'X + Y'Y + g(Z'X + X'Z) + f(Z'Y + Y'Z) + cZ'Z = 0$ for the circle, $X'X/a^2 + Y'Y/b^2 = Z'Z$ for the ellipse and $X'X/a^2 - Y'Y/b^2 = Z'Z$ for the hyperbola. The reason is that these equations become the known equations of the tangents in $\mathbb{E}^2$ at (x', y'), when we put $Z = Z' = 1$, $X = x$, $Y = y$, $X' = x'$, $Y' = y'$.

The generalization is continued by allowing $Z = 0$ (X, Y not both zero) to exist. In Chapter 2 we have already met the concept of a line at infinity, with equation $x + y + z = 0$ in areals. The same thing now takes place with the newly defined homogeneous co-ordinates, and $Z = 0$ is now to be regarded as the line at infinity.

We may now regard the vector space V_3 to be defined over the complex field, thereby creating impossible points as far as the original Euclidean plane is concerned. This allows us to look at conics in relation to the line at infinity. For the parabola, it intersects the line $Z = 0$ at $Y^2 = 0$, the double root $Y = 0$ indicating the parabola touches the line at infinity at $(1, 0, 0)$. Every circle meets the line $Z = 0$ where $X^2 + Y^2 = 0$. That is, every circle passes through $(1, \pm i, 0)$, the so-called circular points at infinity. Neither does the ellipse intersect $Z = 0$ at real points, consistent with the fact that in the real plane it is a finite closed curve. The hyperbola intersects $Z = 0$ at the point $(a, b, 0)$ and $(a, -b, 0)$, that is where the asymptotes intersect the line at infinity. Also, as can be checked, the tangent to the hyperbola at $(a, b, 0)$ is $X/a - Y/b = 0$, showing that the asymptote touches the hyperbola on the line at infinity. This provides supplementary reasoning for having anticipated these results in the classification of conics in Chapter 2 using areal co-ordinates.

At this stage we have a Euclidean plane enhanced with a line at infinity, but after the next generalization there is indeed a really new structure. It is prompted by the simple enquiry as to why $Z = 0$ should be anything special. Just forget Euclidean geometry, forget distance, and put X, Y, Z on an equal footing. Work with $V_3 \setminus \{(0, 0, 0)\}$ over the complex field, call the rays through the origin *points*, call solution sets of homogeneous equations of the first degree *lines* and call solution sets of homogeneous equations of the second degree *conics*. (Curves with equations that are homogeneous of higher degree are, of course, admitted, but we are very seldom concerned with them in this book.) This now defines the *projective plane* $\mathbb{P}^2$, and a study of its properties is called *projective geometry*. For the most part we use only the real field in dealing with the projective plane. Where exceptions occur this should be obvious from the context. See also the Appendix for an explanatory paragraph.

In Euclidean geometry the interest is in distance, angle, shape and area. The underlying group of transformations is the set of similarities and distance preserving transformations. In projective geometry the interest is in incidence, configurations of points and lines, cross-ratio of points on a line, and for conics with the notions of tangency, poles and polars. The underlying group of transformations is the projective general linear group over the complex field, which maps points on to points and lines on to lines, so that a given configuration is mapped on to another with the same fundamental properties.

3.2 Basic ideas

3.2.1 Duality

The equation of a line is of the form $lX + mY + nZ = 0$, where if we regard l, m, n as fixed, then the equation means that the point (x, y, z) lies on the line if, and only if, $lx+my+nz = 0$. In like manner we may define the equation of a point to be of the form $xL+yM+zN = 0$, where if x, y, z are fixed, then the equation means that the line (l, m, n) passes through the point if, and only if, $xl + ym + zn = 0$. It follows that for any configuration of points and lines there corresponds a *dual configuration* of lines and points. Thus a *quadrangle* consists of four points A, B, C, D and six lines AB, AC, AD, BC, BD, CD. The dual figure is the quadrilateral, which consists of four lines a, b, c, d and six points ab, ac, ad, bc, bd, cd. Here ab means the point of intersection of lines a and b. (Sometimes $a \wedge b$ is written for ab.) If AC meets BD at E, AB meets CD at F and AD meets BC at G, then EFG is called the *diagonal point triangle.* If the join of the points ac, bd is the line e, the join of the points ab, cd is f and the join of the points ad, bc is g, then efg is called the *diagonal line triangle.*

The dual interpretation of equations means that for every theorem that holds for a quadrangle there is a dual theorem for a quadrilateral in which the word 'point' and 'line' are exchanged and phrases such as 'the line joining two points' and 'the point that is the intersection of two lines' are exchanged. The same holds true for other dual configurations.

3.2.2 Configurations

We use the standard notation, whereby a set of m points and n lines is called a configuration (m_s, n_t) if s of the lines pass through each of the m points and t of the points lie on each of the n lines. Clearly $ms = nt$. The dual configuration is (n_t, m_s). Thus a quadrangle is the configuration $(4_3, 6_2)$ and the quadrilateral is the configuration $(6_2, 4_3)$. A configuration is self-dual if

$m = n$ and $s = t$. Then we can denote that configuration by (m_s). The triangle is self-dual and is obviously the configuration (3_2). Fig. 3.1 below shows the self-dual configuration (9_3), the existence of which is a consequence of Pappus's theorem, which is proved shortly.

The content of the theorem is that, if A, C, E are distinct points on a line and B, D, F are distinct points on a second line, then $L = CF \wedge ED$, $M = EB \wedge AF$, $N = AD \wedge CB$ are collinear. Thus, as can be seen in Fig. 3.1, there are 9 points with 3 lines through each of them and 9 lines with 3 points on each of them.

3.2.3 The intersection of two lines

Two distinct lines meet in a unique point. Distinct lines with equations $l_1X + m_1Y + n_1Z = 0$ and $l_2X + m_2Y + n_2Z = 0$ meet at the unique point that satisfies both equations and which therefore has co-ordinates

$$(x, y, z) = (m_1n_2 - n_1m_2, n_1l_2 - l_1n_2, l_1m_2 - m_1l_2).$$

This is because the lines are distinct, so not all of the three co-ordinates can vanish simultaneously. Note that as there is no 'line at infinity' there is no concept of parallel or perpendicular lines in projective geometry and there is no concept of the gradient of a line. However, in Chapter 17 we show how these concepts can be simulated, producing generalizations of many of the famous theorems in Euclidean geometry.

3.2.4 The line joining two points

Two distinct points are joined by a unique line. Distinct points with equations $x_1L + y_1M + z_1N = 0$ and $x_2L + y_2M + z_2N = 0$ are joined by the unique line that satisfies both equations and therefore has co-ordinates

$$(l, m, n) = (y_1z_2 - z_1y_2, z_1x_2 - x_1z_2, x_1y_2 - y_1x_2). \tag{3.1}$$

This is because the points are distinct, so not all of the three co-ordinates can vanish simultaneously. The line joining the points has equation $lX + mY + nZ = 0$, where l, m, n are given by Equation (3.1). If the point with co-ordinates (x_3, y_3, z_3) also lies on the line, then

$$y_1z_2x_3 + z_1x_2y_3 + x_1y_2z_3 - z_1y_2x_3 - x_1z_2y_3 - y_1x_2z_3 = 0.$$

3.2.5 The triangle of reference and the unit point

Given any triangle in the projective plane it is possible to choose the vertices to be $A(1,0,0)$, $B(0,1,0)$, $C(0,0,1)$ so that BC has equation $x = 0$, CA has equation $y = 0$, and AB has equation $z = 0$. A fourth point U, lying on none of these lines may be given co-ordinates $U(1,1,1)$. Lines AU, BU, CU have equations $y = z$, $z = x$, $x = y$ respectively. The reason for Section 3.2.5 is that there exists a non-singular real transformation leaving the vertices of the triangle of reference invariant but which transforms the co-ordinates of any other point into the point with co-ordinates $(1,1,1)$. This is, however, the limit of our flexibility. For example, we cannot then specify some line in the diagram as having equation $x + y + z = 0$, unless we know already that two points lying on it have co-ordinates that sum to zero. It is important to understand that ratio of line segments has no meaning in projective geometry, so although the point $R(0,m,n)$ lies on BC, there is no implication (as when using areals) that $BR/RC = n/m$. There is no centroid to the triangle of reference, so although U has co-ordinates $(1,1,1)$ there is no positional inference to be made. U is called the *unit point*. Though it is evident that AU meets BC at $(0,1,1)$, this point must not be interpreted as the midpoint of BC. After having used areal co-ordinates in the Euclidean plane, it is sometimes a bit difficult to readjust when using co-ordinates in the projective plane.

3.2.6 Pappus's theorem

Let $ABCDEF$ be a hexagon with alternate vertices A, C, E on one line and B, D, F on another. Then, if $L = CF \wedge ED$, $M = EB \wedge AF$, $N = AD \wedge CB$, it follows that L, M, N are collinear.

See Fig. 3.1. Take CBF as triangle of reference with $C(1,0,0)$, $B(0,1,0)$, $F(0,0,1)$. Let M be the unit point $(1,1,1)$. It follows that BDF is the line with equation $x = 0$, so we may take D to have co-ordinates $(0,c,1)$. Since BME has equation $x = z$ it follows that E may be taken to have co-ordinates $(1,k,1)$. The line ACE therefore has equation $y = kz$ and since AMF has equation $x = y$, it follows that A has co-ordinates $A(k,k,1)$. The line AND has equation $(k-c)x - ky + kcz = 0$ and since CNB has equation $z = 0$, it follows that N has co-ordinates $(k, k-c, 0)$. The line ELD has equation $(k-c)x - y + cz = 0$ and as CLF has equation $y = 0$, it follows that L has co-ordinates $L(c, 0, c-k)$. It may now be verified that L, M, N are collinear

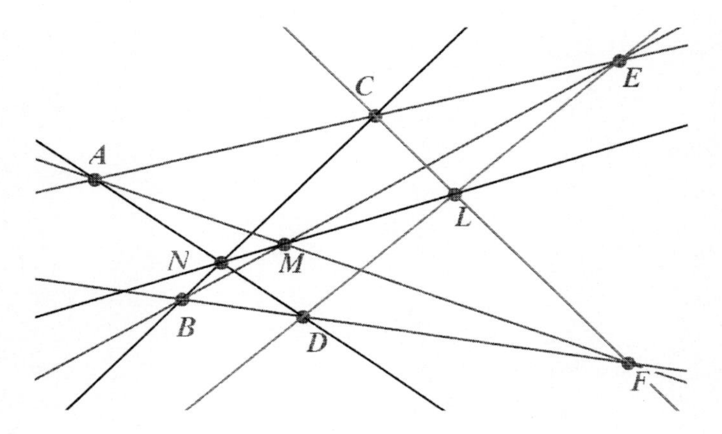

Figure 3.1: Pappus's Theorem

on the line with equation $(c-k)x+ky-cz=0$.

It may be observed that the ability to choose the triangle of reference and the unit point wherever we wish (provided the unit point does not lie on the sides of the triangle of reference) makes a co-ordinate proof in the projective plane easier than in the Euclidean plane. In the latter we would have had to give M arbitrary areal co-ordinates (p,q,r) and although the proof would have gone through, it would have been more involved. The same is true for all theorems involving incidence properties only.

3.2.7 Desargues's theorem

If triangles ABC and DEF are in perspective from a point O (that is AD, BE, CF are concurrent at O, the order of the letters being important), then $L=BC\wedge EF$, $M=CA\wedge FD$, $N=AB\wedge DE$ are collinear. See Fig. 3.2.

The (10_3) configuration depends for its existence on this theorem.

Let ABC be the triangle of reference, with $A(1,0,0)$, $B(0,1,0)$ and $C(0,0,1)$ and let O be the unit point $(1,1,1)$. Since OAD, OBE, OCF are straight lines we may take D,E,F to have co-ordinates $D(d,1,1)$, $E(1,e,1)$, $F(1,1,f)$ for some constants d,e,f. The equation of EF is $(ef-1)x+(1-f)y+(1-e)z=0$. Since BC has equation $x=0$, it follows that L has co-ordinates $L(0,1-e,f-1)$. Similarly M,N have co-ordinates $(d-1,0,1-f),(1-d,e-1,0)$ respectively. These three points lie on the line with equation $(1-e)(1-f)x+(1-f)(1-d)y+(1-d)(1-e)z=0$.

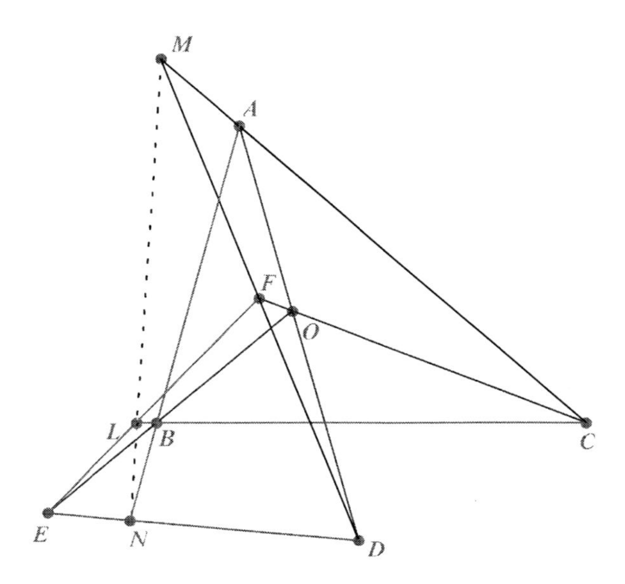

Figure 3.2: Desargues's Theorem

The point O is called the *vertex of perspective* (or *perspector*) and LMN the *axis of perspective* (or *perspectrix*).

Note that the converse of this theorem is the same theorem as its dual and therefore does not require separate proof.

Exercise 3.2.1

1. By varying the hexagon on the two lines in Pappus's theorem how many different Pappus lines, such as LMN, can be drawn?

2. State the dual of Pappus's theorem.

3. ABC is a triangle and P is a point not lying on any of its sides. The line AP meets BC at L; BP meets CA at M; CP meets AB at N. The line MN meets BC at X, NL meets CA at Y and LM meets AB at Z. Prove that X, Y, Z are collinear. If ABC is the triangle of reference and P is the unit point, what is the equation of the line XYZ? Prove the converse of this theorem. The line XYZ is sometimes called the *polar* of P with respect to the triangle ABC.

4. If triangles ABC and TUV are in perspective (regard being paid to the order of the letters) and ABC and UVT are in perspective, prove that ABC and VTU are in perspective.

5. ABC is a triangle and L, M, N are three non-collinear points on BC, CA, AB respectively. A variable line, passing through A, meets LM and LN at P and Q respectively. Prove that BP and CQ meet on a fixed line. State the dual result.

6. ABC is a triangle and D, E, F are points on BC, CA, AB respectively such that AD, BE, CF meet at a point U. P, Q, R lie on EF, FD, DE respectively and are such that DP, EQ, FR meet at a point V. Prove that AP, BQ, CR are concurrent.

7. A line with equation $x + y + z = 0$ meets the sides $x = 0$, $y = 0$, $z = 0$ of the triangle of reference ABC at points X, Y, Z. Let $YB \wedge ZC = P$, $ZC \wedge XA = Q$, $XA \wedge YB = R$. Prove that AP, BQ, CR are concurrent at a point K, whose co-ordinates should be determined.

3.3 Parameters and harmonic conjugates

3.3.1 The line PQ and the parameter of a line

Suppose that P and Q are distinct points with co-ordinates $P(d, e, f)$ and $Q(u, v, w)$, then we define the point $R = mP + nQ$ as the point with co-ordinates $(md + nu, me + nv, mf + nw)$. Because of the linearity of the equation of the line PQ, it follows that R lies on PQ and every point on PQ has co-ordinates in this form, for some values of m and n. See Section 3.2.4. Now put $t = n/m$ (where $t = \infty$ is allowed as we are working in the complex plane) and call t the *parameter of the line* PQ; $t = 0$ corresponds to P, $t = \infty$ corresponds to Q.

3.3.2 Harmonic conjugate points

If R has parameter t, then the point S with parameter $-t$ is called the harmonic conjugate of R with respect to P and Q, whose parameters are 0 and ∞ respectively.

3.3.3 Projective $1-1$ correspondence

Suppose that we have two lines (which may not be distinct) and a parameter s on one line and a parameter t on the other. Then the relationship

$$s = (at+b)/(ct+d), \ ad \neq bc, \tag{3.2}$$

defines a projective $1-1$ correspondence between the points on the two lines. The inverse transformation is given by

$$t = -(ds-b)/(cs-a), ad \neq bc.$$

Clearly the transformation is a bijection between the lines.

3.3.4 Involution

If the lines coincide and the projective $1-1$ correspondence has the property that if, under the transformation (3.2), whenever $U \mapsto V$, it is the case that $V \mapsto U$, then the transformation is called an *involution*. The condition for this is evidently that $d = -a$ in Equation (3.2).

3.3.5 Harmonic conjugate points again

In Section 3.3.2 points R and S with parameters t and $-t$ with respect to points P and Q with parameters 0 and ∞ respectively are defined as harmonic conjugates. In general, points R' and S' are said to be *harmonic conjugates* with respect to points P and Q if $P \leftrightarrow P', Q \leftrightarrow Q', R \leftrightarrow R', S \leftrightarrow S'$ are related by a projective $1-1$ correspondence, where P, Q, R, S are defined as above. For example, if $s = t/(1-t)$, then $t = (0, \frac{1}{2}, 1, \infty) \leftrightarrow s = (0, 1, \infty, -1)$, so points with parameters $\frac{1}{2}$ and ∞ separate harmonically points with parameters 0 and 1.

3.3.6 Cross Ratio

If A, B, C, D are four points on a line with parameters a, b, c, d then the cross-ratio $(A, C; B, D)$ is defined to be

$$\frac{(b-a)(c-d)}{(c-b)(d-a)}.$$

(In Euclidean geometry, if a, b, c, d are the usual (real) co-ordinates on a line $ABCD$ the cross-ratio becomes $(AB/BC) \div (AD/DC)$, where the lengths are signed.)

The results in Sections 3.3.1-3.3.6 have their duals when we deal with lines passing through a point, a so-called *pencil* of lines. We can take two of the lines p and q to have parameters $t = 0$ and $t = \infty$ respectively, and by the same equations, but using line co-ordinates rather than point co-ordinates, we may designate any line t through pq to have that parameter. The lines with parameters t and $-t$ are then said to be harmonic conjugates with respect to p and q.

We now illustrate the above concepts with some examples and exercises.

Example 3.3.1

Let $ABCD$ be a quadrangle and let E, F, G be defined by $E = AC \wedge BD$, $F = AB \wedge CD$, $G = AD \wedge BC$. Let FE meet DA at N and BC at L and let GE meet CD at M and AB at K.

Take ABC to be the triangle of reference and D the unit point and work out the co-ordinates of all the points and the equations of all the lines in the diagram. Set up parameters and find whenever four points on a line are such that two of them are harmonic conjugates with respect to the other two.

The working is easy. We simply state the results. The co-ordinates of the points are $A(1,0,0), B(0,1,0), C(0,0,1), D(1,1,1), E(1,0,1)$, $F(1,1,0)$, $G(0,1,1)$, $L(0,1,-1)$, $M(1,1,2)$, $N(2,1,1)$ and $K(-1,1,0)$. The equations of the lines are $BLCG$, $x = 0$; $KEMG$, $z = x + y$; $ANDG$, $y = z$; $FAKB$, $z = 0$; $FNEL$, $x = y + z$; $FDMC$, $x = y$. On the line $BLCG$ we take the parameter $t = z/y$ and then B, L, C, G have parameters $0, -1, \infty, 1$ respectively so L and G separate B and C harmonically. On $KEMG$ we can take the parameter y/x and then K, E, M, G have parameters $-1, 0, 1, \infty$ respectively, so E and G separate K and M harmonically. On $ANDG$ we can take y/x as parameter and then A, N, D, G have parameters $0, \frac{1}{2}, 1, \infty$ respectively and hence by Section 3.3.5 N and G separate A and D harmonically. Since $FAKB$, $FNEL$, $FDMC$, FG are straight lines with AND, KEM, AND concurrent at G, it is an educated guess that FE and FG are lines of the pencil through F that separate FAB, FDC harmonically. If this is so, and we prove it shortly to be the case, then we have a theorem which says that *pairs of lines of the diagonal point triangle separate harmonically the sides of the quadrangle passing through the same point. Moreover, these*

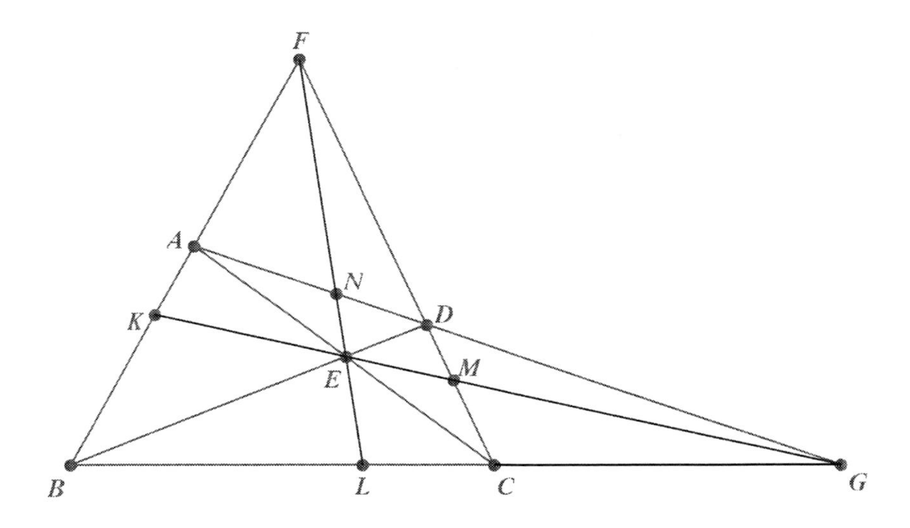

Figure 3.3: The complete quadrangle

lines cut out harmonic ranges, *as sets of such points as* A, N, D, G *are called, on the other pair of sides of the quadrangle.* Now FE, FG have line co-ordinates $(1,-1,-1)$ and $(-1,1,-1)$ respectively. And FAB, FDC have line co-ordinates $(0,0,1)$ and $(1,-1,0)$ respectively. Taking as parameter n/m the first pair have parameters 1 and -1 and the second pair have parameters ∞ and 0, so the theorem holds. It follows that F, A, K, B; F, N, E, L; F, D, M, C are also harmonic ranges. And, if BD and AC intersect FG at U and V, then they separate F and G harmonically.

Exercises 3.3.2.2 and 3.3.2.3 are important results.

Exercise 3.3.2

1. State the dual result to Example 3.3.1 for a complete quadrilateral.

2. Points A, B, C, D on a line have parameters k, l, m, n respectively. Show that the equation satisfied by k, l, m, n if A and C separate B and D harmonically is
$$\frac{(l-k)(m-n)}{(m-l)(n-k)} = -1.$$
That is, the cross ratio $(A, C; B, D) = -1$.

3. Suppose that in a projective $1-1$ correspondence $A \mapsto A'$, $B \mapsto B'$, $C \mapsto C'$, $D \mapsto D'$. Prove that the cross-ratio $(A', C'; B', D') = (A, C; B, D)$. (This justifies, for example, the extension of the definition of harmonic conjugate points in Section 3.3.5 above.)

Example 3.3.3

On the line with equation $x = 0$ we may take any three points A, B, C that lie on it to have co-ordinates $A(0, 1, 0)$, $B(0, 0, 1)$, $C(0, 1, 1)$, where any point on the line has co-ordinates $S(0, 1, s)$, and s is a parameter taking the values $0, \infty, 1$ at A, B, C respectively.

From Section 3.3.1, if $P(x_1, y_1, z_1)$ and $Q(x_2, y_2, z_2)$ are any two points, then the co-ordinates of any point R on PQ may be expressed in the form $R(x_1 + tx_2, y_1 + ty_2, z_1 + tz_2)$, where t is a parameter. $t = 0$ corresponds to P and $t = \infty$ corresponds to Q. Any third point on the line that we desire, provided it is distinct from P and Q, may be given the parameter with value 1. For if we take a point R with parameter $t = \tau$, then by choosing Q to have co-ordinates $(\tau x_2, \tau y_2, \tau z_2)$, we see that, with a parameter s, R has co-ordinates $(x_1 + sx_2, y_1 + sy_2, z_1 + sz_2)$ with $s = 1$. Also $t = 0, \infty$ correspond to $s = 0, \infty$ respectively. Clearly changing scale in this way can only be performed once, so we cannot give an arbitrary fourth point any value of s we like. Now take P to be the point A and Q to be the point B and C to be the point R.

We now explain more fully the idea behind a projective $1-1$ correspondence. It is to set up a correspondence that relates three points on one line to three points on another line, or possibly to three points on the same line. It is sufficiently general to take the first line to have parameter t and the three points on it to be A, B, C with $t = 0, \infty, 1$ respectively and co-ordinates as in Example 3.3.3, and to take the second line to have parameter s and the three points on it to be D, E, F with $s = b/d, a/c, (a+b)/(c+d)$ respectively. (If the two lines are not distinct, we adopt the convention that the point P with parameter $t = k$, also has parameter $s = k$.) When the lines are distinct, then we can choose DEF to have equation $y = 0$ and D, E, F to have co-ordinates $D(d, 0, b)$, $E(c, 0, a)$, $F(c + d, 0, a + b)$, the parameter s being z/x. Note that given any two points on a line it is always possible to normalize their co-ordinates, so that an arbitrary third point has co-ordinates the sum of the other two points. But as in Example 3.3.3 this is a trick that can only be played once and concerns just three points on a line. It must be emphasized that care always has to be taken that the choice of co-ordinates for points in a diagram is sufficiently general. A proof that lacks generality is no proof at all. Note that for D, E, F to be distinct it is necessary and sufficient that $ad \neq bc$. Provided this condition holds, it is possible to choose a, b, c, d so that D, E, F are any three points on the second line. When the lines are the same, all one has to do is to interchange the x and y co-ordinates of D, E, F and use the parameter $s = z/y$. The fact that there are three independent ratios $a : b : c : d$ ensures that the choice of co-ordinates for D, E, F is sufficiently general.

A projective $1 - 1$ correspondence, as defined in Section 3.3.3 above, is often called a *projectivity* and is therefore a relationship between two sets of points with parameters s and t such that $s = (at+b)/(ct+d)$, where a, b, c, d are any complex constants such that $ad \neq bc$.

It is immediately evident that under the projectivity (3.2), A corresponds to D, B corresponds to E and C corresponds to F. A notation commonly used for a projectivity is $(ABC) \wedge (DEF)$, where the order of letters is important.

Note that Equation (3.2) may be written in the form $cst + ds - at - b = 0$. If the two lines are the same, then there are two values of s for which $s = t$. When $c = 0$, one of these points is at $s = t = \infty$. When $b = 0$ one of these points is at $s = t = 0$. When $c = 0$ and $a = d$, then both points are at ∞. When $b = 0$ and $a = d$, then both points are at 0. Such points are called self-corresponding points, for the obvious reason that they are invariant under the projectivity. Their parameters are given by the roots of the quadratic equation $ct^2 + (d - a)t - b = 0$. The two points coincide if, and only if, $(d - a)^2 + 4bc = 0$. If the points are on the same line and $d = -a$, the projectivity is its own inverse and such a projectivity, by the definition in Section 3.3.4 above, is called an involution. Since there are now only two independent ratios $a : b : c$, rather than three independent ratios $a : b : c : d$, it follows that, whereas a projectivity is determined by three pairs of corresponding points, an involution is determined by just two pairs of corresponding points. In an involution there are two self-corresponding points, given by the roots of the quadratic equation $ct^2 - 2at - b = 0$. They cannot coincide since we have excluded the case $bc = ad$, that is, $a^2 + bc = 0$.

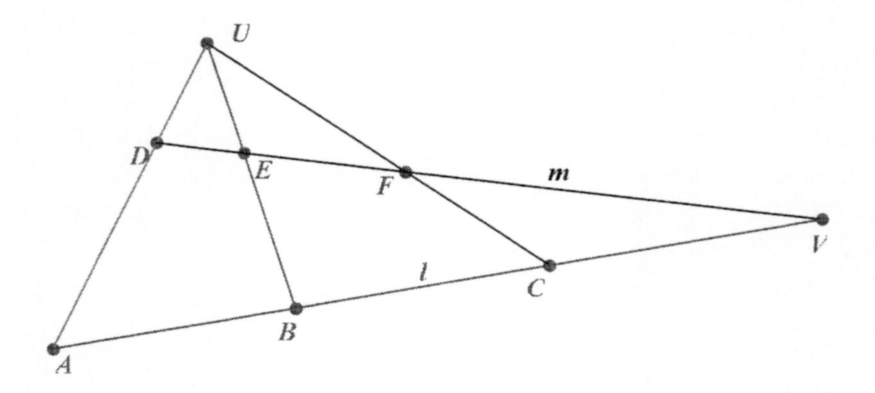

Figure 3.4: Example 3.4.1

3.4 Perspectivity, involution and cross-ratio

Suppose that l is a line passing through three distinct points A, B, C. Suppose that m is another line and that U is a point lying neither on l nor m. Suppose further that AU, BU, CU meet m at D, E, F respectively, then the correspondence between A, D; B, E; C, F is called a perspectivity. See Fig. 3.4. We use the standard notation $(DEF) = U(ABC)$ for a perspectivity through a vertex U. Note that the symbol U could equally well be placed before the symbol (DEF) rather than before the symbol (ABC).

Example 3.4.1

A perspectivity is a projectivity.

Take A, B to be points on $x = 0$ with co-ordinates $A(0, 1, 0)$, $B(0, 0, 1)$, and parameters $t = 0$, ∞, respectively, and take F to be the point $F(1, 0, 0)$. If U is the unit point $U(1, 1, 1)$, then FU meets $x = 0$ at the point $C(0, 1, 1)$ with parameter $t = 1$. Now the equation of UB is $x = y$, so we may take E to have co-ordinates $(1, 1, p)$ for some value of $p \neq 1$. Likewise the equation of UA is $x = z$, and since EF has equation $z = py$, the point D, which is their intersection, has co-ordinates $D(p, 1, p)$. Now we may parameterize DEF with $s = x/y$, so that D, E, F have parameters p, 1, ∞ respectively. Then we have

$$s = (t - p)/(t - 1), \; p \neq 1. \tag{3.3}$$

Comparison with Equation (3.2) shows this to be a projectivity.

Example 3.4.2

A projectivity between the points of two different lines is a perspectivity if, and only if, their point of intersection is an invariant point.

Note that DEF meets ABC at the point V with co-ordinates $(0, 1, p)$. Since V has parameter $t = p$ on $x = 0$, and $s = 0$ on $z = py$, we see from Equation (3.3) that V is an invariant point of the perspectivity. We may therefore write $(DEFV) = U(ABCV)$. Conversely, if there is a projectivity between two lines such that their point of intersection is an invariant point, then $(BCV) \wedge (EFV) \implies (EFV) = U(BCV)$, where U is the intersection of BE and CF.

Example 3.4.3

(a) A projectivity between points on two different lines is expressible as the product of two perspectivities.

(b) A projectivity between points on the same line is expressible as the product of three perspectivities.

We prove these statements in turn.

(a) Suppose that A, B, C lie on l and D, E, F lie on m and $(DEF) \wedge (ABC)$. Suppose that AE meets DB at Y, and AF meets DC at Z. Then draw the line YZ and suppose YZ meets AD at X. Then we have $(DEF) = A(XYZ)$ and $(XYZ) = D(ABC)$. See Fig. 3.5.

(b) If A, B, C and D, E, F lie on the same line, then a perspectivity through some vertex U may be created arbitrarily on some other line, so that $U(ABC) = (PQR)$, and then, since $(PQR) \wedge (DEF)$, part (a) shows that no more than two further perspectivities are required to relate P, Q, R and D, E, F.

The importance of Example (3.4.3) is that it provides the pure geometrical method of handling projectivities. It is done by mapping a range of points on one line to a range of points on another line by means of one or more perspectivities; furthermore we do not miss any projectivities that exist, provided the appropriate perspectivities are used.

As we have seen above an involution is governed by an equation of the form $cst - a(s + t) - b = 0$, where $a^2 + bc \neq 0$. The self-corresponding points

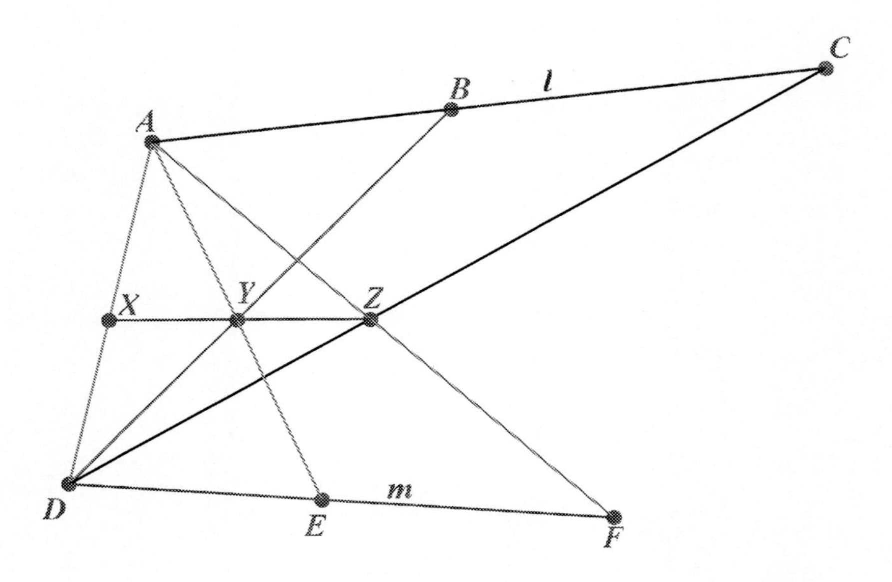

Figure 3.5: Example 3.4.3

are given by $s = t = (a \pm \sqrt{a^2 + bc})/c$. Sometimes these points are called *double points.* If $c = 0$, then one double point is at $t = \infty$. If $b = 0$, then one double point is at $t = 0$. If $b = c = 0$, then the double points are at $t = 0$ and ∞ and the involution has equation $s = -t$. Either ∞ is a double point or its partner in the involution is a/c. Similarly either 0 is a double point or its partner in the involution is $-b/a$.

If (s_k, t_k), $k = 1, 2, 3$ are three pairs in involution, then $cq_k - ap_k - b = 0$, $k = 1, 2, 3$, where $q_k = s_k t_k$ and $p_k = s_k + t_k$. It follows that the determinant of the 3×3 matrix whose k-th row is $(q_k, p_k, 1)$ must vanish. This provides a condition for three pairs to be in involution.

Example 3.4.4

Two distinct involutions amongst points on the same line have a unique common pair.

Suppose that there are two distinct involutions between the points of one line. Then there are two equations $c_1 st - a_1(s + t) - b_1 = 0$ and $c_2 st - a_2(s + t) - b_2 = 0$, one for each involution. If (s, t) is a common pair of these involutions, then s, t will satisfy both these equations and also they must be roots of the quadratic equation $st - (s+t)x + x^2 = 0$. It follows that the determinant with rows (c_1, a_1, b_1), (c_2, a_2, b_2), $(1, x, -x^2)$ must vanish. This is a quadratic equation in x, so provides just one pair (s, t) that belongs to each involution.

A corollary is that if two involutions have two distinct common pairs, then they coincide.

Example 3.4.5

The self-corresponding points of an involution determined by the pairs $A, A'; B, B'$are I, J. We prove that the pairs $A, B'; B, A'; I, J$ are in involution.

Suppose that I, J have parameters $0, \infty$, then the equation of the first involution is $s + t = 0$. Hence we may take the parameters of A, A', B, B' to be $a, -a, b, -b$, respectively. The equation of the second involution must be $s = -ab/t$ and I, J are a pair in this involution.

All that has been said about points on lines in Sections 3.3 and 3.4. has its dual version concerning lines through points. Whereas a set of points on a line is called a range, a set of lines through a point is called a *pencil*. The lines of a pencil may be parameterized as follows. Let the lines pass through the point $(1, 0, 0)$. Then the line co-ordinates of a line through this point are $(0, 1, t)$, where t is a parameter. When $t = n/m$ the line has equation $ny = mz$. Projectivities between three lines p, q, r through one point and three lines f, g, h through a second point are defined by a relation of the form $s = (at + b)/(ct + d)$, $ad \neq bc$. If the two points are the same, then in a projectivity there are two (possibly coincident) self-corresponding lines. A perspectivity is said to exist if a line u exists so that pf, qg, rh are collinear on u. Then we can write $(pqr) = u(fgh)$. Analogues of all the results in Sections 3.3 and 3.4 hold equally well for pencils of lines.

Exercise 3.4.6

1. If $\sigma, \tau$are distinct self-corresponding points of a projectivity on a line, find the most general form for the equation of the projectivity.

2. If σ is the one and only self-corresponding point of a projectivity on a line, find the most general form for the equation of the projectivity.

3. Suppose that A, B, C are three points on a line and $(ABC) \wedge (BCA)$. Prove that the self-corresponding points of the projectivity are distinct.

4. If σ, τ are the self-corresponding points of an involution, find the most general form for the equation of the involution.

5. Prove that a projectivity that interchanges a pair of points is an involution.

6. The pairs of distinct points $A, D; B, E; C, F$ are in involution. Prove that the three involutions, of which $A, D; B, E; C, F$ respectively are the self-corresponding points, have one common pair of points.

The cross-ratio of four points on a line was defined in Section 3.3.6 as follows: If A, B, C, D are four points on a line with parameters a, b, c, d, then the cross-ratio $(A, C; B, D)$ is defined to be $(b-a)(c-d)/(c-b)(d-a)$. We may also define $(a, c; b, d)$ to have the same value.

Since a, b, c, d can be arranged in 24 ways, it is possible that there might arise 24 different values by permuting the symbols of a cross-ratio. However, since

$$(a, c; b, d) = (c, a; d, b) = (b, d; a, c) = (d, b; c, a),$$

there are, in fact, just 6 values in general, which may be evaluated by holding a fixed in the first position and permuting b, c, d. Hence if $(a, c; b, d) = x$, then $(a, c; d, b) = 1/x$, $(a, b; c, d) = 1 - x$, $(a, b; d, c) = 1/(1-x)$, $(a, d; c, b) = (x-1)/x$ and $(a, d; b, c) = x/(x-1)$. Note when three of the values of the parameter are $1, 0, \infty$, then $(a, 1; 0, \infty) = a$, $(a, 1; \infty, 0) = 1/a$, $(a, 0; \infty, 1) = 1/(1-a)$, $(a, 0; 1, \infty) = 1 - a$, $(a, \infty; 1, 0) = (a-1)/a$ and $(a, \infty; 0, 1) = a/(a-1)$. Hence if A, B, C, D are four points on the line $x = 0$ and $A(0, 1, a)$, $C(0, 1, 1)$, $B(0, 1, 0)$, $D(0, 0, 1)$, their parameters z/y are $a, 1, 0, \infty$ respectively, and we may write $(A, C; B, D) = a$.

3.4.1 Cross-ratio is invariant under a projectivity

Suppose that (s_1, t_1), (s_2, t_2), (s_3, t_3) are three distinct pairs of a projectivity, then the equation of the projectivity is

$$(s_1, s_2; s_3, s) = (t_1, t_2; t_3, t).$$

So if (s_4, t_4) is a distinct fourth pair of the projectivity we may put $s = s_4, t = t_4$ to obtain $(s_1, s_2; s_3, s_4) = (t_1, t_2; t_3, t_4)$. Conversely if two cross-ratios are equal and their elements are distinct, there is a projectivity defined by any three pairs, and the fourth pair belongs to that projectivity.

The importance of Section 3.4.1 is emphasized. First it means that a cross-ratio is independent of the parameter system used, which is necessary since otherwise it would not be a useful concept. Secondly it means one can use a perspectivity to say that the cross-ratio of a range of four points on one line is equal to the cross-ratio of four points on another line. We use the notation $O(A, B; C, D)$ for the cross-ratio of the pencil of lines OA, OB, OC, OD formed by O and the range $ABCD$; and $l(a, b; c, d)$ to denote the cross-ratio of the range on l of the points of intersection on l by the pencil of lines $abcd$. These observations imply $O(A, B; C, D) = P(A, B; C, D)$ for distinct vertices O, P and $l(a, b; c, d) = m(a, b; c, d)$ for distinct lines l, m.

In other words the cross-ratio of a range of four points on a line may be transferred to an equal cross-ratio of four points on another line by means of an intermediate pencil $abcd$. Dually the cross-ratio of a pencil of four lines through a point may be transferred to an equal cross ratio of a pencil of four lines through another point by means of an intermediate range $ABCD$.

When $(a, c; b, d) = (a, c; d, b)$, then $x = 1/x$ and $x = \pm 1$. If $x = 1$, and $c \neq d$, then $a = b$, which is not interesting. When $x = -1$ the corresponding range of points $ABCD$ is a *harmonic range*. A and B are harmonic conjugate points with respect to C and D. Also C and D are conjugate points with respect to A and B, and in these statements points A and B may be interchanged and the points C and D may be interchanged. Harmonically conjugate lines are defined in similar fashion. When $x = -1$, the six values of the cross ratio found by permuting the symbols a, b, c, d reduce to 3 values, namely $-1, 2, \frac{1}{2}$. If A and C are fixed and B, D are variable pairs such that $(A, C; B, D) = 1$, then the points A, C are the double points of an involution in which the pairs are B, D. See Example 3.4.7 for an analysis of this.

When $(a, c; b, d) = (a, b; d, c)$ we have $x = 1/(1 - x)$, so that $x = -\omega$ or $-\omega^2$, where ω is a complex cube root of unity. The points $ABCD$ are said to form an *equianharmonic range*. Only two of the six values of the cross-ratios remain unequal, the two values of x already stated. If A is a fixed point and C, B, D are triads such that $(A, C; B, D) = (A, B; D, C)$, then the triads C, B, D are sometimes called *Hessian triads*. The context is often concerned with a projectivity that maps the vertices of a triangle cyclically on to each other.

Example 3.4.7

Prove that, if B, D are any two points on a line and points A_k, C_k on the line are harmonic conjugates with respect to B, D for all k, then the pairs (A_k, C_k) are in involution with B, D as self-corresponding points.

When $(a, c; b, d) = -1$, then $(a + c)(b + d) = 2(ac + bd)$. This equation highlights the symmetries $a \mapsto c, c \mapsto a; b \mapsto d, d \mapsto b; a \mapsto b, b \mapsto c, c \mapsto d, d \mapsto a$. Hence, if a, c are the roots of the quadratic equation $hu^2 - pu + q = 0$ and b, d are the roots of the quadratic equation $ku^2 - ru + m = 0$ we have $pr = 2(qk + mh)$. If $b = 0$ and $d = \infty$, then $k = m = 0$ and hence $p = 0$ and $a, c = \pm\sqrt{q/h}$, that is a and c have equal and opposite values. This means that a, c are pairs in the involution $s + t = 0$, whose self-corresponding points are $0, \infty$.

Suppose that $P_1(x_1, y_1, z_1)$ and $P_2(x_2, y_2, z_2)$ are any two distinct points. Then the points P with co-ordinates $P(x_1 + tx_2, y_1 + ty_2, z_1 + tz_2)$ lie on P_1P_2 and P_1 has parameter $t = 0$ and P_2 has parameter $t = \infty$. Example 3.4.7 shows that the points with parameters $t = k$ and $t = -k$ separate P_1 and P_2 harmonically, and they are pairs in an involution whose self-corresponding points are P_1 and P_2.

Example 3.4.8

The sides BC, CA, AB of a triangle ABC are cut by a line at points P, Q, R respectively. P' is the harmonic conjugate of P with respect to B and C. Q', R' are similarly defined. We prove that AP, BQ, CR are concurrent.

Take ABC to be the triangle of reference and let the line have equation $lx + my + nz = 0$. Then P has co-ordinates $(0, n, -m)$. Using z/y as parameter on BC we have B, C to have parameters $0, \infty$ respectively, and P has parameter $-m/n$. It follows from above that P' has parameter m/n, and co-ordinates $(0, 1/m, 1/n)$. Clearly AP' contains the point $(1/l, 1/m, 1/n)$, which, by symmetry lies on BQ' and CR'.

Exercise 3.4.9

1. Prove that if the pairs of points (A, P), (B, Q), (C, R) on a line are such that P is the harmonic conjugate of A with respect to B and C, Q is the harmonic conjugate of B with respect to C and A, R is the harmonic conjugate of C with respect to A and B, then the three pairs of points are in involution.

2. A, B, C, D are four collinear points. The harmonic conjugates of D with respect to the pairs B, C; C, A; A, B are P, Q, R respectively. Prove that $(A, B; C, D) = (P, Q; R, D)$.

3. $ABCD$ are four collinear points. The harmonic conjugates of B and D with respect to A and C are B' and D' respectively. Prove that $(A, C; B, D) = (A, C; B', D')$.

4. ABC is a triangle and L, M, N are three distinct points in the plane of the triangle not on the sides. The variable point P is chosen on BC and PN meets CA at Q, QL meets AB at R and RM meets BC at P'. Show that, in general there are two choices of the position P for which P' coincides with P. Is it possible to choose L, M, N so that P' always coincides with P?

5. If the pencils $O(A, C; B, D) = O'(A, C; B, D)$ have equal cross-ratios and A, B, C are collinear, prove that D also lies on ABC.

6. If $(A, Q; P, R) = (A, Q'; P', R') = -1$ prove that QQ', PR', RP' are concurrent.

7. Show that if D is a fixed point and DPQ, DRS are two variable lines meeting the fixed lines AB, AC at P, Q and R, S, then the locus of the intersection of PS and QR is a straight line.

8. Two lines FDA, FCB are met by two other lines GBA, GCD. BD meets AC at E. Prove that $F(A, B; E, G) = -1$.

9. A, B, C, D are four given collinear points. Show that, if the cross-ratios $(A, D; B, P)$ and $(A, P; B, C)$ are equal there are two possible positions for P, and that they are harmonically separated by A and B.

10. On a straight line are taken points O, A, B, C, A', B', C' such that

$$(O, C; A, B) = (O, C'; A, B') = (O, C'; A', B) = (O, C; A', B').$$

Prove that each is equal to $(O, C'; A', B')$, and that

$$(O, A'; B, C) = (O, B'; C, A) = (O, C'; A, B) = -1.$$

(JW)

11. Two fixed straight lines intersect in a point O on the side BC of triangle ABC. A variable point P is taken on AO and the straight lines PB, PC meet the two fixed lines at B_1, B_2 and C_1, C_2 respectively. Prove that B_1C_2 and B_2C_1 both meet BC in fixed points. (JW)

12. Points on a line form pairs in involution, and a projectivity exists between these points and points on another line also forming pairs in involution. A_1, A_2 are pairs of points of the former involution and A'_1, A'_2 are corresponding points on the latter. Prove that the locus of intersection of $A_1A'_1, A_2A'_2$ is a straight line. (JW)

3.5 Triangles in multiple perspective

We refer to Exercises 3.2.1 Problem 4. The content of that problem is that if we have two triangles ABC and TUV in perspective and if triangles ABC and UVT are in perspective, then triangles ABC and VTU are also in perspective. Here the ordering of the labels is important. In other words if two triangles are in double perspective, then they are automatically in triple perspective. In this section we consider the possibility of triple and higher order multiple perspectives. The solution to the problem emerges, admittedly rather tangentially, as part of what follows.

Suppose that ABC and TUV are two triangles labelled with the same orientation, say both anticlockwise. Three of the six ways in which these triangles might be in perspective reverse the orientation: when ABC is in perspective with VUT, TVU or UTV. We refer to any of these perspectives as *reverse perspectives*.

3.5.1 Triple reverse perspective

Two triangles ABC and TUV, in which T, U, V are distinct from A, B, C and do not lie on the sides of ABC and in which ABC and TVU are in perspective, with vertex of perspective D, are in triple reverse perspective, with the three vertices of perspective collinear, if, and only if, the vertices of the two triangles lie on a conic.

We give a proof of this result. Take ABC as the reference triangle with $A(1,0,0)$, $B(0,1,0)$ and $C(0,0,1)$ and D to be the point with co-ordinates $D(p,q,r)$, so we may take T, U, V to have co-ordinates $T(1,q,r)$, $U(p,q,1)$, $V(p,1,r)$. Here p, q, r are not all equal and $p, q, r \neq 0$, since T, U, V do not lie on the sides of ABC.

We now find the relation satisfied by p, q, r for ABC and VUT to be in perspective. The equation of BU is $x = pz$, that of CT is $y = qx$ and that of AV is $z = ry$, so the condition for BU, CT, AV to be concurrent is

$pqr = 1$. Since the equations of CV, AU, BT are $x = py$, $y = qz$, $z = rx$ respectively it follows that when $pqr = 1$ it is automatic that triangles ABC and UTV are also in perspective. That is, triangles in double reverse (or direct) perspective are in triple reverse (or direct) perspective.

Suppose now that BU, CT, AV are concurrent at E, then it may be verified that E has co-ordinates $E(p, pq, 1)$. Similarly if CV, AU, BT are concurrent at F, then F has co-ordinates $(1, qr, r)$. The condition that D, E, F are collinear is that the determinant of coefficients of the co-ordinates of D, E, F should be zero, and using the fact that $pqr = 1$, we find this condition to be $p + q + r = 3$.

It is interesting to see what happens if AT, BU, CV are also concurrent, so that we have quadruple perspective. The equations of these three lines are $ry = qz$, $z = x/p$ and $x = py$, and these meet if, and only if, $q = r$. If the vertices of reverse perspective are collinear this means (since $p = q = r = 1$ is forbidden) that $q = r = -\frac{1}{2}$ and $p = 4$. Thus there is quadruple perspective when three vertices of perspective are collinear only in three cases ($q = r = -\frac{1}{2}$, $p = 4$; $r = p = -\frac{1}{2}$, $q = 4$; $p = q = -\frac{1}{2}$, $r = 4$). It follows that in this case there is never more than four perspectives. However, if the three vertices of reverse perspective are not collinear, then the only condition on p, q, r for there to be a fourth perspective is $pq^2 = 1$. And so there is an infinity of such possibilities. If there is a fifth perspective (and hence a sixth), then the condition is $p = q = r$ and $p^3 = 1$. So, in the projective plane over the real field there is never a fifth perspective, but over the complex field there are cases (which cannot be drawn, of course), of sextuple perspective when $p = q = r = \omega$ or ω^2.

We now return to the case when there is triple reverse perspective so that $pqr = 1$, and show the condition $p + q + r = 3$, that D, E, F are collinear is precisely the same condition that A, B, C, T, U, V lie on a conic, where we anticipate the condition that the equation of a conic Σ passing through the vertices of the triangle of reference is of the form $fyz + gzx + hxy = 0$, for some constants f, g, h. Putting in the condition that T, U, V lie on the conic Σ we obtain the three equations

$$qrf + rg + qh = 0, rf + rpq + ph = 0, qf + pg + pqh = 0.$$

The condition for these equations to hold simultaneous for f, g, h not all zero is that the determinant of coefficients is zero, and using $pqr = 1$, this is also seen to be $p + q + r = 3$. The corresponding (unique) equation of Σ is therefore

$$\frac{yz}{1-qr}+\frac{zx}{1-rp}+\frac{xy}{1-pq}=0. \tag{3.4}$$

None of the denominators in Equation (3.4) can vanish, since $p=q=r=1$ is not allowed, and the equation of the line DEF may be verified to be

$$(1-qr)x+(1-rp)y+(1-pq)z=0.$$

3.6 Conics, tangents, poles and polars

3.6.1 General equation of a conic

We define a conic as a curve whose equation is a homogeneous equation of degree two in x, y, z. The general equation of a conic is therefore

$$S \equiv ax^2+by^2+cz^2+2fyz+2gzx+2hxy=0.$$

We refer loosely to the conic $S=0$, or just to the conic S. Note that, in general, a unique conic can be drawn through any 5 points, since there are 5 independent ratios $a:b:c:f:g:h$. However, if three, four or five of the points are collinear, then the conic may be degenerate, non-unique or even indeterminate.

3.6.2 Important expressions

We use the following symbols for key expressions:

$$S_{11} \equiv ax_1^2+by_1^2+cz_1^2+2fy_1z_1+2gz_1x_1+2hx_1y_1,$$

so that $S_{11}=0$ means that the point $P_1(x_1,y_1,z_1)$ lies on the conic $S=0$.

$$S_{12} \equiv ax_1x_2+by_1y_2+cz_1z_2+f(y_1z_2+y_2z_1)+g(z_1x_2+z_2x_1)+h(x_1y_2+x_2y_1),$$

and

$$S_1 \equiv (ax_1+hy_1+gz_1)x+(hx_1+by_1+fz_1)y+(gx_1+fy_1+cz_1)z.$$

$S_1=0$ is the equation of a line, and if P_1 lies on the line $S_1=0$, then it also lies on the conic $S=0$. If $P_2(x_2,y_2,z_2)$ lies on the line $S_1=0$, then $S_{12}=0$ and conversely.

3.6.3 When the conic degenerates into two straight lines

The determinant with rows (a, h, g), (h, b, f), (g, f, c) is denoted by Δ and

$$\Delta \equiv abc + 2fgh - af^2 - bg^2 - ch^2. \tag{3.5}$$

Its minors A, B, C, F, G, H are defined by $A = bc - f^2$, $H = fg - hc$ etc., so that

$$\Delta = aA + hH + gG \text{ etc.}, \; aH + hB + gF = 0 \text{ etc.}$$

The conic S degenerates into a pair of straight lines if, and only if, $\Delta = 0$. If $S = (lx + my + nz)(px + qy + rz)$, then $a = lp$, $b = mq$, $c = nr$, $f = \frac{1}{2}(mr+nq)$, $g = \frac{1}{2}(np+lr)$, $h = \frac{1}{2}(lq+mp)$ and substitution shows that $\Delta = 0$. The converse is less straightforward. If $\Delta = 0$, then there exist constants x_1, y_1, z_1 not all zero such that $ax_1 + hy_1 + gz_1 = 0$, $hx_1 + by_1 + fz_1 = 0$, $gx_1 + fy_1 + cz_1 = 0$. Multiplying by x_1, y_1, z_1 and adding we obtain $S_{11} = 0$, so P_1 lies on S. Multiplying by x_2, y_2, z_2 and adding we obtain $S_{12} = 0$. This means that an arbitrary line through P_1 meets the conic S in two coincident points, and if P_2 actually lies on S, then $S_{22} = 0$ and the entire line lies on the conic. Interpret Equation (3.6) for proof of these remarks. It follows that the conic must be degenerate. It can be shown that if the minors A, B, C, F, G, H vanish, then the conic degenerates into two coincident lines. We assume in what follows that $\Delta \neq 0$.

3.6.4 Joachimstal's equation, tangent to a conic

If P_1 and P_2 are defined as above, then the co-ordinates of any point P on P_1P_2 is $t_1(x_1, y_1, z_1) + t_2(x_2, y_2, z_2)$, where t_1 and t_2 are parameters. If P lies on S, then

$$t_1^2 S_{11} + t_1 t_2 S_{12} + t_2^2 S_{22} = 0. \tag{3.6}$$

This is called Joachimstal's equation, and its two solutions for the ratio t_2/t_1 provide the two points where the line meets the conic S. If P_1 lies on S, then $S_{11} = 0$ and one root of Joachimstal's equation is $t_2/t_1 = 0$. If P_1P_2 is a tangent, then the second root must also be zero, a tangent being defined as a line that meets the conic in two coincident points. Hence $S_{12} = 0$. As the co-ordinates of P_2 vary, then they satisfy the equation $S_1 = 0$. Hence $S_1 = 0$ is the equation of the tangent to the conic S at P_1.

3.6.5 Conjugate points, pole and polar

If a line through P_1 and P_2 meets a conic in Q_1, Q_2 and $(P_1, P_2; Q_1, Q_2) = -1$, then P_1 and P_2 are said to be *conjugate points* with respect to S. Since P_1 and P_2 have parameters 0 and ∞, as defined in Section 3.6.4, and Q_1 and Q_2 are harmonically separated by P_1 and P_2, it follows that the parameters of Q_1 and Q_2 are equal and opposite. Hence $S_{12} = 0$ in Joachimstal's equation. It follows that the locus of P_2 as the line through P_1 varies has equation $S_1 = 0$. This has the same form as the tangent through P_1 when P_1 lies on the conic. It is called the *polar* of P_1, and P_1 is the *pole* of this line. If Q_1 and Q_2 coincide, then P_2 coincides with them. (You may need to check this from the definition of cross-ratio.) It follows that the polar of a point is the chord of contact of the tangents through that point. Note also that the condition for two points to be conjugates is $S_{12} = S_{21} = 0$. It is an immediate consequence that if P_2 lies on the polar of P_1, then P_1 lies on the polar of P_2.

3.6.6 The line equation of a conic

The line with equation $lx + my + nz = 0$ touches S at P_1 if it coincides with $S_1 = 0$. This means that a constant k must exist so that

$$ax_1 + hy_1 + gz_1 = kl, \tag{3.7}$$

$$hx_1 + by_1 + fz_1 = km, \tag{3.8}$$

$$gx_1 + fy_1 + cz_1 = kn, \tag{3.9}$$

and

$$lx_1 + my_1 + nz_1 = 0. \tag{3.10}$$

Multiplying Equations (3.7), (3.8), (3.9) by A, H, G respectively, and adding, one gets

$$\Delta x_1 = k(Al+Hm+Gn), \Delta y_1 = k(Hl+Bm+Fn), \Delta z_1 = k(Gl+Fm+Cn).$$

Substituting these into Equation (3.10) yields the condition

$$Al^2 + Bm^2 + Cn^2 + 2Fmn + 2Gnl + 2Hlm = 0. \tag{3.11}$$

This is the line equation of the conic S.

3.6.7 Duality

Section 3.6.6 means that not only may we consider a conic to be generated as a system of points obeying an equation of the second degree, but also as the envelope of a system of lines obeying an equation of the second degree. Points and tangents become dual concepts, as do poles and polars. In general an envelope of a conic is completely determined by any five of its lines. A line conic may degenerate into two points, or even into a coincident pair of points. Two lines are said to be *conjugate* if the two lines of the envelope through their common point separate them harmonically. The reader is invited to consider what happens in degenerate cases when

$$ABC + 2FGH - AF^2 - BG^2 - CH^2 = 0.$$

3.6.8 Conics through the vertices or touching the sides of the triangle of reference

Since $fyz + gzx + hxy = 0$ is the equation of a conic through the vertices of the triangle of reference, then so $f'mn + g'nl + h'lm = 0$ is the line equation of a conic touching the sides of the triangle of reference. If $f+g+h = 0$, then the first conic also passes through the point $(1, 1, 1)$ and if $f' + g' + h' = 0$, then the second conic also touches the line with equation $x + y + z = 0$.

Example 3.6.1

Find the eight points of contact of common tangents to the conics with equations $x^2 + y^2 + z^2 = 0$ and $ax^2 + by^2 + cz^2 = 0$ and show they lie on a conic. (Here is a case when we need to use the complex field.)

The line equations of the conics are $l^2 + m^2 + n^2 = 0$ and $bcl^2 + cam^2 + abn^2 = 0$ from which we see the common tangents are given by

$$l^2 : m^2 : n^2 = a(b - c) : b(c - a) : c(a - b).$$

It is now a short calculation to show that the four points of contact with the first conic are

$$(\pm\sqrt{a(b-c)}, \pm\sqrt{b(c-a)}, \pm\sqrt{c(a-b)})$$

and that the four points of contact with the second conic are

$$(\pm\sqrt{bc(b-c)}, \pm\sqrt{ca(c-a)}, \pm\sqrt{ab(a-b)}).$$

These points all lie on the conic with equation

$$a(b+c)x^2 + b(c+a)y^2 + c(a+b)z^2 = 0.$$

Example 3.6.2

A and B are two fixed points in the plane of a conic. PA, PB are such that they are conjugate lines with respect to the conic. Show that the locus of P is a conic through A and B.

Suppose that A and B have co-ordinates (d, e, f) and (u, v, w) and that the line equation of the conic is

$$al^2 + bm^2 + cn^2 + 2gmn + 2hnl + 2klm = 0.$$

Let P have co-ordinates (X, Y, Z). If PA has equation $\alpha x + \beta y + \gamma z = 0$ and PB has equation $px + qy + rz = 0$ then, since they are conjugate lines,

$$a\alpha p + b\beta q + c\gamma r + g(\beta r + \gamma q) + h(\gamma p + \alpha r) + k(\alpha q + \beta p) = 0. \tag{3.12}$$

Now $\alpha d + \beta e + \gamma f = 0$ and $\alpha X + \beta Y + \gamma Z = 0$, so

$$\alpha : \beta : \gamma = eZ - fY : fX - dZ : dY - eX.$$

Similarly

$$p : q : r = vZ - wY : wX - uZ : uY - vX.$$

Substituting into Equation (3.12) we get a homogeneous quadratic equation in X, Y, Z, which therefore represents a conic. The equation is satisfied by $X = d$, $Y = e$, $Z = f$ and by $X = u$, $Y = v$, $Z = w$ and so passes through A and B.

Example 3.6.3

A triangle ABC is inscribed in a conic. Prove that the tangents at the vertices meet the opposite sides in collinear points. See Fig. 3.6.

Take ABC to be the triangle of reference and the conic to be $fyz + gzx + hxy = 0$, then the equation of the tangent at $A(1, 0, 0)$ is $hy + gz = 0$. This meets BC at $L(0, -g, h)$, which lies on the line with equation $x/f + y/g + z/h = 0$, and by symmetry this line contains M and N, the points of intersection of CA and AB with the tangents at B and C respectively.

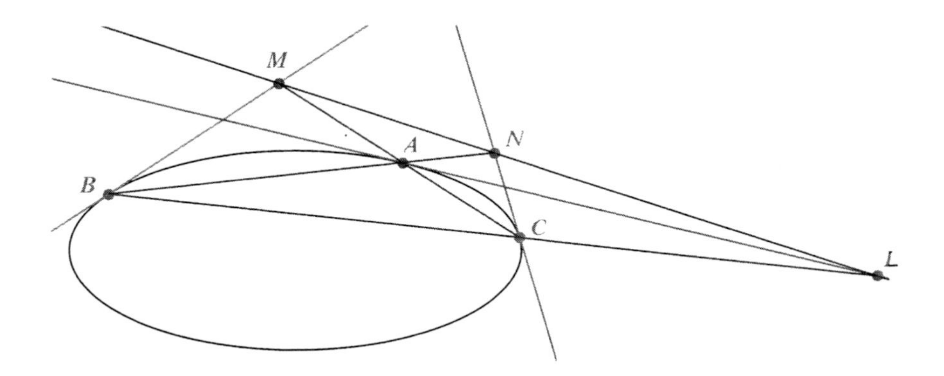

Figure 3.6: Example 3.6.3

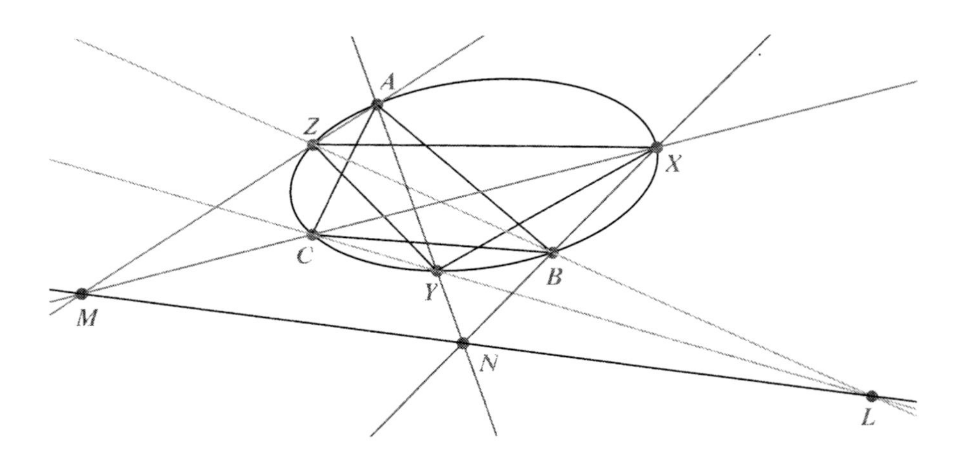

Figure 3.7: Pascal's theorem

3.6.9 Pascal's theorem

Two triangles XYZ, ABC are inscribed in a conic. The points of intersection L of YC, ZB and M of ZA, XC and N of XB, YA are collinear. See Fig. 3.7. Such a line is called a Pascal line.

Take XYZ to be the triangle of reference and the conic to be $fyz + gzx + hxy = 0$. Let points A, B and C have co-ordinates $A(x_1, y_1, z_1), B(x_2, y_2, z_2)$ and $C(x_3, y_3, z_3)$. The equations of YC, ZB are $x/x_3 = z/z_3$ and $x/x_2 = y/y_2$ and so meet at $L(1, y_2/x_2, z_3/x_3)$. Similarly $M(x_1/y_1, 1, z_3/y_3)$ and $N(x_1/z_1, y_2/z_2, 1)$. Now A, B, C lie on the conic so $f/x_k + g/y_k + h/z_k = 0$, $k = 1, 2, 3$. It follows that the 3×3 determinant whose k-th column is $(1/x_k, 1/y_k, 1/z_k)$, $k = 1, 2, 3$ vanishes. The determinant whose rows are the co-ordinates of L, M, N is proportional to this, and so L, M, N are collinear.

Example 3.6.4

If a conic S is inscribed in a triangle ABC and L, M, N are the points of contact, then AL, BM, CN are concurrent. See Fig. 3.8.

This is the dual of Example 3.6.3 but we provide a separate proof. Take the conic to have equation

$$f^2x^2 + g^2y^2 + h^2z^2 - 2ghyz - 2hfzx - 2fgxy = 0.$$

This touches $x = 0$ at $L(0, 1/g, 1/h)$. The line AL contains the point with co-ordinates $(1/f, 1/g, 1/h)$, which, by symmetry, lies on BM and CN.

3.6.10 Brianchon's theorem

This is the dual of Pascal's theorem. If the sides of a hexagon touch a conic, then the lines joining opposite vertices are concurrent. See Fig. 3.9.

This is the dual of Example 3.6.4. A separate proof is not given.

Exercise 3.6.5

1. Prove that there are two tangents to a conic S from an external point P_1 and that their equation (of the second degree) is $S_1^2 = S_{11}S$.

2. ABC is a triangle and S a conic that does not pass through any of A, B, C and does not touch the sides of ABC. The tangents from A to S meet BC in the points P_1, P_2. The points Q_1, Q_2 and R_1, R_2 are similarly defined. Prove that $P_1, P_2, Q_1, Q_2, R_1, R_2$ lie on a conic.

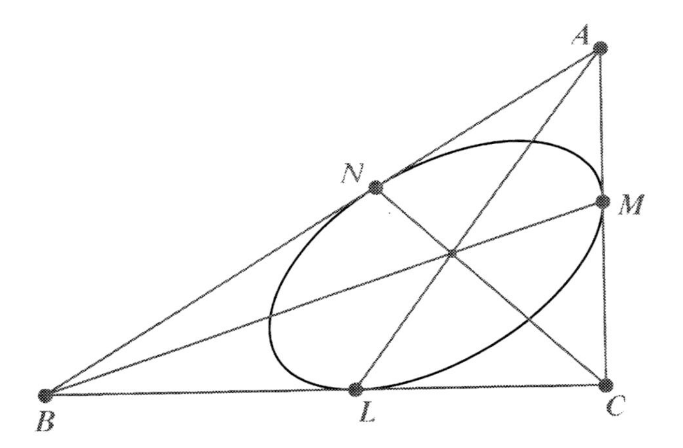

Figure 3.8: Example 3.6.4

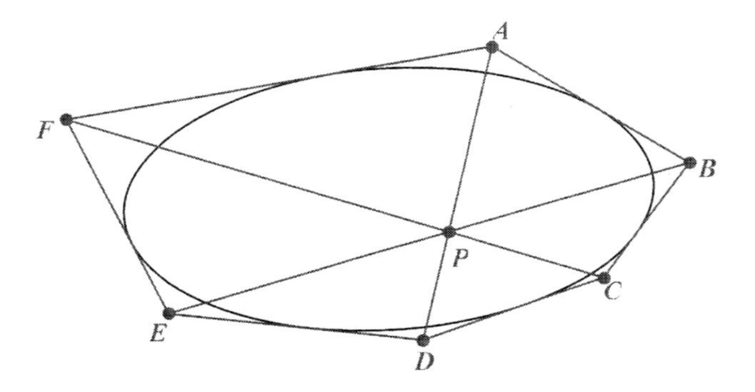

Figure 3.9: Brianchon's theorem

3. Verify that the equation of the conic that passes through the vertices of the triangle of reference, through the point with co-ordinates $(1, 1, 1)$, and through the point with co-ordinates (u, v, w), where u, v, w are not all equal, is

$$u(v-w)yz + v(w-u)zx + w(u-v)xy = 0.$$

4. Find the equation of a point conic touching the sides of the triangle of reference. What does the equation become if it also touches the line with equation $x + y + z = 0$?

5. Prove that the conic touching the sides of the triangle of reference, and the lines with equations $x+y+z = 0$ and $ux+vy+wz = 0$ has equation

$$u^2(v-w)^2x^2 + v^2(w-u)^2y^2 + w^2(u-v)^2z^2$$

$$-2uv(v-w)(w-u)yz - 2vw(w-u)(u-v)zx - 2wu(u-v)(v-w)xy = 0.$$

6. Suppose that you are given a conic with line co-ordinates (l, m, n) and Equation (3.11). Define $a' = BC - F^2$, $h' = FG - HC$ etc., prove that $a' = \Delta a$ and $h' = \Delta h$ etc. What is the geometrical significance of this for a conic with $\Delta \neq 0$?

7. Given the conic $S = 0$, as defined in Section 3.6.1, show that the pole of the line with equation $lx + my + nz = 0$ has co-ordinates

$$(Al + Hm + Gn, Hl + Bm + Fn, Gl + Fm + Cn).$$

8. ABC is a triangle and U, K are distinct points not lying on the sides. Cevians AUL, BUM, CUN, AKD, BKE, CKF are drawn. Prove that a conic passes through L, M, N, D, E, F.

9. Triangle ABC is inscribed in a conic S and circumscribes a conic Σ. The poles with respect to Σ of the tangents at A, B, C to S are denoted by P, Q, R. Prove that AP, BQ, CR are concurrent.

3.7 Two special forms

3.7.1 Conic with the triangle of reference as self-polar triangle

Let X be any point not on S and let Y be a point on the polar of X. The polar of Y passes through X. Suppose that it meets the polar of X at Z. Then XY is the polar of Z, and each side is the polar of the opposite vertex. Such a triangle is called *self-polar* or *self-conjugate*. Suppose now that XYZ is the triangle of reference. The polar of $X(1,0,0)$ is $ax + hy + gz = 0$. If this is to be $x = 0$, then $h = g = 0$. Similarly $f = 0$, and so the conic has an equation of the form $ax^2 + by^2 + cz^2 = 0$. By choice of unit point this equation can be put in the form $x^2 + y^2 + z^2 = 0$. In line co-ordinates the equation of the conic is $l^2/a + m^2/b + n^2/c = 0$ or by choice of unit line $l^2 + m^2 + n^2 = 0$.

3.7.2 The parametric form $(\theta^2, \theta, 1)$ for a conic

The special form $y^2 = zx$ is a perfectly general choice for the equation of a conic and the parameterization that results is often very useful in dealing with properties of a conic, particularly those involving projectivities. To see that it is a general choice, choose the vertices Z, X of the triangle of reference to lie on the conic. Let the tangents at Z and X meet at Y. The tangent at X has equation $ax + hy + gz = 0$ and this passes through X and Y if, and only if, $a = h = 0$. The tangent at Z has equation $gx + fy + cz = 0$, and this passes through Y and Z if, and only if, $c = f = 0$. The equation of S is therefore $by^2 + 2gzx = 0$. If we now choose the unit point to lie on the conic, then its equation is further simplified to $y^2 = zx$, which has the parametric form $(\theta^2, \theta, 1)$. The equation of the chord joining 'θ' and 'ϕ' is easily verified to be

$$x - (\theta + \phi)y + \theta\phi z = 0. \tag{3.13}$$

The equation of the tangent at 'θ' is $x - 2\theta y + \theta^2 z = 0$. The tangents at '$\theta$' and '$\phi$' meet at the point with co-ordinates $(\theta\phi, \frac{1}{2}(\theta + \phi), 1)$, which is therefore the pole of the chord joining these points.

Example 3.7.1

Prove that if two triangles are self-polar with respect to a conic, then their six vertices lie on a conic.

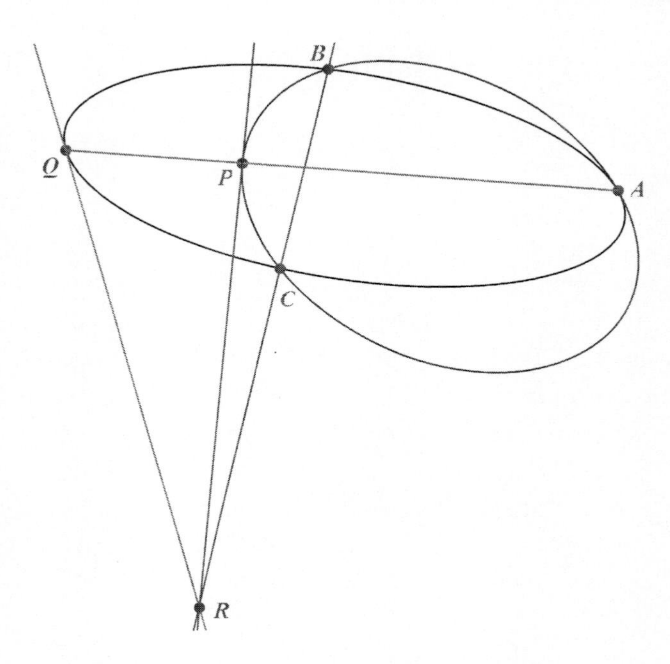

Figure 3.10: Example 3.7.2

Take the conic to have equation $ax^2 + by^2 + cz^2 = 0$, and the triangle of reference to be one of the self-polar triangles. Let $P_k(x_k, y_k, z_k)$, $k = 1, 2, 3$ be the other self-polar triangle. Since P_j and P_k $(j \neq k)$ are conjugate points we have $ax_jx_k + by_jy_k + cz_jz_k = 0$ $(j \neq k)$. Hence the determinant with rows $(1/x_k, 1/y_k, 1/z_k)$, $k = 1, 2, 3$ vanishes. But this is precisely the condition that constants u, v, w exist such that P_1, P_2, P_3 lie on a conic with equation $uyz + vzx + wxy = 0$, a conic that also passes through the vertices of the triangle of reference.

Example 3.7.2

Two conics touch at A and have two further points B and C in common. A line through A meets the conics at P and Q. Prove that the tangents at P and Q meet on BC. See Fig. 3.10.

Take A, B, C as triangle of reference. Since the two conics touch at A, their equations may be taken as $uyz+vzx+wxy = 0$ and $tyz+vzx+wxy = 0$, $t \neq u$, so that the common tangent is $wy+vz = 0$. Let the line through A be $y = kz$, $k \neq -v/w$. This line meets the first conic at $P(-uk, k(v+wk), (v+$

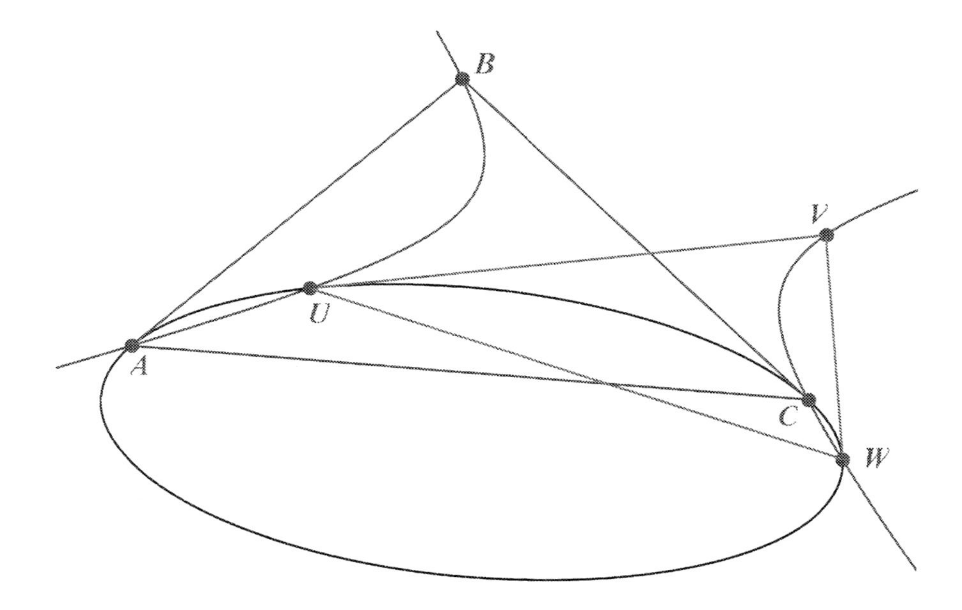

Figure 3.11: Example 3.7.3

$wk)$) and the second conic at $Q(-tk, k(v+wk), (v+wk))$. It is not difficult now to show that the tangent at P to the first conic and the tangent to Q to the second conic meet at the point R on BC with co-ordinates $R(0, -k^2w, v)$.

Example 3.7.3

Two triangles are formed, each by two tangents to a conic and their chord of contact. Prove that their vertices lie on a conic. See Fig. 3.11.

Suppose that the conic has equation $y^2 = zx$ with parameter θ. Take A, B, C to be the triangle of reference, then B is the pole of CA. Suppose that U has co-ordinates $(\theta^2, \theta, 1)$ and W has co-ordinates $(\phi^2, \phi, 1)$, then the pole V of WU has co-ordinates $V(2\theta\phi, \theta+\phi, 2)$. Now it may be verified that the conic with equation $\theta\phi yz - (\theta+\phi)zx + xy = 0$ contains the six points A, B, C, U, V, W.

Exercise 3.7.4

1. A conic is given with four chords AP, BQ, CR, DS all passing through a fixed point K. Prove that the conics $ABCDK$ and $PQRSK$ touch at K. (JW)

2. Two fixed tangents OA, OB are drawn to a conic. C is a fixed point on AB. A variable line is drawn through C to meet OA at P and OB at Q. The other tangents from P and Q meet at R. Prove that the locus of R is a fixed straight line through O. (JW)

3. Two conics circumscribe a triangle ABC. A variable straight line through A meets them again at P and Q. The tangents at P and Q meet BC at R and S. Prove that $(B, C; R, S)$ is constant. (JW)

4. Two conics S and Σ have two points A, B in common. P is a variable point on Σ. The lines PA, PB meet S again at Q, R. Prove that QR envelops a conic.

5. A, B are conjugate points with respect to a conic. R is a variable point on the conic, and RA, RB meet the conic again in P, Q. Show that PQ passes through a fixed point C. Prove also that triangles ABC and QPR are in perspective and find the locus of the vertex of perspective.

6. Three fixed points A, B, C are taken on a conic. Prove that there are infinitely many triangles PQR, self-conjugate with respect to the conic such that P, Q, R lie on BC, CA, AB respectively. Prove further that AP, BQ, CR meet in a point and find the locus of this point.

7. POD, QOE, ROF are three concurrent chords of a conic S and X is any other point of S. Suppose that QR, XD meet in L; RP, XE meet in M; PQ, XF meet in N. Prove that L, M, N lie on a line through O.

8. Suppose that C, D are conjugate points on the polar of P with respect to a given conic. Any line through P cuts the conic in A, B; AD cuts BC in E and AC cuts BD in F. Prove that E, F lie on the conic and that EF passes through P.

9. Suppose that A, B, C are three points in the plane of a conic S. The pole of BC is L, the pole of CA is M and the pole of AB is N. Prove that AL, BM, CN are concurrent.

10. The triangle ABC is inscribed in a conic. The tangents at B and C meet at D, and E, F are similarly defined. Prove that AD, BE, CF are concurrent at a point O. If a line drawn through O meets

BC, CA, AB at L, M, N respectively, prove that the triangle with sides DL, EM, FN is self-conjugate with respect to the conic.

11. Prove that the six points of contact with the sides of a triangle ABC of two inscribed conics S_1, S_2 lie on a conic, which also passes through the two points of contact of the fourth common tangent of S_1, S_2.

12. Prove that if there is one triangle inscribed in a conic S and self-polar with respect to a conic S', then there is an infinite number of such triangles. What is the dual of this result?

13. Prove that if two triangles can be inscribed in a conic S, then there is a conic Σ with respect to which both triangles are self-polar.

14. Let ABC be a self-polar triangle of a conic S. If Q is any point on S and QA meets S again at R, and if RB meets S again at P, prove that PC passes through Q. *Hint: Let S have equation $x^2 + y^2 = z^2$ and let ABC be the triangle of reference. Use a parameter 't' such that $x = (1 - t^2)$, $y = 2t$ and $z = (1 + t^2)$.*

3.8 Projective correspondences on a conic

3.8.1 Introduction

We take the conic S in its canonical form $y^2 = zx$, with parameter 'θ' so that points P on the conic have co-ordinates $P(\theta^2, \theta, 1)$. If A is a fixed point with parameters $(\alpha^2, \alpha, 1)$, then the chord from A to P is uniquely determined by the parameter θ. An arbitrary line through A meets the conic at one and only one point. Also given a point P on the conic there exists only one line of the pencil of lines through A, which passes through P. In other words there is a projective correspondence between the pencil of lines through A and the points of the conic. See Equation (3.13) describing the chord AP, which establishes the correspondence. If C is another point with parameters $(\gamma^2, \gamma, 1)$, there is likewise a projective correspondence between the pencil of lines through C and the pencil of lines through A. In the two pencils the lines AP and CP correspond for each point P on the conic, and the line in each pencil carries the parameter of P. If P is at C, then the line AC in the pencil at A corresponds to the tangent at C in the pencil at C. Similarly the line AC in the pencil at C corresponds to the tangent at A in the pencil at A.

We now prove the converse of these results, and remark that *the following result is often taken as the definition of a conic.*

3.8.2 A projectivity between pencils

If A and C are two distinct points such that there is a projectivity between the pencil of lines through A and the pencil of lines through C, then the locus of points of intersection of corresponding rays is a conic through A and C. Take A and C to be the points $A(1,0,0)$ and $C(0,0,1)$. The equations of the lines of the pencil through A are $y = sz$ and the equations of the lines of the pencil through C are $y = tx$, where s and t are parameters. As in Section 3.3, we take the equation of the projectivity to be $cst + ds - at - b = 0$, $(ad \neq bc)$. The locus of P is therefore $c(y/z)(y/x) + d(y/z) - a(y/x) - b = 0$. That is, $cy^2 + dxy - ayz - bzx = 0$. As this is a homogeneous equation of the second degree it represents a conic Σ. It is easily verified that it passes through A and C; indeed A and C are the intersections of the conic with the line $y = 0$.

The line AC regarded as a line of the pencil through A has parameter $s = 0$, and this corresponds to the line through C with parameter $t = -b/a$. That is, the line through C with equation $bx + ay = 0$. As can be seen from the equation of Σ, this is the tangent at C to Σ. The line AC regarded as a line of the pencil through C has parameter $t = 0$, and this corresponds to the line through A with parameter $s = b/d$. That is, the line through A with equation $dy = bz$. This is the equation of the tangent at A to Σ.

If the line AC through A corresponds to the line AC through C, then $s = t = 0$, so that $b = 0$. The conic Σ is then degenerate consisting of the lines $y = 0$ and $cy + dx = az$. The line AC is then part of the conic. If $b = 0$, then neither a nor d vanishes since $bc \neq ad$, so neither A nor C is the intersection of these two lines. In fact the projectivity becomes one in which the two pencils have a common axis, the line $cy + dx = az$. The other self-corresponding point on the axis has co-ordinates $(c, a - d, c)$.

3.8.3 Cross-ratio and Chasles's theorem

We can now define a cross-ratio of four points A, B, C, D on a conic S. For example, we may suppose the conic has equation $y^2 = zx$ and the points have parameters $\theta = \theta_A, \theta_B, \theta_C, \theta_D$ respectively, then we may define $(A, C; B, D) = (\theta_A, \theta_C; \theta_B, \theta_D)$. Chasles's theorem follows immediately, that if P and Q are any two points on the conic, then $P(A, C; B, D) = Q(A, C; B, D)$. The dual of this result is that if a, b, c, d are four fixed tangents on a conic S and p and q are two distinct tangents of S, then the cross-ratios of the ranges $p(a, c; b, d)$ and $q(a, c; b, d)$ are equal.

3.8.4 Converse of Chasles's theorem

Let A, B, C, D be four points, no three of which are collinear, and suppose that P moves so that the cross-ratio of the pencil $P(A, C; B, D)$ is constant, then the locus of P is a conic through A, B, C, D. Let $A(1,0,0)$, $B(0,1,0)$, $C(0,0,1)$, $D(1,1,1)$, $P(u, v, w)$, then it is easy to show that $P(A, C; B, D) = 1/k$, where $k = u(w-v)/w(u-v)$ and consequently the locus of P has equation $kz(x-y) = x(z-y)$, which is a conic through A, B, C, D.

3.8.5 Properties of a projectivity on a conic

Let us take the conic to have equation $y^2 = zx$ with the parametric representation of points $(t^2, t, 1)$. Suppose that we set up a projectivity $cst + ds - at - b = 0$, $(ad \neq bc)$ between points 't' and 's'.

If the self-corresponding points are distinct we may suppose they are at $s = t = 0$ and $s = t = \infty$, then the projectivity becomes $s = kt$, $k = a/d$ and the double points are at $A(1,0,0)$ and $C(0,0,1)$. We consider what is the envelope of chords joining corresponding points in the projectivity. The equation of such a chord, from Equation (3.13) is $x - yt(1+k) + zkt^2 = 0$. The line co-ordinates of these chords satisfy

$$l : m : n = 1 : -t(1+k) : kt^2$$

and so they envelop the conic S' with equation $km^2 = (1+k)^2 nl$. The point equation of S' is $(1+k)^2 y^2 = 4kzx$. S and S' touch at A and C. When the self-corresponding points coincide, we may suppose that it lies at $s = t = \infty$. Then the projectivity may be written as $s = t + k$. We leave it as an exercise to show that the envelope of chords joining corresponding points is in this case a conic S'' having four-point contact with S at the self-corresponding point.

3.8.6 Harmonic separation of points on a conic

Suppose that we have points P, Q, R, S lying on a conic Σ and we are told that $(P, Q; R, S) = -1$. The obvious question to ask is what does this mean about

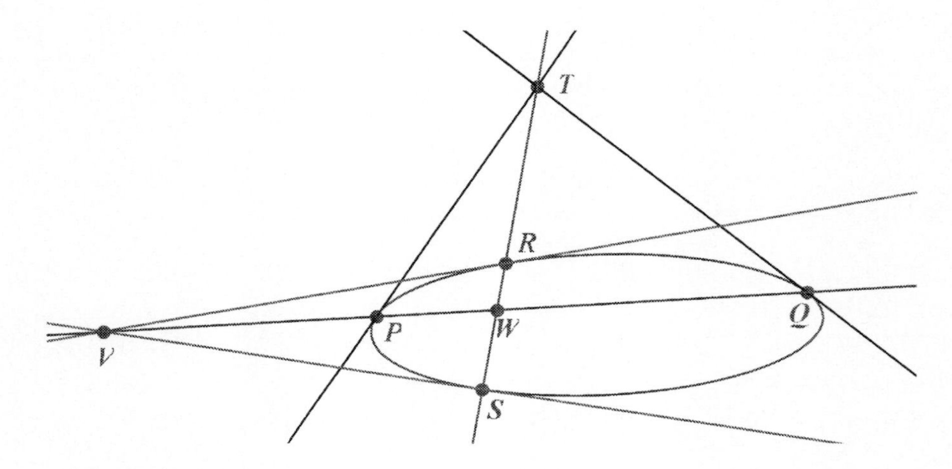

Figure 3.12: Conjugate lines with respect to a conic

the positions of P, Q, R, S. If we suppose that Σ has equation $y^2 = zx$ and that points of Σ have co-ordinates $(\theta^2, \theta, 1)$, then we may choose $P(0,0,1)$, $Q(1,0,0)$, $R(1,1,1)$ and these points have parameters $0, \infty, 1$. Then we know that S must have parameter -1 and hence have co-ordinates $S(1,-1,1)$. PQ has equation $y = 0$ and RS has equation $z = x$. Now the tangent at P has equation $x = 0$ and the tangent at Q has equation $z = 0$, so the pole of PQ is the point $T(0,1,0)$. This lies on RS. It follows that PQ and RS are *conjugate lines*. Similarly the pole of RS is $V(1,0,-1)$, which lies on PQ. PQ and RS meet at $W(1,0,1)$ and on the line $VPWQ$ we have $(P,Q;W,V) = -1$ and on the line $TRWS$ we have $(R,S;W,T) = -1$. This situation is shown in Fig. 3.12.

3.8.7 Involution on a conic

If the projectivity in Section 3.8.5 is an involution, then $d = a$ and if the self-corresponding points are 0 and ∞ the relationship between the parameters is given by $s + t = 0$. Chords joining corresponding points have equations $x = t^2 z$ and these all pass through the point B with co-ordinates $B(0,1,0)$. The self-corresponding points are $A(1,0,0)$ and $C(0,0,1)$. Conversely, if chords of a conic pass through a fixed point it is obvious that pairs at the extremities of these chords are in involution. See Fig. 3.13.

Pairs of points of an involution on a conic are harmonically separated by

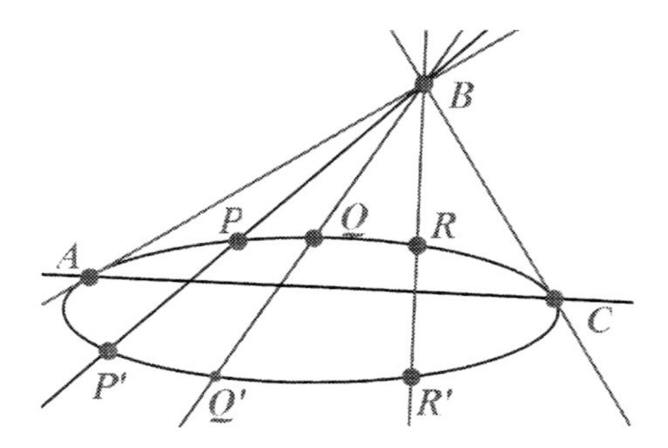

Figure 3.13: Point pairs in involution on a conic

the self-corresponding points of the involution. This is because $(0, \infty; t, -t) = -1$.

3.8.8 Pascal's theorem

We prove Pascal's theorem again, using the methods developed in this section. Let PQR and UVW be two triangles inscribed in a conic, and let QW and RV meet at L, RU and PW meet at M, PV and QU meet at N. Then the points L, M, N are collinear.

Let X, Y be the self-corresponding points of the projectivity $(PQR) \wedge (UVW)$. The cross-ratios $(X, Y; Q, R)$ and $(X, Y; V, W)$ are equal on the conic, and hence $(X, Y; Q, R) = (Y, X; W, V)$. Thus in the projectivity $(XQR) \wedge (YWV)$, Y gives rise to X. The projectivity is therefore an involution. It follows that XY, QW and RV are concurrent; that is, L lies on XY. Similarly M and N lie on XY and so L, M, N are collinear.

Example 3.8.1

Four points A, B, C, D lie on a conic S. A straight line through D meets S again at D' and BC, CA, AB respectively at A', B', C'. Prove that the cross-ratio of the pencil $(A, C; B, D)$ at any point on the conic is equal to the cross-ratio of the range $(A', C'; B', D')$.

Let S have equation $y^2 = zx$ with points represented parametrically by $(\theta^2, \theta, 1)$ and suppose that A, B, C, D, D' are represented by $= s, t, u, \infty, 0$ respectively. Then

$$(A, C; B, D) = (s, u; t, \infty) = (s - t)/(u - t).$$

Now the equation of DD' is $y = 0$ and the equation of BC is $x - (t + u)y + tuz = 0$, so A' has co-ordinates $(-tu, 0, 1)$. Similarly B' and C' have co-ordinates $(-us, 0, 1)$ and $(-st, 0, 1)$ respectively. It follows that

$$(A', C'; B', D') = (-tu, -st; -us, 0) = (s - t)/(u - t).$$

Example 3.8.2

Four points A, B, C, D lie on a conic. AB, CD meet at P, AD, CB meet at Q and the tangents at A and C meet at R. Prove that P, Q, R are collinear.

Take the conic to have equation $y^2 = zx$, with points represented parametrically by $(\theta^2, \theta, 1)$ and A, B, C, D given parameters $\theta = \infty, t, 0, s$ respectively. Then R has co-ordinates $(0, 1, 0)$, since it is the intersection of the tangents at A and C. Now the equations of AB and CD are $y = tz$ and $x = sy$ respectively so P has co-ordinates $(st, t, 1)$. Similarly Q has co-ordinates $(st, s, 1)$. And P, Q, R all lie on the line with equation $x = stz$.

Alternatively, this example may be thought of as a degenerate case of Pascal's theorem, the hexagon being $AABCCD$.

Example 3.8.3

Let triangle ABC be inscribed in a conic, and T be the pole of AB. Any line through T cuts BC, AC in M, N respectively. Prove that M, N are conjugate points with respect to the conic. See Fig. 3.14.

Let TMN meet the conic at X and Y. Now TMN and AB are conjugate lines since T is the intersection of the tangents at A and B. It follows that $(A, B; X, Y)$ is harmonic on the conic. Thus $C(A, B; X, Y)$ is a harmonic pencil. On the range TMN this means that M, N separate X, Y harmonically and hence M, N are conjugate points with respect to the conic.

Example 3.8.4

The tangents at the vertices of a triangle inscribed in a conic meet the opposite sides in collinear points.

See also the solution to Example 3.6.3.

Let ABC be a triangle and suppose that P, Q are the self-corresponding points of the projectivity $(ABC) \wedge (BCA)$. Then on the conic

$$(A, C; P, Q) = (B, A; P, Q) = (A, B; Q, P).$$

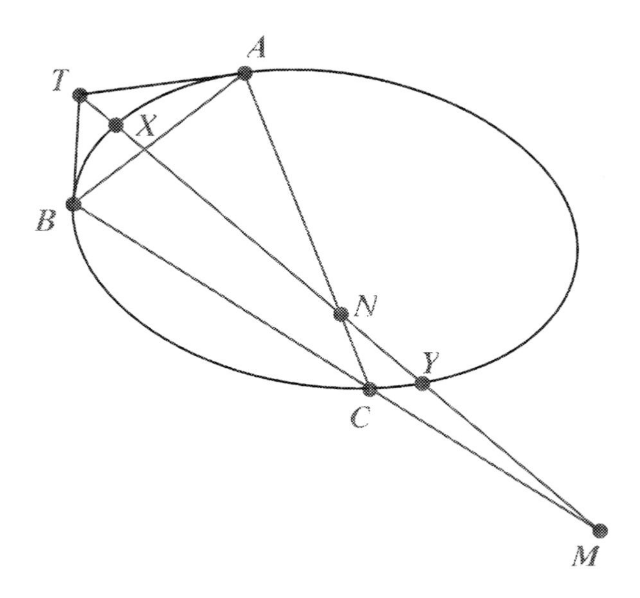

Figure 3.14: Example 3.8.3

Hence A is a self-corresponding point in the involution of which B, C and P, Q are pairs. Hence the tangent at A meets BC at a point of PQ. Similarly the tangent at B meets CA at a point of PQ and the tangent at C meets AB at a point of PQ. So all three points are collinear on PQ.

Exercise 3.8.5

1. The sides of triangle ABC touch a conic S at L, M, N. The tangent at any point O of S meets the sides BC, CA, AB at P, Q, R and the sides MN, NL, LM at U, V, W. Prove that the cross-ratios $(O, P; Q, R)$ and $(O, U; V, W)$ are equal. (JW)

2. Suppose that AOP, BOQ, COR, DOS are four chords of a conic passing through a fixed point O. A conic S' is drawn through O, A, B, C, D and a conic S'' is drawn through O, P, Q, R, S. Prove that S' and S'' touch at O. (JW)

3. Through a given point O a variable line is drawn to meet a given conic S in points Q, R. A point P on QR is selected so that $(O, P; Q, R)$ is

a constant cross-ratio. Prove that the locus of P is an arc of a conic having double contact with S. (JW)

4. A, B, C are three fixed points on a given conic S. P, Q are points on S such that the pencils $(P, B; A, C)$ and $(Q, B; A, C)$ are both harmonic. Prove that PQ, CA and the tangent at B are concurrent.

5. Three fixed points A, B, C lie on a conic S and P is a variable point on S. A line is drawn through P to meet the sides BC, CA, AB at L, M, N so that the range $(P, L; M, N)$ has a fixed cross-ratio. Prove that $PLMN$ passes through a fixed point O of S such that the pencil $(O, A; B, C)$ has the same fixed cross-ratio. (JW)

6. Two conics circumscribe a fixed triangle ABC. A variable line through A meets the conics again at P and Q. Prove that the tangents at P and Q divide BC in a fixed cross-ratio. (JW)

7. Two fixed points A, B are taken on a given conic and a fixed straight line is drawn conjugate to AB. A variable point P is chosen on this second line and chords APQ, BPR are drawn. Prove that QR passes through a fixed point on AB. (JW)

8. Two chords AO and BC of a conic are conjugate and any chord OP meets the sides BC, CA, AB of triangle ABC in points L, M, N. Prove that $(P, L; M, N) = -1$. (JW)

9. A conic S circumscribes a triangle ABC. The tangent at A meets the tangents at C, B in P, Q respectively. O is a fixed point on S. The lines OP, OQ meet BC in points P', Q' respectively. Prove that the two pairs of lines AC, QP' and AB, PQ' meet on the tangent at O.

3.9 Porisms

An interesting situation arises when two fixed conics Σ and Σ' are given and a triangle ABC exists such that ABC is inscribed in Σ and circumscribes Σ'.

3.9.1 Poncelet's porism in Euclidean geometry

Let Σ and Σ' be two circles of radius R and r respectively, and respective centres O and I, where $OI = d$ is given by $d^2 = R^2 - 2Rr$, and let A be *any* point on Σ. Then if AB, AC are drawn to touch Σ' and to meet Σ again at B and C, then BC is also tangent to Σ'. A proof of this result may be found in Bradley and Gardiner [19]. It is also true that if OI does not have the length specified, then no such triangle exists. In other words, if there is one such triangle, there is an infinity of such triangles, with the vertex at any point of the circle Σ. This is why the situation is called a *porism*.

3.9.2 Brocard's porism in Euclidean geometry

Let Σ be a circle and ABC an inscribed triangle, with symmedian point S. Draw AS, BS, CS to meet the sides BC, CA, AB respectively at A', B', C'. Now let Σ' be the ellipse touching triangle ABC internally at A', B', C'. The ellipse Σ' is called *Brocard's ellipse*. Having defined Σ and Σ' in this way, then we may choose any point T on the circle Σ, draw TU, TV to touch Brocard's ellipse and to meet Σ again at U and V, and UV touches Σ'. This porism is called *Brocard's porism*. It turns out that S is also the symmedian point of triangle TUV and hence of any triangle in the porism.

3.9.3 The general porism in the projective plane

Suppose now that Σ and Σ' are two conics and that one triangle exists such that it is inscribed in Σ and circumscribes Σ'. Then there is an infinity of such triangles. A proof of this result appears in most textbooks on projective geometry, for example, Cremona [15]. Such porisms are not restricted to triangles.

What we now show is that Brocard's porism generalizes in the real projective plane and the proof of this may be used to establish Brocard's porism. What follows is a co-ordinate based account of topics similar to those analyzed by Bradley and Smith [8] using synthetic methods. We refer the reader to Section 3.5. The results and much of the notation of that Section are now used without further introduction.

However, we modify the notation as follows. We put $f = 1/(1 - qr)$, $g = 1/(1 - rp)$, $h = 1/(1 - pq)$. With this change of notation, the equation of Σ is

$$fyz + gzx + hxy = 0, \tag{3.14}$$

It will be recalled that ABC and TUV are two triangles inscribed in Σ, that ABC is the triangle of reference, and that triangles ABC and TUV are in triple reverse perspective, with vertices of perspective D, E, F lying on the line with equation

$$\frac{x}{f} + \frac{y}{g} + \frac{z}{h} = 0.$$

Here f, g, h are all non-zero and the line DEF is the polar of the point $K(f, g, h)$ with respect to Σ. So far we have not identified the co-ordinates of T, U, V with f, g, h and we do this shortly by redefining the points D, E, F in terms of a parameter 't' on the co-ordinates of points on the polar of K and hence, by means of the triple reverse perspective, on the points of the conic Σ.

The Cevian lines AK, BK, CK meet BC, CA, AB respectively at the points $A'(0, g, h)$, $B'(f, 0, h)$, $C'(f, g, 0)$. Another conic Σ' is of primary significance and it is the conic that touches the sides of ABC at A', B', C'. The equation of Σ' is

$$\frac{x^2}{f^2} + \frac{y^2}{g^2} + \frac{z^2}{h^2} - \frac{2yz}{gh} - \frac{2zx}{hf} - \frac{2xy}{fg} = 0. \tag{3.15}$$

We take an arbitrary point D on the polar of K with co-ordinates $D(-f, (1-t)g, th)$, where t is some real parameter. The line AD is $thy = (1-t)gz$ and this meets Σ again at T, where T has co-ordinates $T(ft(1-t), -g(1-t), -ht)$. The equation of the line BD is $fz + thx = 0$ and this meets Σ again at V, where V has co-ordinates $V(-f(1-t), -gt, ht(1-t))$. The equation of the line CD is $(1-t)gx + fz = 0$ and this meets Σ again at U, where U has co-ordinates $U(-ft, gt(1-t), -h(1-t))$. A number of results now follow.

3.9.4 Tangents and polars

The tangent at A meets BC on the polar of K and the tangent at B meets CA on the polar of K and the tangent at C meets AB on the polar of K. See Example 3.6.3. The equation of the tangent at A is $gz + hy = 0$ and this meets BC at the point P with co-ordinates $P(0, g, -h)$, which clearly lies on the polar of K. Similarly the tangent at B meets CA on the polar of K at the point Q with co-ordinates $Q(-f, 0, h)$ and the tangent at C meets AB on the polar of K at the point R with co-ordinates $R(f, -g, 0)$.

3.9.5 Tangents and a collinearity

If the tangents at B and C meet at A'', with B'', C'' similarly defined, then A, K, A'' are collinear as are B, K, B'' and C, K, C''. No separate proof is needed as this is the dual of Section 3.9.4. However, as we need them later, the co-ordinates of A'', B'', C'' are $(-f, g, h)$, $(f, -g, h)$, $(f, g, -h)$ respectively.

3.9.6 Triple reverse perspective and the polar of K

The lines AV, BU, CT are concurrent at E with co-ordinates $E(ft, -g, h(1-t))$ and AU, BT, CV are concurrent at $F(f(1-t), gt, -h)$ and E, F as well as D lie on the polar of K. The proof is routine and is omitted. What we have shown is that if we start with any point D on the polar of K and construct the triangle TUV as shown above, then the triangles ABC and TUV are in triple reverse perspective with the vertices of perspective all lying on the polar of K.

3.9.7 Parameters and Hessian triads

Each side of the triangle TUV is tangent to the conic Σ'. T has co-ordinates $T(ft(1-t), -g(1-t), -ht)$ and we may regard 't' either as the parameter of D on the polar of K or the parameter of T on the conic Σ. We now take another point R on Σ with co-ordinates $R(fr(1-r), -g(1-r), -hr)$, where R is eventually going to be either U or V in triangle TUV. The equation of TR is

$$\frac{x}{f} + \frac{rty}{g} + \frac{(1-r)(1-t)z}{h} = 0. \tag{3.16}$$

The condition that TR touches Σ' is found by substituting Equation (3.16) into Equation (3.15) and putting in the condition for equal roots and this condition turns out to be

$$r^2t^2 + 3rt + 1 = (rt + 1)(r + t) + 3.$$

Solving for r for fixed t provides $p = u = 1/(1-t)$ or $p = v = (t-1)/t$, where u and v are the parameters of U and V. With $u = 1/(1-t)$ and $v = (t-1)/t$ it is easy to check that the co-ordinates of U and V, given above in terms of t, are indeed $U(fu(1-u), -g(1-u), -hu)$ and $V(fv(1-v), -g(1-v), -hv)$, as required.

What we have proved is that as the parameter t varies so that triangle TUV moves around the conic Σ, then TUV always touches Σ' and all such triangles are in triple reverse perspective with triangle ABC, the vertices of perspective all lying on the polar of K. Since the reference triangle was chosen as an arbitrary triangle, it follows that any pair of such triangles are in triple reverse perspective. In other words we have established a porism P_K, which means that given Σ and Σ', we can start with any point T on Σ and draw the tangents TU and TV to Σ', where U and V lie on Σ and then UV touches Σ'. We refer to the triangles as belonging to the K-*cycle*.

Observe that $v = 1/(1-u)$ and $t = 1/(1-v)$, which confirms that the vertices T, U, V form a Hessian triad (triads in which a projective transformation exists in which $T \mapsto U, U \mapsto V, V \mapsto T$, the self-corresponding points of which do not exist in the real plane). Thus, starting from any point T on Σ, the vertices U, V are found by the same projective correspondence.

3.9.8 Cross-ratios and intertwining

In the above system of parameters A, B, C have parameters $t = \infty, 0, 1$ respectively. Taking T, U, V to have parameters t, u, v respectively and defining the cross-ratio

$$(d, e; f, g) = \frac{(d-f)(e-g)}{(d-g)(e-f)},$$

then, in terms of their parameters,

$$(T, A; B, C) = (U, B; C, A) = (V, C; A, B) \tag{3.17}$$

In fact we have

$$(T, A; B, C) = (t, \infty; 0, 1) = t/(t-1),$$

$$(U, B; C, A) = (u, 0; 1, \infty) = (1-u) = t/(t-1)$$

and

$$(V, C; A, B) = (v, 1; \infty, 0) = 1/v = t/(t-1).$$

What Section 3.9.8 implies is that given a conic Σ with Equation (3.14), passing through the vertices of an arbitrary triangle ABC as triangle of reference, then if we define a point T on it with co-ordinates given parametrically by $T(ft(1-t), -g(1-t), -ht)$, then $(T, A; B, C) = t/(t-1)$ and U, V on Σ are thereby determined by the cross-ratio Equations (3.17) and have co-ordinates as given above. It then follows that ABC and TUV are in triple reverse perspective with vertices of perspective on the polar of $K(f, g, h)$ with respect to Σ and a porism exists in which ABC and TUV are two triangles in the K-cycle.

Cross-ratio identities ensure that there is nothing special about triangle ABC and it follows that if TUV and $T_0U_0V_0$ are any two triangles in the cycle, then

$$(T_0, T; U, V) = (U_0, U; V, T) = (V_0, V; T, U). \tag{3.18}$$

When Equation (3.18) holds we say that triangles TUV and $T_0U_0V_0$ *intertwine* each other. Thus the vertices of every pair of triangles in the porism intertwine each other.

3.9.9 Coincident polars

The polars of K with respect to the conics Σ and Σ' coincide. This is straightforward and is left as an exercise for the reader.

3.9.10 Nested porisms

Let ABC and TUV be two triangles of the porism, and suppose the tangents at B and C meet at A'', with B'', C'' similarly defined. Suppose that the tangents at U and V meet at T'' with U'', V'' similarly defined. Then A'', B'', C'', T'', U'', V'' lie on a conic Σ''. This follows from the principle of duality. The intersection A'' of the tangents at B and C corresponds to the line BC etc. Now by the porism P_K the sides BC, CA, AB, UV, VT, TU all touch the conic Σ'. It follows that the points $A'', B'', C'', T'', U'', V''$ lie on a conic Σ''. As cross-ratios are unaffected, it follows that the triangles $A''B''C''$, $T''U''V''$ are in triple reverse perspective and that all members of the K-cycle on Σ get mapped on to triangles of a K-cycle on Σ''. This is by virtue of the projective map $ABC \to_K A''B''C''$, which exists by virtue of Section 3.9.5. Now observe that Σ touches the sides of all triangles in the K-cycle on Σ'' and so the triangles $A''B''C'', T''U''V''$ etc. are all inscribed in Σ'' and circumscribed to Σ. Thus we have another porism. This may be regarded as the porism dual to the porism P_K. This means that the polars of K with respect to all of the conics Σ, Σ', Σ'' coincide. Dual porisms occur for porisms in general and are not restricted to those in which the triangles are in triple reverse perspective. However, in Poncelet's porism, the outer conic Σ'' is not a circle.

Note that by means of the projective map $ABC \to_K A'B'C'$, where A', B', C' lie on Σ' then the K-cycles are also transferred to the triangles on Σ'. There is no reason why matters should stop there. These triangles may now be regard as being inscribed in Σ' and by means of the same point K we shall find another conic, which touches all their sides and so on, indefinitely.

We now determine the equation of the conic Σ''. The co-ordinates of U and V are given in the paragraph subsequent to Equation (3.15) and the tangents at these points to Σ are respectively

$$gh(1-t)^2x + hfy + fgt^2z = 0$$

and

$$ght^2x + hf(1-t)^2y + fgz = 0.$$

Solving for x, y, z in terms of t and then eliminating t we find the locus of T'' has equation

$$\frac{x^2}{f^2} + \frac{y^2}{g^2} + \frac{z^2}{h^2} + \frac{3yz}{gh} + \frac{3zx}{hf} + \frac{3xy}{fg} = 0.$$

For Brocard's porism we move from the projective plane to the Euclidean plane and use areal co-ordinates instead of projective co-ordinates. K become the symmedian point S and (f, g, h) is replaced in all the working above by (a^2, b^2, c^2). The conic Σ is the circumcircle of ABC, Σ' is Brocard's ellipse (which has its axes parallel and perpendicular to OS). We call Σ'' the *outer Brocard conic.*

Exercise 3.9.1

1. Determine conditions on the sides of triangle ABC for the outer Brocard conic to be an ellipse, parabola or hyperbola.

3.9.11 A result of Honsberger

If ABC is a triangle and AS, BS, CS meet the circle again at I, J, K respectively, then triangle IJK has S as symmedian point and triangles ABC and IJK are in quadruple perspective. This result, due to Honsberger [23], also follows from our analysis. IJK is a triangle in the Brocard cycle, the fourth vertex of perspective is S itself (the others being on the polar of S). In terms of the parameter t, with A, B, C having parameters $\infty, 0, 1$, then I, J, K have parameters $\frac{1}{2}, 2, -1$ respectively. See Section 3.5.1 where the possibility of a quadruple perspective in the real projective plane is first discussed.

Chapter 4

Vectors, Circumcentre, Orthocentre

4.1 Notation and formulas

4.1.1 Assumed knowledge

We assume knowledge of plane vectors, including their definition and elementary properties such as those covered in secondary school courses. That is, how to add or to subtract them, and how to multiply them by a real scalar. Also assumed known is the meaning of a position vector and how to form the scalar and vector product of two vectors. For the most part we use rectangular Cartesian co-ordinates, so that when we write $\mathbf{p} = (p_1, p_2)$ we mean $\mathbf{p} = p_1\mathbf{i} + p_2\mathbf{j}$, where $\mathbf{i}$ and $\mathbf{j}$ are unit vectors in the x- and y-directions respectively.

4.1.2 Notation

If O is the origin and P is a point of the plane with co-ordinates (p_1, p_2), then the position vector $\mathbf{OP}$ is usually denoted by $\mathbf{p}$. An exception is when dealing with the vertices A, B, C of the triangle of reference, when we write $\mathbf{OA} = \mathbf{x}, \mathbf{OB} = \mathbf{y}, \mathbf{OC} = \mathbf{z}$. Thus we write $\mathbf{AB} = \mathbf{y} - \mathbf{x}$, though normally we write $\mathbf{PQ} = \mathbf{q} - \mathbf{p}$.

4.1.3 Magnitude or length of a vector

For the magnitude or length of a vector we write $|OP| = |\mathbf{p}| = \sqrt{p_1^2 + p_2^2}$. Very often $|\mathbf{PQ}|$ is shortened to PQ. Note, however, that the sides of the triangle of reference are written as $|\mathbf{BC}| = BC = a$, $|\mathbf{CA}| = CA = b$, and $|\mathbf{AB}| = AB = c$.

4.1.4 Scalar product of two vectors

If $\mathbf{p} = (p_1, p_2)$ and $\mathbf{q} = (q_1, q_2)$, then the scalar product $\mathbf{p} \cdot \mathbf{q} = p_1q_1 + p_2q_2$. It follows that $|\mathbf{p}|^2 = \mathbf{p} \cdot \mathbf{p}$ and, by the cosine rule, the angle θ between $\mathbf{p}$ and $\mathbf{q}$ is given by $\cos\theta = \frac{\mathbf{p}\cdot\mathbf{q}}{|\mathbf{p}||\mathbf{q}|}$. In particular, non-zero vectors $\mathbf{p}$ and $\mathbf{q}$ are at right angles if, and only if, $\mathbf{p} \cdot \mathbf{q} = 0$.

4.1.5 Vector product of two vectors

If $\mathbf{p}$ and $\mathbf{q}$ are defined as in Section 4.1.4, then the vector product $\mathbf{p} \times \mathbf{q} = (p_1q_2 - q_1p_2)\mathbf{k}$, where $\mathbf{k}$ is a unit vector normal to the plane OPQ. To be consistent we must have $\mathbf{i} \times \mathbf{j} = \mathbf{k}$, $\mathbf{j} \times \mathbf{k} = \mathbf{i}$, $\mathbf{k} \times \mathbf{i} = \mathbf{j}$, $\mathbf{p} \times \mathbf{q} = -\mathbf{q} \times \mathbf{p}$. The area of the triangle OPQ is given by $\frac{1}{2}|\mathbf{p} \times \mathbf{q}| = \frac{1}{2}|\mathbf{p}||\mathbf{q}|\sin\theta$. Note that $(\mathbf{p} \cdot \mathbf{q})^2 + |\mathbf{p} \times \mathbf{q}|^2 = |\mathbf{p}|^2|\mathbf{q}|^2$.

4.2 The Euler line and nine-point circle

Let ABC be a triangle and O the *circumcentre*, which is the point where the perpendicular bisectors of the sides meet. Take O to be the origin of vectors and write $\mathbf{OA} = \mathbf{x}, \mathbf{OB} = \mathbf{y}, \mathbf{OC} = \mathbf{z}$. Let the circumradius be R, then $|\mathbf{x}| = |\mathbf{y}| = |\mathbf{z}| = R$. In an acute-angled triangle, $\angle BOC = 2A, \angle COA = 2B, \angle AOB = 2C$ and we have $\mathbf{y} \cdot \mathbf{z} = R^2 \cos 2A, \mathbf{z} \cdot \mathbf{x} = R^2 \cos 2B, \mathbf{x} \cdot \mathbf{y} = R^2 \cos 2C$. If the triangle is obtuse at, say, A, then $\angle BOC = 360° - 2A$ and $\mathbf{y} \cdot \mathbf{z}$ still has the same value. So the scalar product relationships hold for all triangles. The point L with position vector $\mathbf{l} = \frac{1}{2}(\mathbf{y} + \mathbf{z})$ is the midpoint of BC. The point G that divides AL in the ratio $2 : 1$ has position vector $(1/3)\mathbf{x} + (2/3)\mathbf{l} = (1/3)(\mathbf{x} + \mathbf{y} + \mathbf{z})$, and so, by symmetry G also lies on BM and CN. The point G, where the medians AL, BM, CN concur is called the *centroid* of ABC. Now consider the point H with position vector $\mathbf{OH} = \mathbf{x} + \mathbf{y} + \mathbf{z}$. We have $\mathbf{AH} = \mathbf{y} + \mathbf{z}$ and $\mathbf{AH} \cdot \mathbf{BC} = (\mathbf{y} + \mathbf{z}) \cdot (\mathbf{z} - \mathbf{y}) = |\mathbf{z}|^2 - |\mathbf{y}|^2 = R^2 - R^2 = 0$, and so AH is perpendicular to BC. Similarly BH

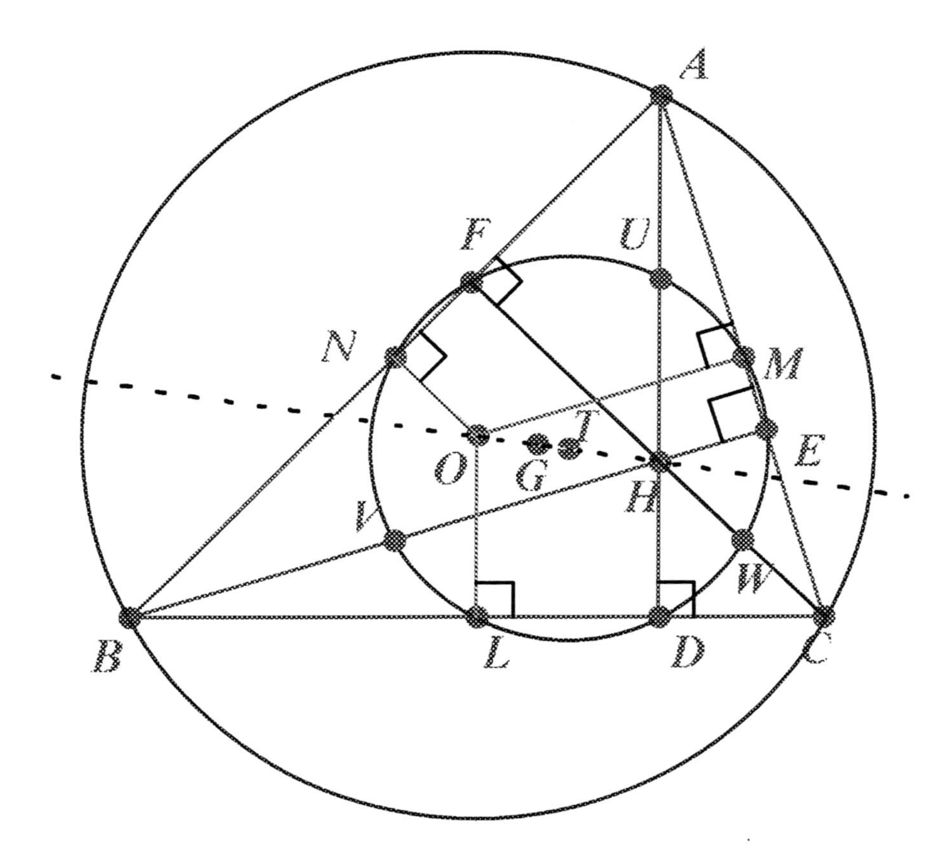

Figure 4.1: The Euler line and the nine-point circle

is perpendicular to CA and CH is perpendicular to AB. It follows that H is the *orthocentre,* the point where the altitudes AD, BE, CF concur. Hence OGH is a straight line and $OG : GH = 1 : 2$. OGH is called the *Euler line* of ABC (or sometimes the *central line*). See Fig. 4.1.

Now consider the point T on the Euler line with vector position $\mathbf{t} = \frac{1}{2}(\mathbf{x}+\mathbf{y}+\mathbf{z})$. T is at distance $R/2$ from L, M, N and is therefore the centre of the circle through the midpoints of the sides. The points U, V, W, the midpoints of AH, BH, CH respectively, have position vectors $\mathbf{u} = \mathbf{x}+\frac{1}{2}(\mathbf{y}+\mathbf{z}), \mathbf{v} = \mathbf{y}+\frac{1}{2}(\mathbf{z}+\mathbf{x}), \mathbf{w} = \mathbf{z}+\frac{1}{2}(\mathbf{x}+\mathbf{y})$, so that U, V, W are also at distance $R/2$ from T. Thus U, V, W also lie on the same circle. Now $HOLD$ is a right-angled trapezium with parallel sides HD and OL perpendicular to

DL, so the midpoint T of OH is equidistant from D and L. Thus D, the foot of the perpendicular from A to BC also lies on the circle. Similarly E, F lie on the circle. It is called the *nine-point circle*, the nine points being $L, M, N, D, E, F, U, V, W$. The point H is the external centre of similitude of the nine-point circle and the circumcircle and G is the internal centre of similitude, the enlargement factor being 2. This means, for example, that if HD is extended to meet the circumcircle at X, then $HD = DX$; and G being the internal centre of similitude is consistent with facts such as $AG = 2GL$.

Exercise 4.2.1

1. Prove that $OH^2 = R^2(3+2(\cos 2A+\cos 2B+\cos 2C))$ and hence show that in any triangle $\cos^2 A + \cos^2 B + cos^2 C \geq \frac{3}{4}$.

2. The point A lies on the circumference of a fixed circle passing through two fixed points B and C. Find the locus of the orthocentre of triangle ABC.

3. H is the orthocentre of triangle ABC. Prove that

$$\frac{BC}{AH} + \frac{CA}{BH} + \frac{AB}{CH} \geq 3\sqrt{3}.$$

4. Let P, Q, R be the reflections of the circumcentre in the sides BC, CA, AB respectively of triangle ABC. Prove that AP, BQ, CR are concurrent.

5. The points diametrically opposite to A, B, C on the circumcircle of triangle ABC are A', B', C'. The midpoints of BC, CA, AB are L, M, N respectively. Prove that $A'L$, $B'M$, $C'N$ are concurrent.

6. In a given plane A is a fixed point. Variable points B, C in the plane are chosen to satisfy

   (i) $AB + AC = k =$ constant;

   (ii) $\angle BAC = 60°$.

   Prove that the locus of the nine-point centre of triangle ABC is a circle, and find its centre and radius.

7. The nine-point centre of a triangle ABC lies on BC. Prove that $|\angle B - \angle C| = 90°$.

8. ABC is a triangle with P an internal point. Through the midpoints of BC, CA, AB lines are drawn parallel to PA, PB, PC respectively. Prove that these lines are concurrent. Locate the point of concurrence when P is the circumcentre O.

9. Consider vectors in a plane. Is it possible to find four non-zero vectors, so that the sum of any two of them is perpendicular to the sum of the other two? Give a geometrical explanation.

10. Six points are given on a circle. From the mean centre of any four of them, the perpendicular is drawn to the chord joining the other two. From the centroid of any three of them, lines are drawn to the orthocentres of the other three. Prove that these thirty-five lines are concurrent.

11. In triangle ABC the lengths of the altitudes are d, e, f. D is a point on the altitude from A, on the same side of A as BC such that $AD = k^2/d$. E and F are similarly defined. Prove that triangles ABC and DEF have the same centroid. *Hint: Prove that* $\mathbf{AD} + \mathbf{BE} + \mathbf{CF} = \mathbf{0}$.

12. The altitudes of triangle ABC are AD, BE, CF. The lines through A, B, C parallel to the opposite sides meet the circumcircle again at U, V, W respectively. Prove that DU, EV, FW are concurrent at the centroid of ABC.

13. ABC is an acute-angled triangle and H is the orthocentre of triangle ABC. AH meets the circle BHC at P. Q and R are similarly defined. Prove that $(HP)(HQ)(HR) \leq 8R^3$, where R is the circumradius.

14. In triangle ABC with circumcentre O, $AB = AC$, N is the midpoint of AB and K is the centroid of triangle ACN. Prove that OK is perpendicular to CN.

15. Let ABC be a triangle with circumcentre O, orthocentre H and nine-point centre T. Prove that $AT^2 + BT^2 + CT^2 \leq AO^2 + BO^2 + CO^2 \leq AH^2 + BH^2 + CH^2$. *Hint: First prove that if P is any point in the plane of ABC, then $AP^2 + BP^2 + CP^2 = AG^2 + BG^2 + CG^2 + 3PG^2$, where G is the centroid of ABC.*

16. In triangle ABC one pair of trisectors of the angles B and C meet at the orthocentre. Show that the other pair of trisectors of these angles meet at the circumcentre.

17. ABC is a triangle, circumcentre O. P, Q, R are the reflections of O in the sides BC, CA, AB respectively. Find the centre and the radius of the circle PQR. If P' is the other point of circle PQR lying on OP, determine $P'O/OP$ in terms of the angles of ABC.

18. ABC is an acute-angled triangle, centroid G and circumradius R. The circle S, centre G and radius $2R/3$ is drawn. Locate, with proof, the six points where S meets the altitudes of the triangle.

19. Locate the points P on the circumcircle of triangle ABC such that $AP^2 + BP^2 + CP^2$ is a maximum or a minimum. *Hint: Use the hint to Problem 15.*

20. ABC is a triangle. X, Y, Z are points such that $BACX$, $CBAY$, $ACBZ$ are parallelograms. Prove that the common chord of the circles on AX and BY as diameters pass through Z. Locate the radical centre of the circles on AX, BY, CZ as diameters.

4.3 Other applications of vectors

There are many problems that may be solved with the help of vectors, other than those in which the circumcentre of a triangle is chosen as origin, and other than those in which Cartesian or areal co-ordinates are appropriate. Some of these appear in later chapters, but we give here two further examples of the use of vectors and an additional exercise.

The first example uses the fact that if $\mathbf{i}$ and $\mathbf{j}$ are unit vectors in the directions $\mathbf{OA}$ and $\mathbf{OB}$ respectively, then $\mathbf{i}+\mathbf{j}$ lies along the internal bisector of $\angle AOB$. For proof consider a rhombus two of whose sides are $\mathbf{i}$ and $\mathbf{j}$ and diagonal $\mathbf{i}+\mathbf{j}$. Since $\mathbf{i}-\mathbf{j}$ is perpendicular to $\mathbf{i}+\mathbf{j}$, $\mathbf{i}-\mathbf{j}$ lies along the external bisector.

Example 4.3.1

In triangle ABC it is given that AB and AC are fixed in direction, with A a fixed point. Also $[ABC]$, the area of triangle ABC, is constant as well as $\angle BAC$. Through L the midpoint of BC, a line is drawn parallel to the internal bisector of $\angle BAC$ to meet AB in P and AC in Q. Prove that $(LP)(LQ)$ is constant and determine the constant in terms of $[ABC]$ and $\angle BAC$.

Take A as origin, And let $\mathbf{i}$ and $\mathbf{j}$ be unit vectors in the directions $\mathbf{AB}$ and $\mathbf{AC}$ respectively. Let $AB = c$ and $AC = b$. A vector parallel to the internal bisector of $\angle BAC$ is $\mathbf{i}+\mathbf{j}$. The vector displacements $\mathbf{LP}$ and $\mathbf{LQ}$ are soon found to be $-\frac{1}{2}b(\mathbf{i}+\mathbf{j})$ and $-\frac{1}{2}c(\mathbf{i}+\mathbf{j})$ respectively. Hence $(LP)(LQ) = \frac{1}{2}bc(1 + \cos A)$, since $\mathbf{i}\cdot\mathbf{j} = \cos A$. And this is equal to $[ABC]\cot\frac{A}{2}$, since $[ABC] = \frac{1}{2}bc\sin A$.

The second example uses the fact that if $ABCD$ is a cyclic quadrilateral and its diagonals AC and BD meet at E, then it is possible to choose E as origin and to define two unit vectors $\mathbf{u}$ and $\mathbf{v}$ in the directions $\mathbf{EA}$ and $\mathbf{EB}$. Then we may write $\mathbf{EA} = l\mathbf{u}, \mathbf{EB} = n\mathbf{v}, \mathbf{EC} = -m\mathbf{u}$, $ED = -p\mathbf{v}$, where $lm = np$. Conversely, if $ABCD$ is a convex quadrilateral whose diagonals meet at E and, if $lm = np$, then $ABCD$ is cyclic. This follows from the intersecting chord theorem $(EA)(EC) = (EB)(ED)$ and its converse.

Example 4.3.2

$ABCD$ is a convex quadrilateral in which AD is not parallel to BC. The diagonals AC and BD meet at E. The points F and G divide AB and DC respectively in the ratio $AD : BC$. Prove that if E, F, G are collinear, then A, B, C, D are concyclic.

Set up and use the notation in the preamble to the example. We need to show that $lm = np$. Let $AD = x$ and $BC = y$. Since F, G divide AB, DC respectively in the ratio $x : y$ we have $\mathbf{EF} = (yl\mathbf{u} + xn\mathbf{v})/(x + y)$ and $\mathbf{EG} = (y(-p\mathbf{v}) + x(-m\mathbf{u}))/(x + y)$. It follows that E, F, G are collinear if, and only if, $yl\mathbf{u} + xn\mathbf{v} = k(yp\mathbf{v} + mx\mathbf{u})$ for some constant k. Hence $y^2/x^2 = mn/lp$. By the cosine rule for triangles AED and BEC, we find, after eliminating $\mathbf{u}\cdot\mathbf{v}$, that $mnx^2 - lpy^2 = mn(l^2 + p^2) - lp(m^2 + n^2) = 0$. Hence $(lm - np)(ln - mp) = 0$. But DA is not parallel to BC, so $ln \neq mp$ and $lm = np$, as required.

Exercise 4.3.3

1. In the hexagon $AYCXBZ$ the three sides AY, CX, BZ are parallel, as are the three sides YC, XB, ZA. Prove that AX, BY, CZ are concurrent.

2. A straight line is drawn through one corner A of a parallelogram to meet the diagonal and the two sides which do not pass through A in P and Q, R respectively. Prove that $AP^2 = (PQ)(PR)$. (JW)

3. G is the centroid of triangle ABC and K is a circle with centre G, which meets the three lines BC, CA, AB. The tangents to K at its intersections with BC meet at P. Points Q and R are defined similarly. Prove that triangle PQR also has centroid G.

4. $UVWXY$ is a plane pentagon. The lines joining V, W, X, Y to the midpoints of UW, VX, WY, XU respectively all pass through a point O. Show that the line joining U to the midpoint of YV also passes through O.

5. In the convex quadrilateral $ABCD$ the midpoints of AB, AC, BD are on the same line. Prove that the midpoint of CD is on the same line.

6. Let the angle bisectors of triangle ABC meet the circumcircle again in P, Q, R. Let $\mathbf{OP} = \mathbf{p}$, $\mathbf{OQ} = \mathbf{q}$, $\mathbf{OR} = \mathbf{r}$, where O is the circumcentre. Prove that $\mathbf{OI} = \mathbf{p} + \mathbf{q} + \mathbf{r}$, where I is the incentre and that $\mathbf{OI}_1 = \mathbf{p} - \mathbf{q} - \mathbf{r}$, where I_1 is the excentre opposite A.

7. Triangle ABC is equilateral; D is the midpoint of BC and DB is extended to E so that $EB = BD$. X is a point of the side AB and Y is taken on the extension of BA so that $XY = BA$. The lines EX and DY meet at Z. Prove that $\angle ZCB = 2\angle ZEB$.

8. Two triangles are so related that straight lines drawn through the vertices of one parallel to the sides of the other meet in a point. Prove that the straight lines drawn through the corresponding vertices of the second parallel to the sides of the first also meet in a point.

9. $ABCDE$ is a convex pentagon. Suppose that the lines from A, B, C, D to opposite sides are concurrent at O. Prove that EO is perpendicular to BC.

10. ABC is a triangle with orthocentre H. Prove that the Euler lines of triangles ABC, BHC, CHA, AHB are concurrent and identify the point of concurrence.

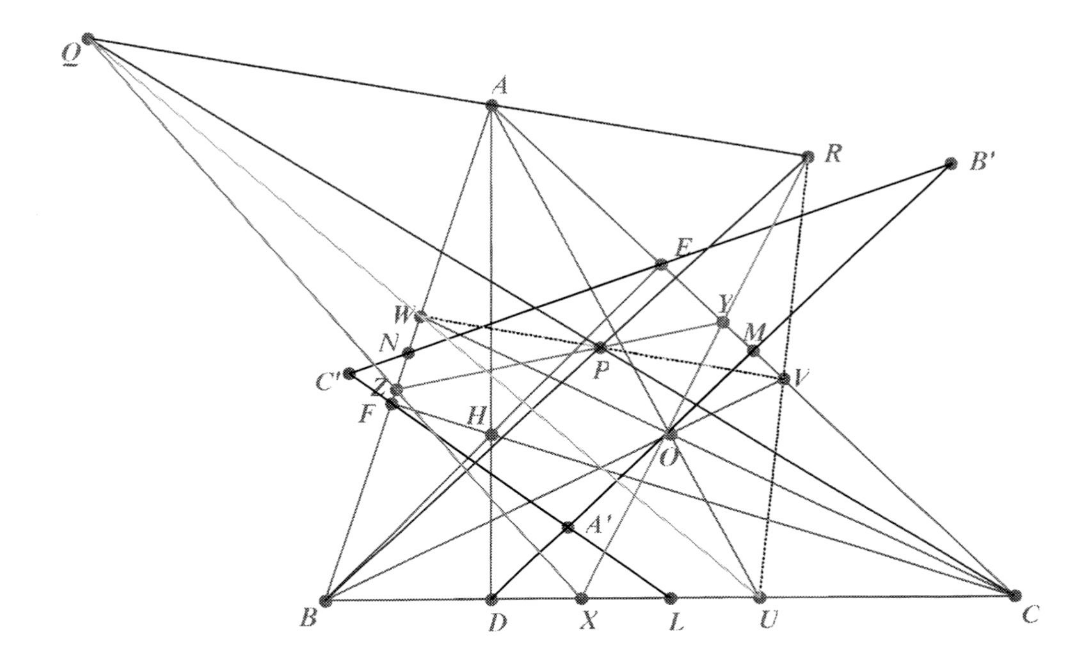

Figure 4.2: First configuration for Part 4.4

4.4 Circumcircle and nine-point circle problems

In this part we use areal co-ordinates.

In Fig. 4.2 the altitudes AD, BE, CF meet at the orthocentre H. The point O is the circumcentre of triangle ABC and AO, BO, CO meet BC, CA, AB at U, V, W respectively. L, M, N are the midpoints of BC, CA, AB respectively. X, Y, Z are the points on BC, CA, AB respectively such that $XL = LU$, $YM = MV$, $ZN = NW$, that is, they are the reflections of U, V, W in the perpendicular bisectors of the sides. Points A', B', C' are defined such that $DM \wedge FL = A'$, $EN \wedge DM = B'$, $FL \wedge EN = C'$, Points P, Q, R are defined such that $YZ \wedge VW = P$, $ZX \wedge WU = Q$, $XY \wedge UV = R$. Two main results hold for this configuration:

4.4.1 Triangles ABC, XYZ, $A'B'C'$ are mutually in perspective at a point J

The point H has areal co-ordinates $(\tan A, \tan B, \tan C)$, so the co-ordinates of D, E, F are $D(0, \tan B, \tan C)$, $E(\tan A, 0, \tan C)$, $F(\tan A, \tan B, 0)$. O has co-ordinates $(\sin 2A, \sin 2B, \sin 2C)$, so U, V, W have co-ordinates

$$U(0, \sin 2B, \sin 2C), V(\sin 2A, 0, \sin 2C), W(\sin 2A, \sin 2B, 0).$$

L, M, N have co-ordinates $L(0,1,1)$, $M(1,0,1)$, $N(1,1,0)$. X, Y, Z are the reflections of U, V, W in the perpendicular bisectors of the sides and so have co-ordinates

$$X(0, \sin 2C, \sin 2B), Y(\sin 2C, 0, \sin 2A), Z(\sin 2B, \sin 2A, 0).$$

The equation of AX is $y \sin 2B = z \sin 2C$, with similar equations by cyclic change for BY and CZ. These three lines are concurrent at the point with co-ordinates $J(\operatorname{cosec} 2A, \operatorname{cosec} 2B, \operatorname{cosec} 2C)$. J is called the *isotomic conjugate* of O. See Chapter 10, where the term isotomic conjugate is defined for points other than O. Now the equation of FL is $x \tan B - y \tan A + z \tan A = 0$ and the equation of DM is $x \tan B + y \tan C - z \tan B = 0$. These meet at $A'$where $y(\tan A + \tan C) = z(\tan A + \tan B)$, that is $y \sin 2B = z \sin 2C$, so A' lies on AX. Similarly B' lies on BY and C' lies on CZ. The result of Section 4.4.1 now follows.

4.4.2 Three collinearities

Q, A, R are collinear, as are R, B, P and P, C, Q. In proving this result we use the abbreviations $l = \sin 2A$, $m = \sin 2B$, $n = \sin 2C$, then we have as co-ordinates

$$U(0, m, n), V(l, 0, n), W(l, m, 0), X(0, n, m), Y(n, 0, l), Z(m, l, 0).$$

The equation of UV is $x/l + y/m - z/n = 0$ and that of XY is $lx + my - nz = 0$. These meet at the point R with co-ordinates $R(l(m^2 - n^2), m(n^2 - l^2), n(m^2 - l^2))$. Similarly Q has co-ordinates $Q(l(m^2 - n^2), m(l^2 - n^2), n(l^2 - m^2))$. The equation of QR is $yn(l^2 - m^2) = zm(l^2 - n^2)$, which passes through A. Similarly RP passes through B and PQ passes through C.

The reader is invited to show that AP, BQ, CR are concurrent, so that triangles ABC and PQR are also in perspective. Whereas Section 4.4.1

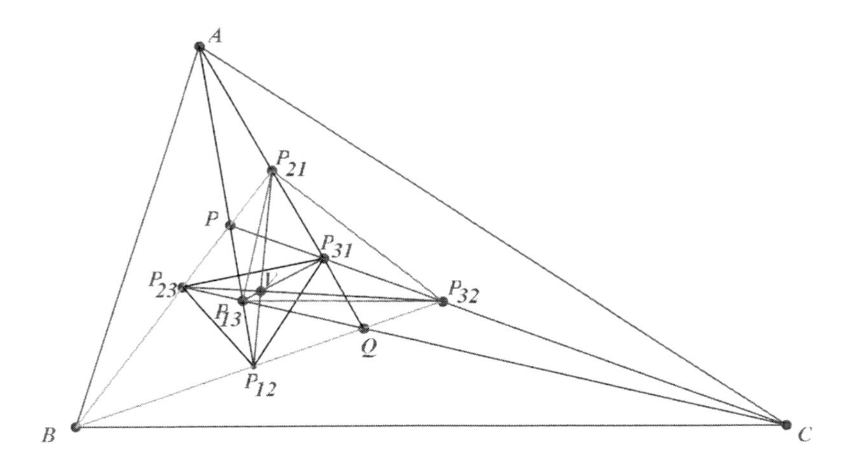

Figure 4.3: Second configuration for Part 4.4

is particular to the configuration involving the orthocentre and the circumcentre, the fact that the result in Section 4.4.2 is true for any values of l, m, n means that it is a result that holds good for isotomic conjugate pairs other than O and J. The result in Section 4.4.1 is, of course, true if the lines DM, EN, FL are replaced by DN, EL, FM forming another triangle $A''B''C''$, so there are in fact four triangles mutually in perspective, all with the same vertex J.

In Fig. 4.3 P and Q might be the circumcentre and the orthocentre of triangle ABC and then P_{12} would be the intersection of AO with BH (1 for the A and 2 for the B). P_{13}, P_{23}, P_{21}, P_{31}, P_{32} would then be similarly defined.

4.4.3 Triangles $P_{12}P_{23}P_{31}$ and $P_{21}P_{32}P_{13}$ are in perspective

The result for this configuration is that triangles $P_{12}P_{23}P_{31}$ and $P_{21}P_{32}P_{13}$ are in perspective. This is a particular case of a general theorem, which says that the result is true if any pair of isogonal conjugate points P and Q, as actually illustrated in Fig. 4.3, replace O and H. See Chapter 10 for the definition of isogonal conjugate pairs. So we prove the more general theorem.

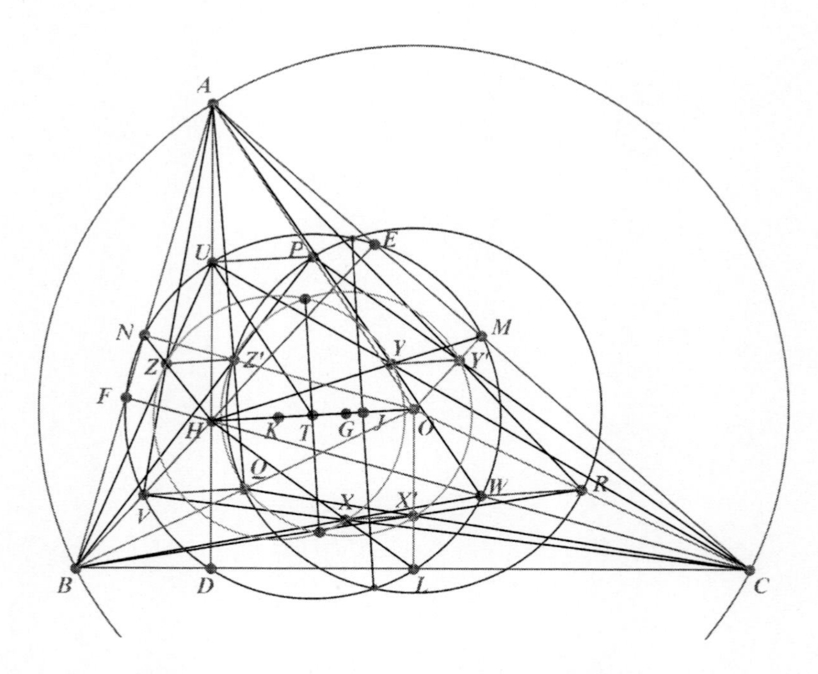

Figure 4.4: The configuration for Part 4.5

Let P have co-ordinates (l, m, n) so Q has co-ordinates $(a^2/l, b^2/m, c^2/n)$, which are the characteristic areal co-ordinates of a pair of isogonal conjugate points. The equation of AP is $ny = mz$ and the equation of BQ is $c^2x/n = a^2z/l$. These meet at P_{12} with co-ordinates $P_{12}(n^2a^2, lmc^2, nlc^2)$. Similarly P_{21} has co-ordinates $P_{21}(lmc^2, n^2b^2, mnc^2)$. The equation of the line $P_{12}P_{21}$ is $nlc^2(m^2c^2 - n^2b^2)x + mnc^2(l^2c^2 - n^2a^2)y + (n^4a^2b^2 - l^2m^2c^4)z = 0$ and those of $P_{23}P_{32}$ and $P_{31}P_{13}$ follow by cyclic change of a, b, c and l, m, n and x, y, z. These three lines share the common point V with co-ordinates $V(a^2l(b^2n^2 + c^2m^2), b^2m(c^2l^2 + a^2n^2), c^2n(a^2m^2 + b^2l^2))$, which is the vertex of perspective of the two triangles.

4.5 Circles allied to the nine-point circle

In Fig. 4.4 O and H are the circumcentre and orthocentre of triangle ABC. The lines AHD, BHE, CHF are the altitudes. G is the centroid and AGL, BGM, CGN are the medians. U, V, W are the midpoints of AH, BH, CH

respectively. We suppose that the circumcircle of ABC has radius R. Points X, Y, Z are the centroids of triangles HBC, HCA, HAB respectively. P, Q, R (which cannot be confused with the circumradius) are the midpoints of OA, OB, OC respectively and X', Y', Z' are the centroids of triangle OBC, OCA, OAB respectively. In this configuration the following results hold:

4.5.1 Points on the nine-point circle

The points $D, E, F, L, M, N, U, V, W$ lie on a circle (the nine-point circle) with centre T (the midpoint of OH) and radius $R/2$. This has been established in Section 4.2.

4.5.2 G lies on OH and is such that $OG = (1/3)OH$

This has also been established in Section 4.2

4.5.3 P, Q, R lie on a circle centre O of radius $R/2$

This result is trivial, since $OP = OQ = OR = R/2$.

4.5.4 The common chord

The common chord of these two circles intersects OH at a point J which is such that $OJ = OH/4$. This circle and the nine-point circle are of equal radius and have centres O and T respectively. Their common chord is therefore cut by the Euler line at a point J such that $OJ = JT$. Thus $OJ = OH/4$.

4.5.5 The circle XYZ

The circle XYZ has centre K and radius $R/3$, where K lies on OH and is such that $OK = (2/3)OH$. X is the centroid of triangle HBC and therefore has position vector $\mathbf{OX} = (1/3)(\mathbf{y} + \mathbf{z} + (\mathbf{x} + \mathbf{y} + \mathbf{z})) = (1/3)(\mathbf{x} + 2\mathbf{y} + 2\mathbf{z})$. Now the point K with position vector $(2/3)(\mathbf{x} + \mathbf{y} + \mathbf{z})$ lies on OH and is such that $OK = (2/3)OH$, and also $\mathbf{XK} = (1/3)\mathbf{x}$. Similarly $\mathbf{YK} = (1/3)\mathbf{y}$ and $\mathbf{ZK} = (1/3)\mathbf{z}$. Hence the circle XYZ has centre K and radius $R/3$.

4.5.6 The circle $X'Y'Z'$

The circle $X'Y'Z'$ has centre G and radius $R/3$. X' is the centroid of triangle OBC and therefore has position vector $\mathbf{OX'} = (1/3)(\mathbf{y}+\mathbf{z})$. It follows that $\mathbf{X'G} = (1/3)\mathbf{x}$ and similarly $\mathbf{Y'G} = (1/3)\mathbf{y}$ and $\mathbf{Z'G} = (1/3)\mathbf{z}$. Hence the circle $X'Y'Z'$ has centre G and radius $R/3$.

4.5.7 The line OH

The line OH bisects the common chord of these last two circles at the point T. These last two circles have equal radius and centres K and G. Their common chord is therefore bisected by KG at its midpoint, which is T, the nine-point centre.

4.5.8 $XX' = YY' = ZZ' = GO$

Finally $\mathbf{XX'} = \mathbf{YY'} = \mathbf{ZZ'} = -(1/3)(\mathbf{x}+\mathbf{y}+\mathbf{z}) = \mathbf{GO}$. Similarly $\mathbf{UP} = \mathbf{VQ} = \mathbf{WR} = -(1/2)(\mathbf{x}+\mathbf{y}+\mathbf{z}) = \mathbf{TO}$.

Exercise 4.5.1

1. Prove that the circle on GH as diameter (the orthocentroidal circle), the nine-point circle and the circumcircle are coaxal.

4.6 The median, halfway and orthic triangles

We investigate properties of the median, halfway and orthic triangles.

In this section we consider only the case in which triangle ABC is acute. When the triangle is obtuse minor changes in the proofs are needed. The text that follows refers to Fig. 4.5.

Triangle DEF is called the *orthic triangle* or *pedal triangle*, and more of its properties are given in Part 4.7.

Triangle LMN is called the *median triangle.* Its centroid is G, the centroid of triangle ABC, and since its circumcentre is T, the nine-point centre of ABC, then by the Euler line property its orthocentre is O, the circumcentre of ABC. Its sides are parallel and half the length of those of ABC and it is homothetic with ABC, the internal centre of similitude being at G.

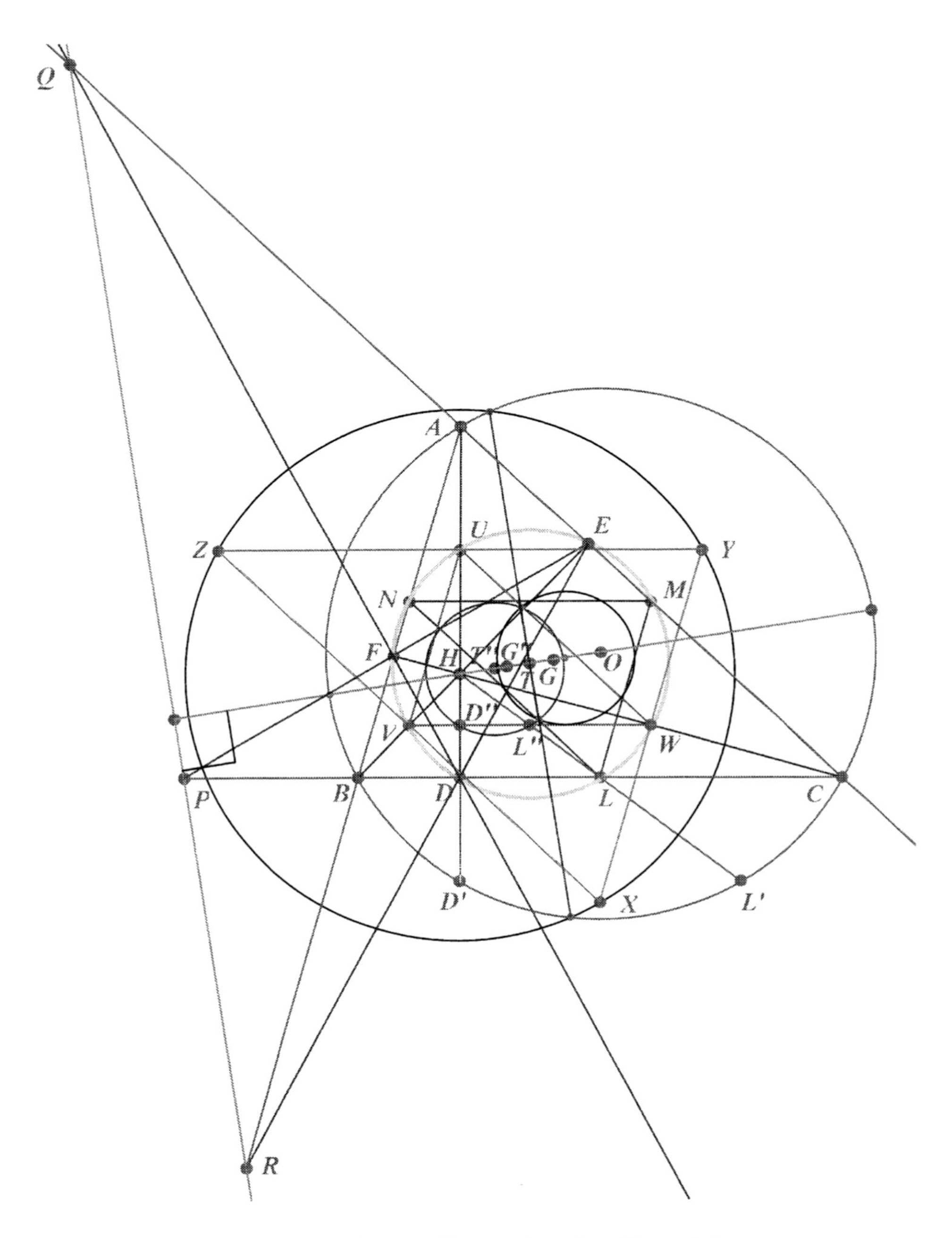

Figure 4.5: The configuration for Part 4.6

Its Euler line coincides with that of ABC. The nine-point circle of triangle LMN is also shown in the diagram, and its radius is $R/4$. Its centre is midway between T and O.

We have taken the liberty of calling triangle UVW the *halfway triangle*, as we have not found a name for it in the literature. Its centroid is at the point G'' with vector position $(2/3)(\mathbf{x}+\mathbf{y}+\mathbf{z})$ and is therefore on the Euler line of ABC at the point such that $OG = G''H$. The orthocentre of triangle UVW is at H and its circumcentre is at T. Like LMN it shares an Euler line with ABC and its nine-point centre, T'', is such that $TT'' = T''H$. Now UVW is congruent to LMN and is similar to and half the size of ABC. Its sides are parallel to those of LMN, so that, for example $LMUV$ is a rectangle. Its nine-point circle of radius $R/4$ is also drawn in the diagram. Triangle UVW is homothetic with ABC with external centre of similitude at H.

Also shown in Fig. 4.5 is triangle XYZ, which is the image of triangle ABC under a half-turn about T. Equivalently, X is the point on AT such that $AT = TX$, with similar definitions for Y on BT and Z on CT. Clearly triangles ABC and XYZ are congruent and homothetic, the internal centre of similitude being T. The position vectors of X, Y, Z are $\mathbf{y}+\mathbf{z}, \mathbf{z}+\mathbf{x}, \mathbf{x}+\mathbf{y}$ respectively. The centroid of XYZ is at G'', its circumcentre is at H, its orthocentre is at O. It shares a nine-point circle with ABC, and its median triangle is UVW. Triangle LMN is to triangle XYZ what triangle UVW is to triangle ABC. The orthic triangle of XYZ is not shown in the figure, but provides three more points on the nine-point circle. If one considers that these twelve points when reflected in the common chord of the nine-point circles of LMN and UVW also lie on the nine-point circle, it is evident that the nine-point circle contains many more than nine interesting points!

Using areal co-ordinates we have $E(\tan A, 0, \tan C)$ and $F(\tan A, \tan B, 0)$ so the equation of EF is $x \tan B \tan C = y \tan A \tan C + z \tan A \tan B$. This meets BC at the point P with co-ordinates $P(0, -\tan B, \tan C)$. Points Q and R defined by $CA \wedge FD$ and $AB \wedge DE$ respectively have co-ordinates $(\tan A, 0, -\tan C)$, $(-\tan A, \tan B, 0)$ respectively, and hence P, Q, R are collinear on the line with equation $\cot Ax + \cot By + \cot Cz = 0$. Now the circumcircle of ABC and the nine-point circle of ABC have Equations (2.11) and (2.14), and their radical axis is found by subtraction to be the line PQR. The line PQR, shown on the diagram, is therefore perpendicular to the Euler line.

We now mention, without proof, certain further properties of the configuration. Some of the results are given in textbooks, such as Durell [17].

4.6.1

The perpendicular bisector of EF passes through L.

4.6.2

DH bisects $\angle EDF$.

4.6.3

H is the incentre of triangle DEF.

4.6.4

The triangles AEF and ABC are similar.

4.6.5

$AH = 2R\cos A$.

4.6.6

$HD = 2R\cos B\cos C$.

4.6.7

If AD meets the circumcircle again at D', then $HD = DD'$.

4.6.8

The perpendicular from A to EF passes through O.

4.6.9

$(AH)(HD) = (BH)(HE) = (CH)(HF)$.

4.6.10

A is the orthocentre of triangle HBC.

4.6.11

If $FB = FE$, then $AE = BD$.

4.6.12

If $AG = BC$, then $\angle BGC = 90^\circ$.

4.6.13

Triangles HBC and ABC have the same nine-point circle.

4.6.14

$\angle FUN = |\angle BAC - \angle ABC|$.

4.6.15

LU, MV, NW are concurrent.

4.6.16

$AH^2 + BC^2 = 4R^2$.

4.6.17

TL is parallel to AO.

4.6.18

UG, HL meet at the point L' on the circumcircle.

4.6.19

UL is parallel to AL'.

Exercise 4.6.1

1. Derive the following formulas for the area of triangle ABC: $[ABC] = \frac{1}{2}bc\sin A = \frac{1}{2}ca\sin B = \frac{1}{2}ab\sin C = abc/(4R) = 2R^2\sin A\sin B\sin C$.

2. Use the formula in Problem 1 for $\sin C$ and the cosine formula for $\cos C$ to derive, by squaring and adding, Heron's formula $[ABC] = \sqrt{s(s-a)(s-b)(s-c)}$ for the area of triangle ABC. Here $s = (a+b+c)/2$ is the semi-perimeter.

3. Prove that the perimeter of the pedal triangle is $2[ABC]/R$, and prove that its area is $[DEF] = \frac{1}{2}R^2\sin 2A\sin 2B\sin 2C$. ($ABC$ is to be taken as acute.)

4. O, H, I, r, R are the circumcentre, orthocentre, incentre, inradius and circumradius of a scalene triangle ABC. D, E, F are the feet of the altitudes. P is the orthocentre of triangle DEF. Prove that the line OI divides the line segment PH internally in the ratio $r : R$. *Hint: In this problem the use of trilinear co-ordinates is advised.*

5. Let P, Q, R be the midpoints of EF, FD, DE respectively. Prove that AP, BQ, CR are concurrent at the symmedian point of ABC, with areals (a^2, b^2, c^2).

6. Prove Apollonius's theorem that $AL^2 = (2b^2 + 2c^2 - a^2)/4$. Prove that if AL meets the circumcircle again at K, then $(AG)(AK) = (a^2 + b^2 + c^2)/3$.

7. Through A, B, C three parallel lines are drawn to meet MN, NL, LM respectively in A^*, B^*, C^*. Prove that A^*B^* passes through C, B^*C^* through A, C^*A^* through B. Prove also that $[A^*B^*C^*] = \frac{1}{2}[ABC]$.

8. Show that if $\tan B\tan C = 3$, then the Euler line is parallel to BC.

9. ABC is an acute-angled triangle. With A, B, C as centres circles are drawn whose radii are $a\cos A, b\cos B, c\cos C$ respectively, and the internal common tangents are drawn to each pair. Prove that three of these pass through O and the other three through H.

10. If $[HOB] = [MOHE]$ prove that $\tan A \tan C = 3$.

11. If G is the centroid of an acute-angled triangle ABC, whose circumradius is R, prove that $AG^2 + BG^2 + CG^2 > 8R^2/3$.

12. ABC is an acute-angled triangle, circumcentre O and circumradius R. AO meets circle BOC again at D. Points E and F are similarly defined. Prove that $(OD)(OE)(OF) \geq 8R^3$.

13. ABC is a scalene triangle with circumcentre O and orthocentre H. Points X, Y, Z are reflections of the vertices A, B, C in the sides BC, CA, AB respectively. BZ meets CY at K. Prove that KX is parallel to OH.

14. A circle centre H cuts the sides of triangle LMN (or the extensions of those sides) at six points: D_1, D_2 on MN, E_1, E_2 on NL and F_1, F_2 on LM. Prove that $AD_1 = AD_2 = BE_1 = BE_2 = CF_1 = CF_2$.

15. Let P be a point on the circumcircle. The line through A parallel to BP meets CH at Q; the line through A parallel to CP meets BH at R. Prove that QR is parallel to AP.

16. Let X, Y, Z be arbitrary points on the sides BC, CA, AB of the obtuse-angled triangle ABC. Prove that the tangents from the orthocentre H to the circles with diameters AX, BY, CZ are equal.

17. Given that $CF = BM$ and $\angle ACF = \angle CBM$, prove that triangle ABC is equilateral.

18. Triangle ABC is acute. The line through H parallel to CA meets BA at X; the line through H parallel to BA meets CA at Y. Prove that triangle OXY is isosceles.

19. ABC is a variable triangle with fixed circumradius R. Find the maximum value of $BC + AL$.

20. Prove that $2(AD + BE + CF) \leq (a + b + c)\sqrt{3}$.

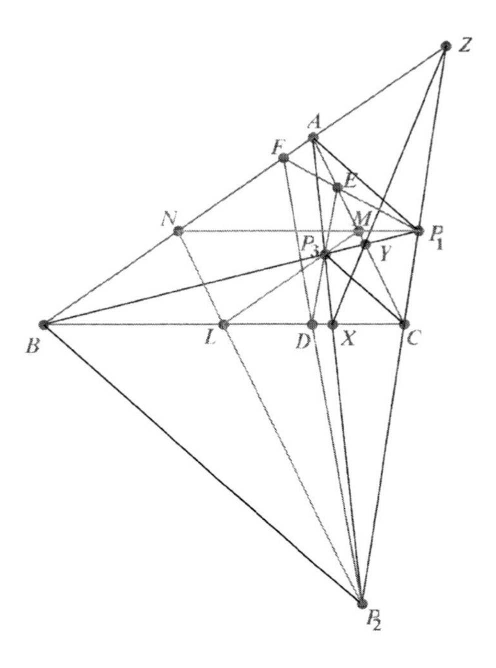

Figure 4.6: Configuration for Exercise 4.7.1

4.7 Median and orthic triangles revisited

We investigate more properties of the median and orthic triangles. In Fig. 4.6 LMN is the median triangle and DEF is the orthic triangle. Points $P_1 = EF \wedge MN$, $P_2 = FD \wedge NL$ and $P_3 = DE \wedge LM$. P_2P_3 meets BC at X, P_3P_1 meets CA at Y and P_1P_2 meets AB at Z. Exercise 4.7.1 is concerned with Fig. 4.6. Use areals with $L(0,1,1)$, $M(1,0,1)$, $N(1,1,0)$, $H(d,e,f)$, where $d = \tan A, e = \tan B, f = \tan C$.

Exercise 4.7.1

1. Prove that AP_1, BP_2, CP_3 are parallel.

2. Prove that A, P_2, P_3 are collinear; B, P_3, P_1 are collinear; and C, P_1, P_2 are collinear.

3. Prove that X, Y, Z are collinear.

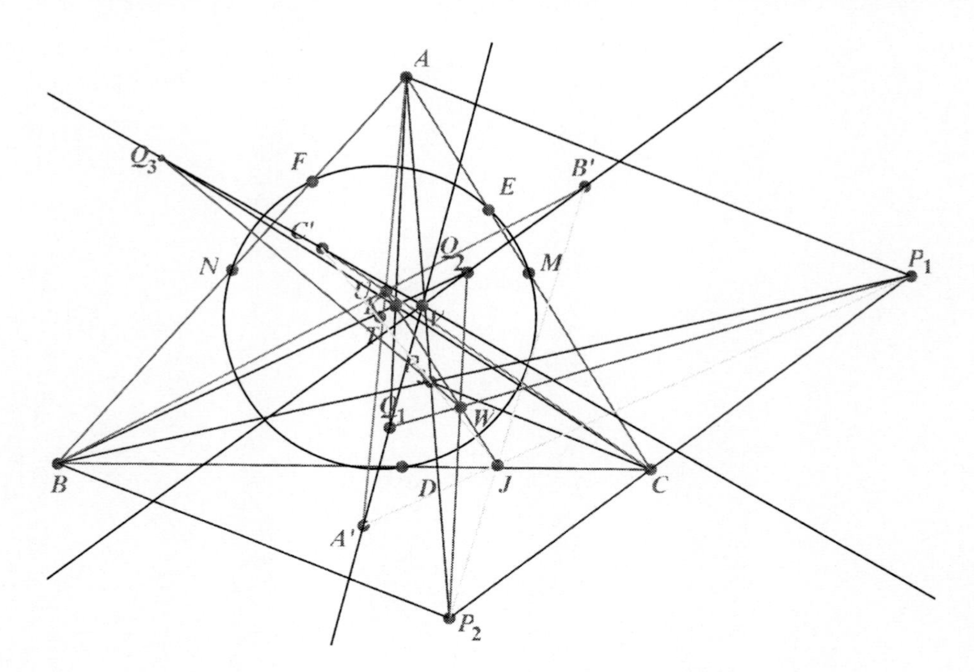

Figure 4.7: Configuration for Exercise 4.7.2

4. Suppose L, M, N remain fixed, but H is replaced by another point K and AKD, BKE, CKF are the three Cevians through K. Are the results of Problems 1–3 still true?

In Fig. 4.7 we show the same configuration as in Fig. 4.6, but with much more detail. In addition to the points P_1, P_2, P_3 we have Q_1, Q_2, Q_3 defined by $Q_1 = FD \wedge LM$, $Q_2 = DE \wedge MN$, $Q_3 = EF \wedge NL$ and also A', B', C' defined by $A' = FL \wedge DM$, $B' = DM \wedge EN$, $C' = EN \wedge FL$. Exercise 4.7.2 is concerned with Fig. 4.7.

Exercise 4.7.2

1. Prove that triangles ABC and $A'B'C'$ are in perspective with vertex at a point U.

2. Prove that triangles $A'B'C'$ and $P_1P_2P_3$ are in perspective with vertex at a point J.

3. Prove that triangles ABC and $Q_1Q_2Q_3$ are in perspective with vertex at a point K.

4. Prove that the points J, K, U defined in Problems 1–3 are collinear.

5. Prove that triangles $A'B'C'$ and $Q_1Q_2Q_3$ are in perspective with vertex at a point V.

6. Prove that triangles $P_1P_2P_3$ and $Q_1Q_2Q_3$ are in perspective with vertex at a point W.

7. Prove that W also lies on the line JKU.

It would be possible to add more detail to this configuration by ordering the letters of the hexagon $DEFLMN$ in different ways. As this hexagon is inscribed in a circle, the conditions of Pascal's theorem apply. Chapter 13 gives a more complete account of results obtained by considering hexagons inscribed in conics.

Chapter 5

Incentre and Excentres

5.1 Elementary facts

5.1.1 Incentre and Excentres

Fig. 5.1 shows a triangle ABC with its internal angle bisectors AIP, BIQ, CIR where P, Q, R are the midpoints of arcs BC, CA, AB of the circumcircle. It also shows the external angle bisectors at A, B, C forming the triangle $I_1I_2I_3$ of excentres. Since I_2I_3, the external bisector of $\angle A$ is at right angles to AI and similarly at B and C, it follows that I is the orthocentre of triangle $I_1I_2I_3$ and that the circumcircle of ABC is the nine-point circle of $I_1I_2I_3$. The nine points are A, B, C, P, Q, R and D, E, F the midpoints of I_2I_3, I_3I_1, I_2I_3 respectively.

It follows from the theory of the nine-point circle that P, Q, R are the midpoints of II_1, II_2, II_3 respectively, and that POD, QOE, ROF are diameters of the circumcircle of ABC perpendicular to the sides BC, CA, AB respectively. Since P is the midpoint of II_1 and $\angle IBI_1 = 90°$, it follows that P is the centre of a circle through the points I, B, I_1 and C, with similar results for Q and R relative to I_2 and I_3. The reader is invited to show that the angles of triangle $I_1I_2I_3$ are $(B + C)/2, (C + A)/2, (A + B)/2$ and that its side lengths are $4R \cos A/2$, $4R \cos B/2, 4R \cos C/2$. The radius of each of the circles $I_1I_2I_3$, II_2I_3, II_3I_1, II_1I_2 is $2R$ (just as the radius of each of the circles ABC, HBC, HCA, HAB is R). The area of triangle $I_1I_2I_3$ is equal to $R(a + b + c)$.

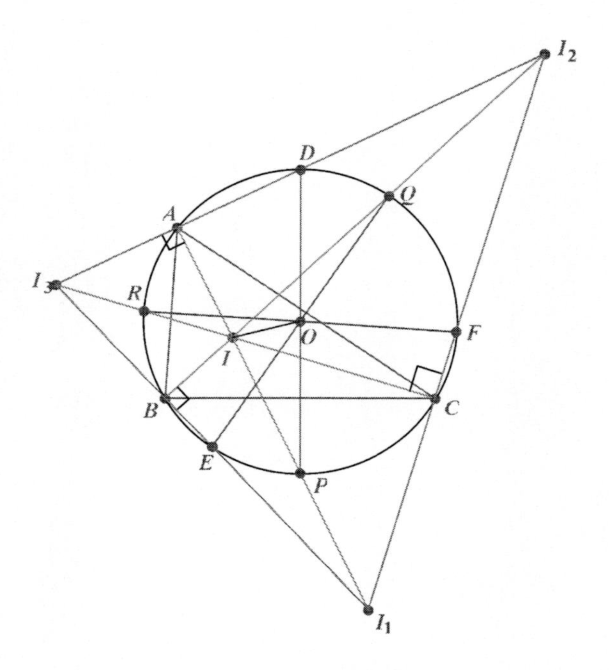

Figure 5.1: Triangle ABC, its incentre and excentres

5.1.2 Points of contact

We investigate points of contact and their distances from the vertices.

Fig. 5.2 shows the incircle and the three excircles of triangle ABC. We use the notation that the points of contact of the incircle with BC, CA, AB are X, Y, Z respectively. The points of contact with the excircle opposite A with BC, CA, AB respectively are denoted by X_1, Y_1, Z_1 respectively. Similarly X_2, Y_2, Z_2 are the points of contact of the excircle opposite B with the sides of triangle ABC and X_3, Y_3, Z_3 are the points of contact of the excircle opposite C with the sides of triangle ABC.

Since $AY = AZ$, $BZ = BX$, $CX = CY$ it follows that $AY = AZ = s-a$, $BZ = BX = s - b$, $CX = CY = s - c$, where $s = \frac{1}{2}(a + b + c)$. Similar reasoning gives

$$AY_1 = AZ_1 = BZ_2 = BX_2 = CX_3 = CY_3 = s$$

and

$$CX_1 = BX, BX_1 = CX, AY_2 = CY, \tag{5.1}$$

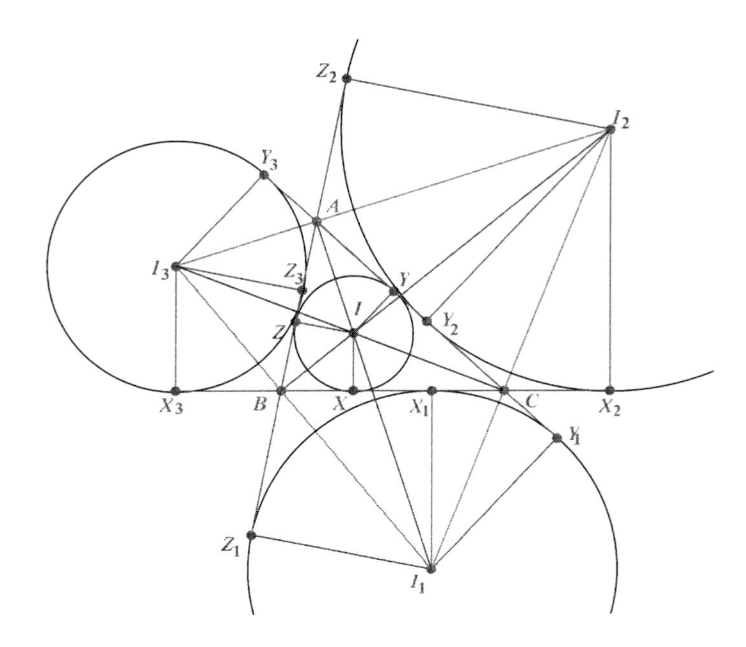

Figure 5.2: The incircle and excircles of triangle ABC

$$CY_2 = AY, BZ_3 = AZ, AZ_3 = BZ.$$

Furthermore, since $[ABC] = [IBC] + [ICA] + [IAB]$, we have $[ABC] = \frac{1}{2}(a + b + c)r = rs$ and similarly $[ABC] = r_1(s - a) = r_2(s - b) = r_3(s - c)$. Here r, r_1, r_2, r_3 are the radii of the incircle and the three excircles.

The above relations account for the form of the areal co-ordinates of I, I_1, I_2, I_3 given in Part 2.2 Section 2.2.9. Note also that $AI = r \operatorname{cosec} A/2$ and $AI_1 = r_1 \operatorname{cosec} A/2$, with similar expressions for BI, BI_2, CI, CI_3.

Exercise 5.1.1

1. Prove that $OI^2 = R^2 - 2Rr$ and hence that $R \geq 2r$.

2. Prove that $r = 4R \sin A/2 \sin B/2 \sin C/2$. Find similar expressions for r_1, r_2, r_3.

3. Prove that in triangle ABC, AI bisects $\angle OAH$.

4. Prove that $[I_1YAZ] = [ABC]$.

5. If X', Y', Z' are the feet of the perpendiculars from X, Y, Z on to YZ, ZX, XY respectively, prove that triangle $X'Y'Z'$ is similar to triangle ABC. What is the scale factor of the similarity?

6. Find, in terms of the angles of ABC, the angles of triangles BIC, CIA, AIB.

7. Show that the centres of the circumcircles of triangles BIC, CIA, AIB lie on AI_1, BI_2, CI_3 respectively.

8. If K is the circumcentre of triangle $I_1I_2I_3$, prove that KI_1 is perpendicular to BC.

9. If M, N are the feet of the perpendiculars from P on to AB, AC respectively, prove that $AM = AN = (b+c)/2$ and that $BM = CN = |b-c|/2$.

10. Prove that $[ABC] = r_1r_2r_3/\sqrt{r_2r_3 + r_3r_1 + r_1r_2}$.

11. Prove that $4R(r_2r_3 + r_3r_1 + r_1r_2) = (r_2 + r_3)(r_3 + r_1)(r_1 + r_2)$.

12. Prove that $[ABC] = \sqrt{rr_1r_2r_3}$.

13. Prove that $8r_1r_2r_3 = (I_2I_3)(I_3I_1)(I_1I_2)\sin A \sin B \sin C$.

14. Prove that

$$8r_1r_2r_3 = (II_1)(II_2)(II_3)(1+\cos A)(1+\cos B)(1+\cos C).$$

15. Prove that $4R = I_2I_3^2/(r_2 + r_3) = II_1^2/(r_1 - r)$ etc.

16. Prove that if $\angle BAC = 90°$, then $r_1 = r_2 + r_3 + r$.

17. Prove that $4R = II_1 \operatorname{cosec} A/2 = II_2 \operatorname{cosec} B/2 = II_3 \operatorname{cosec} C/2$.

5.2 Gergonne's point and Nagel's point

We investigate Gergonne's point, Nagel's point, and the related the ex-Gergonne and ex-Nagel points. In Fig. 5.3 we show the Gergonne point J, the Nagel point N, the ex-Gergonne points J_1, J_2, J_3 and the ex-Nagel points N_1, N_2, N_3. In the sections that follow we use the notation defined in Part 5.1.

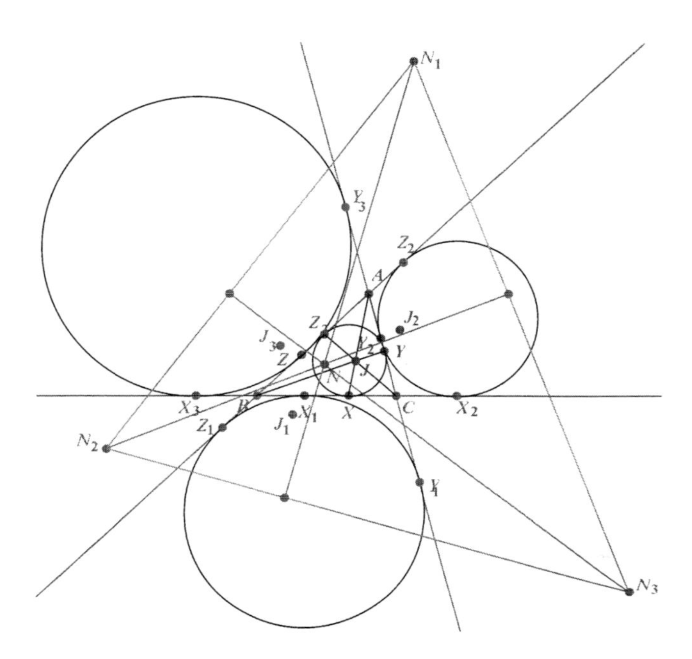

Figure 5.3: The Gergonne and Nagel points

5.2.1 The Gergonne point J

This is the point of concurrence of AX, BY, CZ. Points X, Y, Z have areals $X(0, a+b-c, c+a-b)$, $Y(a+b-c, 0, b+c-a)$, $Z(c+a-b, b+c-a, 0)$. The equation of AX is $(c+a-b)y = (a+b-c)z$, the equation of BY is $(a+b-c)z = (b+c-a)x$ and the equation of CZ is $(b+c-a)x = (c+a-b)y$. These three lines are concurrent at Gergonne's point J with co-ordinates

$$J\left(\frac{1}{b+c-a}, \frac{1}{c+a-b}, \frac{1}{a+b-c}\right).$$

5.2.2 The Nagel point N

This is the point of concurrence of AX_1, BY_2, CZ_3 and in view of Equations (5.1) it is the isotomic conjugate of Gergonne's point J. The areals we require are $X_1(0, c+a-b, a+b-c)$, $Y_2(b+c-a, 0, a+b-c)$, $Z_3(b+c-a, c+a-b, 0)$. We find that AX_1, BY_2, CZ_3 are concurrent at the point $N(b+c-a, c+a-b, a+b-c)$. Note that, as with all isotomic conjugates, the co-ordinates of N are the reciprocals of those of J.

5.2.3 The ex-Gergonne point J_1

This is the point of concurrence of AX_1, BY_1, CZ_1. The points involved have areals $X_1(0, c+a-b, a+b-c)$, $Y_1(b+a-c, 0, a+b+c)$, $Z_1(c+a-b, a+b+c, 0)$. The equation of AX_1 is $(a+b-c)y = (c+a-b)z$, the equation of BY_1 is $(a+b+c)x = (b-a-c)z$ and the equation of CZ_1 is $(a+b+c)x = (c-a-b)y$, and it is easy to see that these three lines are concurrent at

$$J_1\left(\frac{-1}{a+b+c}, \frac{1}{a+b-c}, \frac{1}{c+a-b}\right).$$

5.2.4 The ex-Gergonne points J_2 and J_3

These arise from J_1 by cyclic change of $1, 2, 3$ and x, y, z and a, b, c. Hence J_2 has areals

$$\left(\frac{1}{a+b-c}, \frac{-1}{a+b+c}, \frac{1}{b+c-a}\right)$$

and J_3 has areals

$$\left(\frac{1}{c+a-b}, \frac{1}{b+c-a}, \frac{-1}{a+b+c}\right).$$

5.2.5 The ex-Nagel point N_1

This is the point of concurrence of AX, BY_3 and CZ_2 and its areals are $((a+b+c), -(a+b-c), -(c+a-b))$, as can be seen from Section 5.2.3 since it arises from co-ordinates that are reciprocals of those used in determining J_1. It is therefore the isotomic conjugate of J_1.

5.2.6 The ex-Nagel points N_2 and N_3

These are the isotomic conjugates of J_2 and J_3 and have areals $N_2(-(a+b-c), (a+b+c), -(b+c-a))$ and $N_3(-(c+a-b), -(b+c-a), (a+b+c))$.

5.2.7 N is the orthocentre of triangle $N_1N_2N_3$

It is sufficient to show that N_1N is perpendicular to N_2N_3. In fact the displacement $\mathbf{N_1N} \propto (-b-c, b, c)$, which is parallel to $\mathbf{AI_1}$ and the displacement $\mathbf{N_2N_3} \propto (b-c, -b, c)$, which is parallel to I_3I_2. Then just as I is the orthocentre of triangle $I_1I_2I_3$, so N is the orthocentre of triangle $N_1N_2N_3$.

5.2.8 IGN is a straight line and $GN = 2IG$

The reader is invited to show that triangles $I_1I_2I_3$ and $N_1N_2N_3$ are homothetic and that the scale factor is 2, so that $N_2N_3 = 2I_3I_2$ etc. Also it is easily shown that I_1N_1, I_2N_2, I_3N_3 are concurrent at the centroid G of triangle ABC. It follows at once that the orthocentres, I and N, of the two triangles lie on a line passing through G and satisfying $GN = 2IG$.

Exercise 5.2.1

1. Prove that G and N are the internal and external centres of similitude of the incircles of the triangle ABC and its median triangle.

2. Prove that the midpoint of IN is the centre of mass of a uniform wire framework having the shape of triangle ABC.

Problem 1 of Exercise 5.2.1 implies that the line IGN is analogous to the Euler line OGH, in which $GH = 2OG$ and G and H are the internal and external centres of similitude of the circumcircles of triangle ABC and its median triangle.

5.3 Collinear sets of points

We consider collinear sets of points in the incircle and excircle configuration. It is assumed that triangle ABC is not isosceles, otherwise the figure degenerates to some degree.

In Fig. 5.4, YZ meets BC at P_1 and points P_2, P_3 are similarly defined. Other points are defined as follows: $P_{11} = Y_1Z_1 \wedge BC$, $P_{12} = Z_1X_1 \wedge CA$, $P_{13} = X_1Y_1 \wedge AB$, $P_{22} = Z_2X_2 \wedge CA$, $P_{23} = X_2Y_2 \wedge AB$, $P_{21} = Y_2Z_2 \wedge BC$, $P_{33} = X_3Y_3 \wedge AB$, $P_{31} = Y_3Z_3 \wedge BC$, $P_{32} = Z_3X_3 \wedge CA$.

5.3.1 P_1, P_2, P_3 are collinear

The incircle meets BC at $X(0, a+b-c, c+a-b)$. Similarly $Y(a+b-c, 0, b+c-a)$, $Z(c+a-b, b+c-a, 0)$. The equation of YZ is therefore $y(c+a-b)+z(a+b-c) = x(b+c-a)$. It follows that the point P_1, where YZ meets BC has co-ordinates $P_1(0, -(a+b-c), (c+a-b))$. Similarly $P_2((a+b-c), 0, -(b+c-a))$ and $P_3(-(c+a-b), (b+c-a), 0)$. The points P_1, P_2, P_3 are collinear on the line with equation $(b+c-a)x + (c+$

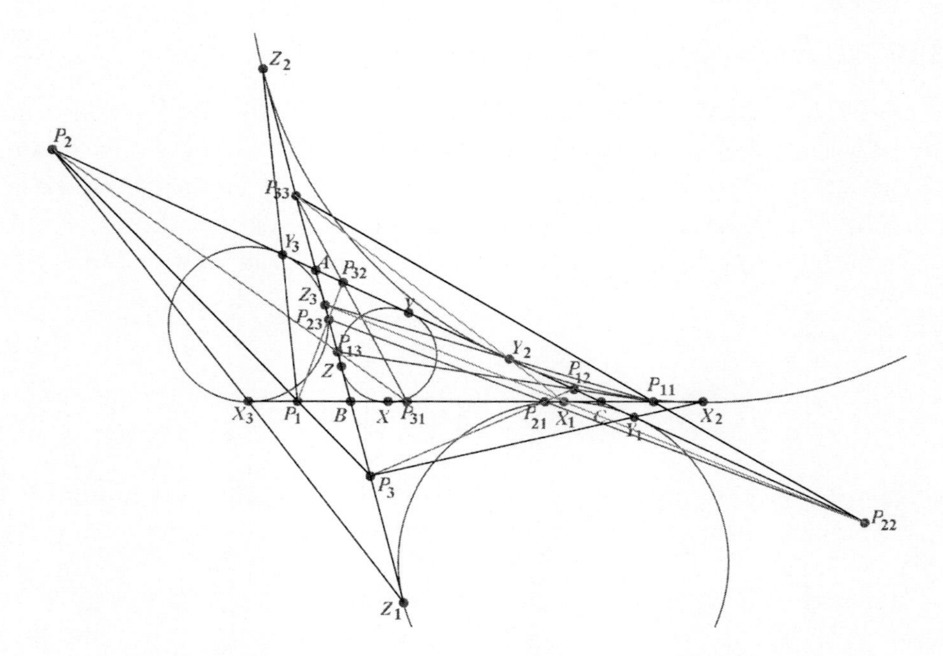

Figure 5.4: Pascal lines for degenerate hexagons

$a-b)y+(a+b-c)z=0$. This is the Pascal line for the degenerate hexagon $XXYYZZ$, where a line such as XX is to be interpreted as the tangent at X.

5.3.2 The points P_{11}, P_{12}, P_{13} are collinear

The excircle opposite A meets BC at $X_1(0, c+a-b, a+b-c)$. It meets CA at $Y_1(b-c-a, 0, a+b+c)$ and AB at $Z_1(c-a-b, a+b+c, 0)$. The equation of Y_1Z_1 is

$$x(a+b+c)+y(a+b-c)+z(c+a-b)=0.$$

This meets BC at the point P_{11} with co-ordinates $P_{11}(0, -(c+a-b), (a+b-c))$. The equation of X_1Y_1 is $y(a+b-c)=x(a+b+c)+z(c+a-b)$ and this meets AB at the point P_{13} with co-ordinates $P_{13}(a+b-c, a+b+c, 0)$. Similarly Z_1X_1 has equation

$$z(c+a-b)=x(a+b+c)+y(a+b-c)$$

and this meets CA at the point P_{12} with co-ordinates $P_{12}(c+a-b, 0, a+b+c)$. The points P_{11}, P_{12}, P_{13} are collinear on the line with equation

$$x(a+b+c) = y(a+b-c) + z(c+a-b).$$

This is the Pascal line for the degenerate hexagon $X_1X_1Y_1Y_1Z_1Z_1$.

5.3.3 The points P_{22}, P_{23}, P_{21} are collinear

Dealing with the excircle opposite B in similar fashion we find the points of tangency to be $Y_2(b+c-a, 0, a+b-c)$, $Z_2(a+b+c, c-a-b, 0)$, $X_2(0, a-b-c, a+b+c)$. Similarly the points P_{22}, P_{21}, P_{23} have co-ordinates $P_{22}((b+c-a), 0, -(a+b-c)$, $P_{21}(0, b+c-a, a+b+c)$, $P_{23}(a+b+c, a+b-c, 0)$ and these three points are collinear on the line with equation $y(a+b+c) = z(b+c-a)+x(a+b-c)$. This is the Pascal line for the degenerate hexagon $X_2X_2Y_2Y_2Z_2Z_2$.

5.3.4 The points P_{33}, P_{31}, P_{32} are collinear

Similar analysis with the excircle opposite C provides points with the following co-ordinates $Z_3(b+c-a, c+a-b, 0)$, $X_3(0, a+b+c, -(b+c-a))$, $Y_3(a+b+c, 0, -(c+a-b))$, $P_{33}(-(b+c-a), (c+a-b), 0)$, $P_{32}(a+b+c, 0, c+a-b)$, $P_{31}(0, a+b+c, b+c-a)$ and the three latter points are collinear on the line with equation $z(a+b+c) = x(c+a-b) + y(b+c-a)$. This is the Pascal line for the degenerate hexagon $X_3X_3Y_3Y_3Z_3Z_3$.

Exercise 5.3.1

1. Prove that in Fig. 5.4 the following sets of points are collinear:

   (i) P_{11}, P_{22}, P_{33} on the line

   $$\frac{x}{b+c-a} + \frac{y}{c+a-b} + \frac{z}{a+b-c} = 0;$$

   (ii) P_1, P_{23}, P_{32} on the line

   $$\frac{y}{a+b-c} + \frac{z}{c+a-b} = \frac{x}{a+b+c};$$

(iii) P_2, P_{31}, P_{13} on the line

$$\frac{z}{b+c-a} + \frac{x}{a+b-c} = \frac{y}{a+b+c};$$

(iv) P_3, P_{12}, P_{21} on the line

$$\frac{x}{c+a-b} + \frac{y}{b+c-a} = \frac{z}{a+b+c};$$

(v) P_1, Z_2, Y_3 on the line

$$\frac{x}{a+b+c} + \frac{y}{a+b-c} + \frac{z}{c+a-b} = 0;$$

(vi) P_2, X_3, Z_1 on the line

$$\frac{x}{a+b-c} + \frac{y}{a+b+c} + \frac{z}{b+c-a} = 0;$$

(vii) P_3, Y_1, X_2 on the line

$$\frac{x}{c+a-b} + \frac{y}{b+c-a} + \frac{z}{a+b+c} = 0;$$

(viii) P_{11}, Z_3, Y_2 on the line

$$\frac{x}{b+c-a} = \frac{y}{c+a-b} + \frac{z}{a+b-c};$$

(ix) P_{22}, X_1, Z_3 on the line

$$\frac{y}{c+a-b} = \frac{z}{a+b-c} + \frac{x}{b+c-a};$$

(x) P_{33}, Y_2, X_1 on the line

$$\frac{z}{a+b-c} = \frac{x}{b+c-a} + \frac{y}{c+a-b}.$$

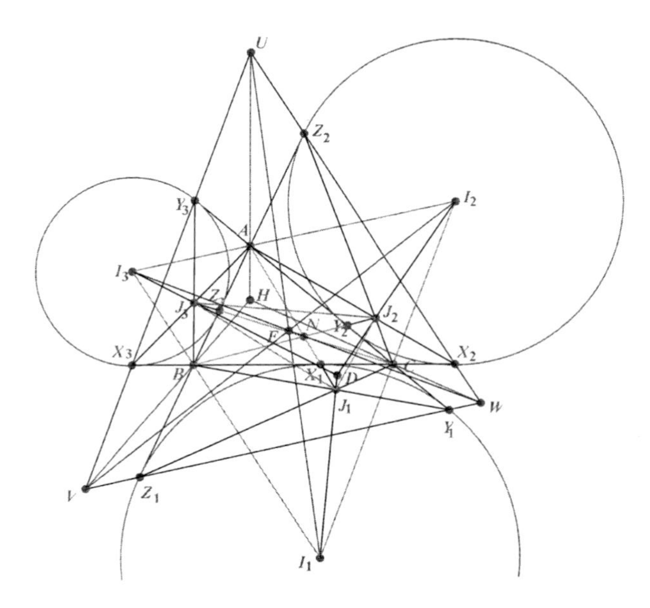

Figure 5.5: Configuration for Part 5.4

5.4 Another configuration with excircles

Let ABC be an acute-angled scalene triangle. Suppose the excircle opposite A, centre I_1, touches BC at X_1, CA at Y_1, AB at Z_1. Suppose the excircle opposite B, centre I_2, touches CA at Y_2, AB at Z_2, BC at X_2. Suppose the excircle opposite C, centre I_3, touches AB at Z_3, BC at X_3 and CA at Y_3. The ex-Gergonne points J_1, J_2, J_3, described earlier, are also shown in Fig. 5.5. So far this is the same labelling as in Part 5.3, but we now introduce some new points. Let Z_2X_2 and X_3Y_3 meet at U, X_3Y_3 and Y_1Z_1 meet at V, Y_1Z_1 and Z_2X_2 meet at W. We now have four triangles ABC, $J_1J_2J_3$, $I_1I_2I_3$, UVW. The purpose of this section is to determine how many pairs of these triangles are in perspective. The vertex of perspective of triangles ABC and UVW is a surprise. Before providing the solutions, some comments are worth making.

Firstly the results are true for acute and obtuse angled triangles, but scalene rules out the right angled case, which is singular and also the case of an isosceles triangle, which is also singular.

Secondly part of this problem is trivial. For example, it is trivial that ABC and $I_1I_2I_3$ are in perspective with the incentre I as the centre of perspective.

Next, the answer to the problem is that five out of six possible pairs of triangles are in perspective. Any analytic proof that triangles $J_1J_2J_3$ and UVW are not in perspective would be technically difficult. However, only one instance is needed as a counter-example and the CABRI construction, which is tremendously accurate, shows these triangles are, in general, far from being in perspective.

Next, being in perspective is not a transitive relation on the set of triangles, so when you get instances in which three triangles are mutually in perspective, it is always interesting. Here ABC, $J_1J_2J_3$, and $I_1I_2I_3$ are three such triangles, as are ABC, $I_1I_2I_3$, and UVW.

Finally the centre of perspective of triangles ABC and UVW is in fact the orthocentre H. This result is intriguing, and provides a curious link between points of the triangle.

This is again a problem in which areal co-ordinates are appropriate. The unnormalised co-ordinates of the following points are known: $I_1(-a, b, c)$, $I_2(a, -b, c)$, $I_3(a, b, -c)$, $X_1(0, s-b, s-c)$, $Y_1(-s+b, 0, s)$, $Z_1(-s+c, s, 0)$, $X_2(0, -s+a, s)$, $Y_2(s-a, 0, s-c)$, $Z_2(s, -s+c, 0)$, $X_3(0, s, -s+a)$, $Y_3(s, 0, -s+b)$, $Z_3(s-a, s-b, 0)$.

5.4.1 Triangles ABC and $J_1J_2J_3$ are in perspective

The equations of BY_1, CZ_1 and AX_1 are therefore $sx+(s-b)z=0$, $sx+(s-c)y=0$ and $(s-b)z=(s-c)y$. These lines meet at $J_1(-1/s, 1/(s-c), 1/(s-b))$. Similarly J_2 and J_3 have co-ordinates $J_2(1/(s-c), -1/s, 1/(s-a))$, $J_3(1/(s-b), 1/(s-a), -1/s)$. Also, by similar working J_2 lies on BY_2 and J_3 lies on CZ_3. Now we know from Section 5.2.2 that AX_1, BY_2, CZ_3 are concurrent at Nagel's point N with co-ordinates $(s-a, s-b, s-c)$, which is therefore the centre of perspective of triangles ABC and $J_1J_2J_3$.

5.4.2 Triangles ABC and $I_1I_2I_3$ are in perspective

The centre of perspective is the incentre $I(a, b, c)$.

5.4.3 Triangles ABC and UVW are in perspective

We now deduce the co-ordinates of U, V, W and show the centre of perspective of triangles ABC and UVW is the orthocentre H. The equation of Z_2X_2, as may be checked by inserting their co-ordinates, is $(s-c)x+sy+(s-a)z=0$.

Similarly the equation of X_3Y_3 is $(s-b)x+(s-a)y+sz=0$ and these two lines meet at the point U with co-ordinates $U(2a(b+c), -(a^2+b^2-c^2), -(a^2-b^2+c^2))$. Similarly V, W have co-ordinates $V(-(b^2-c^2+a^2), 2b(c+a), -(b^2+c^2-a^2))$, $W(-(c^2+a^2-b^2), -(c^2-a^2+b^2), 2c(a+b))$. It is now evident that AU, BV, CW are concurrent at the point with co-ordinates

$$\left(\frac{1}{b^2+c^2-a^2}, \frac{1}{c^2+a^2-b^2}, \frac{1}{a^2+b^2-c^2}\right).$$

Now $b^2+c^2-a^2 = 2bc\cos A = 2abc\cos A/a = abc\cot A/R$, where R is the circumradius of triangle ABC. Hence the unnormalised co-ordinates of the vertex of perspective are $(\tan A, \tan B, \tan C)$, showing it to be the orthocentre H.

It is helpful to modify the co-ordinates of J_1, J_2, J_3 into the following equivalent forms:

$$J_1(-(a+b-c)(a-b+c), (a+b+c)(a-b+c), (a+b+c)(a+b-c)),$$

$$J_2((a+b+c)(b+c-a), -(b+c-a)(b-c+a), (a+b+c)(b-c+a)),$$

$$J_3((a+b+c)(c-a+b), (a+b+c)(c+a-b), -(c+a-b)(c-a+b)).$$

5.4.4 Triangles $I_1I_2I_3$ and $J_1J_2J_3$ are in perspective

The equations of I_1J_1, I_2J_2, I_3J_3 are now found to be

$$I_1J_1: \quad (a+b+c)^2(b-c)x+(c+a)(a+b-c)^2y-(a+b)(a-b+c)^2z=0;$$

$$I_2J_2: \quad -(b+c)(b-c+a)^2x+(a+b+c)^2(c-a)y+(a+b)(b+c-a)^2z=0;$$

$$I_3J_3: \quad (b+c)(c+a-b)^2x-(c+a)(c-a+b)^2y+(a+b+c)^2(a-b)z=0.$$

After some distinctly heavy algebra it is found that these three lines concur at the point with co-ordinates

$$(3a^4-2a^2(b^2+c^2)-(b^2-c^2)^2, 3b^4-2b^2(c^2+a^2)-(c^2-a^2)^2,$$
$$3c^4-2c^2(a^2+b^2)-(a^2-b^2)^2)$$

showing triangles $I_1I_2I_3$ and $J_1J_2J_3$ are in perspective. The vertex of perspective is shown as D in Fig. 5.5.

5.4.5 Triangles $I_1I_2I_3$ and UVW are in perspective

After some simplification it is found that the equations of I_1U, I_2V, I_3W are

$$I_1U: \quad (b-c)x + ay - az = 0;$$

$$I_2V: \quad -bx + (c-a)y + bz = 0;$$

$$I_3W: \quad cx - cy + (a-b)z = 0.$$

I_1U clearly passes through the midpoint of BC and similarly I_2V passes through the midpoint of CA and I_3W passes through the midpoint of AB. Thus triangle $I_1I_2I_3$ and UVW are in perspective. The vertex of perspective E of these triangles has co-ordinates $(a(b+c-a), b(c+a-b), c(a+b-c))$.

Example 5.4.1

Triangle ABC is not isosceles. I and I_1 are the centres of the inscribed circle and of the escribed circle opposite A. II_1 meets BC at U and any straight line (other than BC) through U meets AB, AC at S, T respectively. IS, I_1T meet at P and IT, I_1S meet at Q. Prove that PAQ is a straight line perpendicular to II_1. See Fig. 5.6.

Let I_2, I_3 be the centres of the escribed circles opposite B, C respectively. Since I_2AI_3 is perpendicular to II_1 it is sufficient to show that P and Q lie on I_2I_3. In fact, by symmetry, it is sufficient to show that P lies on I_2I_3. Set up areal co-ordinates with ABC as triangle of reference. Unnormalized co-ordinates of I, I_1, I_2, I_3 are $I(a,b,c)$, $I_1(-a,b,c)$, $I_2(a,-b,c)$, $I_3(a,b,-c)$.

Clearly the equation of I_2I_3 is $cy + bz = 0$. U is the intersection of II_1 and BC and so has co-ordinates $U(0,b,c)$. TUS is a transversal of triangle ABC, so by Menelaus's theorem

$$(BU/UC)(CT/TA)(AS/SB) = -1.$$

But $BU/UC = c/b$ and hence we may take $CT/TA = k/c$ and $AS/SB = -b/k$, for some constant k. (This constant determines the precise line through U which has been chosen.) Hence the co-ordinates of T are $(k,0,c)$ and of S are $(-k,b,0)$. The equation of IS is $b(a+k)z = c(bx+ky)$ and the equation of I_1T is $b(kz-cx) = c(a+k)y$. Eliminating x from these two equations we get $kcy + bkz = (a+k)(cy+bz)$, that is $cy + bz = 0$. This means that P lies on I_2I_3, as required.

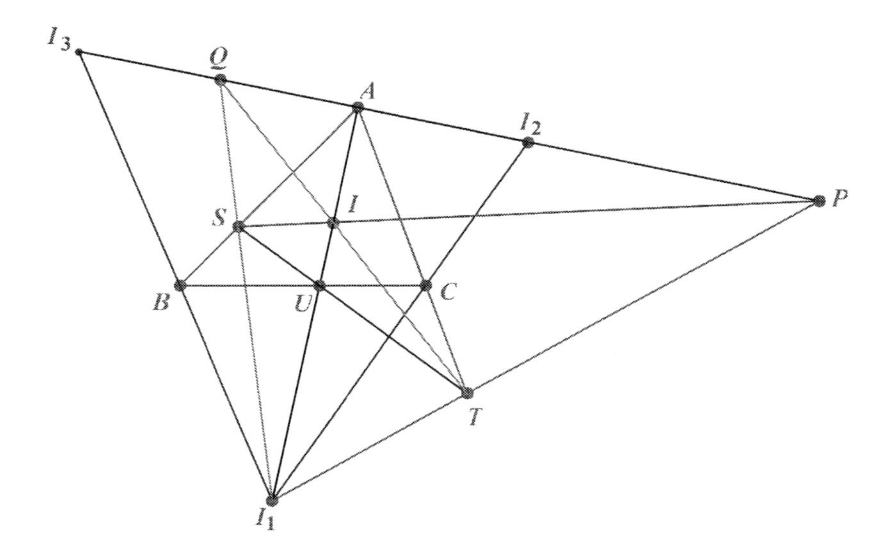

Figure 5.6: Example 5.4.1

Exercise 5.4.2

(Note that parts of Problem 20 are very difficult)

1. Prove that the centroid V of three uniform thin rods in the shape of a triangle is the radical centre of the three excircles of the triangle.

2. Show that if the excircle opposite A is orthogonal to the circumcircle, then $r_1 = 2R$.

3. Prove that the inscribed circles of triangles ABX and ACX touch each other. (JW)

4. Prove that $II_1^2 + I_2I_3^2 = 16R^2$.

5. ABC is a triangle in which $AB \neq AC$. The perpendicular bisector of BC and the internal bisector of $\angle BAC$ meet at K. AK meets BC at U. Prove that A and U are inverse points with respect to the circle centre K and radius KB.

6. A triangle is formed by joining the feet of the perpendiculars of an acute-angled triangle, and the inscribed circle to this triangle touches

its sides at A^*, B^*, C^*, so that A^* is nearer A than B^* or C^* etc. Prove that

$$B^*C^*/BC = C^*A^*/CA = A^*B^*/AB = 2\cos A\cos B\cos C.$$

7. In triangle ABC the incentre is I. BI meets CA at V and CI meets AB at W. What can be said about ABC if $AWIV$ is cyclic?

8. ABC is a triangle with $\angle BCA = 60°$. The internal bisectors of angles A, B meet the opposite sides at P, Q respectively. Prove that $AB = AQ + BP$.

9. Prove that the length of the internal bisector of $\angle BAC$ does not exceed $\frac{1}{2}(AB + AC)$.

10. ABC is a triangle with $BC = a$, $CA = b$, $AB = c$. Describe the locus of P when
$$aAP^2 + bBP^2 + cCP^2 = bc(b + c).$$

11. In triangle ABC the length of the median through A is l and the length of the internal bisector of $\angle BAC$ is u. Prove that $l/u > (b+c)^2/(4bc)$.

12. By considering IH^2, prove that
$$8R^2 + 4r^2 \geq (a^2 + b^2 + c^2).$$
(*Hint: The nine-point circle touches the incircle. See Problem 20 (iii).*)

13. Let u, v, w be the lengths of the angle bisectors and U, V, W the lengths of the angle bisectors extended until they are chords of the circumcircle. Prove that $\sqrt{uvwUVW} = abc$.

14. In triangle ABC, I is the incentre and AI, BI, CI meet the circumcircle at P, Q, R respectively. Prove that I is the orthocentre of triangle PQR.

15. The inscribed circle to triangle ABC touches the sides BC, CA, AB at X, Y, Z respectively. Show that $[XYZ]/[ABC] \leq \frac{1}{4}$.

16. ABC is an acute-angled triangle with incentre I and circumcentre O. Suppose that $AB < AC$ and $IO = \frac{1}{2}(AC - AB)$. Prove that $[IAO] = \frac{1}{2}([BAO] - [CAO])$.

17. ABC is a triangle with inradius r. Points D, E lie on the line segments AB, AC respectively and are varied subject to the condition that DE is always a tangent to the inscribed circle of ABC. Find the angle DE makes with BC when r_0, the radius of the inscribed circle of triangle ADE, is as large as possible. Show that the corresponding value of r_0 is given by $r_0 = r(1 - \sin A/2)/(1 + \sin A/2)$.

18. Prove that I always lies within the circle on GH as diameter.

19. ABC is an acute-angled triangle. Let AI, BI, CI meet BC, CA, AB at U, V, W respectively, and let D, E, F be the feet of the altitudes from A, B, C respectively. Prove that O is an interior point of triangle UVW if, and only if, I is an interior point of triangle DEF.

20. (i) If A, B, C are the angles of a triangle prove, by establishing the identity $OH^2 = R^2(1 - 8\cos A\cos B\cos C)$, that

$$\cos A\cos B\cos C \leq 1/8.$$

(ii) By applying the cosine rule to triangle IAH, prove that

$$IH^2 = 4R^2((1 - \cos A)(1 - \cos B)(1 - \cos C) - \cos A\cos B\cos C)$$

and hence show that if A, B, C are the angles of a triangle, then

$$(1 - \cos A)(1 - \cos B)(1 - \cos C) \geq \cos A\cos B\cos C.$$

(iii) By applying Apollonius's theorem to triangle IOH, using parts (i) and (ii) and the formula $r = 4R\sin A/2\sin B/2\sin C/2$, prove that $IT = \frac{1}{2}(R - 2r)$. (Feuerbach's theorem.)

(iv) Prove that when ABC is not equilateral, then $\angle THI < 90°$, where T is the nine-point centre, H the orthocentre and I the incentre.

(v) Use (iv) to show that if A, B, C are the angles of a triangle, then

$$\sin^2 A + \sin^2 B + \sin^2 C + \frac{\sin^3 A + \sin^3 B + \sin^3 C}{\sin A + \sin B + \sin C} \leq 3.$$

21. Let AU bisect $\angle BAC$ with U on BC. Let V be the point on the line segment BC such that $BV = CU$. Prove that $AV \geq AU$.

22. Let AI, BI, CI met the incircle of triangle ABC at P, Q, R respectively, where P is between A and I, Q between B and I and R between C and I. Let AI, BI, CI meet BC, CA, AB at U, V, W respectively. Prove that $AU + BV + CW \leq 3(AP + BQ + CR)$.

23. Let r be the radius of the inscribed circle of a right-angled triangle, and let h be the height of the triangle drawn to the hypotenuse. This altitude divides the original triangle into two smaller triangles, and let s, t be the radii of the circles inscribed in these triangles. Prove that $r + s + t = h$ and $s^2 + t^2 = r^2$.

24. ABC is a triangle with incentre I. AI meets circle BIC again at I_A, BI meets circle CIA again at I_B and CI meets circle AIB again at I_C. Prove that $(II_A)(II_B)(II_C) \leq 8R^3$, where R is the circumradius. Prove also that $[I_AI_BI_C] \geq 4[ABC]$.

25. In triangle ABC let the radii of the circumcircle and incircle be R and r respectively. Prove that (i) $(abc)^{2/3} \geq 6Rr$; (ii) $[ABC]^2 \geq 27Rr^3/2$; (iii) $(a+b+c)^2 \geq 54Rr$; (iv) $R \geq (1/3)\sqrt{a^2+b^2+c^2} \geq 2r$; (v) $R^2(a+b+c) \geq abc$; (vi) $2R(\sin^2 A+\sin^2 B+\sin^2 C) \geq 9r$; (vii) $a+b+c \geq 6\sqrt{3}r$.

26. ABC is a triangle. Its incircle touches the sides BC, CA, AB at X, Y, Z respectively. Points P, Q, R are the feet of the perpendiculars from X to YZ, Y to ZX, Z to XY respectively. Prove that AP, BQ, CR are concurrent. Prove also that $[ABC] \geq 16[PQR]$.

27. Triangle ABC is inscribed in a circle. The chord AP bisects $\angle BAC$. Suppose that $AB = \sqrt{2}BC = \sqrt{2}AP$. Determine the angles of ABC.

28. ABC is a triangle with incentre I. Lines AI, BI, CI meet the circumcircle at P, Q, R. Prove that $(AI)(BI)(CI) \leq (IP)(IQ)(IR)$.

29. ABC is a triangle with incircle S of radius r. Circles S_A, S_B, S_C are drawn, each touching two sides of ABC and touching S externally. Thus S_A has tangents AB, AC and touches S externally and similarly for S_B and S_C. If the radii of the three circles are denoted by r_A, r_B, r_C respectively, prove that $r \leq r_A + r_B + r_C \leq 2r$.

30. The inscribed circle of triangle ABC touches BC, CA, AB at X, Y, Z respectively. The line through A parallel to YZ meets the line XY at K. Prove that BK passes through the midpoint of YZ.

31. A and B are fixed points and l is a fixed line passing through A. C is a variable point on l, staying on the same side of A. The incircle of triangle ABC touches BC at X and AC at Y. Show that the line XY passes through a fixed point.

32. Triangle ABC is given with $\angle BAC = 72^o$. The perpendicular from B to CA meets the internal bisector of $\angle ACB$ at P. The perpendicular from C to AB meets the internal bisector of $\angle ABC$ at Q. It is found that P, Q, A are collinear. Find the other angles of triangle ABC.

33. Given triangle ABC, altitudes AD, BE, CF and internal angle bisectors AI, BI, CI, let BE meet CI at P, CF meet AI at Q and AD meet BI at R. Prove that AP, BQ, CR are concurrent. Is the result still true if AD, BE, CF are replaced by the medians AL, BM, CN?

34. ABC is a triangle. The internal bisector of $\angle BAC$ meets BC at U. Prove that $AU^2 = (AB)(AC) - (UB)(UC)$. Is the converse true?

35. ABC is a triangle in which $AB \neq AC$. The internal and external bisectors of $\angle BAC$ meet the circumcircle again at L and M respectively. Points L' and M' lie on the extensions of AL and AM respectively and are such that $AL = LL', AM = MM'$. Circles ALM' and $AL'M$ meet again at P. Prove that AP is parallel to BC.

36. The internal bisectors of triangle ABC meet the circumcircle again at P, Q, R. Prove that $[PQR] \geq [ABC]$.

5.5 Geometrical inqualities and side lengths

We are specifically interested in geometrical inequalities involving $s-a, s-b, s-c$. Obviously any inequality that is true for any three positive real numbers is also true when those three numbers are the sides of a triangle. However, there are many inequalities that are true for a, b, c that are not true when a, b, c are replaced by any positive real numbers. The reason for this is that a, b, c must obey the constraints $b+c > a$, $c+a > b$, $a+b > c$. The way round this is to use the lengths $AZ = s-a$, $BX = s-b$, $CY = s-c$ instead, as these can be any positive quantities. To see this, consider the 1-1 transformation $a = m+n$, $b = n+l$, $c = l+m$ with inverse transformation $l = (b+c-a)/2 = s-a$, $m = s-b$, $n = s-c$. The triangle inequality $b+c > a$

now becomes $l > 0$, and similarly $m > 0, n > 0$. The use of l, m, n instead of a, b, c transforms the required inequality into the familiar context of an inequality to be satisfied by any three real numbers, for which, in particular, the Arithmetic Mean/Geometric Mean inequality is often appropriate.

Example 5.5.1

Show that if a, b, c are the sides of a triangle then

$$\frac{1}{b+c-a} + \frac{1}{c+a-b} + \frac{1}{a+b-c} \geq \frac{1}{a} + \frac{1}{b} + \frac{1}{c}.$$

This transforms to

$$\frac{1}{2l} + \frac{1}{2m} + \frac{1}{2n} \geq \frac{1}{m+n} + \frac{1}{n+l} + \frac{1}{n+l}$$

for $l, m, n > 0$. Now $1/m + 1/n \geq 4/(m+n)$ and adding this to two similar inequalities resulting from cyclic change of l, m, n we have the required result.

Exercise 5.5.2

(Not all the problems in this exercise are to be solved by the method above.)

1. Let a, b, c denote the lengths of the sides of a triangle. Prove that

$$\frac{3}{2} \leq \frac{a}{b+c} + \frac{b}{c+a} + \frac{c}{a+b} < 2.$$

2. If a, b, c are the sides of a triangle, prove that

$$\frac{1}{2} < \frac{bc+ca+ab}{a^2+b^2+c^2} \leq 1.$$

3. Let a, b, c be the sides of a triangle and $[ABC]$ its area. Prove that

$$a^2+b^2+c^2 \geq 4\sqrt{3}[ABC].$$

4. If a, b, c are the sides of a triangle, prove that

$$2(a^2b+b^2c+c^2a+ab^2+bc^2+ca^2) \geq a^3+b^3+c^3+9abc.$$

5. If a, b, c are the sides of a triangle and $s = (a + b + c)/2$, prove that
$$abcs \geq 2bc(s-b)(s-c) + 2ca(s-c)(s-a) + 2ab(s-a)(s-b).$$

6. ABC is an acute-angled triangle with sides a, b, c. Prove that
$$\sqrt{b^2+c^2-a^2} + \sqrt{c^2+a^2-b^2} + \sqrt{a^2+b^2-c^2} \leq a+b+c.$$

7. Let a, b, c be the sides of a triangle. Prove that
$$a^3+b^3+c^3+6abc \geq (ab+bc+ca)(a+b+c) > a^3+b^3+c^3+5abc.$$

5.6 Some points in a triangle

It will have been noticed that many new points have been identified in this Chapter, notably Gergonne's point, Nagel's point, the ex-Gergonne points and the ex-Nagel points. Then there have been some vertices of perspective defined. In recent decades a vast number of triangle points have been obtained and catalogued, see Kimberling [24], one of whose particular interests has been in triangle centres (those that coincide when the triangle becomes equilateral). Some points, however, have more interest than others, and in this book I introduce only those that appear to me to be more fundamental. It has not been a deliberate policy to include or exclude certain points, but rather to include those that seemed to appear as a matter of course, as the text has developed.

We first consider the triangle LMN. As its sides are parallel to those of ABC and L, M, N are the midpoints of the sides of ABC, it follows that the orthocentre of triangle LMN is the circumcentre O of triangle ABC. The centroid of triangle LMN is obviously the same as the centroid G of ABC. It follows that the Euler line of triangle LMN coincides with the Euler line of ABC. If we put a co-ordinate on the Euler line of ABC, so that O has co-ordinate 0, G has co-ordinate 1/3 and H has co-ordinate 1, then the circumcentre T of triangle LMN has co-ordinate $\frac{1}{2}$. T is, of course, the nine-point centre of triangle ABC.

5.6.1 de Longchamps's point

This is the point deL that is the image of the half-turn rotation of H about O, so that deL is the point on the Euler line with co-ordinate -1. It follows that the de Longchamps point of triangle LMN is the orthocentre H of ABC.

The next question is where the incentre of triangle LMN is situated. Since the side-lengths of LMN are $a/2, b/2, c/2$ the areal co-ordinates of this point are

$$\frac{1}{2}a(0,1,1) + \frac{1}{2}b(1,0,1) + \frac{1}{2}c(1,1,0) \propto (b+c, c+a, a+b).$$

This point is called the *Spieker centre* and we denote it by Sp.

5.6.2 The Spieker centre

The Spieker centre lies on the line IGN and $ISp = SpN$. From Section 5.2.2 the normalised co-ordinates of N are $(b+c-a, c+a-b, b+c-a)/(a+b+c)$ and those of I are $(a,b,c)/(a+b+c)$. The midpoint of IN is evidently the Spieker centre $Sp(b+c, c+a, a+b)/2(a+b+c)$.

We now consider the triangle $I_1I_2I_3$. Its orthocentre, as we have seen in Part 5.1, is the incentre I, so the Euler line of triangle $I_1I_2I_3$ passes through I. The equation of the I_1L is $(c-b)x = a(y-z)$. Similarly the equation of I_2M is $(a-c)y = b(z-x)$. These meet at the point with co-ordinates $(a(b+c-a), b(c+a-b), c(a+b-c))$, which by symmetry also lies on I_3N. Note that this is the vertex of perspective of triangles $I_1I_2I_3$ and UVW (see Part 5.4).

5.6.3 The Mittelpunkt

The point just defined is called the *Mittelpunkt* and is denoted by Mi.

Exercise 5.6.1

1. Prove that the Mittelpunkt is the Gergonne point of triangle LMN.

2. Prove that in triangle ABC, the incentre, the Gergonne point and the de Longchamps point are collinear.

3. Prove that the Mittelpunkt, the Spieker centre, the de Longchamps point of triangle LMN, and the incentre of triangle $I_1I_2I_3$ are collinear.

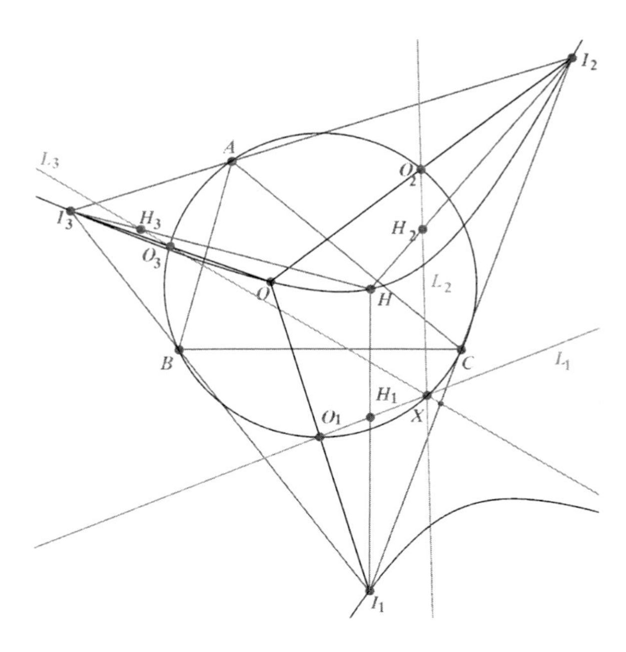

Figure 5.7: The configuration for Part 5.7

5.7 The Euler lines of three triangles

We consider the Euler lines of triangles I_1BC, I_2CA, I_3AB.

In Fig. 5.7 ABC is a triangle with I_1, I_2, I_3 the excentres opposite A, B, C respectively. The triangles I_1BC, I_2CA, I_3AB are denoted by T_1, T_2, T_3 respectively. O_k, H_k, are the circumcentre, orthocentre respectively of triangle T_k, $k = 1, 2, 3$. L_k is the Euler line of T_k, $k = 1, 2, 3$. Define the point $E_k(t)$ on L_k by the equation $E_k(t) = (1-t)O_k + tH_k$, where t is a parameter. Then the following results hold:

5.7.1 L_1, L_2, L_3 are concurrent

L_1, L_2, L_3 are concurrent at a point X with areal co-ordinates (with respect to ABC as triangle of reference) $(a/(b-c), b/(c-a), c/(a-b))$.

5.7.2 $I_1E_1(t), I_2E_2(t), I_3E_3(t)$ are concurrent

$I_1E_1(t), I_2E_2(t), I_3E_3(t)$ are concurrent at a point $E(t)$.

5.7.3 The locus of $E(t)$

The locus of $E(t)$ as t varies is the rectangular hyperbola passing through I_1, I_2, I_3, the orthocentre O and the circumcentre H of triangle $I_1I_2I_3$. Note that the symbols $O \equiv E(0)$ and $H \equiv E(1)$ here are used to be consistent with the notation defined Section 5.7.2. As mentioned in Part 5.1, O coincides with the incentre I of triangle ABC.

Example 5.7.1

1. Prove that the equation of L_1 is $(c^2 - b^2)x + a(c - a)y + a(a - b)z = 0$ with the equations of L_2 and L_3 given by cyclic change of a, b, c and x, y, z.

2. Prove that L_2 and L_3 meet at the point X with co-ordinates $(a/(b - c), b/(c - a), c/(a - b))$ and check that X also lies on L_1.

3. Prove that the co-ordinates of E_1, E_2, E_3 are (X_1, Y_1, Z_1), (X_2, Y_2, Z_2), (X_3, Y_3, Z_3), where $X_1 = a((2a-b-c)t-a)$, $Y_1 = (-a^2+ab-b^2+c^2)t+b(b+c)$, $Z_1 = (-a^2+ca+b^2-c^2)t+c(b+c)$, $X_2 = (-b^2+ab+c^2-a^2)t+a(c+a)$, $Y_2 = b((2b-c-a)t-b)$, $Z_2 = (-b^2+bc-c^2+a^2)t+c(c+a)$, and $X_3 = (-c^2+ca-a^2+b^2)t+a(a+b)$, $Y_3 = (-c^2+bc+a^2-b^2)t+b(a+b)$, $Z_3 = c((2c - a - b)t - c)$.

4. Prove that the equation of the line $I_1E_1(t)$ is

   $$(a + b + c)(b - c)tx + ((a + b - 2c)t + c)ay - ((a - 2b + c)t + b)az = 0$$

   and then the equations of the lines $I_2E_2(t)$ and $I_3E_3(t)$ follow by cyclic change of a, b, c and x, y, z. Verify that these three lines are concurrent at the point $E(t)$ with normalized co-ordinates (X, Y, Z), where

   $$\begin{aligned}\gamma X &= a(c^3 - a^3 + b^3 + c^2a - ca^2 - 3b^2c - 3bc^2 + ab^2 - a^2b + 5abc)t^2 \\ &\qquad +2abc(b + c - 2a)t + a^2bc, \\ \gamma Y &= b(a^3 - b^3 + c^3 + a^2b - ab^2 - 3c^2a - 3ca^2 + bc^2 - b^2c + 5abc)t^2 \\ &\qquad +2abc(c + a - 2b)t + ab^2c, \\ \gamma Z &= c(b^3 - c^3 + a^3 + b^2c - bc^2 - 3a^2b - 3ab^2 + ca^2 - c^2a + 5abc)t^2 \\ &\qquad +2abc(a + b - 2c)t + abc^2,\end{aligned}$$

   where γ is a complicated normalization factor chosen so that $X + Y + Z = 1$.

5. Use a computer algebra package, such as MAPLE or DERIVE, to show that the locus of $E(t)$ as t varies is the conic Σ with equation

$$bc(b-c)x^2 + ca(c-a)y^2 + ab(a-b)z^2 = 0.$$

6. Verify that Σ passes through I, I_1, I_2, I_3 and is therefore a rectangular hyperbola. Prove also that Σ has centre X and that ABC is self-polar with respect to Σ and that the circumcircle of ABC passes through X.

An article by Bradley [4] contains a full account of Sections 5.7.1-5.7.3 and Exercise 5.7.1.

Chapter 6

Circle theorems and Cyclic Quadrilaterals

6.1 Elementary circle theorems

Although this is primarily a textbook of analytic methods in geometry, it would be totally misleading to give the impression that such methods are always appropriate. It may be argued by some that an analytic method is in any case bound to drive the solver back to first principles, as if no pure geometrical theorems existed. We do not subscribe to either point of view, as there is a building process in analytic geometry as in all areas of mathematics. New methods and techniques emerge as a consequence of previously established results. But it would be foolish to ignore the primacy of pure methods in many areas, of which the prime example is that of the elementary circle theorems. Yet as work in a later part of this chapter shows, there are problems on circles that can be solved by analytic methods, for which no catalogue of known pure geometrical theorems seems to be sufficient.

The best advice that can be offered to a reader is to use whichever method seems appropriate, though if you are learning the subject it is sometimes a good idea to try to use the method you are less comfortable with. We concede, however, it is often wise to look for a pure geometrical method first, as such methods are often shorter and more elegant than analytic methods. However, there is a subjective element about the choice. What suits one person does not necessarily suit another. The subjective element about what is aesthetically pleasing has always led me to dispute that there is a best method for any given problem. The reader should be aware that even amongst the exercises set to illustrate analytic methods, there are many that are better solved by synthetic methods. Also there are some problems in which an astute combination of methods is the most direct. Certainly it would not be sensible for me, even bearing in mind the declared purpose of this particular textbook, to contrive to do without certain sections on pure geometry of which this section is an example.

We now tabulate the main elementary theorems on circles and then provide some worked examples, followed by a large number of exercises.

6.1.1 The angle in a semicircle is a right angle

If AB is a diameter of a circle and P is any point on the circumference other than A or B, then $\angle APB = 90°$. The converse is also true, that if APB is a triangle in which $\angle APB = 90°$, then the circumcircle of triangle APB has as its centre the midpoint of AB.

6.1.2 The angle at the centre

The angle subtended at the centre of a circle by an arc is twice the area subtended at a point on the circumference. If AB is a chord, but not a diameter, of a circle, O is the centre of the circle and P is a point on the major arc AB, then $\angle AOB = 2\angle APB$. If P lies on the minor arc AB, then reflex $\angle AOB = 2\angle APB$. (The converse is not true unless it is given that O lies on the perpendicular bisector of the chord AB.) The result of Section 6.1.1 can be thought of as a limiting case of this result.

6.1.3 Angles in the same segment are equal

If AB is a chord, but not a diameter, of a circle, and P and Q are two points on the arc AB (both being on either the major arc or the minor arc), then $\angle APB = \angle AQB$. The converse is also true, that if AB is a line and P, Q are two points, both on the same side of AB and $\angle APB = \angle AQB$, then the four points A, B, P, Q lie on a circle. The points are said to be *concyclic*. If AB is a diameter then Section 6.1.1 applies and it does not matter whether P and Q lie on the same arc or not.

6.1.4 Opposite angles of a cyclic quadrilateral

The opposite angles of a cyclic quadrilateral add up to $180°$. If A, B, C, D are four distinct points, in that order, lying on a circle, then $\angle A + \angle C = \angle B + \angle D = 180°$. The converse is also true that if $ABCD$ is a quadrilateral and $\angle A + \angle C = 180° = \angle B + \angle D$, then the quadrilateral is cyclic.

6.1.5 Radius and tangent

The angle between the radius and the tangent at a point on a circle is equal to $90°$.

6.1.6 The alternate segment theorem

If P is a point on a circle and TP is tangent at P and PQ is a chord of the circle and R is a point in the opposite segment, then $\angle TPQ = \angle PRQ$. See Figs. 6.1(a) and 6.1(b). The converse is also true, that if this angle relationship holds, then TP is tangent to circle PQR at P.

6.1.7 Tangent lengths

The lengths of the two tangents to a circle from an external point are equal.

6.1.8 The intersecting chord theorem

If AB and CD are chords of a circle that intersect at X, then $(AX)(BX) = (CX)(DX)$. Here X may be internal or external to the circle. The converse is also true, that if $(AX)(BX) = (CX)(DX)$, then A, B, C, D are concyclic.

6.1.9 The tangent secant theorem

If AB is a chord of a circle passing through a point X, external to a circle, and XP is tangent to the circle at P, then $(XA)(XB) = (XP)^2$. The converse is also true, that if $(XA)(XB) = (XP)^2$, then XP is tangent to circle APB at P. The converses of the results in Sections 6.1.6 and 6.1.9 are important, as they are the results to use if one is asked to show by synthetic means that a line is tangent to a circle.

It is certain that you will have met these theorems before, but possibly you will have accepted the converses as true, without proving them. If that is the

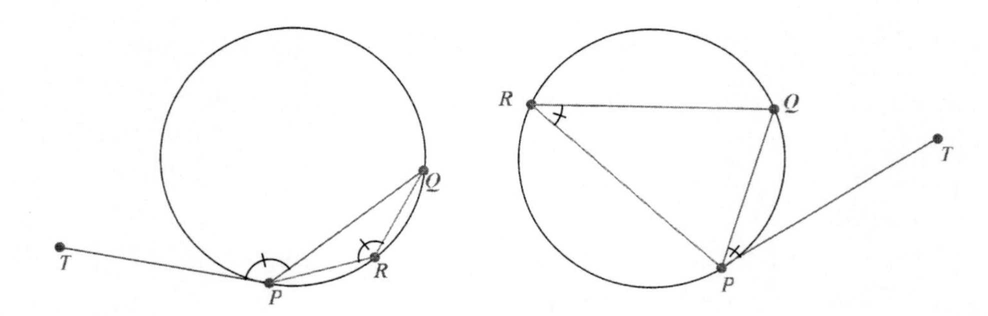

Figure 6.1: The alternate segment theorem

case the converses should be treated as exercises. Proofs are by contradiction, supposing them not to be true and the theorem itself is often used in the proof of the converse. In geometry it is often the case that converses are proved in this way.

Example 6.1.1

A good problem illustrating the theorem that angles in the same segment are equal is as follows:

Two circles intersect at A and B. A variable point T is chosen on one of the circles and lines TA, TB meet the other circle again at P and Q respectively. Then PQ is of constant length.

Fig. 6.2 illustrates the result. Two cases are shown with different sets of subscripts. Four equal angles are shown, proving that $P_1P_2 = Q_1Q_2$. It then follows that arcs P_1Q_1 and P_2Q_2 subtend equal angles at the centre O of the second circle. This means that, as chords, they are of equal length.

Example 6.1.2

Let S_1 and S_2 be two circles intersecting at A and B, and suppose that XBY is a line meeting S_1 at X and S_2 at Y. Find the position of X such that triangle AXY has maximum area. See Fig. 6.3.

The solution is that X lies on S_1 at the opposite end of the diameter through A. The proof is very sweet. $[XAY] = \frac{1}{2}(XA)(YA)\sin\angle XAY$. Now let CBD be the chord, with C on S_1 and D on S_2, such that CD is parallel to the line of centres O_1O_2. Then $\angle XAC = \angle XBC = \angle YBD = \angle YAD$. Thus $\angle CAD = \angle XAY$. But $XA \leq CA$ and $YA \leq DA$, since the diameter of

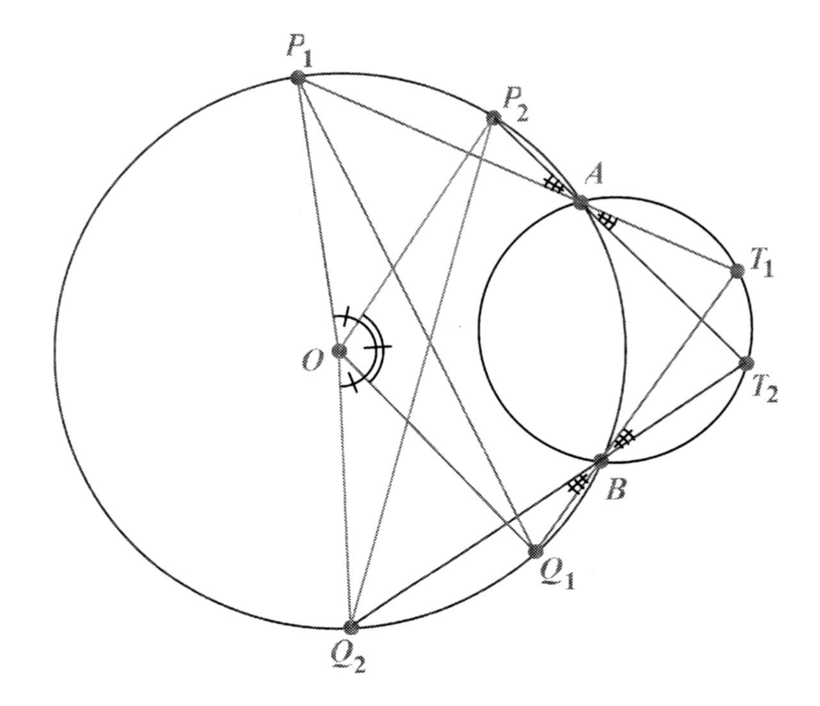

Figure 6.2: Example 6.1.1

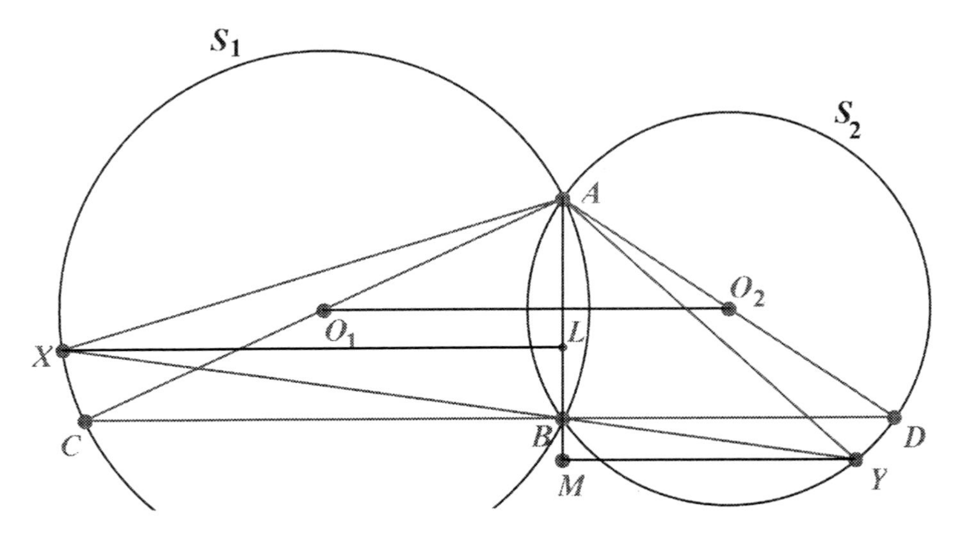

Figure 6.3: Example 6.1.2

a circle is its longest chord, and hence $[XAY] \leq [CAD]$ with equality if, and only if, X coincides with C. In Fig. 6.3 this means that $XL + YM \leq CD$.

There is an interesting consequence of this result. Suppose now that we have three circles S_1, S_2, S_3 with a common point of intersection O. Let S_2 and S_3 meet again at P, with Q and R similarly defined by cyclic change of $1, 2, 3$. Then the following results hold:

1. If you start with a point A on S_1 and draw ARB to meet S_2 at B and AQC to meet S_3 at C then B, P, C are collinear and so ABC is a triangle with A, B, C on S_1, S_2, S_3 respectively and with P, Q, R on BC, CA, AB respectively.

2. From all such triangles ABC, the one with maximum area is when its sides are parallel to the three lines of centres.

The configuration is shown in Fig. 6.4. Result 1 is easily proved by angle properties of cyclic quadrilaterals. Result 2 follows from three applications of the similar result (see Example 6.1.2 for two circles, since $[ABC] = [BOC] + [COA] + [AOB]$).

Example 6.1.3

The tangents from a point O to a circle are bisected by a straight line, which meets a chord PQ of the circle at R. Prove that $\angle ROP = \angle RQO$. See Fig. 6.5.

Let OS, OT be the tangents from O to the circle and let U, V be the midpoints of OS, OT respectively. Let RM be one of the tangents from R. The radical axis of the circle $PSQT$ and the point circle O is the line UV, since $OU = OS$ and $OV = OT$. So UV is the line with the property that from points on it, such as R, $OR =$ length of the tangent from R to the circle $= RM$. By the tangent secant theorem $(OR)^2 = (RM)^2 = (RQ)(RP)$, that is $RQ/OR = RO/PR$. Consider now the triangles RQO and ROP. They have a common angle at R and the ratios of corresponding sides emanating from R are equal. Therefore these triangles are similar, and in particular $\angle ROP = \angle RQO$.

Exercise 6.1.4

1. Two circles C_1 and C_2 touch one another externally at P. Q_1Q_2 is a common tangent (other than the tangent at P) touching C_1 at Q_1 and C_2 at Q_2. What is the value of $\angle Q_1PQ_2$?

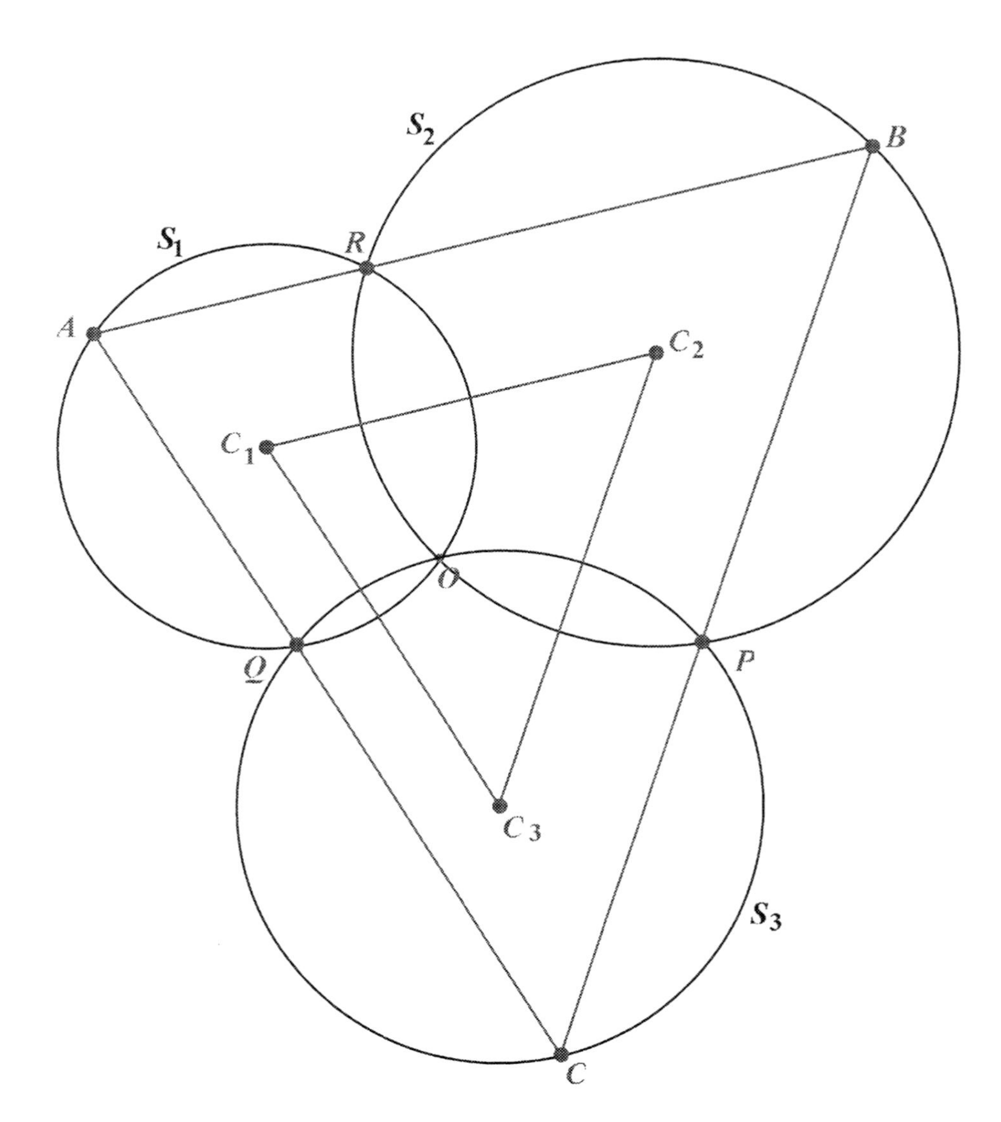

Figure 6.4: Consequence of Example 6.1.2

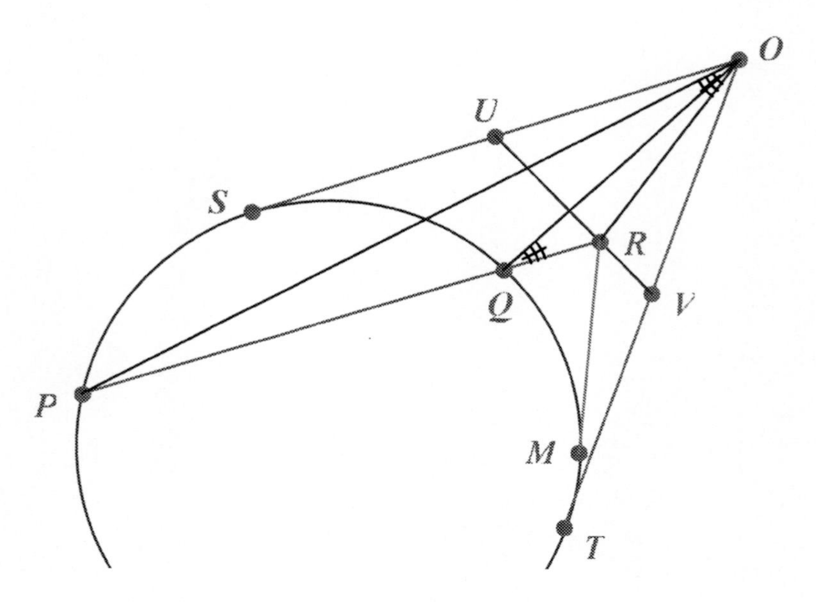

Figure 6.5: Example 6.1.3

2. Two intersecting circles C, D are given. C has the larger radius and passes through the centre of D. The chord PQ of C is tangent to D and the other tangents to D from P, Q meet C again at R, S respectively. Prove that RS is also a tangent to D.

3. A common tangent AB is drawn to two circles, CD is their common chord and tangents are drawn from A to any other circle through C and D. Prove that the chord of contact passes through B.

4. Two circles AP_1Q_1 and AP_2Q_2 cut at A. P_1P_2, Q_1Q_2 are their common tangents. Prove that the circles AP_1P_2 and AQ_1Q_2 touch each other.

5. PAB, QAB are two circles which intersect at A and B. PAQ is a straight line. Prove that BP/BQ is constant.

6. Two circles intersect at A, and through A any two straight lines BAC, $B'AC'$ are drawn terminated by the circles. Prove that the chords BB', CC' of the two circles are inclined at a constant angle.

7. Two circles C_1 and C_2 touch at A. The chord PQ of C_1 touches C_2 at R. PA meets C_2 at S. RA meets C_1 at U. Prove that PU is parallel

to RS.

8. Two circles with centres A, B cut each other at right angles and their common chord meets AB at C. DE is a chord of the first circle passing through B. Prove that A, D, E, C are concyclic.

9. The centre A of a circle lies on another circle, which cuts the former in B, C. AD is a chord of the latter circle meeting BC in E and from D tangents DF, DG are drawn to touch the former circle. Prove that E, F, G are collinear.

10. Two circles C_1 and C_2, centres O_1, O_2 respectively intersect at A and B. O_1A meets C_2 at P. Prove that O_1BO_2P is a cyclic quadrilateral.

11. Circles C_1, C_2 with centres O_1, O_2 and radii r_1, r_2 respectively, are drawn so that the distance between their centres is given by $O_1O_2 = \sqrt{r_1^2 + r_2^2}$. The circles C_1 and C_2 intersect at A and B. From the point P on C_1 furthest from O_2 lines PA, PB are drawn to intersect C_2 at D, E respectively. Prove that DE is the diameter of C_2 perpendicular to O_1O_2. A point Q, distinct from P, is now chosen on the major arc AB of C_1. Lines QA, QB are drawn to intersect C_2 at F, G respectively. Prove that FG is also a diameter of C_2. Locate, with proof, the point where FB and GA meet.

12. Two circles S_1 and S_2 touch each other externally at K; they also touch a circle S internally at A_1 and A_2 respectively. Let P be one point of intersection of S with the common tangent to S_1 and S_2 at K. The line PA_1 meets S_1 again at B_1. The line PA_2 meets S_2 again at B_2. Prove that B_1B_2 is a common tangent to S_1 and S_2.

13. $ABCD$ is a cyclic quadrilateral and O is the centre of the circle $ABCD$. AD and BC meet at P. L, M are the midpoints of AD, BC respectively. S, T are the feet of the perpendiculars from O, P respectively on to LM. Prove that $LS = TM$.

14. P is any point on the circle circumscribing the cyclic quadrilateral $ABCD$. The perpendiculars from P on to AB, BC, CD and DA have lengths s, t, u and v respectively. Prove that $su = tv$.

15. The diagonals of a quadrilateral $ABCD$ meet at O. Circles AOB, BOC, COD, DOA are drawn. Prove that the four centres are the corners of a parallelogram.

16. $ABCD$ is a cyclic quadrilateral. The extensions of AB and DC meet at E. O is the centre of circle ECA. Prove that BD is perpendicular to OE.

17. Three non-intersecting circles SA, SB, SC have centres A, B, C respectively. Prove that A, B, C are collinear if, and only if, there is no unique point P with the property that a circle S_P exists, centre P, such that $\angle PAQ_A = \angle PBQ_B = \angle PCQ_C = 90°$, where Q_A, Q_B, Q_C are one or other of the intersections of S_P with S_A, S_B, S_C respectively. Show further, if more than one point P has the property stated, then all the corresponding circles S_P pass through two fixed points. *Hint: A circle S is said to cut a circle Σ* diametrally *if their common chord is a diameter of Σ. You will find the following lemma indispensable: Let S_A, S_B be circles with centres the distinct points A, B and radii a, b. Then the locus of points P, with the property that there exists a circle S_P, centre P, with suitably chosen radius such that S_P cuts S_A, S_B diametrally, is the line perpendicular to AB through the point N on AB such that $AN : NB = (AB^2 + b^2 - a^2) : (AB^2 + a^2 - b^2)$, due regard being paid to sign.*

18. Two circles C_1 and C_2 of equal radius touch at A. A circle C, of radius twice that of C_1 and C_2 touches C_1 at B so that C_1 lies inside C. Prove that one of the points of intersection of C and C_2 lies on the line BA, and that if P is the other then $PB = 2PA$.

19. Four circles C_1, C_2, C_3, C_4 are external to one another. C_1, C_2 touch at H; C_2, C_3 at K; C_3, C_4 at L; C_4, C_1 at M. Prove that $HKLM$ is a cyclic quadrilateral.

20. A variable circle through two fixed points A and B cuts a fixed circle at P and Q. Prove that AB and PQ meet in a fixed point.

21. ABC is a straight line and O a point not on it. O_1, O_2, O_3 are the centres of circles OBC, OCA, OAB respectively. Prove that O, O_1, O_2, O_3 are concyclic.

22. Two circles touch at T; one lies inside the other. A straight line meets the outer circle at A, D and the inner circle at B, C (in the order $ABCD$). Prove that $AB/CD = ((TA)(TB))/((TC)(TD))$.

23. ABC is an isosceles triangle with $AB = AC$. S is the circumcircle of triangle ABC. The point P lies on the arc BC of S on the opposite side of BC to A. X is the point on AP such that $AX = AC$. Y is the point on BP such that $BY = BC$. Prove that YX, the tangent at B to S and the line through C perpendicular to AB are concurrent. Prove also that C, Y, P, X are concyclic.

24. P is a variable point on the minor arc AB of a circle. AP is extended to Q, making $PQ = PB$. Find the locus of Q as P varies along the arc AB.

25. R, S, T are three points in that order on a circle. The minor arc of RS does not contain T and the minor arc of ST does not contain R. P, Q are the midpoints of the minor arcs RS, ST. The tangent at S meets PQ, RT at D, E respectively and PQ, RT meet at F. Prove that $DE = EF$.

26. Circles S_1 and S_2, with centres O_1 and O_2 respectively, intersect at A and B. A line through A intersects S_1 and S_2 for the second time in C and D respectively. CO_1 and DO_2 meet at P and the line through P perpendicular to CD meets AB at Q. Prove that P, D, Q, C, B are concyclic.

27. D and E are points on the sides AB and AC respectively of triangle ABC and are such that DE is parallel to BC. P is an interior point of triangle ADE. PB and PC meet DE at F and G respectively. Q is the point of intersection, other than P, of the circles PDG and PEF. X and Y are the points of intersection of PB, QD and PC, QE respectively. Prove that P, Y, Q, X lie on a circle. Prove also that APQ is a straight line.

28. ABC is a triangle. L, M, N are the midpoints of the sides BC, CA, AB respectively. What do the circles AMN, BNL, CLM have in common?

29. ABC is a triangle. L and M are the midpoints of AB and AC respectively. The circle through L and M which touches BC does so

at N. D is the foot of the perpendicular from A to BC. Prove that $|BN - NC| = \frac{1}{2}|BD - DC|$.

30. ABC is a triangle. S is a circle, which touches both AB and AC and passes through the centre of BC. Find in terms of a, b, c the possible lengths of the tangents from A to S.

31. The tangent at A to the circumcircle of triangle ABC meets BC at L. Prove that the bisectors of $\angle BAC$ and $\angle ALC$ are at right angles.

32. Three points A, B, C lie on a given circle, centre O. Let P be the midpoint of the arc AB lying on the opposite side of AB to C. Let Q, R be the feet of the perpendiculars from P on to AC, BC respectively. Prove that $OQ^2 + OR^2 = 2OC^2$.

33. O is the circumcentre of an acute-angled triangle ABC. The circle passing through O, B, C has radius R_1, and intersects the sides AB, AC in N, M respectively. Prove that the circle AMN also has radius R_1.

34. Given an isosceles triangle ABC in which $AB = AC$, let AD be the altitude from A and M the midpoint of AB. K is the point of intersection of AD with the circle AMC. Prove that $AK = 3R/2$, where R is the circumradius of ABC.

35. Two circles, centres O_1, O_2, intersect at A and B. Circle O_1BO_2 intersects the second circle at P. Prove that O_1, A, P are collinear.

36. The triangle ABC is right-angled at A. A line through the midpoint D of BC meets AB at X and AC at Y. The point P is taken on this line so that PD and XY have the same midpoint M. The perpendicular from P to BC meets BC at T. Prove that AM bisects $\angle TAD$ (internally or externally).

37. Four fixed points lie on a circle. Two other circles are drawn touching each other, one passing through two fixed points of the four and the other through the other two fixed points. Prove that their point of contact lies on a fixed circle.

38. ABC is an acute-angled triangle. L, M, N are respectively the midpoints of the sides BC, CA, AB. Circles on AL, BM, CN as diameters

are drawn. The common chords of these three circles do, of course, meet in a point. Locate this point with reference to triangle ABC.

39. A chord CD is drawn at right angles to a fixed diameter AB of a given circle and DP is any other chord meeting AB in Q. Prove that $\angle PCQ$ is bisected by either CA or CB.

40. C_1 and C_2 are two circles which touch externally at A. The chord PQ of C_1 meets C_2 at R and S (distinct points). RA meets C_1 at U and SA meets C_1 at V. Prove that $PV = QU$.

41. LMN is a triangle and O is point on the circumcircle of LMN. The line perpendicular to OL through O meets MN at L'. Points M', N' are similarly defined. Prove that O, L', M', N' are collinear.

6.2 Cyclic quadrilaterals

In this section we provide a comprehensive review of some of the more advanced results concerning cyclic quadrilaterals, including a number that may be new. In certain cases the analytic geometry involved is technically complicated and consequently we have used the computer package DERIVE in the proofs. We have been unable to prove one of the results, and this is therefore designated as a conjecture, though the accuracy of CABRI, which allows figures to be moved around, is such that we may be confident that it is true.

We do not include in this section anything to do with side lengths such as Ptolemy's theorem or Brahmagupta's formula for the area. These are mentioned later in Part 6.3. In what follows we use the notation that $ABCD$ is a cyclic quadrilateral, the tangents at A and B meet at P and Q, R, S are similarly defined by cyclic change of letters. O is the centre of circle $ABCD$, E is the intersection of the diagonals AC and BD, F is the intersection of AB and CD, and G is the intersection of AD and BC. Since $PQRS$ is a quadrilateral with an incircle $PQ + RS = QR + SP$. The reader is reminded that two points are said to be conjugate with respect to a circle if each lies on the polar of the other. Some of the proofs use the theorem that the diagonal point triangle of a cyclic quadrilateral is self-conjugate; that is each vertex is the pole of the opposite side with respect to the circle. See Part 1.3.

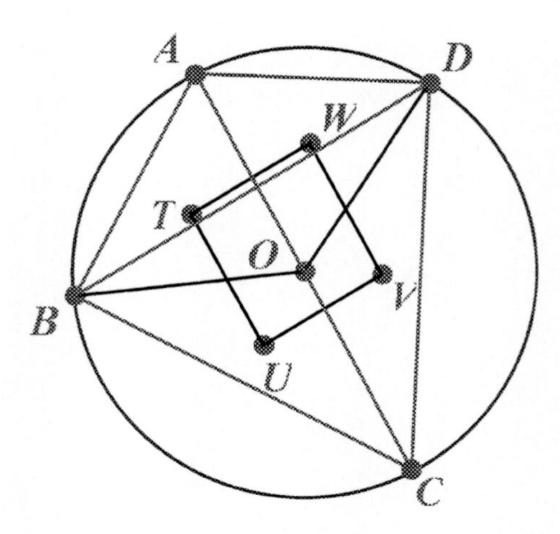

Figure 6.6: Configuration for Section 6.2.1

The contents of this section are an expanded form of the review article by Bradley [3].

6.2.1 Centroids of triangles AOB, BOC, COD, DOA

If T, U, V, W are the centroids of triangles AOB, BOC, COD, DOA respectively, then $TUVW$ is a parallelogram and TU is parallel to AC, and UV to BD. Also $TU = (1/3)AC$ and $UV = (1/3)BD$.

This result is true, as the proof below shows, whether or not $ABCD$ is cyclic, and is true for any point O, either inside or outside $ABCD$. So as O is dragged around $TUVW$ changes position, but not shape or size.

Let O be the origin and write $\mathbf{OA} = \mathbf{a}$ etc.. T is the centroid of triangle OAB, so $\mathbf{t} = (1/3)(\mathbf{a} + \mathbf{b})$. Similarly $\mathbf{u} = (1/3)(\mathbf{b} + \mathbf{c})$, $\mathbf{v} = (1/3)(\mathbf{c} + \mathbf{d})$, $\mathbf{w} = (1/3)(\mathbf{d} + \mathbf{a})$. Then $\mathbf{TU} = \mathbf{u} - \mathbf{t} = (1/3)(\mathbf{c} - \mathbf{a}) = (1/3)\mathbf{AC}$. Similarly $\mathbf{WV} = \mathbf{v} - \mathbf{w} = (1/3)(\mathbf{c} - \mathbf{a}) = (1/3)\mathbf{AC}$. Hence TU is equal and parallel to WV and $TUVW$ is a parallelogram.

Also $TU = WV = (1/3)AC$ and $UV = TW = (1/3)BD$. See Fig. 6.6.

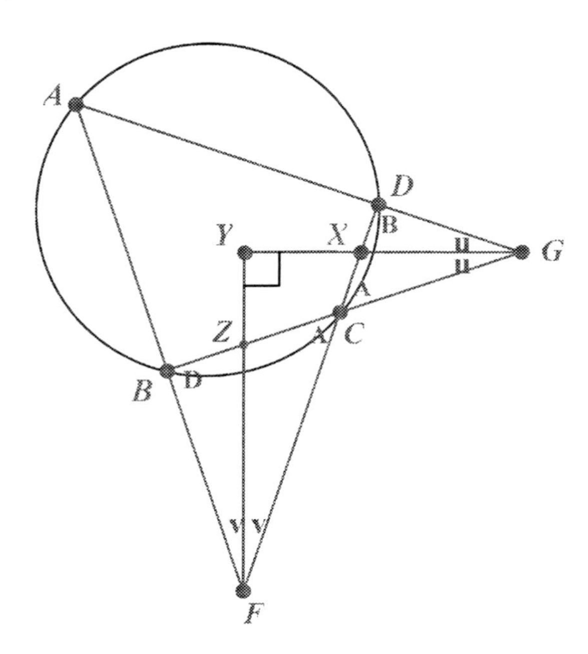

Figure 6.7: Configuration for Section 6.2.2

6.2.2 Internal angle bisectors of the angles at F and G are perpendicular

Let AB, DC meet at F and AD, BC meet at G. If the internal angle bisectors of angles F and G meet at Y, then $\angle FYG = 90°$. This result may be proved using angle theorems only and may be traced back at least as far as Wolstenholme [37]. See Fig. 6.7.

Since $ABCD$ is cyclic we have $A + C = 180°$ and $B + D = 180°$. In quadrilateral $CXYZ$ we have $Z = 180° - D - v$, $C = 180° - A$, $X = 180° - B - u$. Hence $Y = 360° - (180° - D - v) - (180° - A) - (180° - B - u) = A + B + D - 180° + u + v = A + u + v$. Now $u = 90° - A/2 - B/2$ and $v = 90° - A/2 - D/2$, so $\angle FYG = 180° - B/2 - D/2 = 90°$.

6.2.3 The quadrilateral $PQRS$

$PQRS$ is a cyclic quadrilateral if, and only if, AC and BD intersect at right angles. Further, if circle $PQRS$ has centre X, then X, O, E are collinear and PR and QS both pass through E. These results are illustrated in Fig. 6.8.

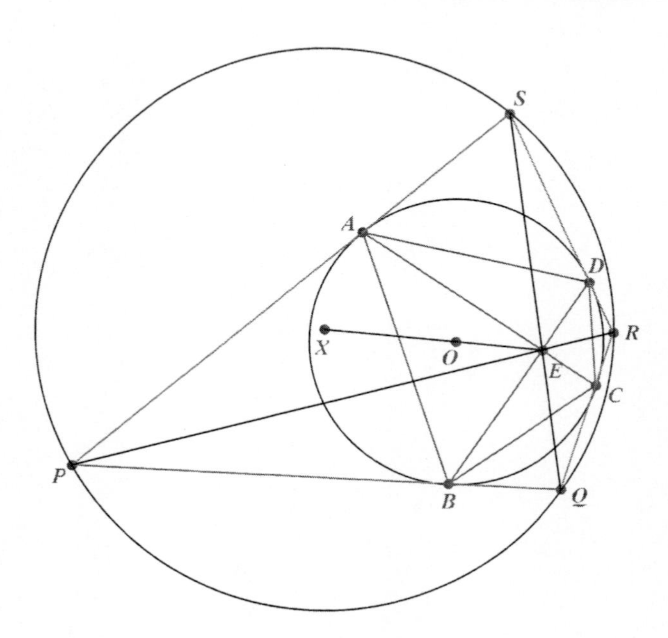

Figure 6.8: Configuration for Section 6.2.3

It is easy to prove that $PQRS$ is cyclic by using angle properties only. We have $\angle APB = 180^o - \angle AOB$, and so $\angle APB + \angle CRD = 360° - \angle AOB - \angle COD = 180°$ if, and only if, $\angle AOB + \angle COD = 180° \Leftrightarrow \angle ADB + \angle CAD = 90° \Leftrightarrow \angle AED = 90° \Leftrightarrow AC$ and BD are at right angles.

The fact that PR and QS pass through E is proved as follows: P is the pole of AB, R is the pole of CD, so PR is the polar of the intersection of AB and CD, which is the point F. But this is GE, since the diagonal point triangle is self-conjugate. Similarly QS is the polar of G, which is EF. So both PR and QS pass through E. The result is an ancient one and appears on p. 99 of Milne [26] but even then it was probably a known result.

The collinearity of X, O, E may be proved using co-ordinate methods. In a general problem we would use parameters as defined in Section 1.3.5, with A having co-ordinates $((1 - a^2)/(1 + a^2), 2a/(1 + a^2))$, and so on. However, in this case, since AC and BD are perpendicular it is possible to choose the axes so that the x-axis lies along the internal bisector of $\angle AOC$ and the y-axis lies along the internal bisector of $\angle BOD$. This limits the number of parameters to two rather than four. Because of the fact that C is the reflection of A in the x-axis and that D is the reflection of B in the y-axis we may take A, B, C, D to have parameters $-t, 1/s, t, s$ respectively. The intersection of AC and BD is therefore $E((1-t^2)/(1+t^2), 2s/(1+s^2))$. Now the tangents at points with parameters h and k meet at the point with co-ordinates $((1-hk)/(1+hk), (h+k)/(1+hk))$. It follows, after some algebra, that P, Q, R, S have the following co-ordinates: $P((s+t)/(s-t), (1-st)/(s-t))$, $Q((s-t)/(s+t), (1+st)/(s+t))$, $R((1-st)/(1+st), (s+t)/(1+st))$, $S((1+st)/(1-st), (s-t)/(1-st))$.

The gradient of PR can now be calculated and turns out to be $\frac{1}{2}(1-s^2)(1+t^2)/(t(1+s^2))$ and so the equation of PR is

$$2t(1+s^2)y - (1-s^2)(1+t^2)x = -(st+s+t-1)(st-s-t-1) \quad (6.1)$$

and by direct verification this contains the point E. Likewise the gradient of QS is the negative of that of PR and the equation of QS is

$$2t(1+s^2)y + (1-s^2)(1+t^2)x = (st-s+t+1)(st+s-t+1) \quad (6.2)$$

and again by direct verification this contains the point E. Now the midpoint of PR has co-ordinates $(s(t^2+1)/((s-t)(1+st)), \frac{1}{2}(1-t^2)(1+s^2)/((s-t)(1+st))$ so the perpendicular bisector of PR has equation

$$\begin{aligned} &2(s-t)(1+st)((1-s^2)(1+t^2)y + 2t(1+s^2)x) \\ &= (st-s+t+1)(st+s-t+1)(1+s^2)(1+t^2). \end{aligned} \quad (6.3)$$

Similarly the perpendicular bisector of QS has equation

$$\begin{aligned} &2(s+t)(st-1)((s^2-1)(1+t^2)y + 2t(1+s^2)x) \\ &= (st-s-t-1)(st+s+t-1)(1+s^2)(1+t^2). \end{aligned} \quad (6.4)$$

These meet at X, the centre of circle $PQRS$, whose co-ordinates are

$$(k(1-t^2)/(1+t^2), 2ks/(1+s^2)),$$

where $k = \frac{1}{4}(1+s^2)^2(1+t^2)^2/((s+t)(s-t)(1+st)(1-st))$. It follows that XOE is a straight line.

6.2.4 $PQRS$ may not be cyclic

Whether or not $PQRS$ is cyclic, PR and QS pass through E. The proof in Section 6.2.3 is still valid. However, for the record, if A, B, C, D have parameters a, b, c, d respectively, then E has co-ordinates (x, y), where

$$x = \frac{(a+c)(bd+1) - (b+d)(ac+1)}{(a+c)(1-bd) - (b+d)(1-ac)},$$

$$y = \frac{2(ac-bd)}{(a+c)(1-bd) - (b+d)(1-ac)}$$

and PR has equation

$$2(cd-ab)y-((a+b)(1+cd)-(c+d)(1+ab))x = (c+d)(1-ab)-(a+b)(1-cd), \tag{6.5}$$

and by direct verification this contains the point E. Similarly QR contains E.

6.2.5 Poncelet's porism

As one of a number of general results under this heading, it is the case that if two circles exist, one within the other, such that a single quadrilateral $PQRS$ both circumscribes the inner circle and is inscribed in the outer circle, then there are an infinity of such quadrilaterals with the vertex P taken to be any point on the outer circle. See Fig. 6.8 again.

During the course of the proof it is found that, if the radius of $PQRS$ is ρ, the radius of $ABCD$ is r and d is the distance between their centres, then

$$\frac{1}{r^2} = \frac{1}{(\rho+d)^2} + \frac{1}{(\rho-d)^2}. \tag{6.6}$$

Suppose that circle $PQRS$, centre $X(0,0)$ has equation $x^2 + y^2 = \rho^2$. Let P, Q, R, S have parameters t, u, v, w, where $x_P = \rho(1-t^2)/(1+t^2)$, $y_P = 2\rho t/(1+t^2)$ etc.. Let the equation of the incircle, centre $O(d, 0)$, be $(x-d)^2 + y^2 = r^2$. The condition for PQ, QR, RS, SP to be tangents to the incircle $ABCD$ is that the perpendicular distance from $(d, 0)$ on to each tangent is r. The chord PQ has equation $(1-tu)x + (t+u)y = \rho(1+tu)$, so the square of the perpendicular distance from $(d, 0)$ on to this line is given by

$$r^2 = \frac{((\rho-d) + tu(\rho+d))^2}{(1+t^2)(1+u^2)}. \tag{6.7}$$

Similarly

$$r^2 = \frac{((\rho - d) + uv(\rho + d))^2}{(1 + u^2)(1 + v^2)}, \tag{6.8}$$

$$r^2 = \frac{((\rho - d) + vw(\rho + d))^2}{(1 + v^2)(1 + w^2)}, \tag{6.9}$$

$$r^2 = \frac{((\rho - d) + wt(\rho + d))^2}{(1 + w^2)(1 + t^2)}. \tag{6.10}$$

From Equations (6.7) and (6.10) u and w must be the roots of the equation

$$r^2(1 + t^2)(1 + x^2) = (\rho - d)^2 + 2xt(\rho^2 - d^2) + x^2t^2(\rho + d)^2. \tag{6.11}$$

Hence

$$u + w = \frac{2t(\rho^2 - d^2)}{(r^2(1 + t^2) - t^2(\rho + d)^2)} \tag{6.12}$$

and

$$uw = \frac{r^2(1 + t^2) - (\rho - d)^2}{r^2(1 + t^2) - t^2(\rho + d)^2}. \tag{6.13}$$

Similarly from Equations (6.8) and (6.9), Equations (6.12) and (6.13) must hold with t replaced by v. Hence

$$\frac{2t(\rho^2 - d^2)}{r^2(1 + t^2) - t^2(\rho + d)^2} = \frac{2v(\rho^2 - d^2)}{r^2(1 + v^2) - v^2(\rho + d)^2}$$

$$\Rightarrow \frac{r^2}{(\rho + d)^2} = \frac{tv}{tv - 1}.$$

Similarly $r^2/(\rho + d)^2 = wu/(wu - 1)$, and so

$$tv = wu. \tag{6.14}$$

Also

$$\frac{r^2(1 + t^2) - (\rho - d)^2}{r^2(1 + t^2) - t^2(\rho + d)^2} = \frac{r^2(1 + v^2) - (\rho - d)^2}{r^2(1 + v^2) - v^2(\rho + d)^2},$$

with a similar equation linking u and w, rather than t and v and since $t \neq v$ this implies that

$$2(d^2 + \rho^2)r^2 = (\rho + d)^2(\rho - d)^2,$$

from which it follows that

$$\frac{1}{r^2} = \frac{1}{(\rho + d)^2} + \frac{1}{(\rho - d)^2}.$$

We now start with an outside circle of radius ρ, an inside circle of radius r, centres distance d apart, with these lengths satisfying Equation (6.6). Start with any point P on the outside circle, parameter t, and choose u to be one of the roots of Equation (6.11) so that PQ touches the inside circle. Now with that value of u, solve Equation (6.11) with u replacing t, one of whose roots is t and the other is v. This value of v defines the point R and we know that QR touches the inside circle. Now with that value of v solve Equation (6.11) with v replacing t, one of whose roots is u and the other is w. This value of w defines the point S and we know that RS touches the inside circle. Now with that value of w solve Equation (6.11) with w replacing t, one of whose solutions we know to be v. Let the other be τ, with τ defining a point Π, and we know that $S\Pi$ touches the inside circle. But

$$tv = \frac{r^2(1+u^2) - (\rho - d)^2}{r^2(1+u^2) - u^2(\rho + d)^2} \tag{6.15}$$

and

$$\tau v = \frac{r^2(1+w^2) - (\rho - d)^2}{r^2(1+w^2) - w^2(\rho + d)^2}. \tag{6.16}$$

Now the right hand sides of Equations (6.15) and (6.16) are equal. Hence $tv = \tau v$ and so $\tau = t$. Thus P and Π coincide. It follows that for every point P on the outer circle the quadrilateral $PQRS$ inscribes the outer circle and circumscribes the inner circle.

6.2.6 Tangents and collinearities

If the tangents at A and C meet at T, and the tangents at B and D meet at U, then T, F, U, G are collinear. It is known that the diagonal point triangle EFG is self-conjugate. AC passes through E so E lies on the polar of T. Similarly it lies on the polar of U, and hence TU is the polar of E, which therefore coincides with FG. Likewise P, E, R, G and S, E, Q, F are collinear. These results are illustrated in Fig. 6.9. This result is proved in Durell [17].

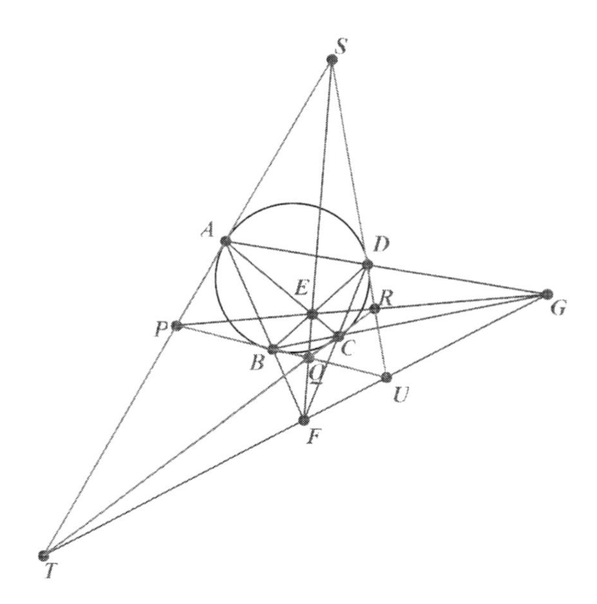

Figure 6.9: Configurations for Sections 6.2.6 and 6.2.7

6.2.7 O is the orthocentre of the diagonal point triangle

Also, using the notation of Section 6.2.6, GO is perpendicular to QS.

See Fig. 6.9, but with O, the centre of circle $ABCD$ added. E is the pole of FG, so OE is perpendicular to FG. Similarly OF is perpendicular to GE, and hence O is the orthocentre of triangle EFG.

6.2.8 Circles and the diagonal point triangle

The circles whose diameters are the sides of the diagonal point EFG are orthogonal to the circle $ABCD$. This is illustrated in Fig. 6.10 and is a development of Sections 6.2.6 and 6.2.7. Let K be the intersection of OF and EG. Since these lines are perpendicular K lies on the circle on FG as diameter. Then $(OK)(OF) = t^2$, where t is the length of the tangent from O to the circle on FG as diameter. But K and F are inverse points with respect to circle $ABCD$, so t is also the radius of this circle. It follows that the two circles are orthogonal. Sections 6.2.7 and 6.2.8 appear as exercises in Durell [17].

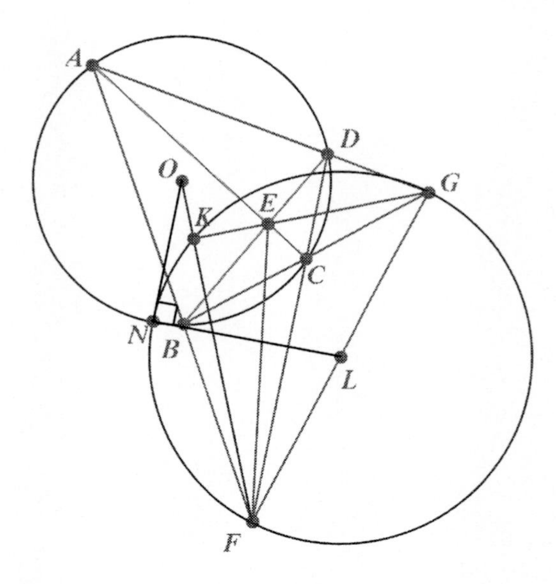

Figure 6.10: Configuration for Section 6.2.8

6.2.9 Circumcentres of AOB, BOC, COD, DOA

Let T, U, V, W be the circumcentres of triangles AOB, BOC, COD, DOA respectively and let TV meet UW at Y, then $TUVW$ is cyclic if, and only if, AC is perpendicular to BD. Further, if X is the centre of circle $TUVW$, then E, Y, O, X are collinear

See Fig. 6.11. T, U, V, W are the circumcentres of triangles AOB, BOC, COD, DOA respectively, and since $\angle PAO = \angle PBO = 90°$, T, U, V, W are also the circumcentres of triangles APB, BQC, CRD, DSA and $OP = 2OT$. P is therefore the point diametrically opposite to O on circle $OAPB$, and similarly for the points Q, R, S. Since $PQRS$ is a cyclic quadrilateral if, and only if, AC and BD are at right angles, the same applies to $TUVW$. When this condition holds we know from Section 6.2.3 that O, X, E are collinear and the centre of circle $PQRS$ also lies on this line. We know that P is the intersection of the tangents at A and B. Q, R, S have like properties. Hence P is the pole of AB with respect to circle $ABCD$, and R is the pole of CD. Therefore PR is the polar of the point of intersection of AB and CD, and hence PR passes through E. Consequently TV passes through the midpoint of OE, and so does UW. Thus Y is the midpoint of OE, and hence X, O, Y, E are collinear.

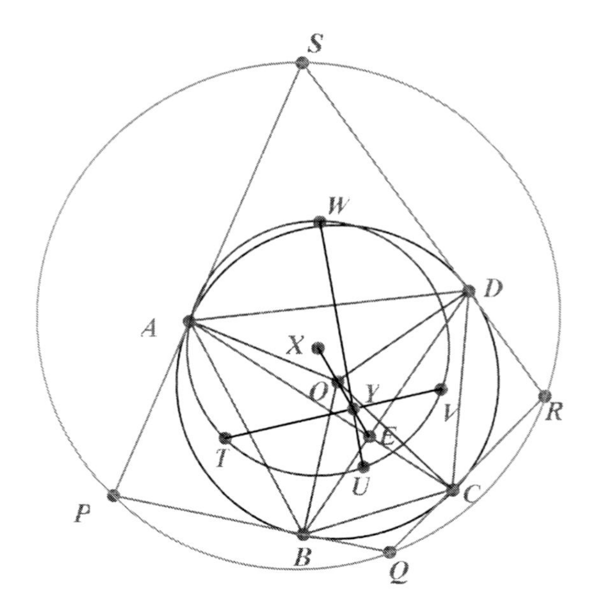

Figure 6.11: Configuration for Section 6.2.9

6.2.10 Orthocentres of AOB, BOC, COD, DOA

Let T, U, V, W be the orthocentres of triangle AOB, BOC, COD, DOA respectively. If AC is perpendicular to BD then T, U, V, W, E are collinear. The same co-ordinate method as for Section 6.2.3 is suitable. When a strange and unexpected result like this occurs it is interesting to look further and in this case to try to identify the line. Monk [28] has in fact shown that the line is tangent at E to the rectangular hyperbola which is the locus of centres of all conics passing through A, B, C, D. If X and Y are the midpoints of AC and BD, the line is perpendicular to XY. This result is illustrated in Fig. 6.12. CABRI indicates the converse is also true.

Take axes as shown in Fig 6.12 so that A and C have parameters a and $-a$ respectively, and B and D have parameters b and $1/b$ respectively. See Section 6.2.3 for the analysis whereby E has co-ordinates

$$\left(\frac{1-a^2}{1+a^2}, \frac{2b}{1+b^2}\right).$$

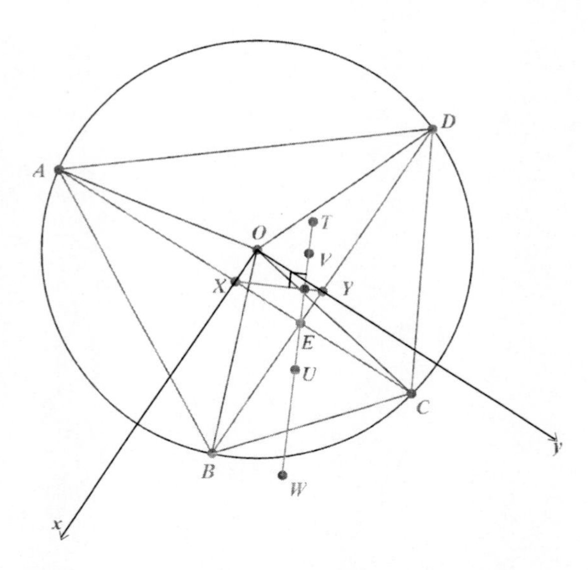

Figure 6.12: Configuration for Section 6.2.10

We omit details of the working, but we find that with

$$k = \frac{(ab+a-b+1)(ab-a+b+1)}{(1+a^2)(1+b^2)}$$

and

$$h = \frac{(ab-a-b-1)(ab+a+b-1)}{(1+a^2)(1+b^2)},$$

T, U, V, W have co-ordinates

$$T(k(1-ab)/(1+ab), k(a+b)/(1+ab)),$$

$$V(-k(a+b)/(b-a), -k(1-ab)/(b-a),$$

$$U(h(1+ab)/(1-ab), h(b-a)/(1-ab)),$$

$$W(-h(b-a)/(b+a), -h(1+ab)/(b+a)).$$

Now the gradient of TV is

$$\frac{(1-a^2)(1+b^2)}{2b(1+a^2)}$$

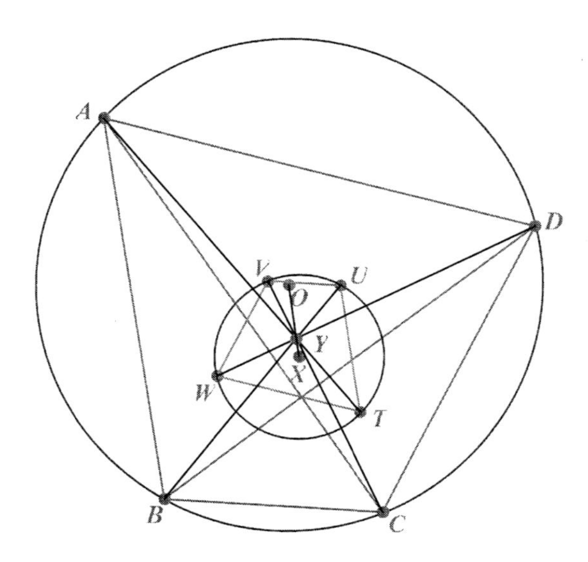

Figure 6.13: The configuration for Section 6.2.11

so its equation is

$$(2b(1+a^2)y-(1-a^2)(1+b^2)x)(1+ab)(1+a^2)(1+b^2)+$$

$$(1+ab)(ab-a+b+1)(ab-a-b-1)(ab+a+b-1)(ab+a+b+1)=0,$$

and by DERIVE this line contains E. Now it is easily shown that UW has the same equation as TV and so coincides with TV.

In the next four sections we consider triangles BCD, ACD, ABD, ABC.

6.2.11 Centroids of BCD, ACD, ABD, ABC

If T, U, V, W are the respective centroids of these four triangles, then $TUVW$ is similar to $ABCD$ with $TU = \frac{1}{3}AB$ etc..

See Fig. 6.13. Take O as origin of vectors, then $\mathbf{t} = \frac{1}{3}(\mathbf{b}+\mathbf{c}+\mathbf{d})$, $\mathbf{u} = \frac{1}{3}(\mathbf{a}+\mathbf{c}+\mathbf{d})$, $\mathbf{v} = \frac{1}{3}(\mathbf{a}+\mathbf{b}+\mathbf{d})$, $\mathbf{w} = \frac{1}{3}(\mathbf{a}+\mathbf{b}+\mathbf{c})$. Then $\mathbf{TU} = \frac{1}{3}(\mathbf{a}-\mathbf{b}) = \frac{1}{3}\mathbf{BA} = -\frac{1}{3}\mathbf{AB}$, with similar expressions for the other sides. It follows that $TUVW$ is homothetic with $ABCD$ and one-third the size. Consider the line AT with equation $\mathbf{r} = \mathbf{a} + s(\frac{1}{3}(\mathbf{b}+\mathbf{c}+\mathbf{d}) - \mathbf{a})$. Choose $s = \frac{3}{4}$, then $\mathbf{r} = \frac{1}{4}(\mathbf{a}+\mathbf{b}+\mathbf{c}+\mathbf{d})$. This point Y lies on AT, BU, CV, DW and is the internal centre of similitude of the circles $ABCD$ and $TUVW$. If

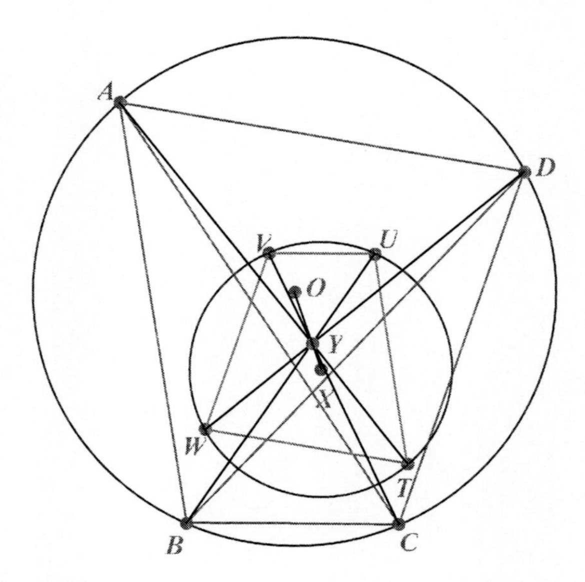

Figure 6.14: The configuration for Section 6.2.12

the centre of the circle $TUVW$ is X, then $\mathbf{OY} = 3\mathbf{YX}$ and X therefore has position vector $\frac{1}{3}(\mathbf{a}+\mathbf{b}+\mathbf{c}+\mathbf{d})$.

6.2.12 Nine-point centres of BCD, ACD, ABD, ABC

If T, U, V, W are the nine-point centres of these four triangles respectively, then $TUVW$ is similar to $ABCD$ and $TU = \frac{1}{2}AB$ etc..

See Fig. 6.14. This is similar to Section 6.2.11, but now $t = \frac{1}{2}(\mathbf{b}+\mathbf{c}+\mathbf{d})$ etc., so $\mathbf{TU} = \frac{1}{2}(\mathbf{a}-\mathbf{b}) = -\frac{1}{2}\mathbf{AB}$ etc.. Thus $TUVW$ is homothetic with $ABCD$ and half the size. The line AT has equation $\mathbf{r} = \mathbf{a}+s(\frac{1}{2}(\mathbf{b}+\mathbf{c}+\mathbf{d})-\mathbf{a})$, and with $s = \frac{2}{3}$ then $\mathbf{r} = \frac{1}{3}(\mathbf{a}+\mathbf{b}+\mathbf{c}+\mathbf{d})$. This point Y lies, by symmetry on all of AT, BU, CV, DW and is the internal centre of similitude of the circles $ABCD$ and $TUVW$. If the centre of the circle $TUVW$ is X, then $\mathbf{OY} = 2\mathbf{YX}$ and X therefore has position vector $\frac{1}{2}(\mathbf{a}+\mathbf{b}+\mathbf{c}+\mathbf{d})$.

6.2.13 Orthocentres of BCD, ACD, ABD, ABC

If T, U, V, W are the orthocentres of these four triangles respectively, then $TUVW$ is congruent to $ABCD$.

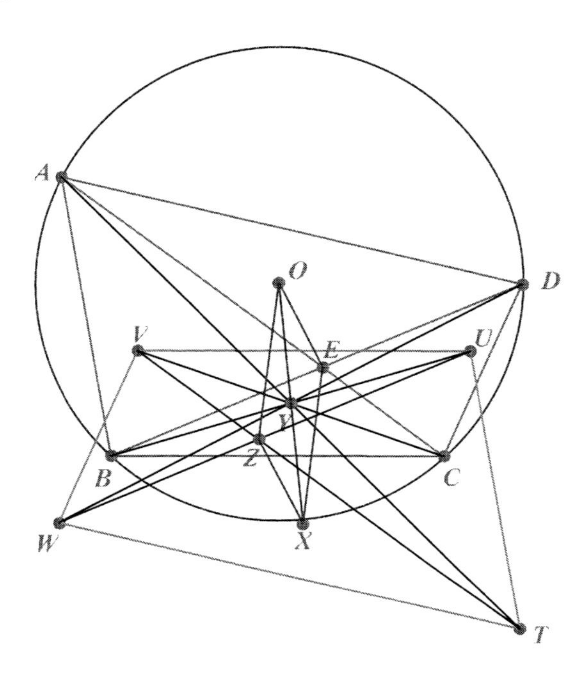

Figure 6.15: Configuration for Section 6.2.13

See Fig. 6.15. Again this is similar to Sections 6.2.11 and 6.2.12. But now $\mathbf{t} = (\mathbf{b} + \mathbf{c} + \mathbf{d})$ etc. and so $TU = (\mathbf{a} - \mathbf{b}) = -\mathbf{AB}$ etc.. Hence $TUVW$ is homothetic with $ABCD$ and congruent to it. The point Y, which is the internal centre of similitude has position vector $\frac{1}{2}(\mathbf{a} + \mathbf{b} + \mathbf{c} + \mathbf{d})$, and so the centre X of circle $TUVW$ has position vector $(\mathbf{a} + \mathbf{b} + \mathbf{c} + \mathbf{d})$. If E is the intersection of AC and BD and Z is the intersection of TV and UW then $EY = YZ$ and so the figure $OZXE$ is a parallelogram.

6.2.14 Incentres of BCD, ACD, ABD, ABC

If T, U, V, W are the respective incentres of these four triangles, then $TUVW$ is a rectangle.

Let K, L, M, N be the midpoints of the arcs DA, AB, BC, CD respectively. Let $\angle DAC = \angle DBC = \angle DLC = 2u$ and $\angle ADB = \angle ACB = \angle ANB = 2v$. By well known properties of incentres of triangles we have $NT = NU = 2R \sin u$, where R is the radius of $ABCD$. So

$$TU = 2R \sin u \sin v = WV.$$

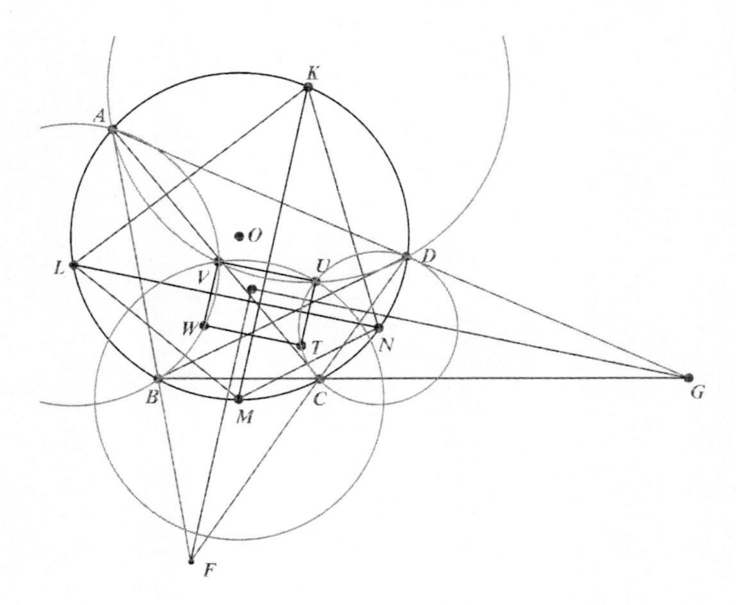

Figure 6.16: The configuration for Section 6.2.14

Similarly $UV = TW$, and $TUVW$ is a parallelogram. That $TUVW$ is in fact a rectangle is linked to the fact that its sides are parallel to the internal bisectors of the angles at F and G. See Fig. 6.16. However the really significant lines are LN and KM. These lines are parallel to the bisectors and they meet at the centre of the rectangle.

6.2.15 Angle between chords of a circle

If XY, ZJ are chords of a circle then the angle θ between them is given by half the sum of the angles subtended at the centre O by arcs ZX, JY, that is $\theta = \frac{1}{2}(\angle ZOX + \angle JOY)$. Angle-chasing easily proves this. For LN and KM it is obvious that the arcs in question amount to half the circumference and so these lines are at right angles. Also in Fig 6.16 note that the quadrilateral $ABWV$ is cyclic, centre L and similarly for quadrilaterals $BCTW$, $CDUT$ and $DAVU$ with centres M, N, K respectively.

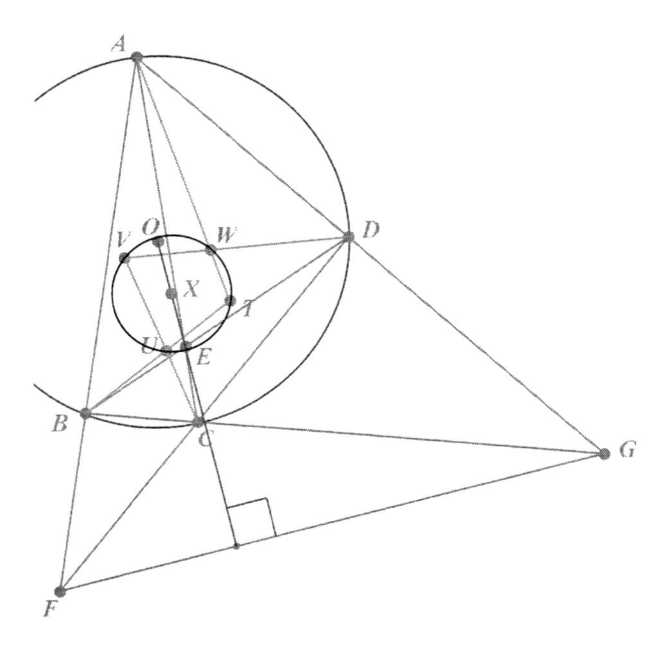

Figure 6.17: Configuration for Section 6.2.16

6.2.16 Internal angle bisectors

Let T be the intersection of the internal bisectors of angles A and B, and let U, V, W be defined similarly by cyclic change of letters. Then $TUVW$ is cyclic and TV and UW intersect at right angles. Further, if X is the centre of circle $TUVW$, then O, X, E are collinear. This result remained a conjecture until Monk [28] supplied the following pure geometrical proof. See Fig. 6.17.

Let $\angle BAD = \alpha, \angle ADC = \delta$. Then $\angle TWV = 180° - \frac{1}{2}\alpha - \frac{1}{2}\delta$. Also $\angle ABC = 180° - \delta$, $\angle BCD = 180° - \alpha$ and so $\angle TUV = 180° - \frac{1}{2}(180° - \alpha) - \frac{1}{2}(180° - \delta) = \frac{1}{2}(\alpha + \delta)$. Thus $\angle TWV + \angle TUV = 180°$ and $TUVW$ is cyclic.

Now W is the incentre of triangle AFD, U is the excentre opposite F of triangle BFC. Hence the line UW is the internal bisector of $\angle AFD$. Similarly TV is the internal bisector of $\angle AGB$. Now triangles CUF and DWF are similar, both having angles $\frac{1}{2}\delta$ and $90° + \frac{1}{2}\alpha$. It follows that $(FU)(FW) = (FC)(FD)$. That is, F has equal powers with respect to circles $TUVW$ and $ABCD$. Similarly G has equal powers with respect to the circles, so FG is their radical axis. Hence FG is perpendicular to the line of centres OX. But triangle EFG is self-conjugate with respect to circle $ABCD$ (inscribed quadrangle), so FG is the polar of E and therefore perpendicular to OE. Thus O, X, E are collinear. Note also that by Section 6.2.2, TV and UW are perpendicular.

6.2.17 External angles bisectors

Let T be the intersection of the external bisectors of angles A and B, and let U, V, W be defined similarly by cyclic change of letters. Then $TUVW$ is cyclic and TV and UW intersect at right angles. Further, if X is the centre of circle $TUVW$, then O, X, E are collinear. The proof of this result is similar to that for Section 6.2.16 and is left as an exercise for the reader.

6.2.18 Pependiculars on to the diagonals

Let T and V be the feet of the perpendiculars from A and C respectively on to BD and let U and W be the feet of the perpendiculars from B and D respectively on to AC, then $TUVW$ is a cyclic quadrilateral. This result can be proved by using angle theorems and is left as an exercise for the reader.

In the next five results we consider triangles ABE, BCE, CDE, DAE.

6.2.19 Centroids of ABE, BCE, CDE, DAE

If T, U, V and W are the centroids of these four triangles respectively, then the figure $TUVW$ is a parallelogram. See Section 6.2.1.

6.2.20 Circumcentres of ABE, BCE, CDE, DAE

If T, U, V, W are the circumcentres of these four triangles respectively, then $TUVW$ is a parallelogram. Furthermore, if X is the intersection of TV and UW, then O, X, E are collinear and $OX = XE$.

See Fig. 6.18. For a start $TW \perp AE$, being its perpendicular bisector and $UV \perp EC$, being its perpendicular bisector. But AEC is a straight line. Therefore TW is parallel to UV. Similarly TU is parallel to WV. Therefore $TUVW$ is a parallelogram. For the second part we prove that $OTEV$ is a parallelogram with one diagonal OXE and the other TXV, so that $OX = XE$. OT is the perpendicular bisector of AB, so it is sufficient to prove that EV is perpendicular to AB. Then similarly TE is parallel to OV and

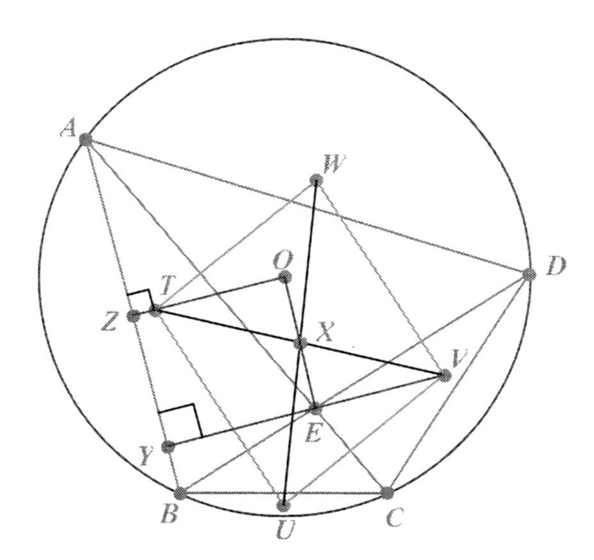

Figure 6.18: The configuration for Section 6.2.20

$OVET$ is a parallelogram. Let VE produced meet AB at Y. Then $\angle YEB = \angle VED = 90^\circ - \angle ECD$ (since V is the circumcentre) $= 90^\circ - \angle EBY$. Hence $EV \perp AB$.

6.2.21 Orthocentres of ABE, BCE, CDE, DAE

If T, U, V, W are the orthocentres of these four triangles respectively, then $TUVW$ is a parallelogram. Since V is the orthocentre of triangle DEC, then DV is perpendicular to AEC. Since W is the orthocentre of triangle ADE, then DW is perpendicular to AEC. So DVW is perpendicular to AC. Similarly BUT is perpendicular to AC. Hence VW is parallel to UT. Similarly WT is parallel to VU and $TUVW$ is a parallelogram.

6.2.22 Nine-point centres of ABE, BCE, CDE, DAE

If T, U, V, W are the nine-point centres of these four triangles respectively, then $TUVW$ is a parallelogram. Since the four circumcentres and the four orthocentres form parallelograms, then since the nine-point centres are midway between the circumcentres and orthocentres, they too form a parallelogram. This follows from the more general result that if $T_kU_kV_kW_k$, $k = 1, 2$ are any two parallelograms and T is the midpoint of T_1T_2 with U, V, W similarly defined, then $TUVW$ is a parallelogram. A proof by vectors is straightforward, bearing in mind that for $PQRS$ to form a parallelogram, it is necessary and sufficient that the position vectors of the vertices satisfy $\mathbf{q} - \mathbf{p} = \mathbf{r} - \mathbf{s}$.

6.2.23 Incentres of ABE, BCE, CDE, DAE

If T, U, V, W are the respective incentres of these four triangles, then $TUVW$ is a quadrilateral in which TV and UW are at right angles. This is trivial since the internal and external bisectors at E are at right angles.

Exercise 6.2.1

1. Let L, M, N, K be the midpoints of AB, BC, CD, DA respectively. Let the perpendicular bisectors of AL, BM meet at U, with V, W, T defined similarly by cyclic change of letters. Let the perpendicular bisectors of LB, MC meet at U', with V', W', T' defined similarly by cyclic change of letters. Then $TUVW$ and $T'U'V'W'$ are cyclic. Further, if circle $TUVW$ has centre X and circle $T'U'V'W'$ has centre X', then X, O, X' are collinear with $XO = OX'$.

2. Triangle UTB is drawn so that $\angle TBC = (1/3)\angle ABC$. Similarly triangle VUC is drawn so that $\angle UCD = (1/3)\angle BCD$. T, U, V, W are defined in this way by cyclic change of letters. Then $TUVW$ is a cyclic quadrilateral.

6.2.24 An eight-point circle

Let L, M, N, K be the midpoints of AB, BC, CD, DA respectively. Let the feet of the perpendiculars from L, M, N, K on to the opposite sides be denoted by T, U, V, W respectively. Then LT, MU, NV, KW are concurrent at a point E. Also the perpendicular from the midpoint of each diagonal AC and BD on to each other passes through E. If further AC and BD are at right angles, then L, M, N, K, T, U, V, W are concyclic. If the centre of this eight-point circle is X, then OXH is a straight line and $OX = XE$. This wonderful result is the analogue of the nine-point circle in a triangle and may be proved using vectors. The result is illustrated in Fig. 6.19. This result appears on p. 66 of Milne [27], but even then was probably a known result.

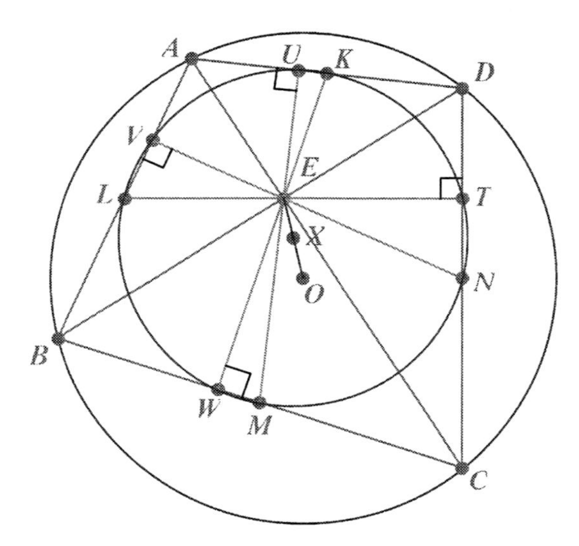

Figure 6.19: The eight-point circle

Take O to be origin of vectors and $\mathbf{OA} = \mathbf{a}$ etc.. Then $|\mathbf{a}| = |\mathbf{b}| = |\mathbf{c}| = |\mathbf{d}| = R$. Let E have position vector $\frac{1}{2}(\mathbf{a}+\mathbf{b}+\mathbf{c}+\mathbf{d})$, then $\mathbf{LE} = (\mathbf{c}+\mathbf{d})/2$ and this is perpendicular to $\mathbf{CD} = (\mathbf{d}-\mathbf{c})$, since $(\mathbf{d}-\mathbf{c}).(\mathbf{d}+\mathbf{c}) = |\mathbf{d}|^2 - |\mathbf{c}|^2 = 0$. Similarly it follows that LT, MU, NV, KW all pass through E. In fact the perpendicular from the midpoint of AC on to BD also passes through E, as does the perpendicular from the midpoint of BD on to AC.

Now consider X, the midpoint of OE. It has position vector $\frac{1}{4}(\mathbf{a}+\mathbf{b}+\mathbf{c}+\mathbf{d})$. We have $\mathbf{XL} = \frac{1}{4}(\mathbf{a}+\mathbf{b}-\mathbf{c}-\mathbf{d})$ and $\mathbf{XM} = \frac{1}{4}(\mathbf{b}+\mathbf{c}-\mathbf{d}-\mathbf{a})$. These are of equal length $\Leftrightarrow$

$$\mathbf{a.b}+\mathbf{c.d}-\mathbf{a.c}-\mathbf{b.c}-\mathbf{a.d}-\mathbf{b.d} = \mathbf{b.c}+\mathbf{a.d}-\mathbf{c.d}-\mathbf{a.c}-\mathbf{b.d}-\mathbf{a.b}$$

$\Leftrightarrow (\mathbf{b}-\mathbf{d}).(\mathbf{a}-\mathbf{c}) = 0 \;\Leftrightarrow BD \perp AC$. Hence a circle, centre X, passes through K, L, M, N if, and only if, AC is perpendicular to BD. Now consider $UEOK$. OK is parallel to EU, since both are perpendicular to DA. So this quadrilateral is a trapezium. X is the midpoint of EO, so $XU = XK$. Thus the circle, centre X, through K, L, M, N also passes through T, U, V, W.

Exercise 6.2.2

1. Let $ABCD$ be a right-angled kite with $AB = AD$ and $BC = DC$ and $\angle ABC = \angle ADC = 90°$. Let T lie on the internal bisectors of $\angle BAC$

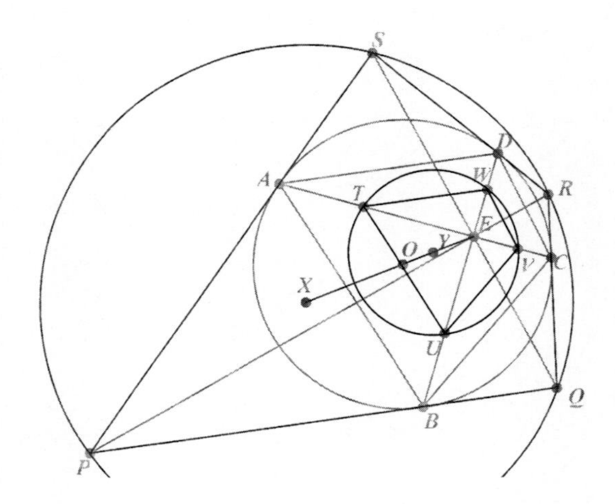

Figure 6.20: Configuration for Section 6.2.25

and $\angle CBD$. Let U, V, W be defined similarly by cyclic change of letters. Prove that $TUVW$ is a cyclic quadrilateral. Let F be the midpoint of TV. Prove that OFE is a straight line.

6.2.25 Incentres of SPE, PQE, QRE, RSE

Suppose that $PQRS$ is cyclic. Let T be the incentre of triangle SPE and let U, V, W be defined similarly by cyclic change of letters. Then $TUVW$ is cyclic and similar to $ABCD$. Further if X is the centre of circle $PQRS$ and Y is the centre of circle $TUVW$, then X, O, Y, E are collinear.

See Fig. 6.20. Let BP and DS meet at Z, then $ZD = ZB$. So $\angle SDE = \angle PBE$. Also $\angle RSQ = \angle RPQ$ (angles in the same segment), so $\angle DSE = \angle BPE$. Hence $\angle SED = \angle PEB$. But $\angle AED = \angle AEB = 90°$ and so $\angle AES = \angle AEP$. Hence T, the incentre of triangle SPE lies on EA. Similarly U, V, W lie on EB, EC, ED respectively. Now since T is the incentre of triangle SPE we have $AT/TE = AP/PE = PB/PE = BU/UE$, since U is the incentre of triangle PQE. It follows that TU is parallel to AB. Similarly UV, VW, WT are parallel respectively to BC, CD, DA. Since AT, BU, CV, DW meet at E it follows that $ABCD$ and $TUVW$ are homothetic, with E the centre of similarity. Their centres O and Y are therefore collinear with E and by Section 6.2.3, X, the centre of circle $PQRS$, also lies on the line OYE.

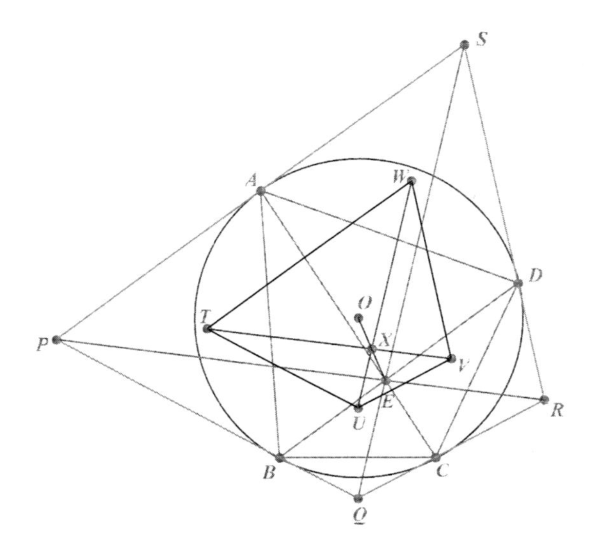

Figure 6.21: Configuration for Section 6.2.26

In the next three results we consider triangles PAB, QBC, RCD, SDA.

6.2.26 Circumcentres of PAB, QBC, RCD, SDA

Let T, U, V, W be the circumcentres of these four triangles respectively, then $TUVW$ is similar to $PQRS$ and half the size. See Fig. 6.21.

See also Section 6.2.9. T, U, V, W are also the circumcentres of triangle AOB, BOC, COD, DOA respectively. This is because $OAPB$ etc. are cyclic quadrilaterals. T is therefore the midpoint of OP, U of OQ, V of OR and W of OS. It follows that $TUVW$ is homothetic with $PQRS$ and half the size. From Section 6.2.4 we know that PR and QS pass through E. So if TV and UW meet at X, it follows from the homothety that X lies on OE and $OX = XE$.

6.2.27 Orthocentres of PAB, QBC, RCD, SDA

Let T, U, V, W be the orthocentres of these four triangles respectively, then $TUVW$ is a parallelogram with TU and WV parallel to AC, and TW and UV parallel to BD. See Fig. 6.22. We give two proofs. First we give the solution using co-ordinate geometry. Let A, B, C, D have parameters a, b, c, d.

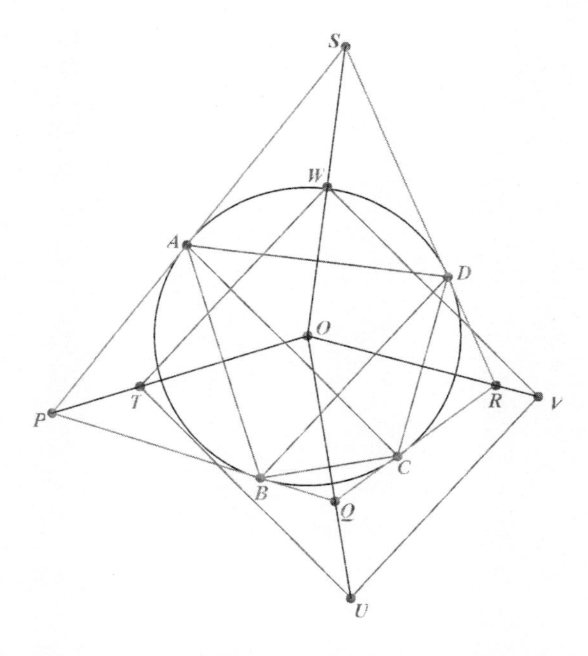

Figure 6.22: Configuration for Section 6.2.27

The tangent at B has equation $(1-b^2)x+2by=1+b^2$. The perpendicular to this through A has equation $(1+a^2)(2bx-(1-b^2)y)=2(b-a)(1+ab)$. Similarly the perpendicular to the tangent at A through B has this equation, but with a and b interchanged. These two lines meet at T with co-ordinates $T(k(1-ab), k(a+b))$, where $k=2(1+ab)/((1+a^2)(1+b^2))$. The co-ordinates of W are similar but with a and d rather than a and b. It now follows that the gradient of TW is $(bd-1)/(b+d)$, which is the same gradient as that of BD. Similarly UV is parallel to BD and both TU and WV are parallel to AC. It follows that $TUVW$ is a parallelogram.

Now we give the following pure geometrical proof by David Monk [28]. Since triangle DRC is isosceles

$$DV = \frac{1}{2}DC/\cos\angle CDV = DC/(2\sin\angle DCR) = DC/(2\sin\angle DAC)$$

by the tangent-chord theorem, but by the sine rule this final expression is R, the radius of circle $ABCD$. Similarly $BU=R$. Also

$$\angle BDV = \angle BDC + \angle CDV = \angle BDC + 90^\circ - \angle DCR = 90^\circ + \angle BDC - \angle DBC$$

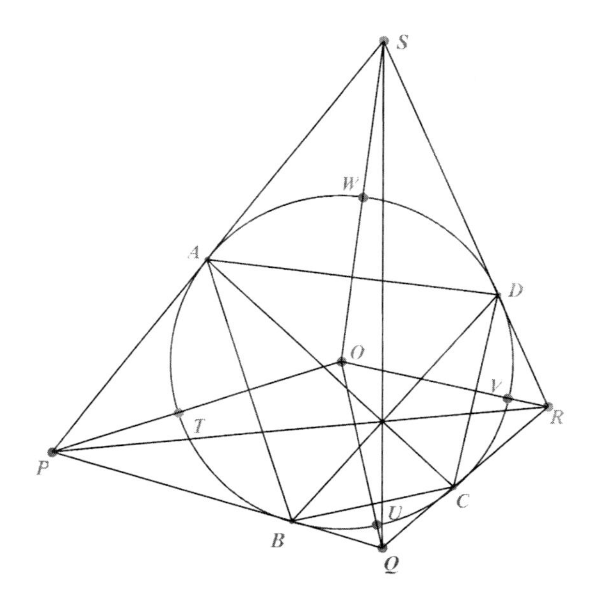

Figure 6.23: Configuration for Section 6.2.28

(tangent-chord theorem). Similarly $\angle DBU = 90^\circ + \angle DBC - \angle BDC$, so $\angle BDV + \angle DBU = 180^\circ$ and BU is parallel to DV. Thus $BUVD$ is a parallelogram and so UV is parallel to BD etc..

6.2.28 Incentres of PAB, QBC, RCD, SDA

Let T, U, V, W be the incentres of these four triangles respectively, then $TUVW$ is a cyclic quadrilateral with $\angle WTU = \frac{1}{2}(\angle DAB + \angle ABC)$. In fact T, U, V, W turn out to lie on circle $ABCD$. See Fig. 6.23.

Obviously SO bisects $\angle ASD$ so W lies on OS etc.. Let W^* be the midpoint of the arc DA. Suppose $\angle AOD = 2x$, then $\angle SAD = x$ (alternate segment theorem). Similarly $\angle W^*AD = \frac{1}{2}x$. Hence W^* coincides with W. That is W lies on circle $ABCD$. Similarly T, U, V all lie on circle $ABCD$, so obviously $TUVW$ is cyclic. Now $\angle WOT = \frac{1}{2}\angle DOB = \angle DCB$. Similarly $\angle TOU = \angle ADC$, so $\angle WVU = \frac{1}{2}(\angle ACB + \angle ADB)$ with similar expressions for the other angles of $TUVW$.

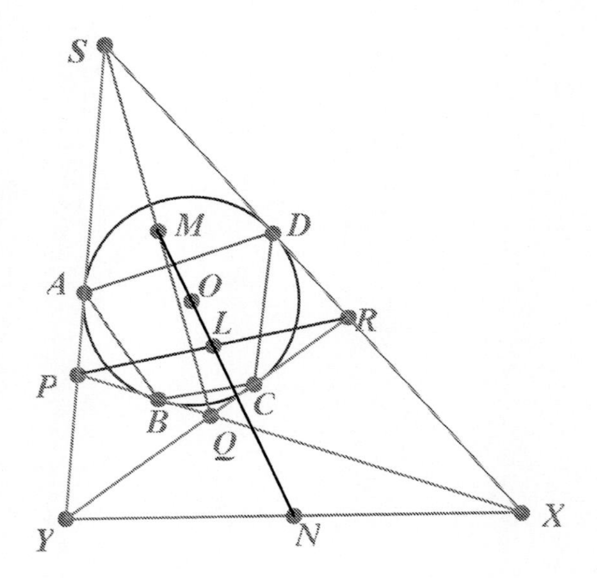

Figure 6.24: Configuration for Section 6.2.29

6.2.29 Collinearity of midpoints of diagonals

Let PQ meet RS at X and let PS and QR meet at Y. Denote the midpoints of PR, QS, XY by L, M, N respectively. Then L, M, N, O are collinear. See Fig. 6.24. The proof is given in Durell [17]. The method is to consider four forces $\mathbf{PS}, \mathbf{PQ}, \mathbf{RS}, \mathbf{RQ}$ and to show that the resultant passes through each of L, M, N, O. In the next three results we consider triangles POQ, QOR, ROS, SOP.

6.2.30 Centroids of POQ, QOR, ROS, SOP

If T, U, V and W are the centroids of these four triangles respectively, then $TUVW$ is a parallelogram. This is a consequence of Section 6.2.1.

6.2.31 Circumcentres of POQ, QOR, ROS, SOP

If T, U, V, W are the circumcentres of these four triangles then $TUVW$ is a cyclic quadrilateral and TV, UW pass through O. See Fig. 6.25.

Both AB and WT are perpendicular to OP so WT is parallel to AB. Similarly WV is parallel to AD, VU to DC and UT to CB. It follows that

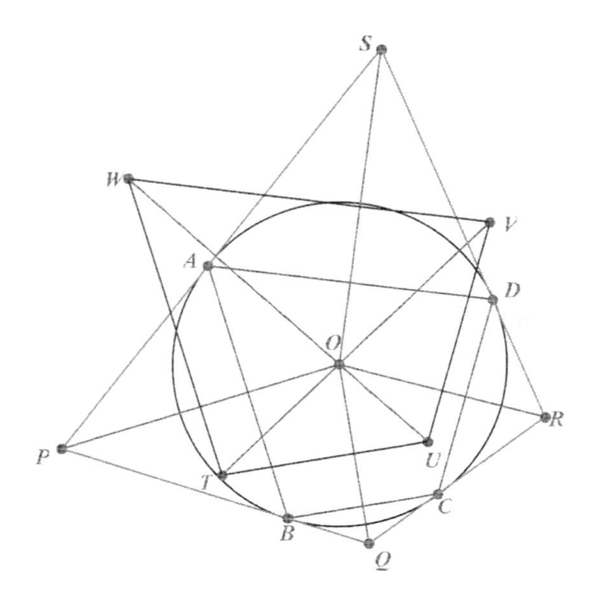

Figure 6.25: Configuration for Section 6.2.31

$WTUV$ is cyclic, having the same angles as $ABCD$. We use a co-ordinate method to show that TV and UW pass through O. We take A, B, C, D to have parameters a, b, c, d.

The equation of OP is $(1-ab)y = (a+b)x$, so TW, which is its perpendicular bisector, has equation

$$2(1+ab)(a+b)y + 2(1+ab)(1-ab)x = (1+a^2)(1+b^2).$$

TU has a similar equation with c replacing a. T therefore has co-ordinates $(k(1-ac), k(a+c))$, where $k = (1+b^2)/(2(1+ab)(1+bc))$. Evidently the gradient of OT is $(a+c)/(1-ac)$ which shows that OT is perpendicular to AC. Similarly OV is perpendicular to AC. Hence TOV is a straight line passing through O. Similarly UOW is a straight line through O.

6.2.32 Orthocentres of POQ, QOR, ROS, SOP

If T, U, V, W are the orthocentres of these four triangles respectively, then T, U, V, W, E are collinear. This can be proved using analytic methods. It turns out that if a, b, c, d are the parameters of A, B, C, D then the equation

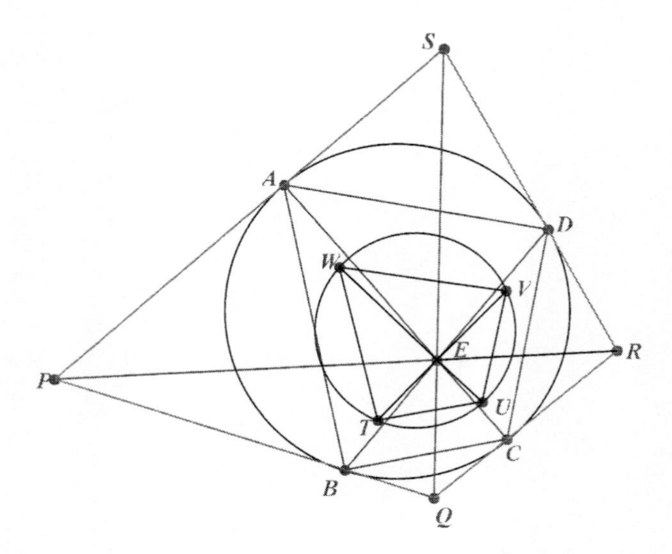

Figure 6.26: Configuration for Section 6.2.33

of $TUVWE$ is

$$(bcd+acd+abd+abc+a+b+c+d)y+2(1-abcd)x = 2(1+ac)(1+bd). \quad (6.17)$$

6.2.33 Conjecture concerning incentres

Conjecture: if T, U, V, W are the incentres of triangles PQE, QRE, RSE, SPE respectively then $TUVW$ is a cyclic quadrilateral. See Fig. 6.26.

Example 6.2.3

Two line segments AC and BD intersect at right angles at a point X, thereby forming what we refer to subsequently as a *cross*. Prove there is a unique cyclic quadrilateral $TUVW$ with the properties that its diagonals TV and UW pass through X, A lies on TU, B on UV, C on VW, D on WT, and AC and BD are the internal and external bisectors of $\angle TXU$ respectively.

Prove further that TW is parallel to UV if, and only if, $XA = XC$, and that TV and UW are at right angles if, and only if, $1/XA + 1/XC = 1/XB + 1/XD$. The configuration is shown in Fig. 6.27.

First we consider how a cross $ABCD$ arises given a cyclic quadrilateral $TUVW$. We then show that given the cross $ABCD$ it can arise from no other

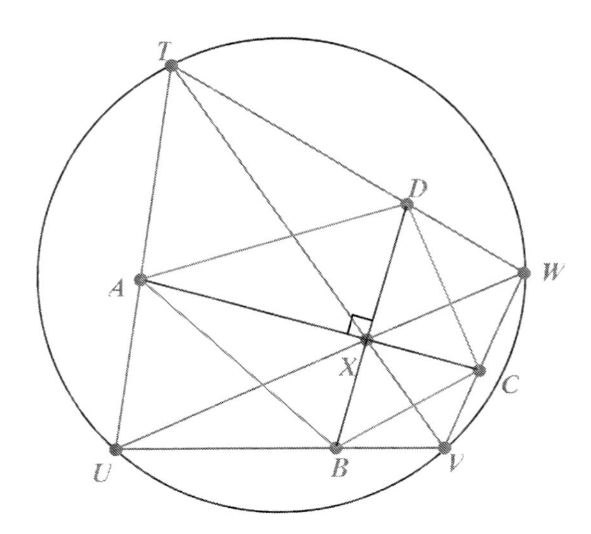

Figure 6.27: Example 6.2.3

cyclic quadrilateral, and that for any cross such a cyclic quadrilateral always exists.

Given $TUVW$ with its diagonals TV and UW intersecting at X, let $\mathbf{XT} = t\mathbf{i}$, $\mathbf{XU} = u\mathbf{j}$, $\mathbf{XV} = -v\mathbf{i}$, $\mathbf{XW} = -w\mathbf{j}$, where $\mathbf{i}$ and $\mathbf{j}$ are unit vectors. The condition for $TUVW$ to be cyclic is $tv = wu$. Let $\mathbf{i}.\mathbf{j} = \cos\theta$. Since AC is the internal bisector of the angle at X, it follows that $TA/AU = TX/XU = t/u$ and hence, by the section theorem of Section 1.2.2,

$$\mathbf{XA} = a\mathbf{e} = tu(\mathbf{i}+\mathbf{j})/(t+u), \tag{6.18}$$

where $\mathbf{e}$ is a unit vector parallel to $\mathbf{i}+\mathbf{j}$. Similarly

$$\mathbf{XB} = b\mathbf{f} = uv(\mathbf{j}-\mathbf{i})/(u+v), \tag{6.19}$$

$$\mathbf{XC} = -c\mathbf{e} = -vw(\mathbf{i}+\mathbf{j})/(v+w), \tag{6.20}$$

$$\mathbf{XD} = -d\mathbf{f} = -wt(\mathbf{j}-\mathbf{i})/(w+t), \tag{6.21}$$

where $\mathbf{f}$ is a unit vector parallel to $(\mathbf{j}-\mathbf{i})$. Note that $|\mathbf{i}+\mathbf{j}| = 2\cos\theta/2$ and $|\mathbf{j}-\mathbf{i}| = 2\sin\theta/2$.

We now reverse the procedure and show that given $ABCD$, that is, given $a, b, c, d, \mathbf{e}, \mathbf{f}$ the values of $t, u, v, w,$ and hence $\mathbf{i}$ and $\mathbf{j}$ are uniquely determined (and always exist). In view of the riders a constructive proof is necessary.

At first we consider only the ratios a/c and b/d and find t, u, v, w such that $tv = uw = 1$. Later we scale the results to reflect the actual size and shape of the cross. So first we look for solutions for t, u, v, w satisfying $(tu(v+w))/((t+u)vw) = a/c$ and $(uv(t+w))/((u+v)tw) = b/d$ and $tv = uw = 1$. The first two equations may be written in the form $c(V+W) = a(T+U)$ and $d(T+W) = b(U+V)$, where $T = 1/t, U = 1/u, V = 1/v, W = 1/w$ and the third then becomes $TV = UW = 1$. Putting $V = 1/T$ and $W = 1/U$ we obtain $c(T+U) = aTU(T+U)$ and $dT(1+UT) = bU(1+UT)$. Now $U, T > 0$ so we can cancel the factors $(U+T)$ and $(1+UT)$ to give $tu = a/c$ and $u/t = b/d$ and hence the solution. In view of the homogeneity of the first two equations, when we scale so that $tv = uw = k^2$, we have an explicit solution

$$t = k\sqrt{\frac{ad}{bc}}, u = k\sqrt{\frac{ab}{cd}}, v = k\sqrt{\frac{bc}{ad}}, w = k\sqrt{\frac{cd}{ab}}. \tag{6.22}$$

It remains to show that the actual size and shape of the cross determine k and θ uniquely. To that end we have to satisfy the Equations (6.18) and (6.19), which give $2\cos\theta/2 = a(1/t + 1/u)$ and $2\sin\theta/2 = b(1/u + 1/v)$. Using Equations (6.22) this leads to $\tan\theta/2 = (bd(a+c))/(ac(b+d))$ and $4k^2 = ac(b+d)^2/bd + bd(a+c)^2/ac$. Since k is positive and $\frac{1}{2}\theta$ is acute we have constructed a solution for all $a, b, c, d, \mathbf{e}, \mathbf{f}$ which is unique.

TW parallel to UV means that a constant λ exists such that $t\mathbf{i} + w\mathbf{j} = \lambda(v\mathbf{i} + u\mathbf{j})$, which holds if, and only if, $t/w = v/u$, which is so if, and only if, $a = c$, that is $XA = XC$.

TV and UW are at right angles if, and only if, $\theta = 90°$, which holds if, and only if, $\tan\theta/2 = 1$, that is if, and only if, $bd(a+c) = ac(b+d)$ or $1/a + 1/c = 1/b + 1/d$, that is if, and only if, $1/XA + 1/XC = 1/XB + 1/XD$.

Exercise 6.2.4

1. Let P, Q, R, S be the reflections of E in AB, BC, CD, DA respectively and let T, U, V, W be the reflections of E in SP, PQ, QR, RS respectively. Suppose PR, QS meet at Y and TV, UW meet at Z. Prove that $TUVW$ is cyclic, with its sides parallel to $ABCD$, circle $TUVW$ has centre E and $EY = YZ$.

2. $ABCD$ is a cyclic quadrilateral. AB, CD meet at F; BC, AD meet at G; AC, BD meet at E. The angle bisector $FNKM$ of $\angle BFC$ meets BC, AD at N, M respectively, and the angle bisector $GUKV$ of $\angle AGB$

meets AB, CD at U, V respectively. Prove that (i) $MUNV$ is a rhombus; (ii) $1/MU = 1/AC + 1/BD$; (iii) K lies on the line joining the midpoints of AC, BD, FG.

3. $ABCD$ is a cyclic quadrilateral inscribed in a circle centre O. L, M are the midpoints of AD, BC. AD and BC meet at V. The feet of the perpendiculars from O, V to LM are P, Q respectively. Prove that $LP = QM$.

4. $ABCD$ is a quadrilateral circumscribing a circle. Prove that the incircles of triangles ABC and ADC touch each other.

5. $ABCD$ is a convex quadrilateral inscribed in a circle S_1. Circle S_2 touches S_1 externally at D. AP, BQ, CR are the tangents from A, B, C to the circle S_2. Prove that

$$(AB)(CR) + (BC)(AP) = (CA)(BQ).$$

6.3 Ptolemy's Theorem

6.3.1 A proof of Ptolemy's theorem

This states that if $ABCD$ is a cyclic quadrilateral (with vertices in that order) then

$$(AB)(CD) + (BC)(AD) = (CA)(BD). \tag{6.23}$$

If in Fig. 6.28, the circumradius $= \frac{1}{2}$, then, for Equation (6.23) to hold, it is sufficient to show

$$\sin u \sin v + \sin w \sin t = \sin(t + v) \sin(v + w). \tag{6.24}$$

Bearing in mind that $u = 180° - v - w - t$, it is now possible to show that Equation (6.24) is an identity in v, w, t. The converse of Ptolemy's theorem is also true.

It is also the case that if A, B, C, D are not concyclic then the equality sign in Equation (6.23) becomes a greater than symbol.

We complete this chapter with an exercise of miscellaneous questions on cyclic quadrilaterals, including some problems about areas. Both Ptolemy's theorem and its converse are proved, using complex numbers, in Example 14.4.1.

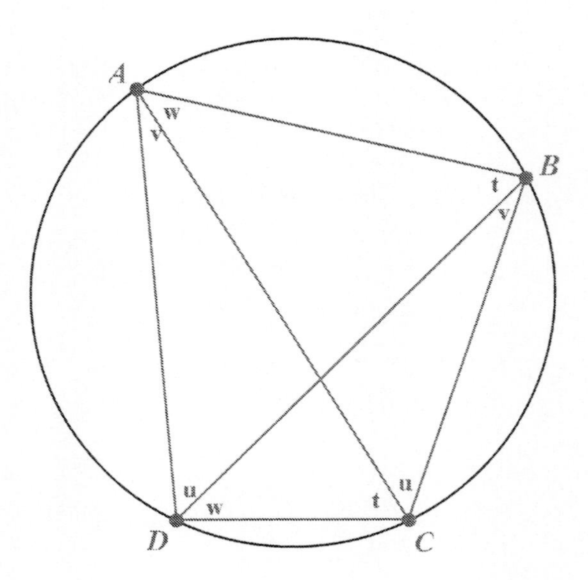

Figure 6.28: Ptolemy's theorem

Example 6.3.1

1. ABC is an equilateral triangle of side a. P is a point on the minor arc BC of the circumcircle of ABC. Prove (i) $PA = PB + PC$; (ii) $PA^2 + PB^2 + PC^2 = 2a^2$.

2. D is a point on the circumference of the circumcircle of an equilateral triangle ABC. AD, BD, CD meet BC, CA, AB at P, Q, R respectively. The circle with centre D and radius DP meets BC again at X. Points Y, Z are defined similarly on CA, AB respectively. Prove that X, Y, Z are collinear.

3. Prove Brahmagupta's formula for the area of the cyclic quadrilateral $ABCD$ with sides a, b, c, d, that

$$[ABCD] = \sqrt{(s-a)(s-b)(s-c)(s-d)},$$

where $s = (a+b+c+d)/2$.

4. ABC is a triangle with $a > b > c$. D is a point on the side BC and E is a point on the extension of BA beyond A, so that $BD = BE = CA$. Let P be a point on the side AC such that E, B, D, P are concylic, and

let Q be the second intersection of BP with the circumcircle of triangle ABC. Prove that $AQ + CQ = BP$.

5. From the point of intersection of the diagonals of a cyclic quadrilateral, perpendiculars are drawn to the sides. Prove that the sum of the two opposite angles formed by straight lines joining the feet of these perpendiculars is double one of the angles between the diagonals.

6. $ABCD$ is a cyclic quadrilateral, which also has an incircle. Let $AB = a$, $BC = b$, $CD = c$, $DA = d$ and let R, r be the radii of the circumscribed and inscribed circles. Prove that

   (i)
   $$r = \frac{2\sqrt{abcd}}{a+b+c+d};$$

   (ii)
   $$[ABCD] = \sqrt{abcd};$$

   (iii)
   $$R = \frac{1}{4}\sqrt{\frac{(ab+cd)(ac+bd)(ad+bc)}{abcd}};$$

   (iv)
   $$R \geq \sqrt{2}r.$$

7. Quadrilateral $ABCD$ is inscribed in a circle with centre O. The diagonals AC and BD are perpendicular. Prove that the distance from O to AD is half the length of BC.

8. A convex cyclic quadrilateral $ABCD$ possesses an incircle centre I. The line parallel to AB through I meets AD in P and BC in Q. Prove that the length of PQ is a quarter of the perimeter of $ABCD$.

9. A hexagon $ABCDEF$ is inscribed in a circle. Prove that AD, BE, CF are concurrent if, and only if, $(AB/BC)(CD/DE)(EF/FA) = 1$.

10. $ABCD$ is a convex quadrilateral with consecutive sides equal to a, b, c and d. Prove that $[ABCD] \leq \frac{1}{4}(a+c)(b+d)$. When does equality hold?

11. A cyclic quadrilateral $ABCD$ also has an inscribed circle, centre I. Prove that the four line segments parallel to its sides passing through I and terminating on opposite sides of the quadrilateral are equal in length.

12. $ABCD$ is a convex quadrilateral and O is the intersection of its diagonals. Given that

$$OA \sin A + OC \sin C = OB \sin B + OD \sin D,$$

prove that $ABCD$ is cyclic.

13. $ABCDE$ is a regular pentagon and P is a point on the circumcircle between A and E. Show that $PA + PC + PE = PB + PD$.

Chapter 7

Pencils of Conics and the Wallace-Simson Line

7.1 The harmonic quadrangle and quadrilateral

Example 7.1.1

We refer to Example 3.3.1 in which a preliminary treatment is given of the harmonic properties of the quadrangle. The analysis there is carried out using projective co-ordinates. In Chapter 6 we stated without proof that the diagonal point triangle is self-conjugate with respect to the circumscribing circle of a cyclic quadrangle. In Section 7.3.1 we extend the treatment to prove that the diagonal point triangle is self-conjugate with respect to any non-degenerate conic circumscribing the quadrangle, and to show that the set of such conics produces a natural example of what is meant by a *pencil of conics*. In Section 7.3.3 we generalize the idea of pencils of conics. The later part of this section reverts to the Euclidean plane, where we use areal co-ordinates, and provides an account of how the harmonic properties of the quadrangle can be described by cross-ratios involving ratios of line segments, rather than the values of parameters. Note that dual results involving quadrilaterals also hold good, the corresponding *tangential pencils of conics* now being those that touch the four sides of the quadrilateral.

Using the notation of Example 3.3.1 the quadrangle $ABCD$ may be taken to have projective co-ordinates $A(1,0,0), B(0,1,0), C(0,0,1), D(1,1,1)$. The diagonal point triangle EFG has vertices given by $E = AC \wedge BD$, $F = AB \wedge CD$, $G = AD \wedge BC$. The co-ordinates of E, F, G are thus $E(1,0,1)$, $F(1,1,0)$, $G(0,1,1)$. In Example 3.3.1 we showed that several pencils, such as $F(BECG)$ are harmonic, cutting out harmonic ranges on lines intersecting the pencil.

The equation of AB is $z = 0$, the equation of CD is $x = y$. Together they form a degenerate line pair with equation $z(x-y) = 0$. The equation of BC is $x = 0$ and the equation of AD is $y = z$. Together they form the degenerate line pair with equation $x(y-z) = 0$. We now claim that any conic through A, B, C, D has an equation of the form $z(x-y) = kx(y-z)$ for some value of k (including $k = 0$ and $k = \infty$). The case $k = -1$ is exceptional as it degenerates into the line pair AC and BD. The single parameter k ensures that the conic can be made to pass through any distinct fifth point by choosing k appropriately. This is the most often quoted example of a *pencil of conics*.

We now show that EFG is self-conjugate with respect to any of these conics ($k \neq -1, 0, \infty$). It is sufficient, by symmetry, to show that FG is the polar of E. The conic has equation $(1+k)zx - kxy - yz = 0$ and the polar of E has equation $(1+k)(z+x-y) = 0$; that is $y = x+z$, since $k \neq -1$. This is the equation of the line FG, as required.

The dual result follows automatically, using line co-ordinates, and shows that the diagonal line triangle is self-conjugate with respect to any conic circumscribed by the four lines of the quadrilateral. The tangential equation of such a conic is $(1+k)nl - klm - mn = 0$, which in point co-ordinates becomes $f^2x^2 + g^2y^2 + h^2z^2 - 2ghyz - 2hfzx - 2fgxy = 0$, where $f = -1$, $g = (1+k), h = -k$. Note that such a conic not only touches the sides of the triangle of reference, but also the unit line, with equation $x + y + z = 0$.

The above calculation transfers in detail to the Euclidean plane only when D happens to be the centroid of triangle ABC, though the analysis is sufficient to assure us that the diagonal point triangle is self-conjugate in the Euclidean plane. The analysis is not much more complicated when we use areal co-ordinates, with ABC the triangle of reference and with D having co-ordinates (p, q, r), where none of p, q, r vanishes (so that D does not lie on any of the lines AB, BC, CA).

As the working is very similar to the above, we merely state the results. The co-ordinates of E, F, G are $E(p,0,r)$, $F(p,q,0)$, $G(0,q,r)$. The pencil of conics through A, B, C, D has equation $pyz - q(k+1)zx + krxy = 0$. (Here $k = -1$ gives the degenerate line pair of AC and BD.) The equation of FG

is $pqz + qrx = rpy$ and the polar of E is $(k+1)(qrx - rpy + pqz) = 0$ and for $k \neq 1$, this is the equation of FG.

We complete the example by appreciating how to identify a harmonic range in the Euclidean plane by means of a cross-ratio involving the ratios of the lengths of line segments. Let FE meet BC at L. Now the equation of FE is $rpy + pqz = qrx$ and this meets BC, $x = 0$, at $L(0, q, -r)$. Now $G(0, q, r)$, so we have $BG/GC = r/q$ and $BL/LC = -r/q$ and we have

$$(BL/LC)/(BG/GC) = -1. \tag{7.1}$$

This is the same as the cross-ratio $(b, c; l, g) = (l-b)(c-g)/(g-b)(c-l) = -1$, where b, c, l, g are the parameters z/y of the points B, C, L, G. Note that when $q = -r$, L is the midpoint of BC and G recedes to infinity, because AD and BC are parallel. This provides the concept in the Euclidean plane of a harmonic range in which two points are harmonically separated by the midpoint between them and the point at infinity on the line.

Exercise 7.1.2

1. Find the equation in areals of an ellipse with major axis $BC = 2a$ and minor axis $AD = 2t$, where ABC is the reference triangle and D has co-ordinates $(-1, 1, 1)$.

7.2 Cross-ratio in the Euclidean plane

From Example 7.1.1 we see that the definition in the Euclidean plane, consistent with the definition of cross-ratio in the projective plane, is

$$(A, C; B, D) = (AB/BC)/(AD/DC), \tag{7.2}$$

where line segments are signed.

The importance of the cross-ratio Equation (7.2) is that it is an invariant under Euclidean and affine transformations, since ratios on line segments are unaltered by these transformations. This is, of course, only to be expected, in view of the invariance of cross-ratio under projective transformations.

Exercise 7.2.1

1. Prove that for all real values of k ($k \neq 0, -1, \infty$) the conic with equation

$$\tan Ayz + k \tan Cxy = (k+1) \tan Bzx$$

is a rectangular hyperbola passing through the vertices of the triangle of reference and its orthocentre. Find the locus of the centre as k varies.

2. Let ABC be a triangle and K a general point in the plane of ABC. The lines AKL, BKM, CKN are the Cevians through K and NM meets BC at L'. Prove that $(B, C; L, L') = -1$. Prove conversely, that if L'' lies on BC and $(B, C; L, L'') = -1$, then L'' lies on MN.

3. Prove that if A, B, C lie on a line, then $(A, C; B, \infty) = -AB/BC = AB/CB$.

4. A variable line meets four fixed planes, which have a common line of intersection, in points A, B, C, D. Prove that $(A, C; B, D)$ is constant.

5. Prove that if $(A, C; B, D) = (A', C'; B', D') = -1$ and AA', BB', CC' are concurrent at a point P, then DD' passes through P.

6. Let S be a conic and P an external point. Give a ruler only construction to draw the pair of tangents to S passing through P.

7. A variable conic passes through four fixed points A, B, C, D. If two fixed lines AX, BY cut the conic at P, Q, prove that PQ passes through a fixed point O.

8. P is a variable point on the circumcircle of triangle ABC. AP meets BC at Q and PC meets AB at R. Show that QR passes through a fixed point.

9. A, B, C, D are the four vertices of a quadrangle, AC, BD meet at E; AB, CD meet at F; AD, BC meet at G. CD meets GE at Q, DB meets FG at P and BC meets EF at R. Prove that P, Q, R are collinear.

7.3 Pencils of conics

7.3.1 Conics passing through four points

Let $S = 0$ and $S' = 0$ be two distinct non-degenerate conics, both of which pass through four given points A, B, C, D then the equation $S = kS'$, for any value of k, represents a conic passing through A, B, C, D. This is because the equation is of the second degree and is satisfied by the co-ordinates of A, B, C, D. It represents the pencil of conics passing through A, B, C, D. For the conic in the pencil that also passes through $P(u, v, w)$ put $k = S(u, v, w)/S'(u, v, w)$. This is true for intersecting circles also, when two of the four points are the circular points at infinity.

Let l be a fixed line. We prove that it cuts the pencil of conics at pairs of points in involution. For if l meets one of the conics at P, it must intersect that conic at a second point P'. Similarly, as the conic is equally well defined by P', it follows that P' gives rise to P. Such pairs of points on l are therefore pairs of points in involution. As an involution has two self-corresponding points it follows that two conics of the pencil touch an arbitrary line.

Since the pairs of opposite sides of the quadrangle $ABCD$ are degenerate line pairs of the pencil of conics, it follows that an arbitrary line cuts the opposite sides of a quadrangle in three pairs of points in involution.

7.3.2 Eleven-point conic

Consider the polar of the point $P_1(x_1, y_1, z_1)$ with respect to the conics $S = kS'$. The polar is given by the equation $S_1 = kS'_1$. For the notation see Section 3.2.5. These polars all pass through the intersection of the lines $S_1 = 0$ and $S'_1 = 0$. Denote this point by Q_1. We have proved that the polars of a fixed point with respect to the pencil of conics through four given points all pass through a fixed point. Suppose now that P_1 moves on a given line $L : lx + my + nz = 0$, so that $lx_1 + my_1 + nz_1 = 0$, then eliminating $x_1 : y_1 : z_1$ between this equation and the equations $S_1 = 0$ and $S'_1 = 0$, we obtain a homogeneous quadratic in x, y, z which shows the locus of Q_1 is a conic $\Sigma_Q = 0$. It can be shown that this conic contains eleven key points. These are:

(i) the diagonal points E, F, G of the quadrangle $ABCD$,

(ii) the points on L where two conics of the pencil touch L and

(iii) the harmonic conjugate on BC of the point where L meets the side BC and the five similarly defined points on CA, AB, AD, BD, CD.

Proof: We prove the result in the projective plane, where the pencil of conics have the equation given in Example 7.1.1, which is $yz-(k+1)zx+kxy=0$. The polar of the point P_1 lies on the line $S_1=0$, that is $-x_1z+y_1z+z_1(y-x)=0$ and also on the line $S_1'=0$, that is $x_1(y-z)+y_1x-z_1x=0$. Eliminating $x_1:y_1:z_1$ between these equations and $lx_1+my_1+nz_1=0$, we get the equation of the locus of Q_1 as

$$\Sigma_Q \equiv lx^2+my^2+nz^2-(m+n)yz-(n+l)zx-(l+m)xy=0. \qquad (7.3)$$

It is easy to check that $E(1,0,1), F(1,1,0)$ and $G(0,1,1)$ lie on this conic. Also it meets BC again where $my=nz$, that is at the point $(0,n,m)$. Now L meets BC where $my+nz=0$, that is at the point $(0,-n,m)$. These points are harmonic conjugates with respect to B and C, as stated. Similar considerations hold for the five points that are harmonic conjugates on CA, AB, AD, BD, CD with respect to the points where L meets those lines. The other two points, each being a point of tangency of L with respect to a member of the pencil, obviously lie on the conic $\Sigma_Q=0$ as the pole of a tangent is its point of contact.

7.3.3 Conics touching four lines

The dual of Section 7.3.1 is that there exists a tangential pencil of conics touching four given lines. Two conics of a tangential pencil pass through an arbitrary point and the lines joining an arbitrary point to the opposite vertices of a quadrilateral are three pairs of lines in involution.

7.3.4 Particular forms of conics

Let A, B, C be the vertices of the triangle of reference and let D be the unit point and let S be a non-degenerate conic through A, B, C, D. We may suppose its equation is $S \equiv (q-r)yz+(r-p)zx+(p-q)xy=0$, where no two of p, q, r are zero. The equation of AD is $y=z$ and the equation of BC is $x=0$. The equation of the tangent at A is $(r-p)z+(p-q)y=0$ and the equation of the tangent at B is $(q-r)z+(p-q)x=0$. The following results now hold:

(i) The conic with equation $S = kx(y - z)$ is an arbitrary conic through A, B, C, D.

(ii) The conic with equation $S = kx((r - p)z + (p - q)y)$ is an arbitrary conic touching S at A and passing through B and C.

(iii) The conic with equation $S = k(y-z)((r-p)z+(p-q)y)$ is an arbitrary conic having triple point contact at A and passing through D.

(iv) The conic with equation $S = k((r - p)z + (p - q)y)^2$ is an arbitrary conic having four point contact with S at A.

(v) The conic with equation $S = k(y - z)^2$ is an arbitrary conic having double contact with S at A and D. The line AD is called the *chord of contact.*

(vi) The conic with equation $S = k((r-p)z+(p-q)y)((q-r)z+(p-q)x)$ is an arbitrary conic having double point contact with S at A and B.

The reader is invited to treat these statements as exercises.

Exercise 7.3.1

1. Given the conics with equations $S + L^2 = 0, S + M^2 = 0, S + N^2 = 0$, having respectively L, M, N as common chords of contact with S, identify the six lines with equations $M \pm N = 0$, $N \pm L = 0$, $L \pm M = 0$.

2. A variable conic passes through four fixed points A, B, C, D. Prove that the tangents to the conic at A, B, C meet BC, CA, AB respectively at three points on a line, whose envelope is the conic which touches the sides of the triangle ABC at the diagonal points of the quadrangle $ABCD$.

3. Two conics S, S' have triple point contact at P. They meet again at D. The other common tangent meets S at E and S' at F. Prove that PT, PD are harmonically conjugate to the lines PE, PF.

7.4 The Wallace-Simson line

Fig. 7.1 shows the classical Wallace-Simson line configuration, in which P lies on the circumcircle of triangle ABC, and PL, PM, PN are the perpendiculars on the sides BC, CA, AB respectively.

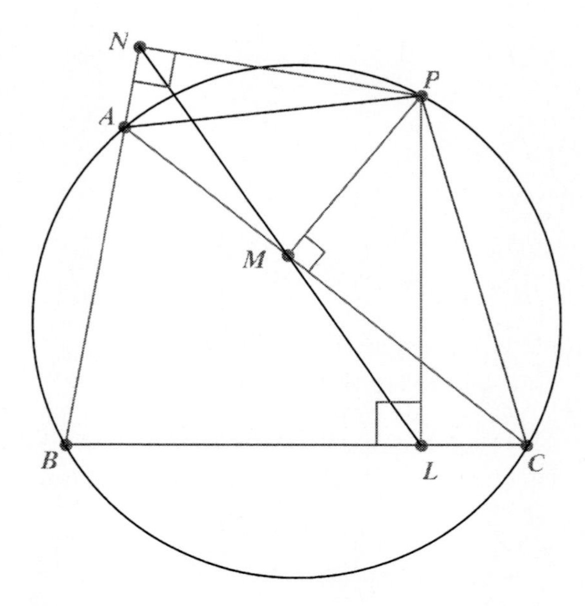

Figure 7.1: The Wallace-Simson line

7.4.1 Wallace-Simson line

The points L, M, N are collinear on what has become known as the *Wallace-Simson* line of P. We have $\angle NMA = 90° - \angle PMN$. But $PNAM$ is cyclic, because of the right angles, so this is equal to $90° - \angle PAN = 90° - \angle PCB$, since $PACB$ is cyclic. This is equal to $90° - \angle PCL = \angle CPL = \angle CML$, as $PMLC$ is cyclic. Hence $\angle NMA = \angle CML$ and since AMC is a straight line it follows that NML is a straight line, it having been proved that the two angles at M are vertically opposite.

Exercise 7.4.1

1. LMN is the Wallace-Simson line of a point P with respect to triangle ABC. Prove that the circles AMN, BNL, CLM meet at P. The centres of these three circles are denoted by A', B', C' respectively. Prove that triangle $A'B'C'$ is homothetic with triangle ABC and half the size. X, Y, Z are the reflections of P in the lines BC, CA, AB respectively. Prove that circles AYZ, BZX, CXY meet at a point Q. If the centres of these circles are denoted by A'', B'', C'' prove that $Q, A'', B'', C''$are concyclic.

2. Let ABC be a triangle and P any other point on the circumcircle of ABC. The perpendiculars from P on to BC, CA, AB, when extended meet the circumcircle again at D, E, F respectively. Prove that AD, BE, CF are all parallel to the Wallace-Simson line of P. Prove also that the axis of perspective of the (congruent) triangles ABC and DEF is at right angles to the Wallace-Simson line of P.

3. Lines AX, BY, CZ are drawn parallel to each other, through A, B, C meeting the circumcircle again at X, Y, Z. Prove that the lines through X perpendicular to BC, through Y perpendicular to CA and through Z perpendicular to AB are concurrent. Locate the point of concurrency.

The first step in generalizing the theorem is to replace the words 'PL, PM, PN are the perpendiculars to the sides BC, CA, AB respectively' by the text 'PL, PM, PN are drawn parallel to AH, BH, CH respectively. Let these lines meet the sides BC, CA, AB in points L, M, N respectively'. The fact that $\mathbf{GH} = 2\mathbf{OG}$ is indicative, as is the fact that P lies on the circle through A, B, C with centre O.

7.4.2 Generalization in the Euclidean plane of the Wallace-Simson line property

See Bradley and Bradley [5]. Let ABC be a triangle and Q a general point in the plane of ABC (that is, not on a side or on a line through a vertex parallel to an opposite side). Let P be a variable point in the plane and let L, M, N be the points $BC \wedge l, CA \wedge m, AB \wedge n$, where l, m, n are the lines through P parallel to AQ, BQ, CQ respectively. Then L, M, N are collinear if, and only if, P lies on the unique conic through A, B, C with centre R, such that $\mathbf{GQ} = 2\mathbf{RG}$. See Fig. 7.2.

The conic is called the *Wallace-Simson conic* of Q and is denoted by S_Q and the line LMN is called the Wallace-Simson line of P with respect to Q. The Wallace-Simson conic S_Q is not always an ellipse, but is a hyperbola when Q lies in certain positions external to the triangle.

We prove the result of Section 7.4.2 using areal co-ordinates. Suppose that Q has co-ordinates (u, v, w) and that P has co-ordinates (X, Y, Z). We require the locus of P so that L, M, N are collinear. $\mathbf{AQ} = (u-1, v, w)$ so points on PL have co-ordinates $(X, Y, Z) + t(u-1, v, w)$ for varying t. Now $u-1 = -(v+w)$ and L lies on BC, so we choose $t = X/(v+w)$ and hence L has

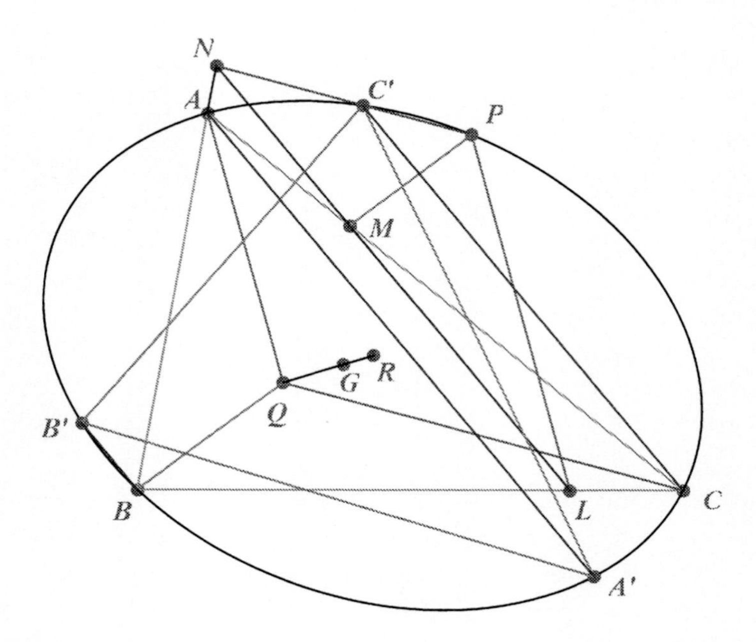

Figure 7.2: The Wallace-Simson line of P with respect to Q

co-ordinates $(0, Y+Xv/(v+w), Z+Xw/(v+w))$ or $(0, Yv+Yw+Xv, Zv+Zw+Xw)$. Similarly M has co-ordinates $(Xw+Xu+Yu, 0, Zw+Zu+Yw)$ and N has co-ordinates $(Xu+Xv+Zu, Yu+Yv+Zv, 0)$. L, M, N are collinear if, and only if, the determinant whose rows are the co-ordinates of L, M, N vanishes. $(X+Y+Z) \neq 0$ is a factor of this determinant, and it may be divided out, leaving the condition for collinearity to be

$$u(v+w)YZ + v(w+u)ZX + w(u+v)XY = 0. \tag{7.4}$$

Equation (7.4) may be interpreted to mean that for fixed Q, LMN is a straight line if, and only if, P lies on the conic through A, B, C with equation

$$u(v+w)yz + v(w+u)zx + w(u+v)xy = 0. \tag{7.5}$$

We say that P lies on the Wallace-Simson conic of Q. Since this equation may be rewritten in the form $x(y+z)vw + y(z+x)wu + z(x+y)uv = 0$, it is automatically the case that the next property holds.

7.4.3 Wallace-Simson reciprocity

If P lies on the Wallace-Simson conic of Q, then Q lies on the Wallace-Simson conic of P,

7.4.4 A midpoint on the Wallace-Simson line

The midpoint of PQ lies on the Wallace-Simson line LMN. This follows immediately from the fact that if you add together the rows of the above determinant and use $X + Y + Z = 1$ and $u + v + w = 1$ you get $(u + X, v + Y, w + Z)$, which are the unnormalised co-ordinates of the midpoint of PQ. To complete the proof of Section 7.4.2 note that the conic with Equation (7.5) has centre R with co-ordinates $\frac{1}{2}((1 - u), (1 - v), (1 - w))$. The reason for this is that the conic passes through the points D, E, F with co-ordinates

$$D(-u, (1 - v), (1 - w)), E((1 - u), -v, (1 - w)), F((1 - u), (1 - v), -w)$$

and R is the midpoint of AD, BE, CF. Also we have $\mathbf{GQ} = 2\mathbf{RG}$, since the co-ordinates of Q, R, G satisfy $2R + Q = (1, 1, 1) = 3G$.

We next establish the condition that the line with equation $px + qy + rz = 0$ should be a Wallace-Simson line of a point lying on the Wallace-Simson conic of $Q(u, v, w)$. Since L, M, N lie on the line we have $q/r = -(vZ + w(1 - Y))/(wY + v(1 - Z))$, $r/p = -(wX + u(1 - Z))/(uZ + w(1 - X))$, $p/q = -(uY + v(1 - X))/(vX + u(1 - Y))$. That is

$$qv + rw + (q - r)(wY - vZ) = 0, \tag{7.6}$$

$$rw + pu + (r - p)(uZ - wX) = 0, \tag{7.7}$$

$$pu + qv + (p - q)(vX - uY) = 0. \tag{7.8}$$

The compatibility condition for a solution for X, Y, Z to exist is

$$p(q - r)^2 vw + q(r - p)^2 wu + r(p - q)^2 uv = 0. \tag{7.9}$$

A dual interpretation may be placed on this equation.

7.4.5 The condition for a line to be a Wallace-Simson line

The fixed line with equation $px + qy + rz = 0$ is a Wallace-Simson line whenever Q lies on the conic circumscribing A, B, C with equation

$$p(q - r)^2 yz + q(r - p)^2 zx + r(p - q)^2 xy = 0. \tag{7.10}$$

Secondly if Q is a fixed point only those transversals with p, q, r satisfying Equation (7.9) are Wallace-Simson lines of some point P on the Wallace-Simson conic of Q.

The original Wallace-Simson line theorem is retrieved when Q coincides with H and then the result is that $px + qy + rz = 0$ is a Wallace-Simson line of some point on the circumcircle if, and only if, p, q, r satisfy the equation

$$p(q-r)^2 \cot A + q(r-p)^2 \cot B + r(p-q)^2 \cot C = 0. \tag{7.11}$$

Another special case is when Q coincides with G, the centroid of ABC. Then R also coincides with G and the Wallace-Simson conic of G is the outer Steiner ellipse, which passes through A, B, C and has centre G and equation $yz + zx + xy = 0$.

Exercise 7.4.2

1. Prove that the centres of the circles circumscribing the four triangles formed by four straight lines are concyclic.

2. I is the incentre of triangle ABC. Perpendiculars through I to IA, IB, IC meet any tangent to the incircle at P, Q, R. By determining the poles of AP, BQ, CR, prove that AP, BQ, CR are concurrent.

3. Show that the locus of the midpoint of PQ as P varies on the Wallace-Simson conic of Q is a conic and show that it passes through the midpoints of BC, CA, AB, AQ, BQ, CQ. Where else does this conic meet BC, CA, AB?

4. Using the same notation as in Section 7.4.2, let the line through P parallel to AQ meet BC at L, CA at $M'$and AB at N'', let the line through P parallel to BQ meet CA at M, AB at N', BC at L'', and let the line through P parallel to CQ meet AB at N, BC at L' and CA at M''. We know that L, M, N are collinear if, and only if, P lies on the Wallace-Simson conic of Q, with Equation (7.5). Prove that L', M', N' are collinear if, and only if, P lies on the *right Wallace-Simson conic* of Q, with equation $wuyz + uvzx + vwxy = 0$. Prove also that L'', M'', N'' are collinear if, and only if, P lies on the *left Wallace-Simson conic* of Q, with equation $uvyz + vwzx + wuxy = 0$.

5. Show that if P lies on the right Wallace-Simson conic of Q, then Q lies on the left Wallace-Simson conic of P.

6. Prove that points common to the right Wallace-Simson conic of Q and the left Wallace-Simson conic of Q also lie on the Wallace-Simson conic of Q.

7. Prove that the left Wallace-Simson conic of the Brocard point Ω with co-ordinates $(1/b^2, 1/c^2, 1/a^2)$ and the right Wallace-Simson conic of the Brocard point Ω' with co-ordinates $(1/c^2, 1/a^2, 1/b^2)$ both coincide with the circumcircle of ABC. (For this problem the co-ordinates given may be taken as the definition of the Brocard points. See also Chapter 10.)

8. Deduce from Problem 7 that if P lies on the circumcircle of triangle ABC and the line through P parallel to $A\Omega'$meets CA at M', the line through P parallel to $B\Omega'$ meets AB at N' and the line through P parallel to $C\Omega'$ meets BC at L', then L', M', N' are collinear.

9. If P is a general point on the circumcircle of a triangle ABC, prove there is a parabola inscribed in ABC, with focus P and the Wallace-Simson line of P as tangent at its vertex.

10. Now let P in Problem 9 be the point on the circumcircle such that the Wallace-Simson line of P is BC. Locate the point where the perpendicular to BC through P meets BC and also that this line passes through de Longchamps point (the reflection of H in O). Prove further that the line through B parallel to AC and the line through C parallel to AB meet at a point Q on this line. Prove finally that the line joining the points of contact with AB and AC of the parabola of Problem 9 also passes through Q.

11. Prove that if P is a point on the circumference of triangle ABC and L, M, N are the feet of the perpendiculars from P on to BC, CA, AB respectively and PL, PM, PN meet the circumcircle again at D, E, F respectively, then the following results hold:

    (i) AD, BE, CF and the directrix of the parabola in Problem 9 are parallel to the Wallace-Simson line LMN;

    (ii) Triangle DEF is congruent to triangle ABC;

    (iii) The line of collinearity of the midpoints of AL, BM, CN is perpendicular to the Wallace-Simson line and parallel to the axis of the parabola in Problem 9.

12. See Fig. 7.2. Is it the case that AA', BB', CC' are parallel to LMN, where A', B', C' are the intersections of PL, PM, PN with the Wallace-Simson conic Σ_Q?

Chapter 8

Three-triangle and Six-triangle configurations

8.1 Three-triangle configurations

A *three-triangle configuration* is made up as follows: start with a triangle ABC (which we assume is not isosceles) and a point K neither on the sides of ABC nor on lines through the vertices parallel to the opposite sides. Then join AK, BK, CK to form the *three* triangles BKC, CKA, AKB. Now complete the configuration by letting P, Q, R be the same triangle centre with reference to each of the three triangles.

For example, if the centroid is chosen as the triangle centre, then P, Q, R are the centroids of triangles BKC, CKA, AKB respectively. We call such a configuration the *KG configuration* (G standing for centroid). The problem is to find out as much as possible about the configuration, such questions as whether AP, BQ, CR are concurrent or not.

We consider altogether four possible triangle centres, the centroid, the circumcentre, the orthocentre and the incentre. If K coincides with a triangle centre of triangle ABC, we wish to make this clear. For example, if K coincides with the orthocentre H of triangle ABC and P, Q, R are still the centroids of their respective triangles, then we call the configuration the *HG configuration.* K, however, may be a general Cevian point.

We also use the following additional notation: O for the circumcentre of ABC, T for the nine-point centre of ABC, I for the incentre of ABC and G for the centroid of ABC. L, M, N always denote the intersections of

KP, KQ, KR with BC, CA, AB respectively. If AP, BQ, CR are concurrent then we call the point of concurrence J. When this happens the intersections of AP, BQ, CR with BC, CA, AB are denoted by X, Y, Z respectively. If AL, BM, CN are concurrent then the point of concurrence is denoted by S. The midpoints of AK, BK, CK are denoted by U, V, W respectively. The midpoints of BC, CA, AB are denoted by D, E, F respectively. All such labels are used without further introduction and are also used when K is a triangle centre of ABC. Areal co-ordinates are used throughout Part 8.1.

8.1.1 The KG configuration

See Fig. 8.1. Triangle PQR is homothetic with triangle ABC and $\mathbf{RQ} = (1/3)\mathbf{BC}$ etc.. Also S coincides with the centroid G of triangle ABC so that L, M, N coincide with D, E, F respectively. K, J, G are collinear and $KJ = 3JG$. The six points L, M, N, X, Y, Z lie on a conic.

Let $K(l, m, n)$, then $P(l/3, (1+m)/3, (1+n)/3)$ etc. and hence $\mathbf{RQ} = (0, -1/3, 1/3) = (1/3)\mathbf{BC}$. The equation of AP is $(1+n)y = (1+m)z$ etc., so AP, BQ, CR are concurrent at $J(\frac{1}{4}(1+l), \frac{1}{4}(1+m), \frac{1}{4}(1+n))$. Triangle PQR is therefore homothetic with triangle ABC and is one-third the size, the (inverse) centre of enlargement being at J. KP has equation $(m-n)x - ly + lz = 0$ and this meets BC at $L(0, \frac{1}{2}, \frac{1}{2})$, which therefore coincides with D. X, Y, Z have unnormalized co-ordinates

$$(0, 1+m, 1+n), (1+l, 0, 1+n), (1+l, 1+m, 0)$$

respectively and it may be verified that X, Y, Z, D, E, F lie on the conic with equation

$$(1+m)(1+n)x^2 + (1+n)(1+l)y^2 + (1+l)(1+m)z^2$$
$$-(1+l)(2+m+n)yz - (1+m)(2+n+l)zx - (1+n)(2+l+m)xy = 0. \quad (8.1)$$

The co-ordinates of K, J, G satisfy $J = \frac{1}{4}(K + 3G)$ so K, J, G are collinear and $KJ = 3JG$.

8.1.2 The OG configuration

The property additional to the KG configuration is that J now lies on the Euler line and is such that $OJ = 3JG$ and hence $OJ = \frac{1}{4}OH$. If we place an x-co-ordinate axis on the Euler line with O at $x = 0$ and H at $x = 1$, then G is at $x = 1/3$, T is at $x = \frac{1}{2}$ and J is at $x = \frac{1}{4}$. See Fig. 8.2.

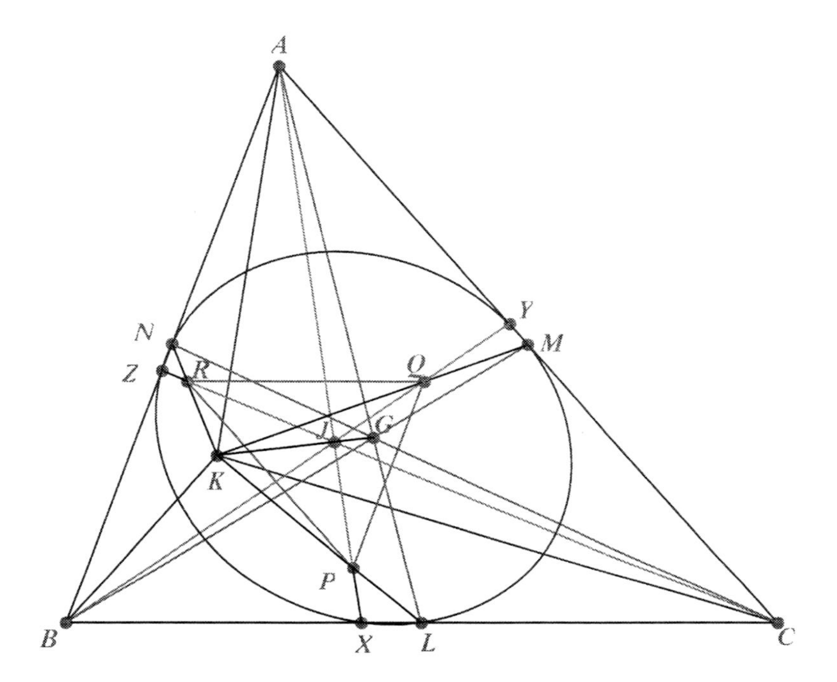

Figure 8.1: The KG configuration

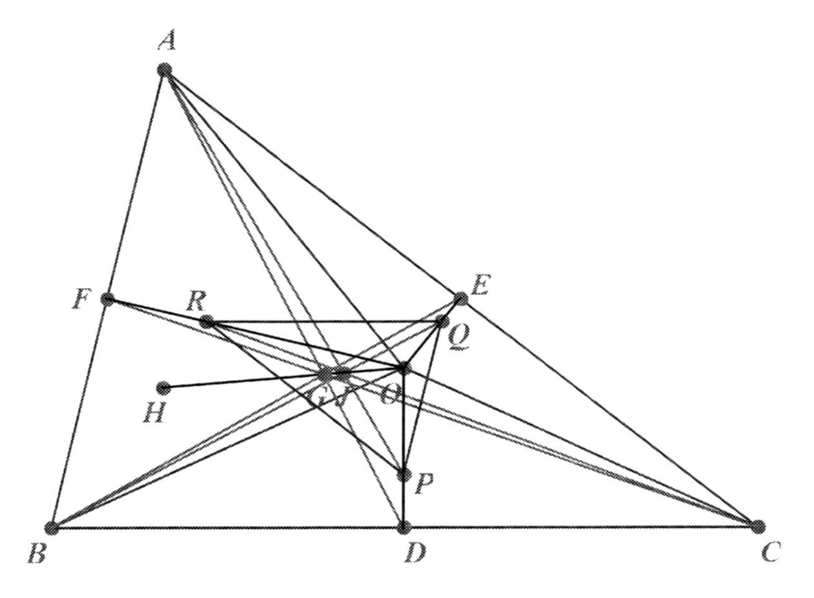

Figure 8.2: The OG configuration

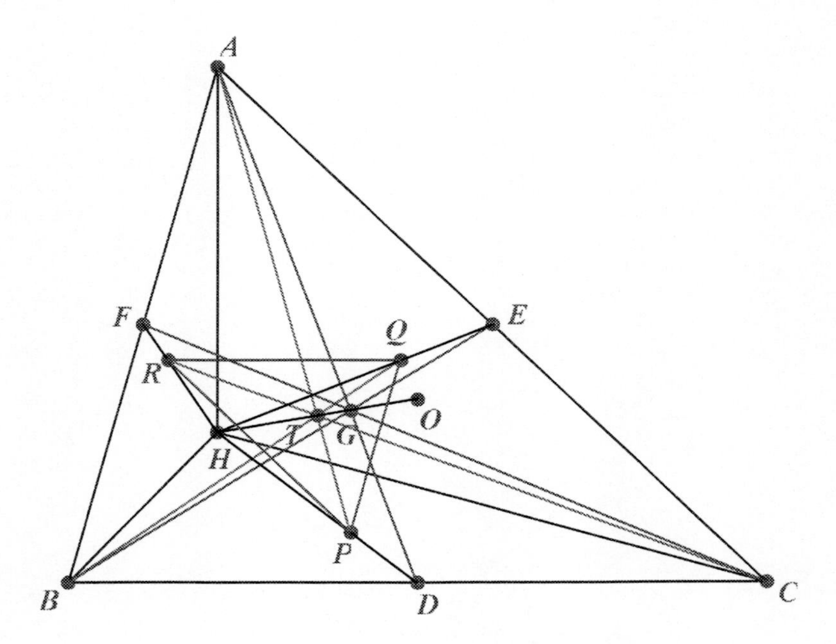

Figure 8.3: The HG configuration

8.1.3 The HG configuration

For the location of J we now have $HJ = 3JG$, so that J coincides with the nine-point centre T with co-ordinate $x = \frac{1}{2}$. Also HP, HQ, HR meet BC, CA, AB respectively at the midpoints of the sides. See Fig. 8.3.

8.1.4 The IG configuration

For the location of J we now have $IJ = 3JG$. The line IG in a triangle is analogous to the Euler line OG and contains two other well-known points. See Section 5.6.2. These are, first, the Spieker centre, Sp, whose position satisfies $IG = 2GSp$. And secondly, Nagel's point Na (labelled N in Chapter 5), which is the point of concurrence of AX_1, BY_2, CZ_3, where X_1 is the point where the excircle opposite A touches BC, with Y_2 and Z_3 similarly defined, and whose position satisfies $INa = 2ISp$.

If we place an x-co-ordinate axis on IG with I at $x = 0$ and Na at $x = 1$, then G is at $x = \frac{1}{3}$, Sp is at $x = \frac{1}{2}$ and hence J is at $x = \frac{1}{4}$. G and Na are respectively the internal and external centres of similitude of the incircles of triangles ABC and DEF, just as G and H are of their circumcircles.

8.1.5 The GG configuration

The points G, J, S now coincide, as do $X, D; Y, E$; and Z, F and the conic in this limiting case becomes the conic that touches the sides of the triangle at its midpoints. This is not drawn in Fig. 8.4, which does, however, contain eight other conics. We now give a description of these conics and proofs of their properties.

$PQRVW$ is a conic that touches AB, AC at points we denote by U_1, U_2 respectively. $PQRWU$ is a conic that touches BC, BA at points we denote by V_1, V_2 respectively, $PQRUV$ is a conic that touches CA, CB at points we denote by W_1, W_2 respectively and the six points $U_1, U_2, V_1, V_2, W_1, W_2$ lie on a conic.

$UVWQR, UVWRP, UVWPQ$ are parabolas. These parabolas meet the sides BC, CA, AB at points we denote by P_1, P_2; Q_1, Q_2; R_1, R_2 and these six points lie on a conic.

The unnormalized areal co-ordinates of P, Q, R, V, W are $(1,4,4)$, $(4,1,4)$, $(4,4,1)$, $(1,4,1)$, and $(1,1,4)$ respectively. It may be verified that these five points lie on the conic with equation

$$16x^2 + 4y^2 + 4z^2 - yz - 16zx - 16xy = 0. \tag{8.2}$$

This conic meets AB, with equation $z = 0$ where $(2x - y)^2 = 0$, the double root $y = 2x$ indicating tangency at U_1, which is evidently the point of trisection of AB nearer B. The remainder of the first part of the result follows by symmetry. The six points of trisection of the sides $U_1, U_2, V_1, V_2, W_1, W_2$ lie on the conic with equation

$$2x^2 + 2y^2 + 2z^2 - 5yz - 5zx - 5xy = 0. \tag{8.3}$$

We leave it as an exercise to show this is an ellipse with centre G.

U, V, W, Q, R have unnormalized co-ordinates $(4,1,1)$, $(1,4,1)$, $(1,1,4)$, $(4,1,4)$, $(4,4,1)$ respectively and it may be verified that these five points lie on the conic with equation

$$x^2 + 4y^2 + 4z^2 - 16yz - zx - xy = 0. \tag{8.4}$$

The reason that this conic is a parabola is that it meets the line at infinity with equation $x + y + z = 0$ where $(y - z)^2 = 0$; that is, it touches the line at infinity, and conics with this property are all parabolas. Its axis is parallel to AG since it meets the line at infinity where $y = z$. The $y \leftrightarrow z$ symmetry

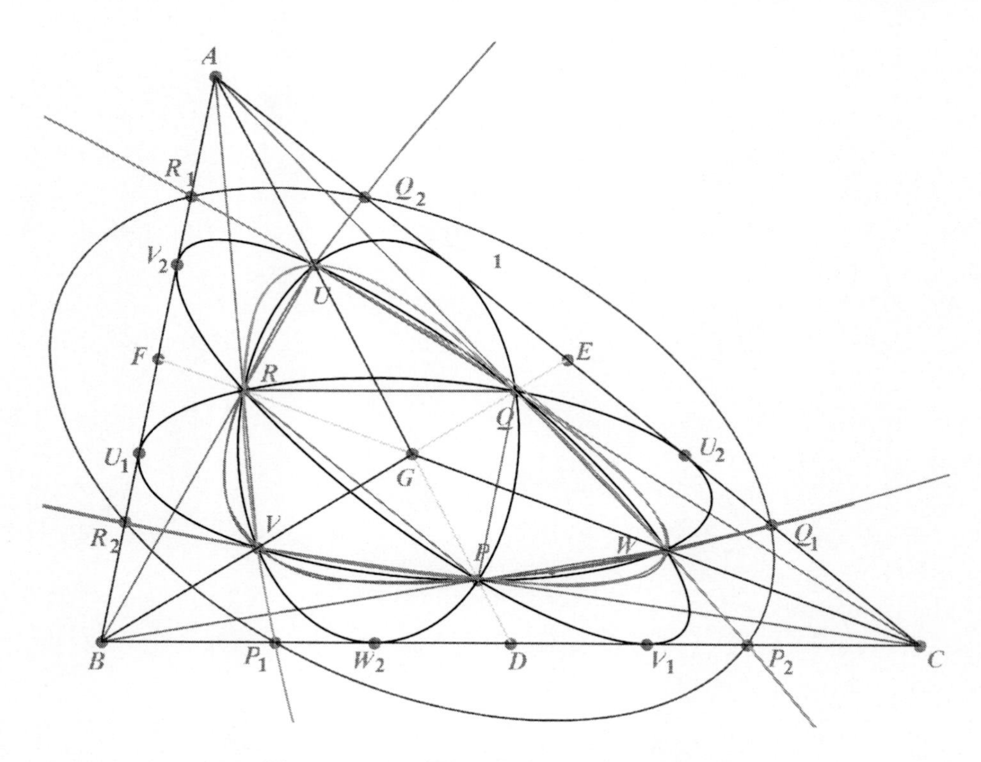

Figure 8.4: The GG configuration

of the equation of this parabola means that the tangent at U is parallel to the line BC. The parabola meets BC where $y^2 + z^2 - 4yz = 0$; that is where $y = (2 \pm \sqrt{3})z$, defining the positions of P_1 and P_2 on BC. $Q_1, Q_2; R_1, R_2$ are similarly situated on CA, AB and all six points lie on the conic with equation

$$x^2 + y^2 + z^2 - 4yz - 4zx - 4xy = 0. \tag{8.5}$$

We leave it as an exercise to show this is also an ellipse with centre G.

Exercise 8.1.1

1. In triangle ABC denote the de Longchamps point by deL. Show that in the $deLG$ configuration AP, BQ, CR are concurrent at the circumcentre O.

8.1.6 The KO configuration

This does not appear to have any interesting properties.

8.1.7 The OO configuration

The lines AP, BQ, CR are concurrent. See Fig. 8.5.

Use vectors with O as origin and $\mathbf{OA} = \mathbf{a}$ etc.. Since P lies on the perpendicular bisector of BC its position vector $\mathbf{p} = k(\mathbf{b}+\mathbf{c})$ for some constant k. The value of k is to be determined from the fact that P is the circumcentre of triangle BOC so that $OP = PB$. This means that $|k(\mathbf{b}+\mathbf{c})| = |(k-1)\mathbf{b}+k\mathbf{c}|$. After some algebra, using $\mathbf{b}.\mathbf{c} = R^2\cos 2A$, we find $k = \frac{1}{4}\sec^2 A$. We have $BX/XC = [ABP]/[ACP]$. Now

$$[ABP] = |\mathbf{BP}\times\mathbf{BA}| = |((\frac{1}{4}\sec^2 A - 1)\mathbf{b} + \frac{1}{4}\sec^2 A\mathbf{c})\times(\mathbf{a}-\mathbf{b})|$$

and again after some more algebra, using $\mathbf{a}\times\mathbf{b} = \sin 2C\mathbf{u}$, where $\mathbf{u}$ is a vector perpendicular to the plane of ABC, we find

$$[ABP] = \sin 2C + \frac{1}{4}\sec^2 A(\sin 2A + \sin 2B - \sin 2C)$$

$$= \sin 2C + \sec A\sin C\cos B = \sin C(2\cos C + \sec A\cos B).$$

Similarly $[ACP] = \sin B(2\cos B + \sec A\cos C)$. Hence

$$\frac{BX}{XC} = \frac{\sin C(2\cos C\cos A + \cos B)}{\sin B(2\cos A\cos B + \cos C)}$$

and by symmetry we have $(BX/XC)(CY/YA)(AZ/ZB) = 1$. Since X, Y, Z are internal points of the sides, it follows by the converse of Ceva's theorem that AX, BY, CZ are concurrent, which is what we wish to prove.

It is worth noting that whenever the lines AP, BQ, CR are concurrent, triangles ABC and PQR are in perspective. By Desargues's theorem, this means that $BC\wedge QR$, $CA\wedge RP$, $AB\wedge PQ$ are collinear. The six other points of intersection of the two triangles lie on a conic, as shown in Fig. 8.5. The reason for this is that the two triangles are degenerate cubics with nine points of intersection. No more than three of them can be collinear. See Silvester [31]. When the triangles are in perspective three are collinear. It follows that the other six points lie on a conic. The line and the conic together form a degenerate cubic through the nine points.

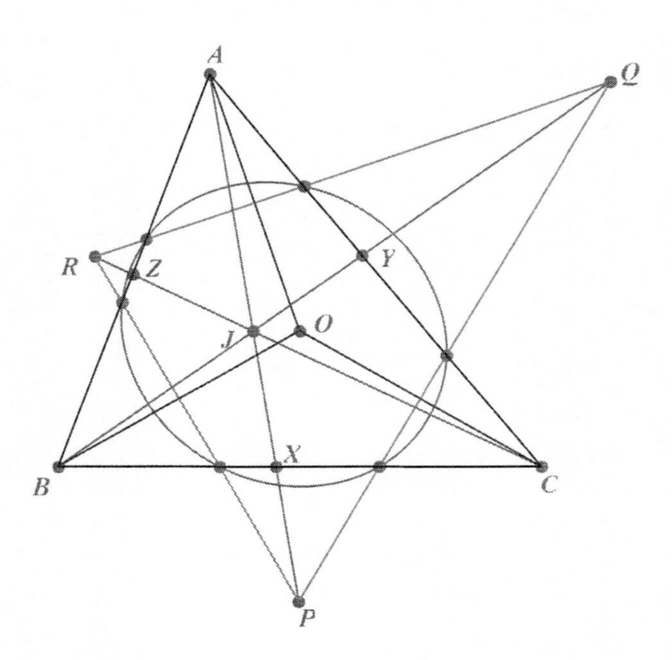

Figure 8.5: The OO configuration

8.1.8 The HO configuration

AP, BQ, CR are concurrent at the nine-point centre T. AL, BM, CN are concurrent at a point S that we identify as the isotomic conjugate of O. The triangle PQR is congruent and homothetic to triangle ABC, the (inverse) centre of similitude being T. Points A, B, C, P, Q, R lie on a conic and L, M, N, X, Y, Z lie on a conic. See Fig. 8.6. The conic that touches the sides of ABC at L, M, N is the Euler inconic with foci at H and O.

Take the circumcentre O as the origin of vectors, with $\mathbf{OA} = \mathbf{a}$ etc.. Then the position vector of H is $\mathbf{h} = \mathbf{a} + \mathbf{b} + \mathbf{c}$. Since $PH = PB = PC$ and $|\mathbf{a}| = |\mathbf{b}| = |\mathbf{c}|$ it follows that $\mathbf{OP} = \mathbf{b} + \mathbf{c}$. Similarly $\mathbf{OQ} = \mathbf{c} + \mathbf{a}$ and $\mathbf{OR} = \mathbf{a} + \mathbf{b}$. Hence $\mathbf{RQ} = \mathbf{c} - \mathbf{b} = \mathbf{BC}$. Similarly $\mathbf{PR} = \mathbf{CA}$ and $\mathbf{QP} = \mathbf{AB}$. Thus triangles ABC and PQR are congruent and homothetic. The midpoint of AP has position vector $\frac{1}{2}(\mathbf{a} + \mathbf{b} + \mathbf{c})$, which, by symmetry, is also the midpoint of BQ, CR. Hence AP, BQ, CR are concurrent at T, the nine-point centre of triangle ABC. Furthermore T is the centre of an ellipse passing through A, B, C, P, Q, R.

Denote the foot of the perpendicular from H on BC by A' and note that

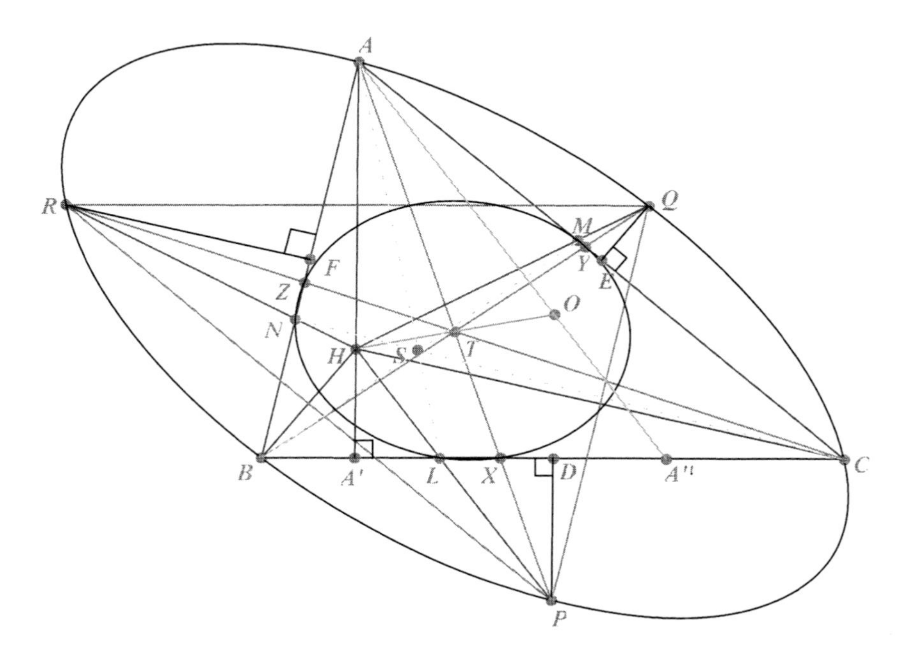

Figure 8.6: The HO configuration

the foot of the perpendicular from P on to BC is D, the midpoint of BC. Clearly triangles $HA'L$ and PDL are similar and $HA' = 2R\cos B\cos C$ and $PD = OD = R\cos A$. Now $A'D = \frac{1}{2}a - c\cos B = \frac{1}{2}(b\cos C - c\cos B) = R(\sin B\cos C - \sin C\cos B) = R\sin(B-C)$. It follows that

$$A'L = \frac{2R\cos B\cos C\sin(B-C)}{2\cos B\cos C + \cos A} = 2R\cos B\cos C\tan(B-C).$$

Thus $BL = A'L + c\cos B = 2R\cos B\cos C\tan(B-C) + 2R\sin C\cos B = R\sin 2B/\cos(B-C)$. Hence $BL/LC = \sin 2B/\sin 2C$, with similar expressions by cyclic change of A, B, C for CM/MA and AN/NB. Thus by the converse of Ceva's theorem AL, BM, CN are concurrent at a point S. Now if AO meets BC at A'' then $BA''/A''C = \sin 2C/\sin 2B$. It follows that $LD = DA''$ etc., and so S is the isotomic conjugate of the circumcentre O. Finally since AX, BY, CZ and AL, BM, CN are two sets of Cevians, the six points X, Y, Z, L, M, N lie on a conic.

8.1.9 The IO configuration

This is a configuration that is featured in most geometry books on the triangle. P, Q, R lie on the circumcircle of triangle ABC. The points J and S coincide with I. No proofs are needed here as they are given in Section 5.1.1.

8.1.10 The GO configuration

This does not appear to have any interesting properties.

8.1.11 The KH configuration

$ABCPQR$ is a rectangular hyperbola passing through K and H. See Fig. 8.7.

Any conic that passes through A, B, C, H is known to be a rectangular hyperbola. Hence the conic defined by A, B, C, H, K is a rectangular hyperbola. Now use Cartesian co-ordinates, supposing that its equation is $xy = c^2$. Let $(ct, c/t)$ be the parametric representation and suppose A, B, C, K have parameters t_1, t_2, t_3, s then it is known that H, P, Q, R have parameters $-1/t_1t_2t_3, -1/st_2t_3, -1/st_3t_1, -1/st_1t_2$ and all lie on the rectangular hyperbola. It can also be shown that $[PQR] = [ABC]$, which is a very intriguing result.

8.1.12 The OH configuration

As in the KH configuration, a rectangular hyperbola passes through A, B, C, O, H, P, Q, R. In addition, since $OA = OB = OC$ then OP meets BC a at its midpoint. Thus L, M, N coincide with D, E, F respectively and AL, BM, CN are concurrent at G. See Fig. 8.8.

8.1.13 The IH configuration

As in the KH configuration, a rectangular hyperbola passes through A, B, C, I, H, P, Q, R. In addition AL, BM, CN are concurrent at S and S also lies on the rectangular hyperbola. See Fig. 8.9.

The conic with equation

$$\sin A(\cos B - \cos C)yz + \sin B(\cos C - \cos A)zx + \sin C(\cos A - \cos B)xy = 0$$

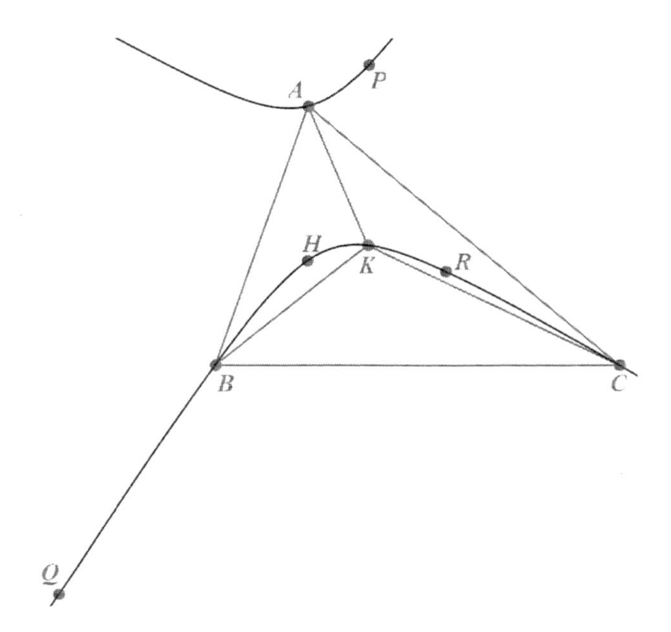

Figure 8.7: The KH configuration

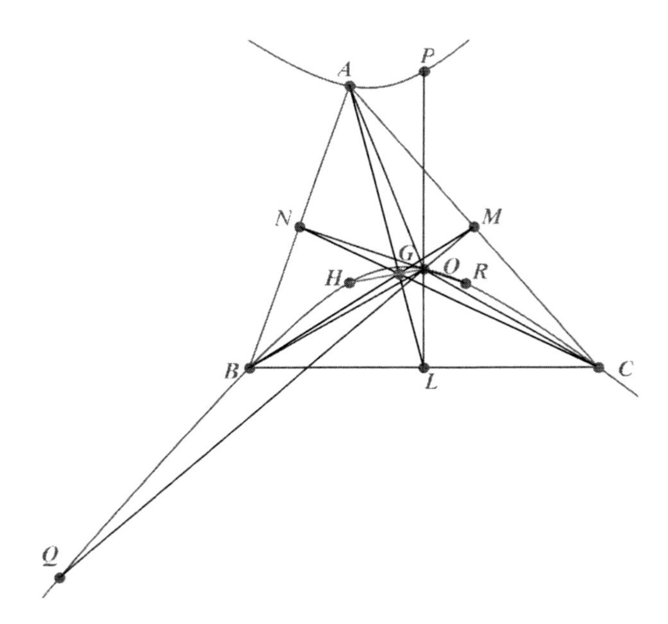

Figure 8.8: The OH configuration

clearly passes through the vertices $A(1,0,0), B(0,1,0), C(0,0,1)$ and the in-centre $I(\sin A, \sin B, \sin C)$. Since

$$\cos A(\cos B - \cos C) + \cos B(\cos C - \cos A) + \cos C(\cos A - \cos B) = 0$$

it also passes through $H(\tan A, \tan B, \tan C)$. It is therefore the equation of the rectangular hyperbola containing these five points and, from the result holding for the KH configuration, it contains P, Q, R as well.

Now the equation of the altitude through A is $y \cot B = z \cot C$. Lines parallel to this have equations of the form $kx + (k + \cot B)y + (k - \cot C)z = 0$ for some non-zero constant k. Such a line passes through $I(\sin A, \sin B, \sin C)$ when $k = (\cos C - \cos B)/(\sin A + \sin B + \sin C)$. Putting in this value of k and setting $x = 0$ we get the co-ordinates of L, which after some algebra, is $(0, \tan \frac{1}{2}B, \tan \frac{1}{2}C)$. The co-ordinates of M and N follow by cyclic change, and hence AL, BM, CN are concurrent at $S(\tan \frac{1}{2}A, \tan \frac{1}{2}B, \tan \frac{1}{2}C)$. Since $\sin A \cot \frac{1}{2}A = 2\cos^2 \frac{1}{2}A$ etc. and

$$cos^2 \frac{1}{2}A(\cos B - \cos C) + \cos^2 \frac{1}{2}B(\cos C - \cos A) + \cos^2 \frac{1}{2}C(\cos A - \cos B) = 0$$

it follows that S lies on the rectangular hyperbola.

8.1.14 The GH configuration

There are no interesting properties, apart from the usual rectangular hyperbola.

Exercise 8.1.2

1. In triangle ABC denote the de Longchamps point by deL. Prove that in the $deLH$ configuration AL, BM, CN are concurrent at a point S. See Fig. 8.10.

8.1.15 The II configuration

Whenever I is the triangle centre AL, BM, CN are concurrent. This is set as Exercise 8.1.3 Problem 3. But in the II configuration it is also the case that AP, BQ, CR are concurrent and L, M, N, X, Y, Z lie on a conic. See Fig. 8.11.

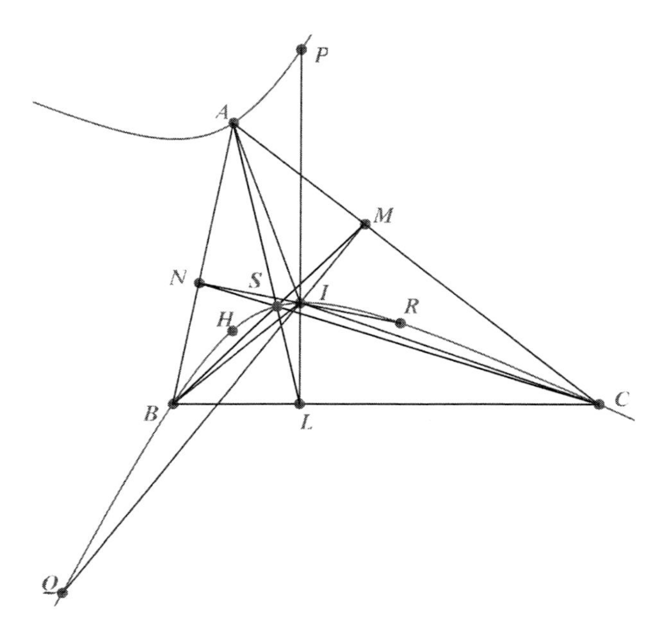

Figure 8.9: The IH configuration

The unnormalized areal co-ordinates of P are $(a^2, ab + (a+b+c)CI, ac + (a+b+c)BI)$. Now $ab + (a+b+c)CI = ab + (a+b+c)r \operatorname{cosec} \frac{1}{2}C$, where r is the inradius. But $ab = 2[ABC] \operatorname{cosec} C$ and $r(a+b+c) = 2[ABC]$, where $[ABC]$ is the area of ABC. Hence $ab + (a+b+c)CI = 2[ABC](\operatorname{cosec} C + \operatorname{cosec} \frac{1}{2}C)$. Similarly $ac + (a+b+c)BI = 2[ABC](\operatorname{cosec} B + \operatorname{cosec} \frac{1}{2}B)$.

It follows that $BX/XC = (\operatorname{cosec} B + \operatorname{cosec} \frac{1}{2}B)/(\operatorname{cosec} C + \operatorname{cosec} \frac{1}{2}C)$ with similar expressions by cyclic change for CY/YA and AZ/ZB. Thus, by the converse of Ceva's theorem, AX, BY, CZ are concurrent, which shows that AP, BQ, CR are concurrent. For the conic $LMNXYZ$ note that the points concerned are the feet of two sets of Cevians.

Exercise 8.1.3

1. Let ABC be a triangle with symmedian point S, and let P, Q, R be the symmedian points of triangle BSC, CSA, ASB respectively. Prove that AP, BQ, CR are concurrent. Suppose now that SP, SQ, SR meet BC, CA, AB at L, M, N respectively. Prove that AL, BM, CN are concurrent. See Fig. 8.12.

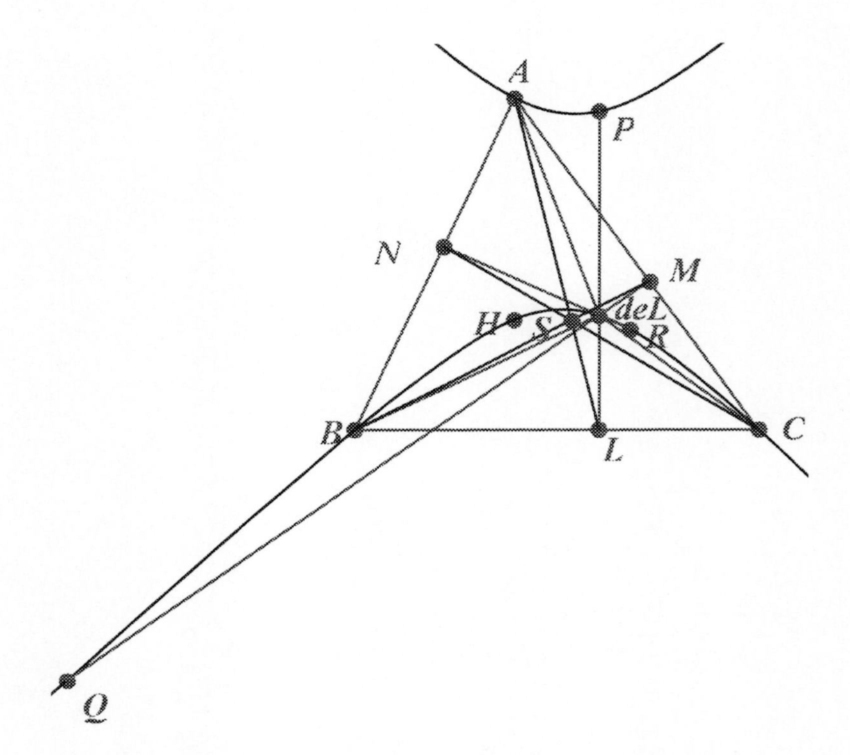

Figure 8.10: The $deLH$ configuration

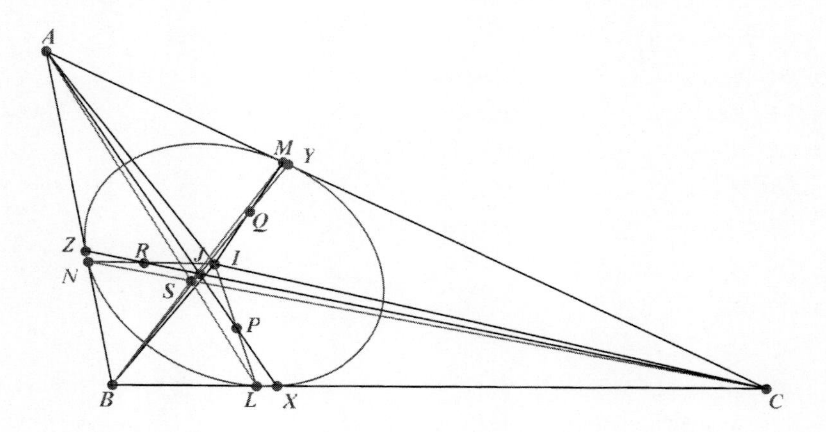

Figure 8.11: The II configuration

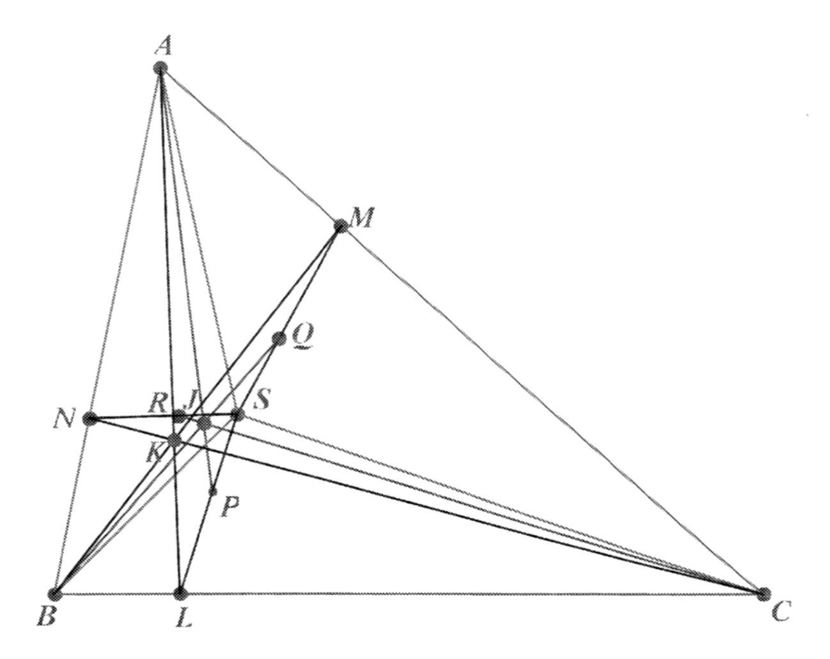

Figure 8.12: The SS configuration

2. What happens in the HH configuration?

3. Prove that in the KI configuration AL, BM, CN are concurrent. (This is true whenever I is the triangle centre.)

4. In the TT configuration, where T is the nine-point centre, let AT, BT, CT meet BC, CA, AB respectively at A', B', C'. Prove that AP, BQ, CR are concurrent. Prove also that $A'B'C'XYZ$ lie on a conic. See Fig. 8.13.

5. In triangle ABC let deL be the de Longchamps point. Prove that in the $deLdeL$ configuration the point J lies at the orthocentre of triangle ABC.

6. Let I be the incentre of triangle ABC. Prove that the Euler lines of triangle ABC, IBC, ICA, IAB are concurrent. (Schliffer's problem)

7. Repeat Problem 6 when I is replaced by O.

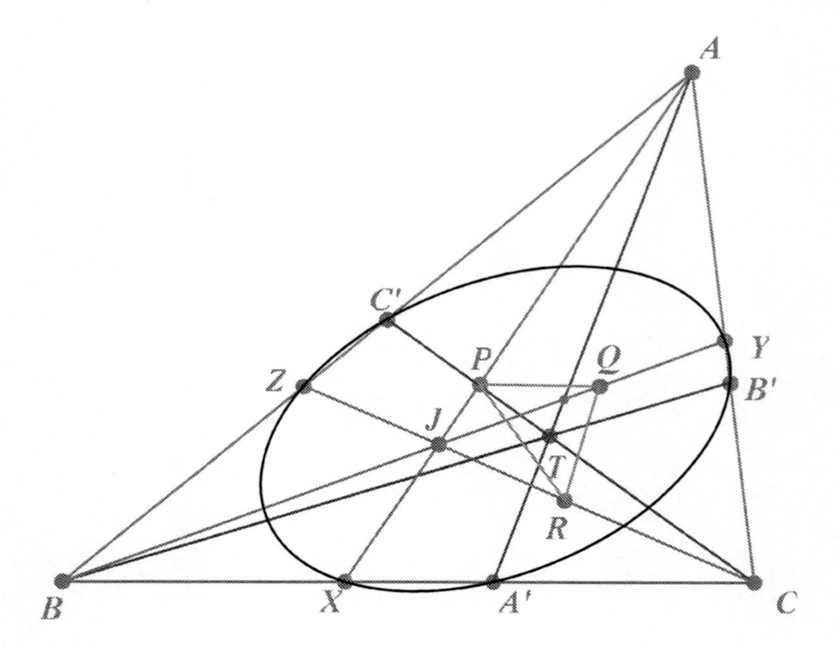

Figure 8.13: The TT configuration

8. Repeat Problem 6 when I is replaced by H and show the point of concurrence is T, the nine-point centre of ABC.

9. Prove that if P is a point such that the Euler lines of triangles PBC, PCA, PAB are concurrent then the point of concurrence lies on the Euler line of triangle ABC. (A conjecture that is CABRI indicated.)

8.2 Six-triangle configurations

A *six-triangle configuration* is formed when you take a point K, internal to triangle ABC, draw the Cevians AKP, BKQ, CKR and then the six triangles concerned are triangles $T_{12} \equiv PKB$ (1 for P, 2 for B), $T_{13} \equiv PKC$, $T_{23} \equiv QKC$, $T_{21} \equiv QKA$, $T_{31} \equiv RKA$, $T_{32} \equiv RKB$. The configuration is made complete by taking a triangle centre, such as the circumcentre, and forming the hexagon created by taking that particular key point in each of the six triangles. So, if the circumcentre is the triangle chosen, then the hexagon consists of the points O_{ij} $(i, j = 1, 2, 3, i \neq j)$, where O_{ij} is the circumcentre of triangle T_{ij}. Various properties of this figure may then be

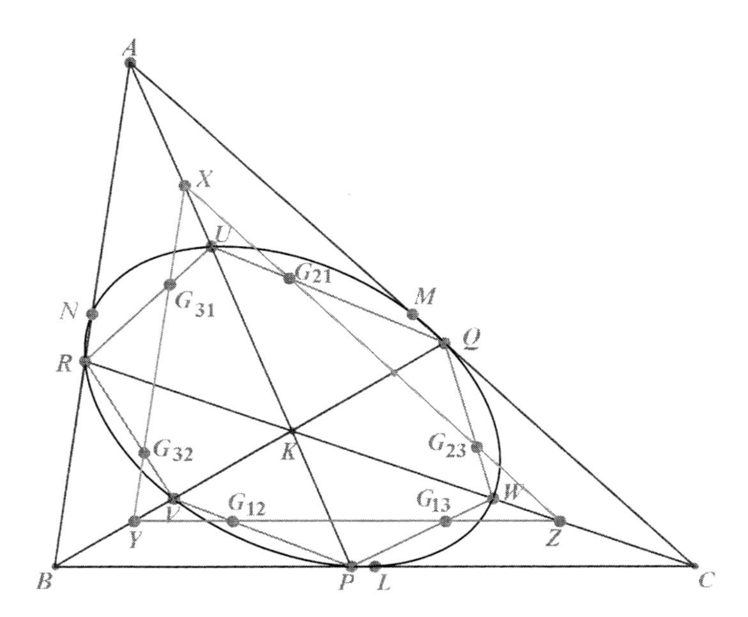

Figure 8.14: The KG configuration

investigated. If K is chosen to be the centroid and O is the triangle centre, then we call the configuration the GO configuration. Unless otherwise stated K is a general point. Areal co-ordinates are used throughout this section.

8.2.1 The KG configuration

Fig. 8.14 shows a general point K in the plane of a triangle ABC, not lying on its sides (or their extensions) and the Cevians AKP, BKQ, CKR. L, M, N are the midpoints of the sides BC, CA, AB respectively. The Cevians divide triangle ABC into six triangles $PKB, PKC, QKC, QKA, RKA, RKB$ and the centroids of these triangles are denoted by $G_{12}, G_{13}, G_{23}, G_{21}, G_{31}, G_{32}$ respectively. U, V, W are the midpoints of AK, BK, CK respectively. The six centroids form three lines $G_{12}G_{13}, G_{23}G_{21}, G_{31}G_{32}$ and these three lines form a triangle XYZ as shown in the figure.

The following results hold for this configuration:

(i) P, L, Q, M, R, N lie on a conic which also passes through U, V, W;

(ii) X, Y, Z lie on AK, BK, CK respectively;

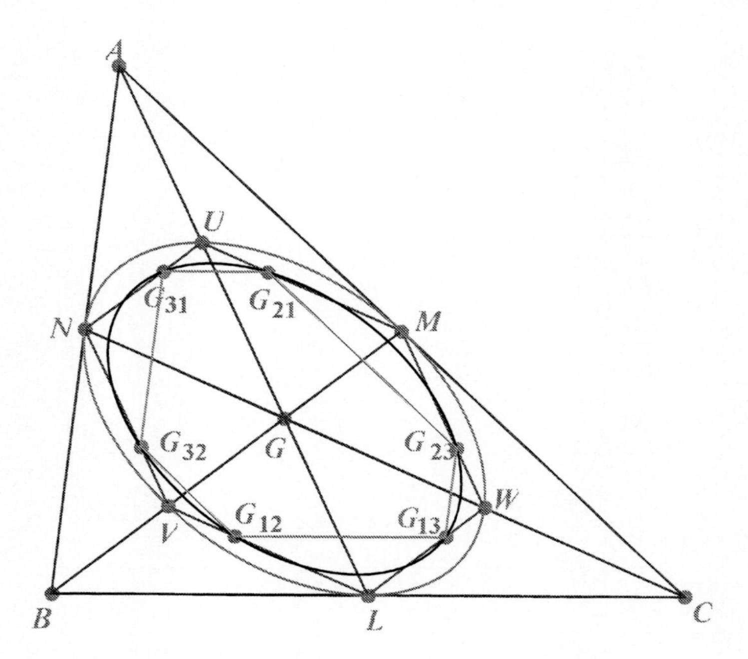

Figure 8.15: The GG configuration

(iii) $AX = (1/3)AK$ etc.;

(iv) $\mathbf{G_{12}G_{13}} = (1/3)\mathbf{BC}$ etc.;

(v) $\mathbf{YZ} = (2/3)\mathbf{BC}$;

(vi) if K coincides with G, the centroid of ABC, then the six centroids lie on a conic.

Result (vi) is illustrated in Fig. 8.15, in which $P, L; Q, M$ and R, N coincide and the conic of Result (i) is the conic touching the sides at its midpoints.

(i) Let $K(l, m, n)$, then $P(0, m, n)/(m+n)$ with similar expressions for Q and R. L, M, N are, of course the points with co-ordinates

$$L\,(0, 1/2, 1/2)\,, M\,(1/2, 0, 1/2)\,, N\,(1/2, 1/2, 0)\,.$$

U, V, W being the midpoints of AK, BK, CK respectively have co-ordinates

$$U\left(l + \frac{m+n}{2}, \frac{m}{2}, \frac{n}{2}\right), V\left(\frac{l}{2}, m + \frac{n+l}{2}, \frac{n}{2}\right), W\left(\frac{l}{2}, \frac{m}{2}, n + \frac{l+m}{2}\right).$$

It may now be verified by substitution that $U, V, W, P, Q, R, L, M, N$ lie on the conic with equation

$$\frac{x^2}{l}+\frac{y^2}{m}+\frac{z^2}{n}-yz\left(\frac{1}{m}+\frac{1}{n}\right)-zx\left(\frac{1}{n}+\frac{1}{l}\right)-xy\left(\frac{1}{l}+\frac{1}{m}\right)=0. \quad (8.6)$$

The centre of this particular conic has co-ordinates $\frac{1}{4}(1+l, 1+m, 1+n)$.

(ii)–(v) Using the fact that a centroid is the centre of mass of equal masses at each vertex of a triangle, we now deduce the co-ordinates of the centroids as

$$G_{12}(1/3)(l, m+(2m+n)/(m+n), n+n/(m+n)),$$
$$G_{13}(1/3)(l, m+m/(m+n), n+(m+2n)/(m+n))$$

and similarly by cyclic change for the other four centroids. It follows that $\mathbf{G_{12}G_{13}} = (1/3)(0,-1,1) = (1/3)\mathbf{BC}$. Since all points on YZ have x-co-ordinate equal to $(1/3)l$, the co-ordinates of X, Y, Z are now easily calculated to be $X(2/3+l/3, m/3, n/3)$, $Y(l/3, 2/3+m/3, n/3)$, $Z(l/3, m/3, 2/3+n/3)$, from which we see that X lies on AK with $AX/AK = 1/3$ and $\mathbf{YZ} = (0,-2/3,2/3) = (2/3)\mathbf{BC}$.

(vi) If K coincides with G, the centroid of ABC, then the 6 centroids have co-ordinates $(2,11,5)$, $(2,5,11)$ etc. and these all lie on the conic, centre G, with equation

$$29(x^2+y^2+z^2)-50(yz+zx+xy)=0. \quad (8.7)$$

8.2.2 The *GO* configuration

ABC is a triangle with centroid G, circumcentre O and nine-point centre T, and with L, M, N the midpoints of the sides, thus providing a configuration with six triangles LGB, LGC, MGC, MGA, NGA, NGB. The circumcentres of these triangles are denoted by $O_{12}, O_{13}, O_{23}, O_{21}, O_{31}, O_{32}$ respectively. See Fig. 8.16.

The following results hold for this configuration:

(i) The hexagon $O_{12}O_{13}O_{23}O_{21}O_{31}O_{32}$ has opposite sides parallel;

(ii) The six circumcentres lie on a circle, with centre at a point marked X in the figure;

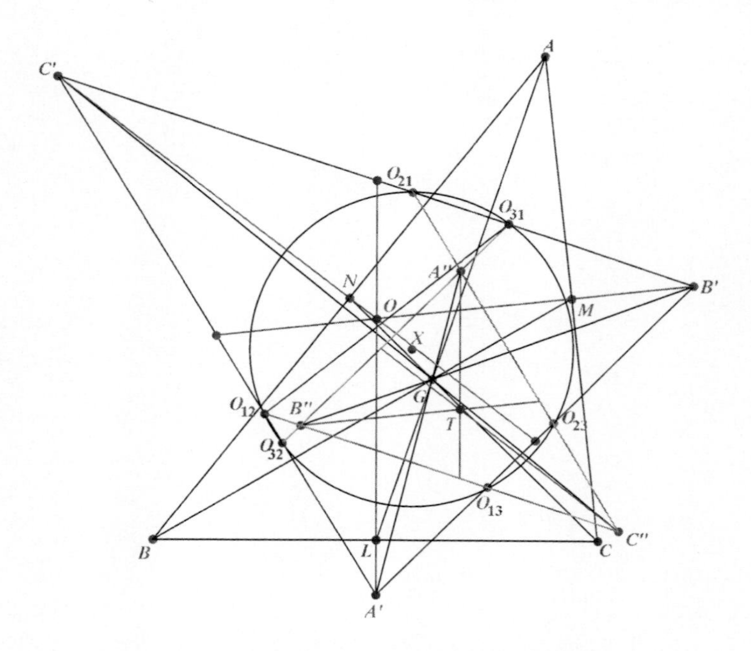

Figure 8.16: The GO configuration

(iii) $O_{21}O_{12} = O_{32}O_{23} = O_{13}O_{31}$;

(iv) Results (i) and (iii) imply that figures such as $O_{12}O_{13}O_{31}O_{21}$ form isosceles trapezia;

(v) The lines $O_{32}O_{12}, O_{13}O_{23}, O_{21}O_{31}$ form a triangle $A'B'C'$ and the perpendicular bisectors of BC, CA, AB pass through A', B', C' respectively;

(vi) The lines $O_{23}O_{21}, O_{31}O_{32}, O_{12}O_{13}$ form a triangle $A''B''C''$ and the triangles $A'B'C'$ and $A''B''C''$ are homothetic with scale factor 2, the internal centre of similitude being the point G itself;

(vii) The centroid of triangle $A'B'C'$ is the circumcentre O of triangle ABC;

(viii) The centroid of triangle $A''B''C''$ is T, the nine-point centre of triangle ABC.

The proof of some of these results is technically quite difficult and the algebra computer package DERIVE was therefore used both to provide and to verify results. We give an outline of the proof, providing details of the methods used and quoting key results, such as expressions for the co-ordinates of the circumcentres and the lengths of lines stated to be equal.

(i)-(iv) We concentrate first on the derivation of the co-ordinates of the point O_{12}. Note that the perpendicular to BC is in the same direction as the altitude through A, which is $(-a, b\cos C, c\cos B)$. Since O_{12} lies on the perpendicular to BC through $(0, \frac{3}{4}, \frac{1}{4})$, the midpoint of BL, we see its co-ordinates must be of the form $(ta, \frac{3}{4} - tb\cos C, \frac{1}{4} - tc\cos B)$. Here the constant t is to be determined from the fact that O_{12} is equidistant from B and G. DERIVE produced a value $t = a(5c^2 - a^2 - b^2)/\Gamma$, where

$$\Gamma = 6(a+b+c)(b+c-a)(c+a-b)(a+b-c) = 6[ABC]^2.$$

It is now a matter of careful bookkeeping to obtain the (unnormalised) co-ordinates of O_{12}, which in terms of a, b, c are

$$O_{12}(a^2(5c^2 - a^2 - b^2), -4a^4 + 2a^2(5b^2 + 3c^2) - 2(b^2 - c^2)(2b^2 - c^2),$$
$$-a^4 + a^2(3b^2 + c^2) - 2(b^2 - 2c^2)(b^2 - c^2)).$$

To normalize these co-ordinates it is necessary to divide by Γ.

In order to write down the co-ordinates of the point O_{13} exchange the y- and z-co-ordinates and exchange b and c. Co-ordinates of O_{23} and O_{31} follow from O_{12} by cyclic change of $1, 2, 3$; x, y, z and a, b, c. The co-ordinates of O_{21}, O_{32} follow from those of O_{13} by the same set of cyclic changes. So, for example, the z-co-ordinate of O_{32} is $c^2(5a^2 - c^2 - b^2)$.

Result (i) follows immediately from the expressions for the co-ordinates. For example

$$(b^2 + c^2 - 2a^2)\mathbf{O_{12}O_{13}} = 3a^2\mathbf{O_{21}O_{13}}, \tag{8.8}$$

with similar relations between the other two pairs of opposite sides of the hexagon.

Result (ii) follows from the fact that the point X with unnormalised x-co-ordinate

$$(1/3)(-10a^4 - 4b^4 - 4c^4 + 10b^2c^2 + 13c^2a^2 + 13a^2b^2)$$

and y- and z-co-ordinates obtained by cyclic change of a, b, c is equidistant from each of the six circumcentres. Again DERIVE was used to verify this. To normalize the co-ordinates of X, divide by Γ.

Use of the areal distance function, see Equation (2.5), shows that

$$\begin{aligned}\Gamma^2(O_{21}O_{12})^2 &= 4(a^{10}+b^{10}+c^{10}) - 14(a^8b^2+b^8a^2+c^8a^2+a^8c^2+b^8c^2+c^8a^2)\\ &+10(a^6b^4+b^6a^4+c^6a^4+a^6c^4+b^6c^4+c^6b^4) \qquad (8.9)\\ &+13(a^6b^2c^2+b^6c^2a^2+c^6a^2b^2)+18(a^4b^4c^2+b^4c^4a^2+c^4a^4b^2).\end{aligned}$$

The symmetry of this expression in a, b, c establishes Result (iii) and Result (iv) is an immediate consequence.

(v)-(viii) First note that C' lies on the perpendicular bisectors of BG and AG and is therefore equidistant from A and B. Similarly A' is equidistant from B and C, and B' is equidistant from C and A. It follows from the three isosceles triangles created that the perpendiculars from C' to AB, from A' to BC and from B' to CA pass through the midpoints N, L, M respectively and hence are concurrent at O, the circumcentre of triangle ABC. This establishes Result (v).

Bearing in mind that $GB = 2GM$ and that $C'A'$ is the perpendicular bisector of GB and $C''A''$ is the perpendicular bisector of GM, and similarly for the other sides of triangles $A'B'C'$ and $A''B''C''$, we see that these two triangles are homothetic. The centre of similitude is G and the scale factor is 2. This proves Result (vi).

It is immediate from Result (v) that O is the centroid of triangle $A'B'C'$ and from the homothety, the centroid T of triangle $A''B''C''$ must lie on OG and be such that $\mathbf{OG} = 2\mathbf{GT}$. T is therefore the nine-point centre of triangle ABC. These are Results (vii) and (viii).

8.2.3 The KO configuration

We now state without proof how these results appear to alter when G is replaced by a general internal point K. As no algebraic proof has been attempted, these results remain conjectures. See Fig. 8.17.

Conjectures

Result (i) is true. Instead of Result (ii) we expect that the six circumcentres lie on a conic. Result (iii) no longer holds, but is dependent on the position of K and on a, b, c. Result (iv) is no longer true. Result (v) still holds, and for the same reason as in the proof in Section 8.2.2. If Result (i) is true, then it follows that the sides of triangles $A'B'C'$ and $A''B''C''$ are parallel, and hence the triangles are in perspective. The vertex of perspective is, however, no longer at K, but at the point J marked in the figure. So Result (vi) is true in a restricted sense. Results (vii) and (viii) are replaced by the following. The centroids X and Y of triangles $A'B'C'$ and $A''B''C''$ lie on a line through J, but this line is not in general the Euler line of triangle ABC.

When K coincides with H, the six circumcentres now coincide in pairs at the midpoints of AH, BH, CH. These points are well known to lie on the nine-point circle of ABC. As far as I am aware no other position of K, other than G or H, produces a case in which the six circumcentres are concyclic.

There are no interesting results for six-triangle configurations when the key points are the orthocentres of the six triangles.

8.2.4 The II configuration

In Fig. 8.18 I is the incentre of triangle ABC and AP, BQ, CR are the Cevians through I. The incentres of triangles $PIB, PIC, QIC, QIA, RIA, RIB$ are $I_{12}, I_{13}, I_{23}, I_{21}, I_{31}, I_{32}$ respectively. II_{12} meets BC at L_2, with L_3, M_3, M_1, N_1, N_2 similarly defined. AIP meets the hexagon formed by the six incentres at X_1, Y_1 with X_1 nearer A than Y_1. X_2, Y_2 on BIQ and X_3, Y_3 on CIR are similarly defined.

The following result now holds: L_2, L_3, M_3, M_1, N_1, N_2 lie on an ellipse.

The following conjectures, prompted by CABRI, appear to be true: I_{12}, I_{13}, I_{23}, I_{21}, I_{31}, I_{32} lie on an ellipse and X_1, Y_3, X_2, Y_1, X_3, Y_2 lie on an ellipse.

First we catalogue some key distances. It is well known that $AI = r \operatorname{cosec} A/2$, $BI = r \operatorname{cosec} B/2$, $CI = r \operatorname{cosec} C/2$, where r is the inradius of ABC. Now, since I is the incentre, $AI = (b+c)/(a+b+c)AP$ etc. and $IP = a/(a+b+c)AP$. It follows that $IP/AI = a/(b+c)$ and $IP = (ar \operatorname{cosec} A/2)/(b+c)$. Similarly $IQ = (br \operatorname{cosec} B/2)/(c+a)$ and $IR = (cr \operatorname{cosec} C/2)/(a+b)$. Thus, using the internal bisector theorem (Example 2.2.2), we have $BL_2/L_2P = BI/IP = ((b+c) \operatorname{cosec} \frac{1}{2}B)/(a \operatorname{cosec} \frac{1}{2}A)$. Now the co-ordinates of B are $(0,1,0)$ and those of P are $(0,b,c)/(b+c)$ and hence by the Section Theorem (see Section 1.2.2), the unnormalised co-ordinates of L_2 are

$$((a \operatorname{cosec} \frac{1}{2}A(0,1,0) + b \operatorname{cosec} \frac{1}{2}B(0,1,0) + c \operatorname{cosec} \frac{1}{2}B(0,0,1))$$

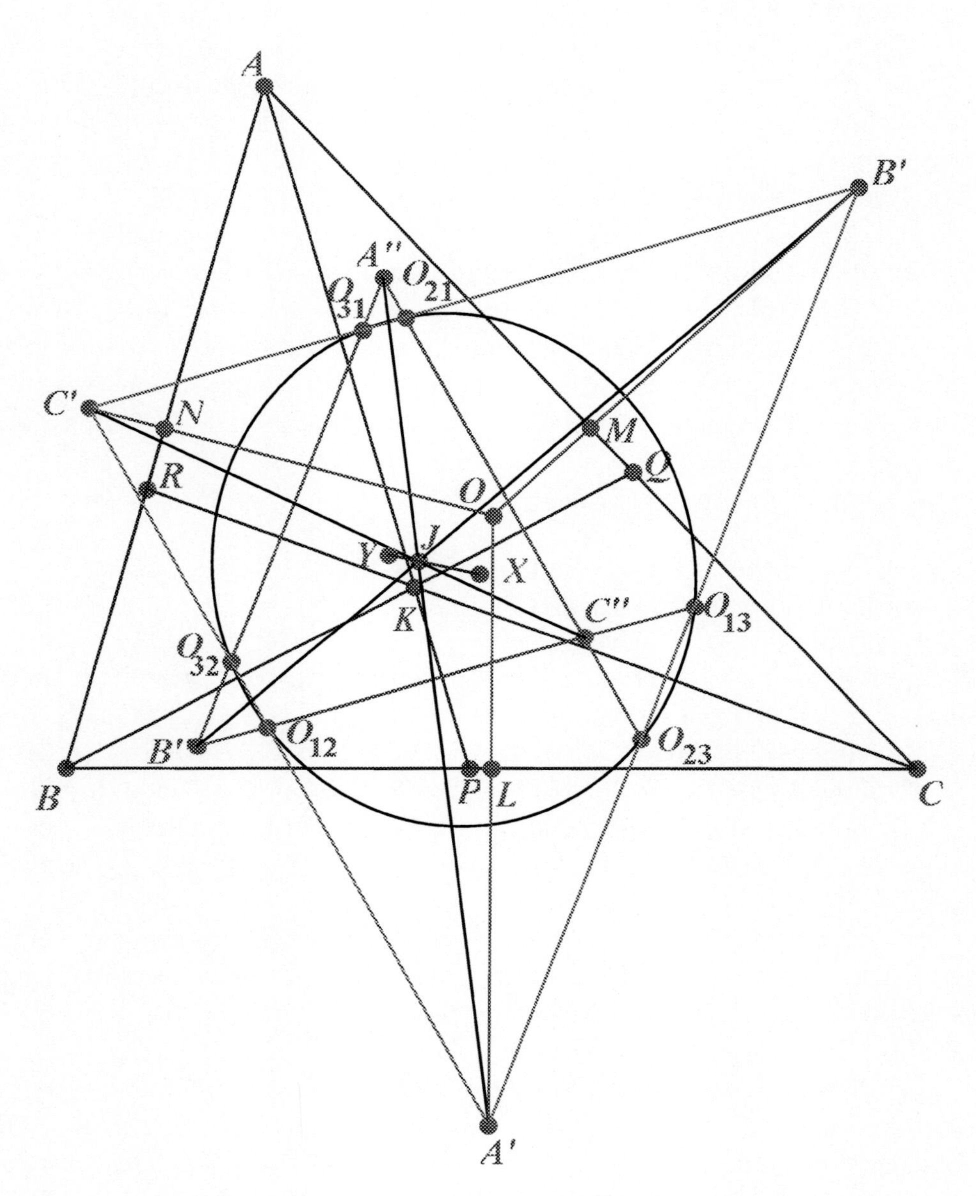

Figure 8.17: The KO configuration

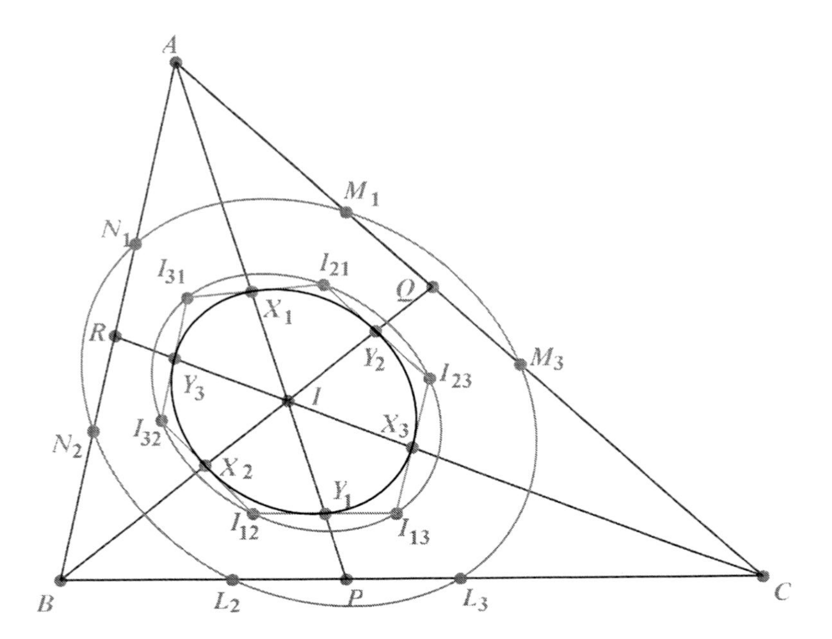

Figure 8.18: The II configuration

$$\doteq (0, a \operatorname{cosec} \frac{1}{2}A + b \operatorname{cosec} \frac{1}{2}B, c \operatorname{cosec} \frac{1}{2}B).$$

Similarly the unnormalised co-ordinates of L_3 are $(0, b \operatorname{cosec} \frac{1}{2}C, a \operatorname{cosec} \frac{1}{2}A + c \operatorname{cosec} \frac{1}{2}C)$. The co-ordinates of M_3, N_1 follow from those of L_2 by cyclic change of $1, 2, 3$; a, b, c; A, B, C and x, y, z. By a similar cyclic change, the co-ordinates of M_1, N_2 follow from those of L_3. We now have

$$\frac{BL_2}{L_2C} = \frac{c \operatorname{cosec} \frac{1}{2}B}{a \operatorname{cosec} \frac{1}{2}A + b \operatorname{cosec} \frac{1}{2}B},$$

$$\frac{BL_3}{L_3C} = \frac{a \operatorname{cosec} \frac{1}{2}A + c \operatorname{cosec} \frac{1}{2}C}{b \operatorname{cosec} \frac{1}{2}C},$$

with similar expressions by cyclic change of a, b, c and A, B, C for CM_3/M_3A, CM_1/M_1A, AN_1/N_1B, AN_2/N_2B. Hence

$$\frac{BL_2}{L_2C} \cdot \frac{BL_3}{L_3C} \cdot \frac{CM_3}{M_3A} \cdot \frac{CM_1}{M_1A} \cdot \frac{AN_1}{N_1B} \cdot \frac{AN_2}{N_2B} = +1,$$

and by the converse of Carnot's theorem, see Equation (12.8), since all the points involved lie internally on the sides of ABC we deduce that L_2, L_3, M_3, M_1, N_1, N_2 lie on an ellipse.

The remaining properties, though undoubtedly correct, as CABRI verifies them as correct to a high degree of accuracy must remain as conjectures, as they have not yet been proved algebraically.

When I is replaced by a general point K, then the main result remains true and the second conjecture appears to be true, but not the first conjecture.

Exercise 8.2.1

In Problems 1-8 the Cevians AKP, BKQ, CKR are drawn through a point K and triangle PQR *(the pedal triangle of the Cevians)* is drawn. Along with triangles AQR, BRP, CPQ this produces four triangles. Properties of these four triangles are considered.

1. Let L, M, N be the midpoints of the sides of triangle PQR. Prove that AL, BM, CN are concurrent at a point J. Show that if K is the orthocentre H, then J is the symmedian point of triangle ABC.

2. Show that if K is a general point and L, M, N are the centroids of triangles AQR, BRP, CPQ, then AL, BM, CN are concurrent.

3. Show that if K in Problem 2 coincides with the centroid G, then the conic that touches ABC at P, Q, R passes through L, M, N and that L, M, N are the midpoints of AG, BG, CG.

4. Let K coincide with G. Let L, M, N be the circumcentres of triangle AQR, BRP, CPQ respectively. Prove that AL, BM, CN are concurrent at O, the circumcentre of triangle ABC, that PL, QM, RN are concurrent at a point W and that G, W, O are collinear with $OW = 3WG$. Prove further that P, Q, R, L, M, N lie on a conic with centre W.

5. Let K coincide with G. Let L, M, N be the orthocentres of triangles AQR, BRP, CPQ respectively. Prove that AL, BM, CN are concurrent at H, the orthocentre of triangle ABC, that PL, QM, RN are concurrent at the nine-point centre T and that G, T, H are collinear with $HT = 3TG$. Prove further that P, Q, R, L, M, N lie on the nine-point circle.

6. Let K coincide with G. Let L, M, N be the incentres of triangles AQR, BRP, CPQ respectively. Prove that AL, BM, CN are concurrent at I, the incentre of triangle ABC, that PL, QM, RN are concurrent at a point W and that G, W, I are collinear with $IW = 3WG$. Prove further that P, Q, R, L, M, N lie on a conic and this conic passes through the points of intersection of AI, BI, CI with BC, CA, AB respectively.

7. Let K coincide with H. Let L, M, N be the orthocentres of triangles AQR, BRP, CPQ respectively. Prove that AL, BM, CN are concurrent at O, the circumcentre of triangle ABC. Prove also that PL, QM, RN are concurrent at a point W.

8. Let K be a general point. Let L, M, N be the orthocentres of triangles AQR, BRP, CPQ respectively. Prove that AL, BM, CN are concurrent.

   In Problems 9–13 ABC is a triangle with excentres I_1, I_2, I_3 and we consider the triangles $T_1 \equiv I_1BC, T_2 \equiv I_2CA, T_3 \equiv I_3AB$ and we consider properties of key points in these triangles.

9. Let G_1, G_2, G_3 be the centroids of triangles T_1, T_2, T_3. Prove that I_1G_1, I_2G_2, I_3G_3 are concurrent at the Mittelpunkt with areal co-ordinates

$$(a(b+c-a), b(c+a-b), c(b+c-a)).$$

10. Let O_1, O_2, O_3 be the circumcentres of triangles T_1, T_2, T_3. Prove that I_1O_1, I_2O_2, I_3O_3 are concurrent at I, the incentre of triangle ABC.

11. Let H_1, H_2, H_3 be the orthocentres of triangles T_1, T_2, T_3. Prove that I_1H_1, I_2H_2, I_3H_3 are concurrent at the circumcentre of triangle $I_1I_2I_3$.

12. Let S_1, S_2, S_3 be the symmedian points of triangles T_1, T_2, T_3. Prove that I_1S_1, I_2S_2, I_3S_3 are concurrent. (A conjecture that is CABRI indicated.)

13. Prove that a similar result to that in Problem 12 holds if the symmedian points are replaced (a) by the nine-point centres and (b) the de Longchamps points of the three triangles. (Conjectures that are CABRI indicated.)

Chapter 9

Pedal triangles and the orthocentroidal disc

9.1 Basic results on pedal triangles

If P is a point in the plane of a triangle ABC, not at a vertex, then the pedal triangle of P is the triangle LMN, where L, M, N are the feet of the perpendiculars from P on to the sides BC, CA, AB respectively. See Figs. 9.1 and 9.2. (If one just speaks of the pedal triangle, it is assumed that P coincides with H, the orthocentre, but this triangle is also called the *orthic triangle*.) If P lies on a side, then the foot of the perpendicular from P on to that side is taken to coincide with P.

9.1.1 Condition for LMN to be a pedal triangle

LMN is a pedal triangle of a point P if, and only if,

$$BL^2 + CM^2 + AN^2 = LC^2 + MA^2 + NB^2. \tag{9.1}$$

If LMN is the pedal triangle of a point P, then by Pythagoras's theorem $BL^2 - LC^2 = PB^2 - PC^2$. Similarly $CM^2 - MA^2 = PC^2 - PA^2$ and $AN^2 - NB^2 = PA^2 - PB^2$, and adding these three equations we obtain Equation (9.1).

Conversely, suppose L, M, N lie on the sides BC, CA, AB respectively and Equation (9.1) holds. Erect perpendiculars at L, M to BC, CA respectively to meet at a point P. Drop the perpendicular from P on to AB to meet it

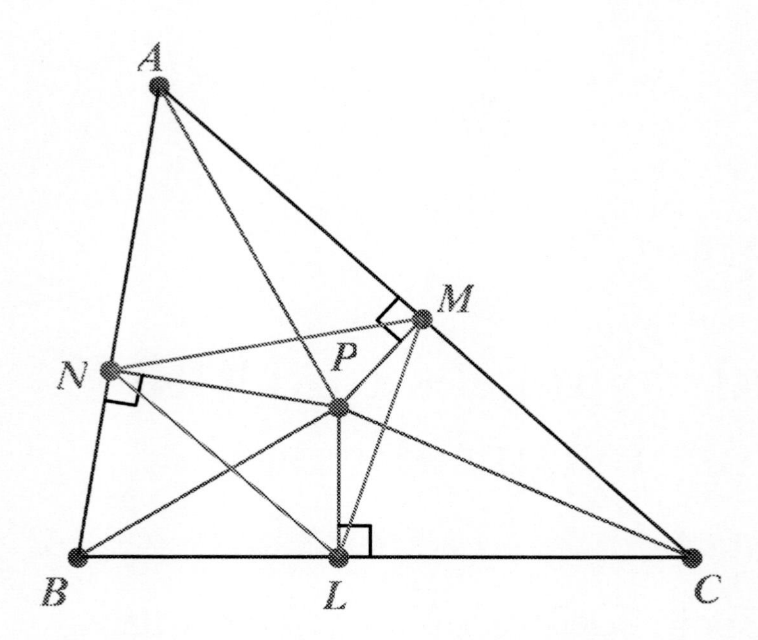

Figure 9.1: The pedal triangle of an internal point P

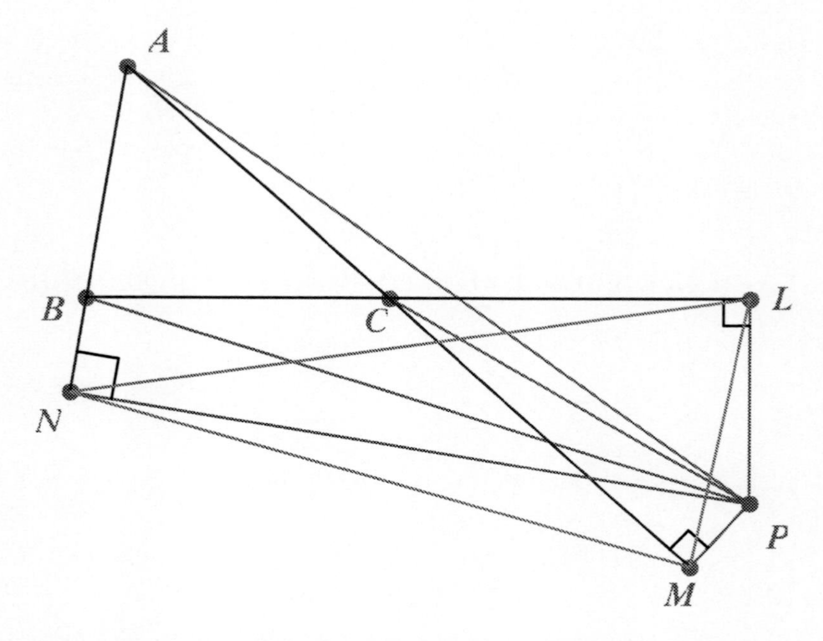

Figure 9.2: The pedal triangle of an external point P

at N'. Then LMN' is the pedal triangle of P, and so Equation (9.1) also holds with N' replacing N. From these two equations we get $AN^2 - NB^2 = AN'^2 - N'B^2$, and hence N and N' coincide.

The above proof is for when P is internal to triangle ABC. It is left as an exercise to show that the same proof holds wherever P is situated relative to ABC.

9.1.2 The angles of a pedal triangle

When P is an internal point of triangle ABC and LMN is the pedal triangle of P, $\angle NLM = \angle BPC - \angle BAC, \angle LMN = \angle CPA - \angle CBA, \angle MNL = \angle APB - \angle ACB$.

To see this, observe that

$$\angle NLM = \angle NLP + \angle MLP = \angle NBP + \angle MCP,$$

since $PNBL$ and $PMCL$ are cyclic,

$$= \angle ABC + \angle ACB - \angle PBC - \angle PCB = \angle BPC - \angle BAC.$$

Similar arguments apply in the other two cases.

Other relationships hold between the angles when P is external to triangle ABC, but they vary depending on which of the six external regions P lies. We do not investigate these cases further.

In Chapter 10 we show that, for a triangle all of whose angles are less than 120°, there is a unique point P internal to triangle ABC for which the pedal triangle is equilateral. It is called the *isodynamic point* or the *first Hessian point*.

9.1.3 Generalization of the sine rule

When P lies internal to triangle ABC, then

$$\frac{(AP)(BC)}{\sin \angle NLM} = \frac{(BP)(CA)}{\sin \angle LMN} = \frac{(CP)(AB)}{\sin \angle MNL}. \tag{9.2}$$

By the sine rule for triangle LMN, we have $NM \sin \angle NLM = 2R_0$, where R_0 is the circumradius of triangle LMN. Now $NM = 2R_A \sin \angle BAC$, where R_A is the circumradius of triangle ANM. But $AMPN$ is cyclic and

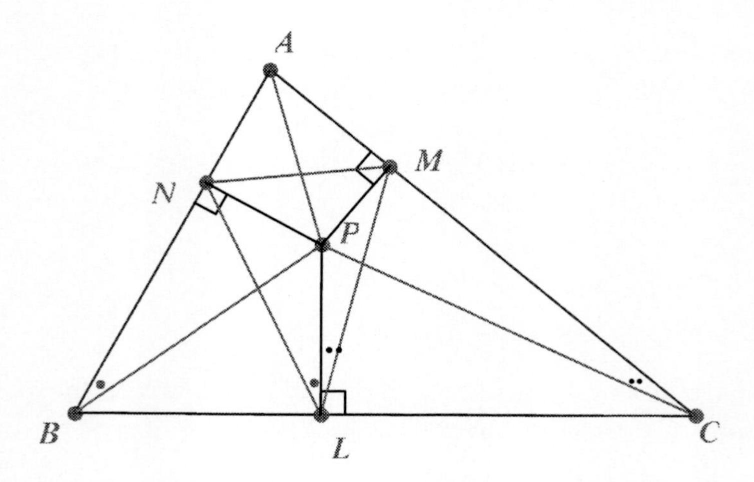

Figure 9.3: The generalized sine rule

$AP = 2R_A$, because $\angle ANP = 90°$. Hence

$$NM = AP \sin \angle BAC = \frac{(AP)(BC)}{2R},$$

where R is the circumradius of triangle ABC. Hence

$$\frac{(AP)(BC)}{\sin \angle NLM} = 4RR_0$$

and the result now follows by cyclic change of A, B, C and L, M, N. See Fig. 9.3.

When P is the circumcentre, Equation (9.2) reduces to the sine rule for triangle ABC. When P is the incentre

$$\angle BPC = 180° - \frac{1}{2}\angle ABC - \frac{1}{2}\angle ACB = 90° + \frac{1}{2}\angle BAC$$

and Equation (9.2) becomes

$$\frac{(AP)(BC)}{\cos \frac{A}{2}} = 4RR_0,$$

giving $AP \sin A/2 = R_0$. But we know that $R_0 = r$, the radius of the incircle, so we just get the known equation $AP = r \operatorname{cosec} A/2$.

We determine the areal co-ordinates of the vertices of the pedal triangle LMN, supposing that the areal co-ordinates of P are (l, m, n), and hence we determine the distances BL, LC, CM, MA, AN, NB.

Since H has co-ordinates $(\tan A, \tan B, \tan C)$ the equation of AH is $y \tan C = z \tan B$. Lines parallel to AH have equations of the form $(y \tan C - z \tan B) - k(x + y + z) = 0$. If such a line passes through $P(l, m, n)$ then $m \tan C - n \tan B = k(l + m + n)$ and hence the equation of PL is

$$(l + m + n)(y \tan C - z \tan B) = (m \tan C - n \tan B)(x + y + z) = 0.$$

This meets BC, $x = 0$, at L, where

$$y((n + l) \tan C + n \tan B) = z((l + m) \tan B + m \tan C).$$

Hence L has areal co-ordinates $(0, m + lb \cos C/a, n + lc \cos B/a)$ and so

$$BL = na + lc \cos B \text{ and } LC = ma + lb \cos C, \tag{9.3}$$

where we have used $l + m + n = 1$; this means that the formulas (9.3) are valid only when the co-ordinates of P are given in normalised form. CM, MA, AN, NB may now be obtained by cyclic change of a, b, c and l, m, n.

Exercise 9.1.1

1. Suppose that L, M, N lie on the sides BC, CA, AB of a triangle ABC. It is known that the circles AMN, BNL, CLM meet at a point P. Prove that lines through A, B, C parallel to PL, PM, PN respectively are concurrent if, and only if, LMN is the pedal triangle of P.

2. L, M, N are the feet of the perpendiculars from a point K to the sides BC, CA, AB respectively of a triangle ABC. Prove that the perpendiculars from A, B, C to MN, NL, LM respectively are concurrent. (The point of concurrence is called the isogonal conjugate of K.) *Hint: A proof using Pythagorass theorem is preferable to an analytic proof. See also Chapter 10 for an extended account of the isogonal conjugate of a point.*

3. From any point inside an equilateral triangle perpendiculars are drawn to the sides. Prove that the sum of the lengths of those perpendiculars

is constant. Is the same true if the triangle is replaced by any regular polygon? Is it true, for a circle of radius a, that the average distance of any point inside the circle from the tangents to the circle is equal to a (where the average is taken on the assumption that points are uniformly distributed around the circumference)?

4. Prove that in the notation of Section 9.1.2

$$[BPC](\cot \angle BAC - \cot \angle BPC) = [CPA](\cot \angle CBA - \cot \angle CPA)$$
$$= [APB](\cot \angle ACB - \cot \angle APB).$$

(JW)

5. Let P be an internal point of triangle ABC, and let $AP = x$, $BP = y$, $CP = z$. Show that given $f(x, y, z)$, then the stationary value of f when P varies is given by the equations

$$\partial f/\partial x : \partial f/\partial y : \partial f/\partial z = \sin \angle BPC : \sin \angle CPA : \sin \angle APB.$$

Hence find the points P when

(i) $f(x, y, z) = x + y + z$;
(ii) $f(x, y, z) = x^2 + y^2 + z^2$;
(iii) $f(x, y, z) = (u^2 + v^2)x + 2uvy + (u^2 - v^2)z$

in a triangle in which $AB = u + v$, $BC = \sqrt{u^2 + v^2}$, $CA = u\sqrt{2}$, and $u > v > 0$.

6. In the notation of Problem 5, prove that $ax + by + cz \geq 4[ABC]$.

7. ABC is an acute-angled triangle, and from an internal point P, perpendiculars PL, PM, PN are drawn to the sides BC, CA, AB respectively. Show that the position of P that maximises the area of triangle LMN is the circumcentre, and calculate the maximium value. Suppose now that the position of P is varied so that the area $[LMN]$ is constant, but less than its maximum value. Find the locus of P.

8. For a point P varying inside an acute-angled triangle ABC let u, v, w be the perpendicular distances of P from BC, CA, AB respectively. Prove that $avw + bwu + cuv$ has its maximum value $abc/4$ when P is at the circumcentre of ABC.

9. ABC is an acute-angled triangle. Establish the existence of a unique point P internal to ABC such that

$$(BC)(AP) = (CA)(BP) = (AB)(CP).$$

What is the property possessed by the corresponding pedal triangle LMN?

10. Using the notation of Part 9.1, prove that

    (i) $BL^2 + CM^2 + AN^2 \geq \frac{1}{4}(a^2 + b^2 + c^2)$;
    (ii) $[ABC]^2 \leq (BL^2 + CM^2 + AN^2)(PL^2 + PM^2 + PN^2)$.

    In each case determine when equality holds.

11. P is a point inside triangle ABC and LMN is the pedal triangle of P. Find the maximum value of $(PL)(PM)(PN)$.

12. ABC is a triangle. Locate, with proof, the point P in the plane of the triangle such that

$$(AP)\cos A/2 + (BP)\cos B/2 + (CP)\cos C/2$$

is a minimum. What is the value of the minimum?

13. Let ABC be a triangle with unit area. Prove there is a point P in the plane of ABC such that $PL^2 + PM^2 + PN^2 = 1/\sqrt{3}$, where LMN is the pedal triangle of P. For which triangles is there only one such point? (BG)

14. ABC is an equilateral triangle of side 2. Prove that there are infinitely many points P on the circumcircle so that the Wallace-Simson line LMN has the property that BL, CM, AN are all rational.

15. Let LMN be the pedal triangle of the incentre I. IL meets MN at U and V, W are similarly defined. Prove that AU, BV, CW are concurrent at the centroid G.

16. ABC is a triangle and P is a point in the plane of ABC. Points L, M, N are defined on BC, CA, AB respectively such that LP, MP, NP are perpendicular to AP, BP, CP respectively. Prove that L, M, N are collinear.

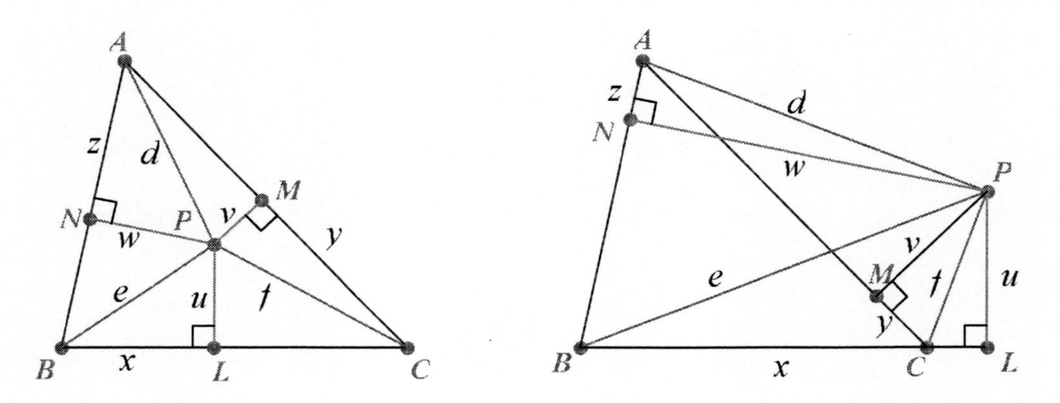

Figure 9.4: Configurations for Part 9.2

9.2 Further results

Let $BL = x, CM = y, AN = z, PL = u, PM = v, PN = w, AP = d, BP = e, CP = f$ as shown in the Fig. 9.4. We work out $x, y, z, u, v, w, d, e, f$ in terms of $l, m, n, a, b, c, [ABC]$, where l, m, n are the normalised areal coordinates of P. Note that in the left picture of Fig. 9.4 $l, m, n > 0$, but in the right picture $l > 0, m < 0, n > 0$. The results that follow are independent of the position of P relative to ABC.

First we have already obtained x, y, z in Equation (9.3). The results may be expressed as

$$
\begin{aligned}
x &= na + \frac{l(c^2 + a^2 - b^2)}{2a}, \\
y &= lb + \frac{m(a^2 + b^2 - c^2)}{2b}, \\
z &= mc + \frac{(b^2 + c^2 - a^2)}{2c}.
\end{aligned} \tag{9.4}
$$

Clearly $ua = 2[PBC]$ and $[PBC]/[ABC] = l$, so

$$
u = \frac{2l[ABC]}{a}, v = \frac{2m[ABC]}{b}, w = \frac{2n[ABC]}{c}. \tag{9.5}
$$

For d, e, f we use the areal distance function, see Equation (2.5), and obtain

$$
d^2 = AP^2 = c^2m^2 + b^2n^2 + (b^2 + c^2 - a^2)mn,
$$

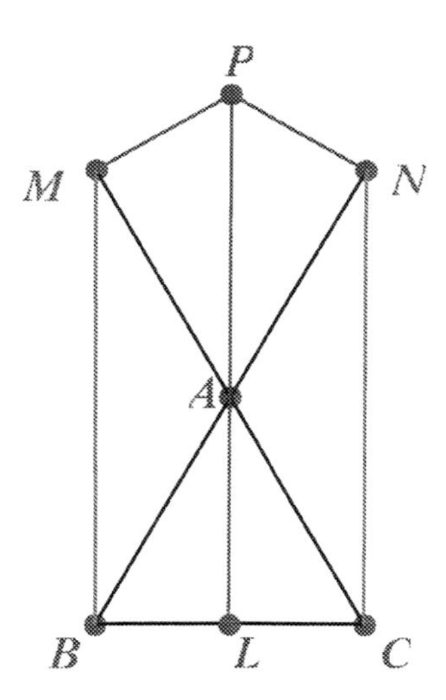

Figure 9.5: Example 9.2.2

$$e^2 = BP^2 = a^2n^2 + c^2l^2 + (c^2 + a^2 - b^2)nl, \tag{9.6}$$
$$f^2 = CP^2 = b^2l^2 + a^2m^2 + (a^2 + b^2 - c^2)lm.$$

Example 9.2.1

(Viviani's theorem) Show that if ABC is an equilateral triangle of side $2k$, then $x + y + z = 3k$ and $u + v + w = k\sqrt{3}$.

From Equations (9.4) we have $x = 2nk + lk$. Similarly $y = 2lk + mk$, $z = 2mk + nk$ and adding and using $l + m + n = 1$, we get $x + y + z = 3k$. From Equations (9.5) we have $u = lk\sqrt{3}$, $v = mk\sqrt{3}$, $w = nk\sqrt{3}$ and adding we get $u + v + w = k\sqrt{3}$.

Example 9.2.2

Show that if ABC is equilateral, then AL, BM, CN are concurrent or parallel if, and only if, P lies on one of the medians or their extensions.

Again let the side be $2k$. By the converse of Ceva's theorem AL, BM, CN are concurrent or parallel if, and only if, $xyz = (2k-x)(2k-y)(2k-z)$. This results in the equation $(2n+l)(2l+m)(2m+n) = (2m+l)(2n+m)(2l+n)$, which reduces to $(m-n)(n-l)(l-m) = 0$. Now $m = n$ if, and only if, P lies on the median through A etc.. Incidentally the parallel case can occur, for example when $l = 7/3$, $m = n = -2/3$. See Fig. 9.5.

Example 9.2.3

ABC is a triangle, with no two sides equal and with incentre I and circumcentre O. LMN is the pedal triangle of P. Prove that $BL + CM + AN = LC + MA + NB$ (where lengths are signed) if, and only if, P lies on OI.

From Equation (9.4) the given equation becomes

$$na+\frac{l(c^2+a^2-b^2)}{2a}+lb+\frac{m(a^2+b^2-c^2)}{2b}+mc+\frac{n(b^2+c^2-a^2)}{2c} = \frac{(a+b+c)(l+m+n)}{2}. \tag{9.7}$$

This is a linear equation in l, m, n so the locus of P is a straight line and, after some algebra, the equation of the locus reduces to

$$xbc(b-c)(b+c-a)+yca(c-a)(c+a-b)+zab(a-b)(a+b-c) = 0. \tag{9.8}$$

With $x : y : z = a : b : c$ Equation (9.8) becomes an identity, so I lies on the locus. From geometrical considerations O obviously lies on the locus, since L, M, N are then the midpoints of the sides. Thus the locus of P is the line OI.

Example 9.2.4

Prove that the vertices of a pedal triangle of a point P are also the feet of a set of Cevians if, and only if, P lies on a certain cubic curve passing through $A, B, C, H, O, I, I_1, I_2, I_3$.

By Ceva's theorem, the necessary and sufficient condition for this is

$$(BL)(CM)(AN) = (LC)(MA)(NB).$$

Using the expressions in Part 9.2 this reduces, after some algebra, to the condition that P should lie on the cubic curve with equation

$$\begin{aligned} &x^2ybc^2(\cos B - \cos C\cos A) + y^2zca^2(\cos C - \cos A\cos B) \\ &+z^2xab^2(\cos A - \cos B\cos C) + x^2zb^2c(\cos A\cos B - \cos C) \\ &+y^2xc^2a(\cos B\cos C - \cos A) + z^2ya^2b(\cos C\cos A - \cos B) = 0. \end{aligned} \tag{9.9}$$

It can be shown that in addition to the nine points stated this cubic curve also contains the points X, Y, Z where the lines $AdeL, BdeL, CdeL$ meet BC, CA, AB respectively, where deL is deLongchamps point. This cubic is called the Darboux cubic, see

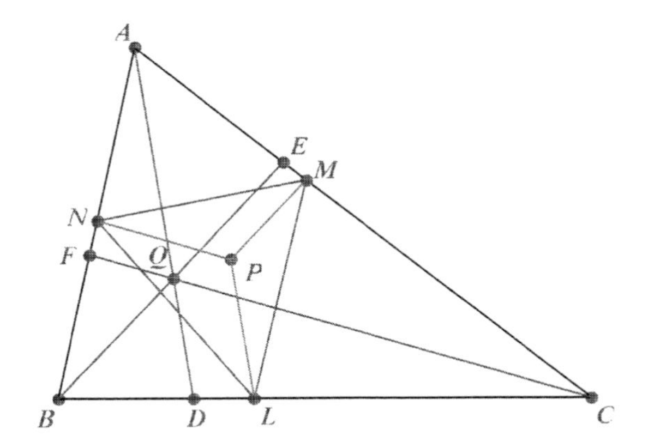

Figure 9.6: Configuration for Section 9.2.1

http://perso.wanadoo.fr/bernard.gibert/index.html

(a website that contains details of many cubic curves associated with a triangle, a topic that we do not cover in this book).

Exercise 9.2.5

1. Let LMN be the pedal triangle of the centroid G. Prove that $BL + CM + AN = (a+b+c)/2$ if, and only if, ABC is isosceles. Prove also that AL, BM, CN are concurrent if, and only if, ABC is isosceles.

2. Using the notation of Section 9.2 prove that, if P is an internal point of ABC, (i) $ad < be + cf$; (ii) $ad \geq bv + cw$; (iii) $ud + ve + wf \geq 2(vw + wu + uv)$; (iv) $def \geq 8uvw$; (v) $d + e + f \geq 2(u + v + w)$. (Erdos, Mordell)

3. Prove that the vertices of a Cevian triangle of a point P are also the feet of a pedal triangle if, and only if, P lies on a cubic curve passing through A, B, C, G, H, the Nagel and ex-Nagel points, the Gergonne and ex-Gergonne points. (The Lucas cubic)

9.2.1 A generalisation of the pedal triangle

The pedal triangle of a point P may be thought of as being the triangle LMN constructed by drawing lines through P parallel to the altitudes AD, BE, CF

to meet BC, CA, AB respectively in L, M, N. This process may be generalised by taking AD, BE, CF to be any three Cevians. See Fig. 9.6. In the figure AQD, BQE, CQF are a set of Cevians and P is any point. Lines through P parallel to AQ, BQ, CQ meet BC, CA, AB respectively at L, M, N. We call triangle LMN the pedal triangle of P with respect to Q. The pedal triangles of Parts 9.1 and 9.2 may then be thought of as the pedal triangles of P with respect to H.

Suppose that Q has areal co-ordinates (u, v, w), then as in the generalization of the Wallace-Simson line property, we find the co-ordinates of L to be $L(0, mv + mw + lv, nv + nw + lw)$, where (l, m, n) are the normalised areal co-ordinates of P. It follows that

$$BL = \frac{(lw + n(v + w))a}{v + w}, \quad LC = \frac{(lv + m(v + w))a}{v + w},$$

where we have used $l + m + n = 1$. These may be simplified as

$$\frac{BL}{a} = n + \frac{lw}{v + w}, \frac{LC}{a} = m + \frac{lv}{v + w}, \tag{9.10}$$

with similar expressions, by cyclic change, for $CM/b, MA/b, AN/c, NB/c$.

Exercise 9.2.6

1. Using the notation of Part 9.3, prove that AL, BM, CN are Cevians if, and only if, P lies on the cubic curve with equation

$$\begin{aligned} &vwzx(vz + wx + wz - vx) + wuxy(wx + uy + ux - wy) \\ &+uvyz(uy + vz + vy - uz) = uvzx(uz + ux + vx - vz) \qquad (9.11) \\ &vwxy(vx + vy + wy - wx) + wuyz(wy + wz + uz - uy). \end{aligned}$$

   (Call this cubic C_Q, the cubic of Q. Then the symmetry of Equation (9.11) under the transformation $u \leftrightarrow x, v \leftrightarrow y, w \leftrightarrow z$ means that, if P lies on C_Q, then Q lies on C_P.)

2. ABC is a triangle that is not equilateral. P is a point in the plane of triangle ABC. AD, BE, CF are the three medians. The lines through P parallel to AD, BE, CF meet the sides BC, CA, AB respectively in the points L, M, N. Prove that the locus of P such that $BL + CM + AN = LC + MA + NB$, where the usual sign convention is used for directed line segments, is the line GI.

3. ABC is a triangle that is not equilateral. I is the incentre of triangle ABC. Lines through I parallel to the medians of ABC meet the sides in L, M, N respectively. Points X, Y, Z are the feet of the perpendiculars from I on to BC, CA, AB respectively. Prove that $LX+MY+NZ = 0$, where in this equation the lengths are signed, so that, for example, LX is positive if the order of the points on BC is B, L, X, C and LX is negative if the order of the points is B, X, L, C.

4. ABC is a triangle that is not equilateral. P is a point in the plane of ABC. AD, BE, CF are the medians. Lines through P parallel to AD, BE, CF meet the sides BC, CA, AB respectively at L, M, N. G is the centroid and S is the symmedian point of ABC. Prove that

   (i) $BL^2 + CM^2 + AN^2 = LC^2 + MA^2 + NB^2$ if, and only if, P lies on GS.

   (ii) Prove that whatever the position of P

   $$\frac{BL^2}{a^2} + \frac{CM^2}{b^2} + \frac{AN^2}{c^2} = \frac{LC^2}{a^2} + \frac{MA^2}{b^2} + \frac{NB^2}{c^2}. \tag{9.12}$$

   (iii) Show that, if Equation (9.12) holds, then L, M, N are such that lines drawn through L, M, N parallel to AD, BE, CF respectively are concurrent.

5. Let ABC be a triangle and P and Q internal points. Draw QL, QM, QN parallel to AP, BP, CP to meet BC, CA, AB respectively in L, M, N. Let D, E, F be the midpoints of sides BC, CA, AB, and X, Y, Z the points on the sides such that D, E, F are the midpoints of LX, MY, NZ respectively. Prove that lines through X, Y, Z parallel to AP, BP, CP respectively are concurrent.

9.3 The *GH* disc

Given a triangle ABC the interior of the orthocentroidal circle on diameter GH is called the GH disc. In this section we give details of some results for what is called the scaled GH disc. By this we mean that given a non-equilateral triangle, it is enlarged so that the GH disc always has the same linear dimensions. The actual value of the length GH does not matter, but we treat it as fixed. Obviously when G and H coincide, which is the case only for an equilateral triangle, there cannot be a GH disc. It is sometimes helpful to think of the equilateral triangle as being a limiting case in which its side is infinite. Obviously, once GH is fixed in length, so are all segments on the Euler line, and in particular the circumcentre O is such that $GH = 2OG$. We denote the centre of the GH disc by J, and the nine-point centre by T, these being the key points on the Euler line within the disc. Along the Euler line we may give O, G, T, J, H the co-ordinates $0, 1/3, 1/2, 2/3, 1$ respectively.

In a classic paper Euler [18] showed that given the positions of O, G and the incentre I, it is possible to reconstruct the lengths of the sides of the corresponding triangle. Recently Smith [32] has shown how to do this more efficiently, with recourse to additional known formulas, such as the length of IT. Guinand [20] has shown that the incentre I always lies within the GH disc, punctured at T (which is the limiting position of I as the triangle becomes equilateral). Furthermore he proved by calculation that the incentre can lie anywhere within the GH disc (except T). Smith [32] gives a short conceptual proof of this fact. Guinand also showed that the excentres lie outside the GH disc. Várilly [35] simplified the work of Guinand and extended the work to cover the Fermat points (see Chapter 10 for their definition), one of which always lies within the GH disc and one always outside.

More recently still Bradley and Smith [6] have extended work on the GH disc to cover the symmedian point S and Gergonne's point Ge, both of which lie within the GH disc, and in particular, they have shown that the positions of O, G and S determine the side lengths of the triangle. Finally, they show how to obtain the joint loci of the Brocard points [7], relative to the scaled GH disc. In this section we review the work on the symmedian point S and prove that it can range freely over the GH disc, punctured at its midpoint J and show how a, b, c may be recovered from the positions of O, G and S.

We shall use the condition that X lies in the GH disc if, and only if, $\angle GXH > \pi/2$ and consequently that X lies in the orthocentroidal disc if, and only if,

$$\mathbf{XG}.\mathbf{XH} \leq 0 \tag{9.13}$$

with equality if, and only if, X is on the boundary.

If we use Cartesian co-ordinates with $\mathbf{OA} = \mathbf{x}$, $\mathbf{OB} = \mathbf{y}$, $\mathbf{OC} = \mathbf{z}$ and

$|\mathbf{x}| = |\mathbf{y}| = |\mathbf{z}| = 1$, then, from the cosine rule we have

$$\mathbf{y}.\mathbf{z} = \cos 2A = \frac{a^4 + b^4 + c^4 - 2a^2(b^2 + c^2)}{2b^2c^2}. \tag{9.14}$$

If X has unnormalised areals (u, v, w), so that

$$\mathbf{OX} = \frac{u\mathbf{x} + v\mathbf{y} + w\mathbf{z}}{u + v + w},$$

then in the Cartesian frame

$$3\mathbf{XG} = \frac{1}{u + v + w}((v + w - 2u), (w + u - 2v), (u + v - 2w)) \tag{9.15}$$

and

$$\mathbf{XH} = \frac{1}{u + v + w}(v + w, w + u, u + v) \tag{9.16}$$

If we now use Equation (9.13), we find (after some algebra) that the condition for $X(u, v, w)$ to lie in the GH disc is

$$(b^2 + c^2 - a^2)u^2 + (c^2 + a^2 - b^2)v^2 + (a^2 + b^2 - c^2)w^2 - a^2vw - b^2wu - c^2uv < 0. \tag{9.17}$$

and the equation of the orthocentroidal circle is

$$S_{GH} \equiv (b^2 + c^2 - a^2)x^2 + (c^2 + a^2 - b^2)y^2 + (a^2 + b^2 - c^2)z^2 - a^2yz - b^2zx - c^2xy = 0. \tag{9.18}$$

9.3.1 Important coaxal circles

The orthocentroidal circle, the polar circle, the circumcircle and the nine-point circle are coaxal The equations of these circles are given by Equations (9.18)–(9.21) respectively.

$$S_P \equiv (b^2 + c^2 - a^2)x^2 + (c^2 + a^2 - b^2)y^2 + (a^2 + b^2 - c^2)z^2 = 0, \tag{9.19}$$

$$S_C \equiv a^2yz + b^2zx + c^2xy = 0, \tag{9.20}$$

$$S_N \equiv (b^2 + c^2 - a^2)x^2 + (c^2 + a^2 - b^2)y^2 + (a^2 + b^2 - c^2)z^2 - 2a^2yz - 2b^2zx - 2c^2xy = 0. \tag{9.21}$$

Evidently $S_{GH} - S_C = S_P$ and $S_N + 2S_C = S_P$, showing the four circles are coaxal.

9.3.2 The symmedian point S lies inside the GH disc

Substituting $u = a^2$, $v = b^2$, $w = c^2$ in Equation (9.17) proves this, since it results in a standard geometrical inequality on the sides a, b, c of a non-equilateral triangle.

9.3.3 The Brocard points and the GH disc

One Brocard point lies in the GH disc and the other lies outside the GH disc, unless the triangle is isosceles, when both lie on the boundary. The Brocard points have unnormalised co-ordinates (a^2b^2, b^2c^2, c^2a^2) and (c^2a^2, a^2b^2, b^2c^2) with the same normalization factor $1/(a^2b^2 + b^2c^2 + c^2a^2)$. It is left as an exercise now to show that the sum of the powers of these two points with respect to S_{GH} is zero, and that both points lie on the boundary of S_{GH} if, and only if, $a = b$, $b = c$ or $c = a$.

We next show how to determine a cubic polynomial which has roots a^2, b^2, c^2, given the positions of O, G and S. This is done by expressing formulas for OS^2, GS^2 and JS^2 in terms of $u = a^2+b^2+c^2$, $v^2 = b^2c^2+c^2a^2+a^2b^2$ and $w^3 = a^2b^2c^2$. We provide only key equations, bearing in mind the details are worked out elsewhere [6].

First, however, recall some familiar results:

9.3.4 $16[ABC]^2 = 4v^2 - u^2$

$16[ABC]^2 = 4v^2 - u^2$ This is Heron's formula in terms of u and v.

9.3.5 $R^2 = w^3/(4v^2 - u^2)$

This comes from squaring the formula $R = abc/(4[ABC])$.

9.3.6 $OG^2 = w^3/(4v^2 - u^2) - u/9$

This comes from the known equation $OG^2 = R^2 - (a^2 + b^2 + c^2)/9$. The following formulas may be obtained using the areal distance function.

$$OS^2 = \frac{4w^3(u^2 - 3v^2)}{u^2(4v^2 - u^2)} \tag{9.22}$$

$$GS^2 = \frac{6uv^2 - u^3 - 27w^3}{9u^2} \tag{9.23}$$

Note the interesting geometrical inequality this provides.

$$JS^2 = OG^2\left(1 - \frac{48[ABC]^2}{(a^2+b^2+c^2)^2}\right) \tag{9.24}$$

$$= \frac{4(u^3+9w^3-4uv^2)(u^2-3v^2)}{9u^2(4v^2-u^2)}$$

The attractive formula (9.24) leads to the famous inequality $a^2+b^2+c^2 \geq 4\sqrt{3}[ABC]$, which is now seen to be equivalent to the fact that S lies in the GH disc.

If we now put $u = p$, $4v^2 - u^2 = q$ and $w^3 = r$, we may now establish the following results:

$$OG^2 = r/q - p/9, \tag{9.25}$$

$$OS^2 = r(1/q - 3/p^2), \tag{9.26}$$

$$GS^2 = p/18 + q/(6p) - 3r/p^2, \tag{9.27}$$

$$OS^2/JS^2 = 1 - (pq)/(9r). \tag{9.28}$$

A final substitution is now made by putting $p = x, q/p = y, r/q = z$ and $r/p^2 = s$, all of these being homogeneous of degree 1 in a^2, b^2, c^2 and from Equations (9.25) – (9.28) we have $OG^2 = z - x/9$, $OS^2 = z - 3s$, $GS^2 = x + 6y - 3s$ and $OS^2/JS^2 = 1 - x/(9z)$. Now u, v and w are known unambiguously and hence the equations determine a^2, b^2 and c^2 and therefore a, b and c.

Exercise 9.3.1

1. Prove that Gergonne's point lies in the GH disc.

2. Prove that Nagel's point lies outside the GH disc.

3. Prove that Feuerbach's point (the point where the incircle and the nine-points circle touch) lies outside the GH disc.

Chapter 10

Isogonal and Isotomic conjugates

10.1 Properties of isogonal conjugate points

Let P have normalised areal co-ordinates $P(l, m, n)$, where $l, m, n \neq 0$, so that P does not lie on a side (nor the extension of a side) of triangle ABC. We define the isogonal conjugate of P as the point Q with normalised areal co-ordinates $Q(a^2/l, b^2/m, c^2/n)$. From this definition it is evident that the isogonal conjugate of Q is the point P, so the relationship is an involution on the set (l, m, n) with $l, m, n \neq 0$. Note that it is possible for the isogonal conjugate of a point P to lie on the line at infinity. In fact the locus of such points is the circumcircle, see Fig. 10.2. This is because the sum of the co-ordinates of Q vanishes when $P(l, m, n)$ lies on the circumcircle, with equation $a^2/x + b^2/y + c^2/z = 0$.

Let LMN be the pedal triangle of P, with L, M, N on BC, CA, AB respectively and let DEF be the pedal triangle of Q with D, E, F on BC, CA, AB respectively. See Fig. 10.1.

The main properties of pairs of isogonal points, which we now prove, are as follows:

10.1.1 A common pedal circumcircle

Provided that P does not lie on the circumcircle, L, M, N, D, E, F lie on a circle, centre T, the midpoint of PQ. Since P has normalised co-ordinates

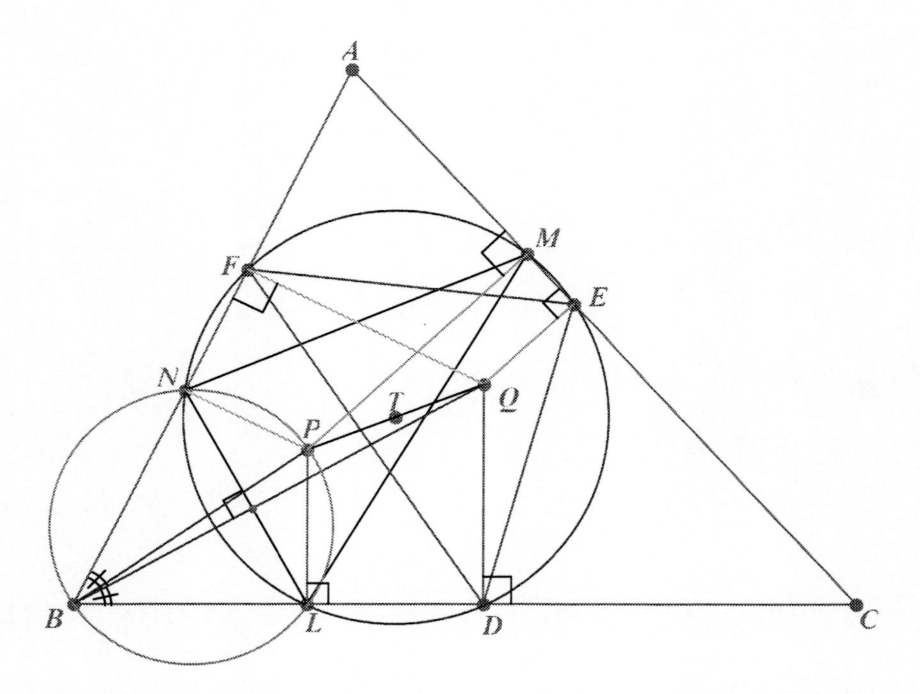

Figure 10.1: Isogonal conjugate points and the common pedal circle

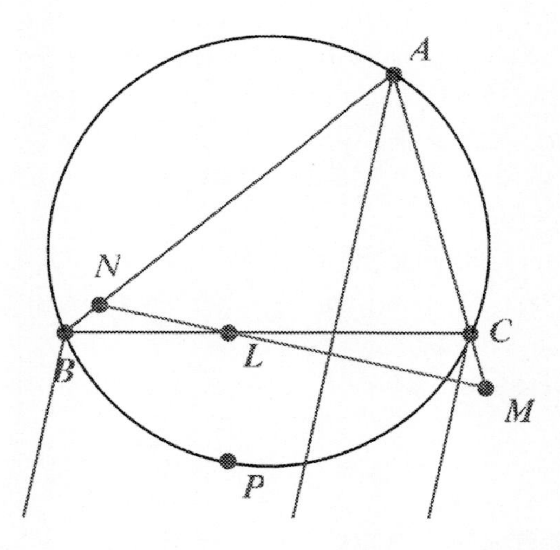

Figure 10.2: The isogonal conjugate of a point on the circumcircle

(l, m, n) the following distances are known. See Equations (9.3). $BL = an + cl\cos B$, $LC = am + bl\cos C$, $CM = bl + am\cos C$, $MA = bn + cm\cos A$, $AN = cm + bn\cos A$, $NB = cl + an\cos B$. Replacing l by a^2/lk, where k is a normalization factor for the co-ordinates of Q etc., we deduce that $BD = ac^2/nk + (ca^2/lk)\cos B$ and $FB = ca^2/lk + (ac^2/nk)\cos B$. It follows that

$$(BL)(BD) = (BN)(BF) = (ac/nlk)(cl\cos B + an)(an\cos B + cl) \quad (10.1)$$

and hence, by the converse of the intersecting chord theorem the points N, L, F, D lie on a circle. Now $LPQD$ and $NPQF$ are right-angled trapezia, so the centre of this circle is T, the midpoint of PQ. Similarly T is the centre of a circle through L, M, D, E. As these circles have the same centre and the same radius, it follows that all six points L, M, N, D, E, F lie on a circle centre T.

10.1.2 Isogonal conjugation and angles

$\angle BAP = \angle CAQ, \angle ABP = \angle CBQ$ (shown in the Fig. 10.1), $\angle BCP = \angle ACQ$, that is the line segments from A to P and Q make equal angles with the sides through A, and similarly for the vertices B, C. We prove, as representative of the angle relationships in this section, that $\angle NBP = \angle DBQ$, for which it is sufficient to prove that $PN/QD = BN/BD$. Now it is known from Equation (9.3) that $PL = (-l, (bl/a)\cos C, (cl/a)\cos B)$. Using the areal distance function given by Equation (2.5), we find

$$PL^2 = -bcl^2\cos B\cos C + (b^2l^2c/a)\cos B + (c^2l^2b/a)\cos C$$

$$= (bcl^2/a)(b\cos B + c\cos C - (b\cos C + c\cos B)\cos B\cos C)$$

$$= (bcl^2/a)(b\cos B\sin^2 C + c\cos C\sin^2 B) = bcl^2\sin B\sin C.$$

It follows by cyclic change that $PN^2 = abn^2\sin A\sin B$ and similarly $QD^2 = (bca^4/l^2k^2)\sin B\sin C$. Using the sine rule we get $PN/QD = knl/ac$. Now $NB = cl + an\cos B$ and $BD = ac^2/nk + (ca^2/kl)\cos B$ and hence $NB/BD = knl/ac = PN/QD$, as required. Similar analysis, by cyclic change of a, b, c and l, m, n, produces the other angle relationships of Section 10.1.2.

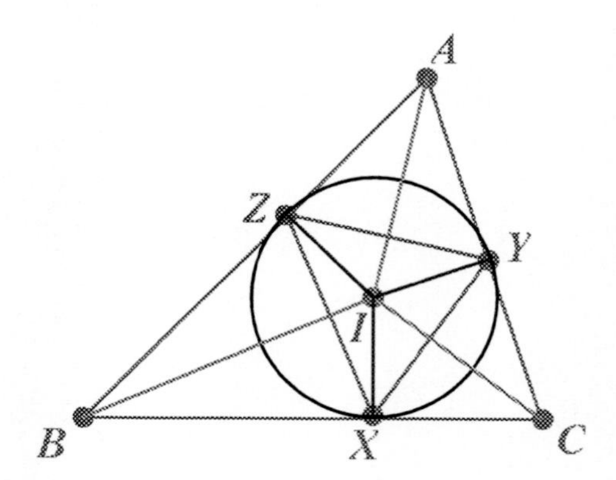

Figure 10.3: The incentre is self-conjugate

10.1.3 Perpendiculars to sides of pedal triangles

The Cevian lines AQ, BQ, CQ are perpendicular to the sides MN, NL, LM of the pedal triangle of P and similarly AP, BP, CP are perpendicular to the sides EF, FD, DE of the pedal triangle of Q. This result follows as an immediate consequence of the results in Section 10.1.2. In fact, since $NPLB$ is a cyclic quadrilateral, with a pair of right angles at N and L, we have $\angle NLB = \angle NPB = 90° - \angle NBP = 90° - \angle LBQ$ (from Section 10.1.2) and hence BQ is at right angles to NL. Similarly AQ is at right angles to MN and CQ is at right angles to LM. And because of the involution the lines AP, BP, CP are perpendicular to the sides of triangle DEF.

Sections 10.1.1 and 10.1.3 provide the means of constructing, with straight edge and compasses only, the isogonal conjugate of a given point P. The first step is to construct the pedal triangle LMN of P. Then, if using the result of Section 10.1.1, construct the circle LMN to meet the sides again at D, E, F. Then erect perpendiculars to the sides from D, E, F and these lines are concurrent at Q. If using the result in Section 10.1.3, the perpendiculars from the vertices to the sides of LMN are concurrent at Q. These constructions do, of course, rely on the truth of the converses of the stated properties, but these are not difficult to establish and are left as exercises.

10.2 Pairs of isogonal conjugate points

The involution has four fixed points, which are when P and Q coincide with (i) the incentre I, or (ii) any of the excentres. If X, Y, Z are the feet of the perpendiculars from I on to the sides, then it is obvious from the isosceles triangles in Fig. 10.3 that AI is perpendicular to YZ and bisects $\angle A$ etc.. The circle, centre I, which passes through X, X, Y, Y, Z, Z touches the sides at X, Y, Z and is the incircle. The results in Sections 10.1.1-10.1.3 in this case are trivial. Similar considerations apply to the excentres.

10.2.1 O and H are isogonal conjugates

The isogonal conjugate of $O(\sin 2A, \sin 2B, \sin 2C)$ is $H(\tan A, \tan B, \tan C)$. This is because

$$\frac{a^2}{\sin 2A} = \frac{R^2 \sin^2 A}{\sin 2A} = \frac{R^2}{2} \tan A.$$

The circle through D, E, F, L, M, N (the feet of the altitudes and the mid-points) is the nine-point circle, the centre T being the nine-point centre, which is the midpoint of OH. The result is Section 10.1.2 is equivalent to the set of well-known angle relationships $\angle HAI = \angle OAI$ and so on. The result in Section 10.1.3, that AO, BO, CO are at right angles to the sides of the pedal triangle (of the altitudes), is less well known, but is not difficult to prove independently using angle relations.

10.2.2 G and S are isogonal conjugates

The isogonal conjugate of $G(1,1,1)$ is $S(a^2, b^2, c^2)$, known as the symmedian (or Lemoine) point. The symmedian point has a number of interesting properties, some of which we have already met and some of which we now review.

The configuration of Fig. 10.4 is often used as the definition of the symmedian point. It shows the circumcircle of ABC and the tangents at the vertices. The tangents at B and C meet at X, with Y and Z defined similarly. The lines AX, BY, CZ are concurrent at the symmedian point S.

The equation of the circumcircle is $a^2yz+b^2zx+c^2xy = 0$, so the equations of the tangents at B and C are $a^2z + c^2x = 0$ and $a^2y + b^2x = 0$. Their intersection is $X(-a^2, b^2, c^2)$ and the equation of AX is therefore $b^2z = c^2y$.

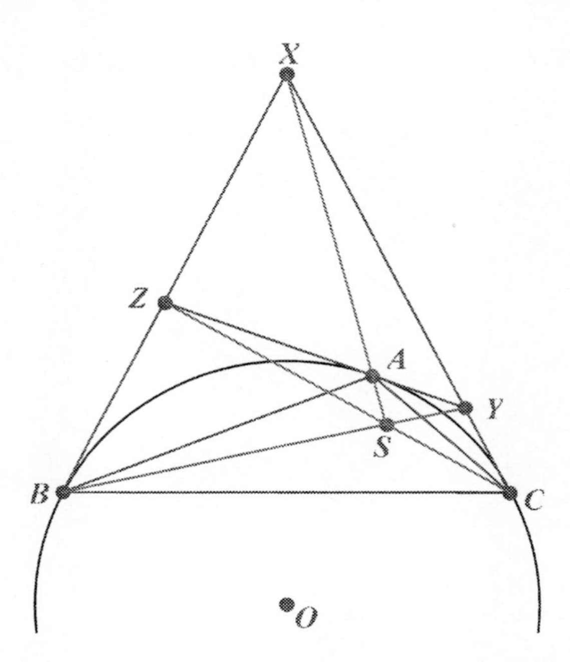

Figure 10.4: One definition of the symmedian point

This line contains the point $S(a^2, b^2, c^2)$, which by symmetry lies on BY and CZ.

This construction requires amplification if one of the angles of ABC is a right angle. For example if $\angle BAC = 90°$, then $BZYC$ is a right-angled trapezium with BZ parallel to CY. The lines BY and CZ meet at S, the point X moves to infinity and AS is parallel to BZ and CY, all three lines may be regarded as passing through X at infinity.

The result in Section 10.1.2 is sometimes used as a definition. What it implies is that AS is the reflection of AG in the internal bisector AI, and as AG is called a median, so AS is called a *symmedian* and S is the point where the three symmedians meet.

Fig. 10.5 illustrates Sections 10.1.2 and 10.2.2 for the symmedian point. It shows the symmedians as the reflection of the medians in the internal bisectors and also the circle through the feet of the pedal triangles of G and S. It also illustrates, in general, that symmedians are no longer than the corresponding medians.

Fig. 10.6 shows another property of the symmedian point. The points D, E, F are the feet of the altitudes and X, Y, Z are the midpoints of EF, FD,

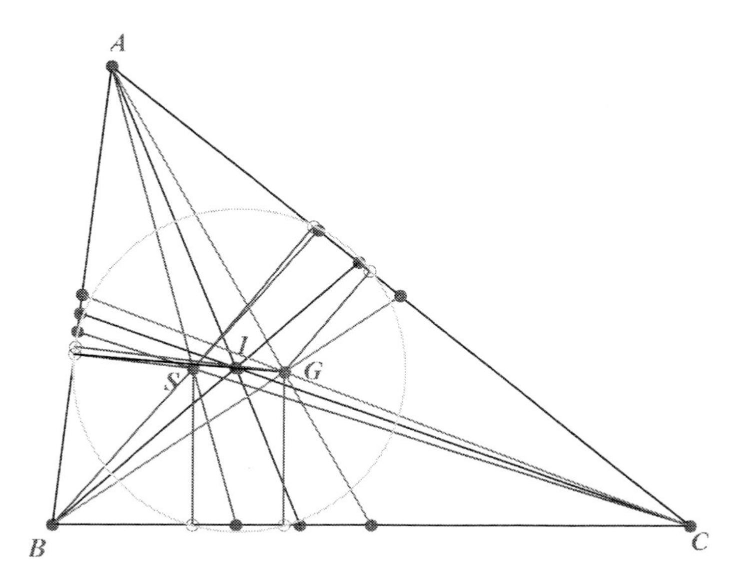

Figure 10.5: The relationship between medians and symmedians

DE respectively. Then AX, BY, CZ meet at S. As proof of this, note that the normalised co-ordinates of E and F are $E(\sin A\cos C, 0, \sin C\cos A)/\sin B$ and $F(\sin A\cos B, \sin B\cos A, 0)/\sin C$. Thus X has y- and z-co-ordinates proportional to $\sin B/\sin C$ and $\sin C/\sin B$. The equation of AX is therefore $y\sin^2 C = z\sin^2 B$, which passes through $S(a^2, b^2, c^2)$. By symmetry this point also lies on BY and CZ.

There is another property of the symmedian point, which is quite interesting. If P is a point internal to a triangle ABC and u, v, w are the perpendicular distances from P on to the sides BC, CA, AB respectively, then it is a reasonable question to ask for the position of P that minimises the expression $u^2 + v^2 + w^2$.

By the Cauchy-Schwarz inequality we have

$$4[ABC]^2 = (au + bv + cw)^2 \leq (a^2 + b^2 + c^2)(u^2 + v^2 + w^2),$$

with equality if, and only if, $u/a = v/b = w/c$ and hence $u^2 + v^2 + w^2$ is minimized when P is the symmedian point and the minimum value is value $4[ABC]^2/(a^2 + b^2 + c^2)$. Thus we have a *variational characterization* of the symmedian point as the point which minimizes a natural function.

Further properties of the symmedian point are given later in this section, when its links with the Brocard points are discussed. It turns out to have

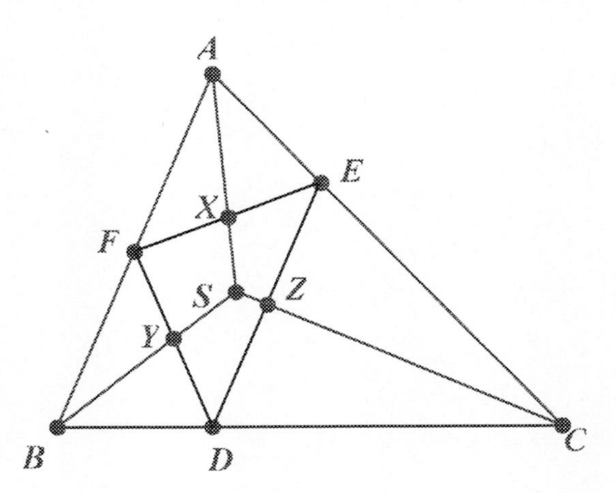

Figure 10.6: Another property of the symmedian point

an important role in the content of Chapter 13 on hexagons. It is also very significant point in the theory of porisms of a particular type.

10.2.3 Ω and Ω' are isogonal conjugates

The Brocard points Ω and Ω' are also a pair of isogonally conjugate points. These points are illustrated in Fig. 10.7.

For the construction of the point Ω' take the following three circles:

(i) The circle that touches AB at B and passes through C;

(ii) The circle that touches BC at C and passes through A;

(iii) The circle that touches CA at A and passes through B.

If the first two circles meet at Ω', then by the alternate segment theorem and its converse the third circle also passes through Ω' and the following angles are equal:

$$\angle AB\Omega' = \angle BC\Omega' = \angle CA\Omega' = w',$$

say.

For the construction of the point Ω take the following three circles:

(i) The circle that touches AB at A and passes through C;

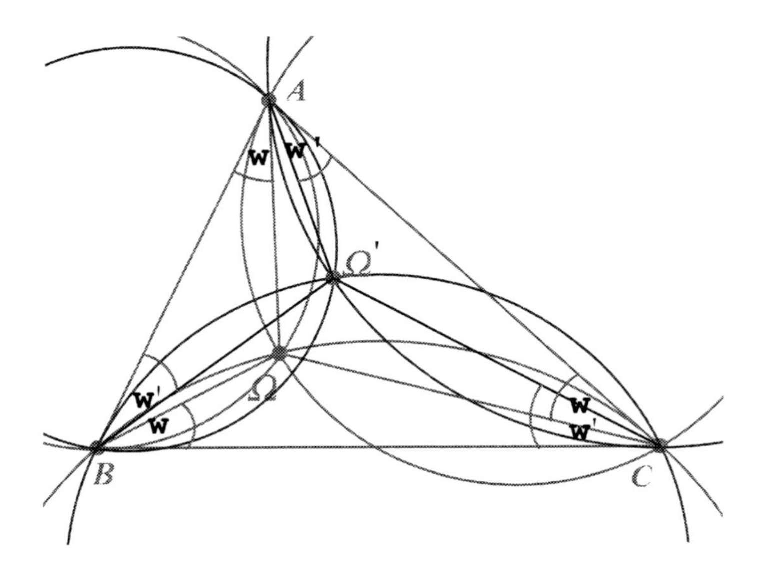

Figure 10.7: The Brocard points

(ii) The circle that touches BC at B and passes through A;

(iii) The circle that touches CA at C and passes through B.

If the first two circles meet at Ω then by the alternate segment theorem and its converse the third circle also passes through Ω and the following angles are equal

$$\angle BA\Omega = \angle CB\Omega = \angle AC\Omega = w,$$

say.

If we can prove that $w = w'$, then, by Section 10.1.2, Ω and Ω' are isogonal conjugate points. We now prove certain properties of the Brocard points, including the required fact that $w = w'$.

By the sine rule on triangles $AB\Omega', BC\Omega', CA\Omega'$ we have

$$\begin{aligned}
\frac{A\Omega'}{\sin w'} &= \frac{B\Omega'}{\sin(A - w')}, \\
\frac{B\Omega'}{\sin w'} &= \frac{C\Omega'}{\sin(B - w')}, \\
\frac{C\Omega'}{\sin w'} &= \frac{A\Omega'}{\sin(C - w')}.
\end{aligned}$$

It follows that

$$\sin^3 w' = \sin(A - w')\sin(B - w')\sin(C - w'). \tag{10.2}$$

By repeating the argument for the triangles involving the point Ω we find that w satisfies the same equation and so, as there is only one acute angle satisfying the equation, it must be the case that $w = w' = \omega$, and hence the Brocard points are isogonally conjugate points. It can, in fact, be shown, by some rather dull trigonometry (or some nice geometry), that

$$\cot\omega = \cot A + \cot B + \cot C, \tag{10.3}$$

thus providing the value of ω in terms of the angles A, B, C. The derivation of Equation (10.3) is left as an exercise. The angle ω is called the *Brocard angle.* It is desirable to work out the areal co-ordinates of the points Ω and Ω'. To do this we use the generalized sine rule, see Section 9.1.3. Now, from Fig. 10.7, it is easy to see that

$$\angle B\Omega'C = C + A, \angle C\Omega'A = A + B, \angle A\Omega'B = B + C$$

and hence the generalized sine rule in this case states that

$$\frac{A\Omega'\sin A}{\sin C} = \frac{B\Omega'\sin B}{\sin A} = \frac{C\Omega'\sin C}{\sin B}. \tag{10.4}$$

Now, by definition, the areal co-ordinates of Ω' are proportional to

$$((B\Omega')(C\Omega')\sin(C + A), (C\Omega')(A\Omega')\sin(A + B), (A\Omega')(B\Omega')\sin(B + C))$$

$$\propto \left(\frac{\sin B}{A\Omega'}, \frac{\sin C}{B\Omega'}, \frac{\sin A}{C\Omega'}\right) \propto \left(\frac{\sin B\sin A}{\sin C}, \frac{\sin C\sin B}{\sin A}, \frac{\sin A\sin C}{\sin B}\right)$$

$$\propto \left(\frac{1}{c^2}, \frac{1}{a^2}, \frac{1}{b^2}\right).$$

Similarly the areal co-ordinates of Ω are proportional to $(1/b^2, 1/c^2, 1/a^2)$.

Example 10.2.1 *(Triplicate ratio circle and seven-point circle)*

Let ABC be a triangle and S the symmedian point of ABC. Through S draw lines parallel to the sides of the triangle, as shown Fig. 10.8, to meet BC at D, D', CA at E, E', and AB at F, F'. The line $E'SD$ is parallel to AB,

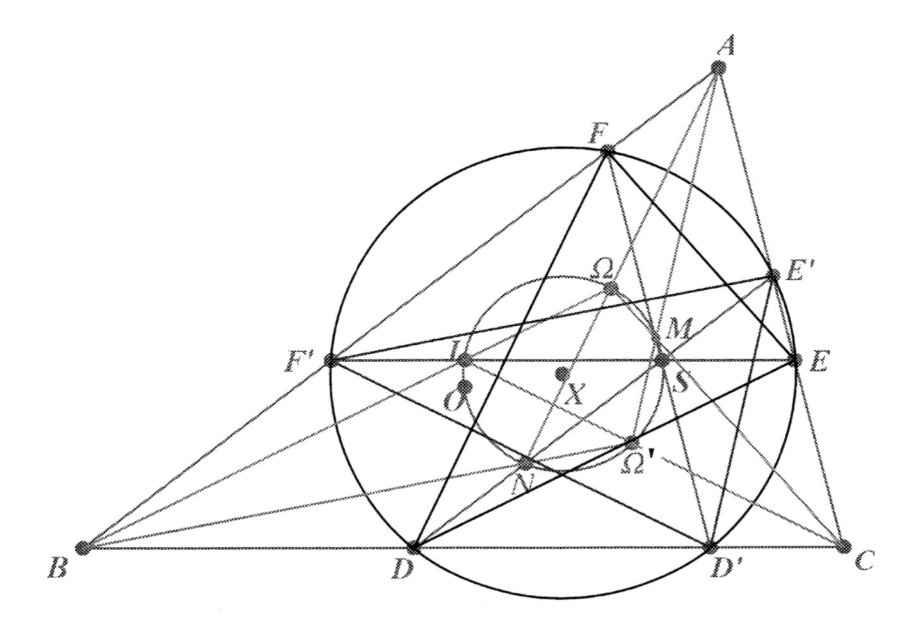

Figure 10.8: Triplicate ratio and Brocard circles

$F'SE$ is parallel to BC and $D'SF$ is parallel to CA. Then D, D', E, E', F, F' are concyclic on the *triplicate ratio circle.*

Lines through A, B, C parallel to the sides FD, DE, EF respectively concur at the Brocard point Ω and lines through A, B, C parallel to the sides $D'E', E'F', F'D'$ concur at the Brocard point Ω'. Let $B\Omega \wedge C\Omega' = L, C\Omega \wedge A\Omega' = M, A\Omega \wedge B\Omega' = N$. Let O be the circumcentre of ABC. Then the seven points $O, L, \Omega, M, S, \Omega', N$ are concyclic on the *Brocard circle* or *seven-point circle* with diameter OS. Furthermore the triplicate ratio circle and the seven-point circle are concentric.

The side BC has equation $x = 0$, so a line parallel to it has equation $x+y+z = kx$, where k is some constant. If the line passes through $S(a^2, b^2, c^2)$, then $k = (a^2+b^2+c^2)/a^2$. Hence the line $F'SE$ has equation $a^2(y+z) = (b^2+c^2)x$. It follows that F' and E have unnormalised co-ordinates $F'(a^2, b^2+c^2, 0)$ and $E(a^2, 0, b^2+c^2)$. The co-ordinates of the other points may be found by cyclic change and are $F(c^2+a^2, b^2, 0)$, $D(0, a^2+b^2, c^2)$, $D'(0, b^2, c^2+a^2)$, $E'(a^2+b^2, 0, c^2)$. Thus $BD = c^2a/(a^2+b^2+c^2)$ and $BD' = (c^2+a^2)a/(a^2+b^2+c^2)$ and

$$(BD)(BD') = \frac{c^2a^2(c^2+a^2)}{a^2+b^2+c^2} = (BF')(BF).$$

By the converse of the intersecting chord theorem D, D', F, F' lie on a circle. Similar results for the vertices A and C ensure that all six points lie on a circle. By substituting their co-ordinates into the equation of a general conic we find the equation of the triplicate ratio circle in areals is

$$\begin{aligned} \frac{(b^2+c^2)x^2}{a^2} + \frac{(c^2+a^2)y^2}{b^2} + \frac{(a^2+b^2)z^2}{c^2} \\ -yz\left(2+\frac{a^2(a^2+b^2+c^2)}{b^2c^2}\right) - zx\left(2+\frac{b^2(a^2+b^2+c^2)}{c^2a^2}\right) \\ -xy\left(2+\frac{c^2(a^2+b^2+c^2)}{a^2b^2}\right) = 0. \end{aligned} \tag{10.5}$$

Now the equation of EF is

$$y(c^2+a^2)(b^2+c^2) + za^2b^2 = xb^2(b^2+c^2).$$

The line $CM\Omega$ parallel to this through C has equation $b^2x = c^2y$. Similarly the lines $AN\Omega$ and $BL\Omega$ have equations $c^2y = a^2z$ and $a^2z = b^2x$ respectively and these meet at $\Omega(1/b^2, 1/c^2, 1/a^2)$. Similarly the equations of $AM\Omega'$, $BN\Omega'$, $CL\Omega'$ are $a^2y = b^2z$, $b^2z = c^2x$, $c^2x = a^2y$ respectively. These meet at $\Omega'(1/c^2, 1/a^2, 1/b^2)$. Ω and Ω' are the Brocard points, defined in Section 10.2.3. It was in this context that Brocard first discovered the points named after him. Now $CM\Omega$ and $AM\Omega'$ have equations $b^2x = c^2y$ and $a^2y = b^2z$, so M has co-ordinates (c^2, b^2, a^2) and similarly L and N have co-ordinates $L(a^2, c^2, b^2)$, $N(b^2, a^2, c^2)$. By substituting in the co-ordinates of the points S, L, M, N into the general equation of a circle, given by Equations (2.12) and (2.13) we find the equation of the seven-point circle in areals is

$$b^2c^2x^2 + c^2a^2y^2 + a^2b^2z^2 - a^4yz - b^4zx - c^4xy = 0. \tag{10.6}$$

Exercise 10.2.2

1. Using the notation of Example 10.2.1, prove that triangles $D'E'F'$ and EFD are congruent.

2. Using the notation of Example 10.2.1 verify that the points O, Ω, Ω' lie on the circle with Equation (10.6).

3. Show that the centre of the circle with Equation (10.6) is the midpoint of the line segment joining the symmedian point and the circumcentre.

4. Put the Equations (10.5) and (10.6) in the standard form of the equation of a circle (2.12) and deduce that the triplicate ratio circle and the seven-point circle are concentric.

5. Prove that the feet of the Cevians through Ω and Ω' lie on a conic.

6. Let $d^4 = a^2b^2 + b^2c^2 + c^2a^2$. Establish the following formulas involving the Brocard points and the Brocard angle:

   (i) $\Omega\Omega'^2 = a^2b^2c^2(a^4 + b^4 + c^4 - d^4)/d^8 = 4R^2\sin^2\omega(4\cos^2\omega - 3)$;

   (ii) $\cos\omega = (a^2 + b^2 + c^2)/(2d^2)$;

   (iii) $\sin\omega = 2[ABC]/d^2$;

   (iv) $A\Omega' = bc^2/d^2$, $A\Omega = b^2c/d^2$, with similar formulas for $B\Omega$, $B\Omega'$, $C\Omega$, $C\Omega'$ by cyclic change of a, b, c;

   (v) $(A\Omega)(B\Omega)(C\Omega) = (A\Omega')(B\Omega')(C\Omega') = 8R^3\sin^3\omega$; and $\Omega\Omega' \leq R/2$.

7. Prove that $\angle SO\Omega = \angle SO\Omega' = \omega$;

8. Prove that if J is the centre of the GH disc, then $JS^2/OG^2 = 1 - 3\tan^2\omega$;

9. Prove that $J\Omega^2 + J\Omega'^2 = 2OG^2$.

10.2.4 The Fermat and isodynamic points are isogonal conjugates

In a triangle ABC with angles all less than 120°, the Fermat point F is defined to be the point inside ABC such that $\angle BFC = \angle CFA = \angle AFB = 120^o$. Its areal co-ordinates are

$$\propto ((BF)(CF)\sin 120^o, (CF)(AF)\sin 120^o, (AF)(BF)\sin 120^o)$$

$\propto (1/AF, 1/BF, 1/CF)$. The isogonal conjugate of F, which we have denoted by V in Fig. 10.9, has areal co-ordinates $\propto (a^2(AF), b^2(BF), c^2(CF))$. V is called the *Hessian point* or the *isodynamic point* of ABC. Now, by the generalized sine rule

$$\frac{a(AF)}{\sin(120^o - A)} = \frac{b(BF)}{\sin(120^o - B)} = \frac{c(CF)}{\sin(120^o - C)}.$$

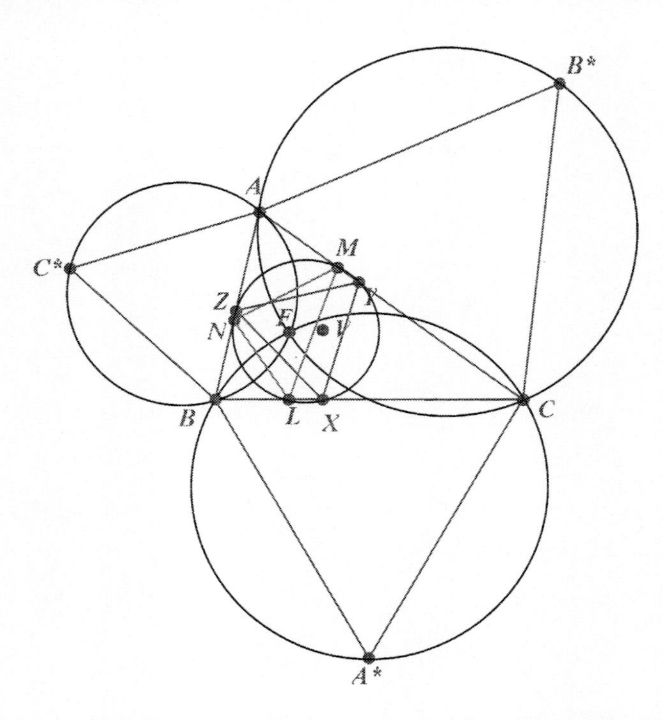

Figure 10.9: The Fermat point and the isodynamic point

So the areal co-ordinates of V are

$$\propto (a\sin(120^o - A), b\sin(120^o - B), c\sin(120^o - C))$$

$$\propto \left(\frac{\sin\angle BVC}{AV}, \frac{\sin\angle CVA}{BV}, \frac{\sin\angle AVC}{CV}\right).$$

But, by the generalized sine rule, we have $a(AV)/\sin\alpha = b(BV)/\sin\beta = c(CV)/\sin\gamma$ where $\alpha = \angle BVC - \angle BAC$ etc. The only values of the angles $\angle BVC, \angle CVA, \angle AVB$ compatible with these relations are $60^o + A$, $60^o + B$, $60^o + C$ with $\alpha = \beta = \gamma = 60^o$ and the values of these angles characterize the point V geometrically. Note that, as a result, the pedal triangle of V is equilateral, and it is in fact the equilateral triangle of minimum size that can be inscribed in triangle ABC. See Fig. 10.9.

Example 10.2.3

ABC is a triangle whose angles are all less than 120°. Find the point F inside ABC such that $PA + PB + PC$ is minimised when P is at F.

Draw equilateral triangles BCA^*, CAB^*, ABC^* external to the sides BC, CA, AB respectively. Draw the circles BCA^*, CAB^* to meet at F. Since $\angle BA^*C$ and $\angle CB^*A = 60^o$, then it follows that $\angle BFC = \angle CFA = 120^o$, so that F is an internal point. (Hence the need for the triangle to have angles less than 120°.) Clearly it follows that $\angle AFB = 120^o$ also, and so AC^*BF is cyclic and circle AC^*B passes through F. Now join FA^*, FB^*, FC^*. In circle BA^*CF we have $\angle BFA^* = \angle BCA^* = 60^o$. It follows that AFA^* is a straight line, as also are BFB^* and CFC^*. All the six angles at F are therefore 60°. Now since BCA^* is equilateral it follows by Ptolemy's theorem for BA^*CF that $A^*F = BF+CF$. It follows that $AA^* = BB^* = CC^* = FA+FB+FC$. By the extension of Ptolemy's theorem (see Example 14.4.1) if P does not lie on circle BA^*CF, then $PB+PC > PA^*$ and $PA+PB+PC > PA+PA^*$. Furthermore if P does not lie on AA^*, we have $PA + PA^* > AA^*$ and $PA + PB + PC > AA^* = FA + FB + FC$. Hence F is the point P inside ABC which minimizes $AP + BP + CP$.

10.2.5 The second Fermat and isodynamic points are isogonal conjugates

In a general triangle acute, right-angled or obtuse, it is still possible to define the Fermat point by means of drawing equilateral triangles BCA^*, CAB^*, ABC^* external to the sides of ABC and then the point of concurrency of AA^*, BB^*, CC^* is the Fermat point. If, say, angle $BAC = 120^o$, then F coincides with A. If an angle in the triangle is greater than 120°, then F lies outside triangle ABC. But then there is no minimum property of $FA+FB+FC$. By drawing the equilateral triangles inwards instead of outwards, it is possible to define the second Fermat point F'. Its isogonal conjugate, V'. is called the *second Hessian point* or the *second isodynamic point.*

Exercise 10.2.4

1. Let F, F' be the Fermat points, and J the centre of the GH disc. Prove that J, F, S, F' are collinear and that $(J, S; F, F) = -1$. Also prove that $JF/FS = (a^2 + b^2 + c^2)/4\sqrt{3}[ABC]^2$.

2. Prove that O, V, S, V' are collinear and that $V, V'$are inverse points with respect to the circle on OS as diameter.

3. An acute-angled triangle ABC is given. Prove that the following is a construction of the equilateral triangle LMN of maximum perimeter circumscribing triangle ABC. The construction is to erect equilateral triangles BCA^*, CAB^*, ABC^* external to the triangle. Join AA^*, BB^*, CC^* to meet at F (the Fermat point). Draw MN through A perpendicular to FA, NL through B perpendicular to FB and LM through C perpendicular to FC.

4. Prove that in triangle ABC, with all its angles less than 120°, the minimum value of $PA + PB + PC$, as P varies over the interior of ABC, is given by d, where $2d^2 = (a^2 + b^2 + c^2) + 4\sqrt{3}[ABC]$.

5. ABC and PQR are two triangles and the perpendiculars from P, Q, R respectively on to the sides BC, CA, AB are concurrent. Prove that the perpendiculars from A, B, C respectively on to the sides QR, RP, PQ are concurrent. The isogonal conjugate of a curve C in the plane of a triangle is defined as the locus of points Q that are the isogonal conjugates of points $P \in C$.

   (i) Prove that the isogonal conjugate of a line is a hyperbola through A, B, C;

   (ii) Prove that the isogonal conjugate of a line through the circumcentre O is a rectangular hyperbola through A, B, C.

   (iii) What is the isogonal conjugate of the circumcircle?

6. Use Problem 5 part (ii) to show that the general equation in areals of a rectangular hyperbola through A, B, C is

$$\sin^2 A(v\sin 2C - w\sin 2B)yz + \sin^2 B(w\sin 2A - u\sin 2C)zx$$
$$+ \sin^2 C(u\sin 2B - v\sin 2A)xy = 0.$$

10.3 Isotomic conjugate points and their properties

The isotomic conjugate of a point P lying in the plane of ABC, but not on its sides nor on lines through a vertex parallel to an opposite side, is defined as follows. Draw the Cevians APL, BPM, CPN with L, M, N on BC, CA, AB

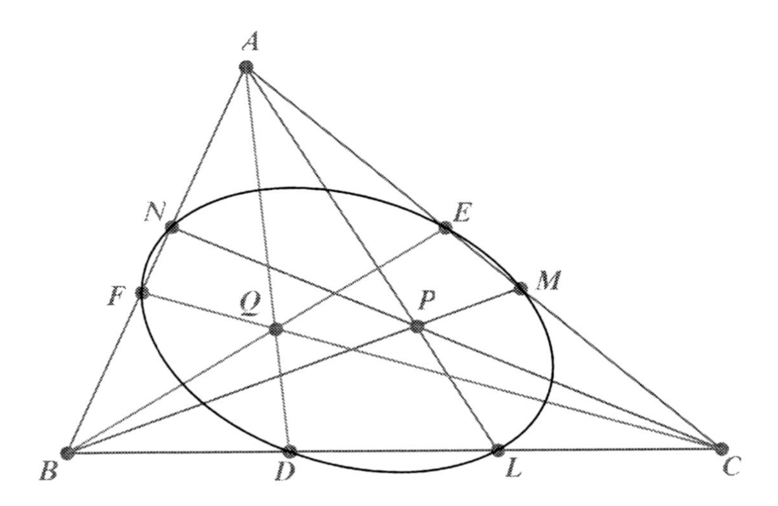

Figure 10.10: Isotomic conjugate points

respectively. Rotate L, M, N through $180°$ about the midpoints of their respective sides to obtain points D, E, F. Then AD, BE, CF are concurrent at a point Q, which is called the isotomic conjugate of P. Clearly the isotomic conjugate of Q takes us back to the point P, so the mapping of points on to their isotomic conjugates is an involution. The fixed points of the involution are the centroid G and the points on the medians with co-ordinates $(-1, 1, 1)$, $(1, -1, 1)$, $(1, 1, -1)$, The proof that the construction produces a well-defined point is straightforward. Let P have areal co-ordinates (l, m, n), where $l, m, n \neq 0$ since P does not lie on a side of ABC. Then L has unnormalised co-ordinates $L(0, m, n)$ and since $BL/LC = CD/DB$, the unnormalised co-ordinates of D are $D(0, n, m)$. The equation of AD is therefore $my = nz$, and the point $Q(1/l, 1/m, 1/n)$ lies on this line. By symmetry it also lies on BE, CF, so the three lines are concurrent. Note that the formula for the co-ordinates of the isotomic conjugate allows one to extend the definition of the isotomic conjugate to a point on a line through a vertex parallel to an opposite side.

Since L, M, N and D, E, F are the feet of two sets of Cevians then they lie on a conic (see Chapter 12 and Exercise 10.3.1 Problem 2) and it may be verified that its equation is

$$x^2 + y^2 + z^2 - (n/m + m/n)yz - (l/n + n/l)zx - (m/l + l/m)xy = 0. \quad (10.7)$$

Exercise 10.3.1

1. Prove that points (other than A, B, C) on the ellipse with equation $xy + yz + zx = 0$ have isotomic conjugates on the line at infinity. This is the equation of the *outer Steiner ellipse* circumscribing ABC, with centre G.

2. Prove that, in general, if AKL, BKM, CKN, AJD, BJE, CJF are two distinct sets of Cevians and $K(l, m, n)$ and $J(d, e, f)$ then a conic passes through the points L, M, N, D, E, F and has equation

$$\frac{x^2}{ld}+\frac{y^2}{me}+\frac{z^2}{nf}-yz\left(\frac{1}{ne}+\frac{1}{mf}\right)-zx\left(\frac{1}{lf}+\frac{1}{nd}\right)-xy\left(\frac{1}{md}+\frac{1}{le}\right)=0. \tag{10.8}$$

3. Prove that if J, K are as defined in Problem 2, where the co-ordinates are supposed normalised, and K is fixed, then the mapping $J \leftrightarrow J'$ given by $(d, e, f) \leftrightarrow (2l - d, 2m - e, 2n - f)$ is an involution, with invariant point K. What does this represent geometrically?

4. Let S be a conic circumscribing triangle ABC and P be a variable point on S. Find the locus of the isotomic conjugate of P.

5. Let L be a line, which is not a side of triangle ABC, and P a variable point on that line. Find the locus of the isotomic conjugate of P. What happens if the line passes through the centroid G? Identify the line whose isotomic conjugate is the circumcircle of ABC.

6. LMN is the pedal triangle of a point P. Points L', M', N' are the reflections of L, M, N in the midpoints of the respective sides. Prove that $L'M'N'$ is a pedal triangle of a point P'. Find the co-ordinates of P' in terms of P. Prove that the mapping is an involution.

7. ABC is a triangle and AD is the altitude through A. The reflection of D through the midpoint of BC is X. Prove that the perpendicular at B to AB, the perpendicular at C to AC and the perpendicular at X to BC are concurrent.

8. LMN is a transversal of triangle ABC. L', M', N' are the reflections of L, M, N in the midpoints of the sides BC, CA, AB respectively. Prove

that $L'M'N'$ is also a transversal. If the equation of LMN is $px+qy+nz=0$, what is the equation of $L'M'N'$? What can be said if LMN is the Wallace-Simson line of a point P on the circumcircle? (Such lines may be called *isotomic conjugate lines.*)

9. Prove that the Nagel and Gergonne points are isotomic conjugates. See Section 4.4.2. Also prove that the ex-Gergonne and ex-Nagel points are isotomic conjugates.

Chapter 11

Inequalities

11.1 Introduction

This chapter is not designed to be anything like a comprehensive review of the topic of geometrical inequalities, which has been one of the main areas of development in geometry during the twentieth century. We propose, however, to show how co-ordinate methods, vector methods and trigonometrical methods may be applied to help solve certain types of inequalities. The first two methods we outline are both linked to the use of areal co-ordinates.

11.2 Distance formula

11.2.1 The parallel axis formula

The key result is the following: let K have areal co-ordinates (l, m, n) (normalized so that $l + m + n = 1$), with ABC as triangle of reference, and let P be any point in the plane of the triangle. Then

$$lPA^2 + mPB^2 + nPC^2 = lKA^2 + mKB^2 + nKC^2 + PK^2 \qquad (11.1)$$

Let A, B, C have position vectors $\mathbf{x}, \mathbf{y}, \mathbf{z}$, with respect to an arbitrary origin O, and let P and K have position vectors $\mathbf{p}$ and $\mathbf{k}$ respectively. The data tell us that $\mathbf{k} = l\mathbf{x} + m\mathbf{y} + n\mathbf{z}$. The left-hand side of Equation (11.1) is

$$l(\mathbf{x} - \mathbf{p})^2 + m(\mathbf{y} - \mathbf{p})^2 + n(\mathbf{z} - \mathbf{p})^2 \qquad (11.2)$$

$$= lOA^2 + mOB^2 + nOC^2 + OP^2 - 2\mathbf{OP}.\mathbf{OK}$$

The right-hand side of Equation (11.1) is

$$l(\mathbf{x}-\mathbf{k})^2+m(\mathbf{y}-\mathbf{k})^2+n(\mathbf{z}-\mathbf{k})^2+(\mathbf{k}-\mathbf{p})^2 \tag{11.3}$$

$$= lOA^2+mOB^2+nOC^2+OK^2+PK^2-2\mathbf{k}.(l\mathbf{x}+m\mathbf{y}+n\mathbf{z}).$$

The left-hand side minus the right-hand side is therefore

$$OP^2-2\mathbf{OP}.\mathbf{OK}+OK^2-PK^2=0.$$

Example 11.2.1

Show that if P is any point in the plane of triangle ABC and O is the centre of the circumcircle, then

$$(PA)^2\sin 2A+(PB)^2\sin 2B+(PC)^2\sin 2C-4(OP)^2\sin A\sin B\sin C \tag{11.4}$$

$$=2[ABC].$$

Note that the unnormalized areal co-ordinates of the circumcentre are $(\sin 2A, \sin 2B, \sin 2C)$ and their sum is $4\sin A\sin B\sin C$. The parallel axis formula tells us, therefore, that

$$(PA)^2\sin 2A+(PB)^2\sin 2B+(PC)^2\sin 2C-4(OP)^2\sin A\sin B\sin C$$

$$=R^2(\sin 2A+\sin 2B+\sin 2C)=4R^2\sin A\sin B\sin C$$

$$=\frac{abc}{2R}=2[ABC],$$

as required. This may be converted into a geometrical inequality by noting that

$$(PA)^2\sin 2A+(PB)^2\sin 2B+(PC)^2\sin 2C\geq 2[ABC],$$

with equality if, and only if, P coincides with O.

Exercise 11.2.2

1. P is a point in the plane of triangle ABC. Find the position of P that minimizes $PA^2+PB^2+PC^2$, and find the minimum value.

2. P is a point in the plane of triangle ABC. Find the position of P that minimizes $aPA^2+bPB^2+cPC^2$, and find the minimum value.

3. Prove that in an obtuse- or acute-angled triangle

$$\tan A(PA)^2 + \tan B(PB)^2 + \tan C(PC)^2$$

is a maximum or minimum respectively when P is at H, and that then the stationary value is $8R^2 \sin A \sin B \sin C$.

4. Prove that if P is a point on the circumcircle of triangle ABC, then

$$(PA)^2 \sin 2A + (PB)^2 \sin 2B + (PC)^2 \sin 2C = 4[ABC].$$

5. Prove that if P is any point on the nine-point circle, then

$$(PA)^2(\sin 2B+\sin 2C)+(PB)^2(\sin 2C+\sin 2A)+(PC)^2(\sin 2A+\sin 2B)$$

$$= 8R^2 \sin A \sin B \sin C(1 + 2\cos A \cos B \cos C).$$

6. ABC is an acute-angled triangle and P is a point inside it. Prove that $a(AP) + b(BP) + c(CP) \geq 4[ABC]$. When does equality hold?

7. Prove that if ABC is a triangle, with orthocentre H and nine-point centre T, then

$$AT^2 + BT^2 + CT^2 \leq 3R^2 \leq AH^2 + BH^2 + CH^2.$$

8. Locate the points P on the circumcircle of triangle ABC such that $PA^2 + PB^2 + PC^2$ is a maximum or a minimum.

9. Repeat Problem 8 for points P lying on the outer Steiner ellipse of triangle ABC with centre the centroid G. Find the maximum and minimum, supposing $a > b > c$.

11.3 Area inequalities

11.3.1 How such inequalities arise

Suppose that points D, E, F have normalised areal co-ordinates $D(d_1, d_2, d_3)$, $E(e_1, e_2, e_3)$, $F(f_1, f_2, f_3)$, then $[DEF]/[ABC]$ is equal to the modulus of the determinant with these sets of co-ordinates as rows (or columns). If the value of the modulus of the determinant is Δ, then $[DEF] = \Delta[ABC]$.

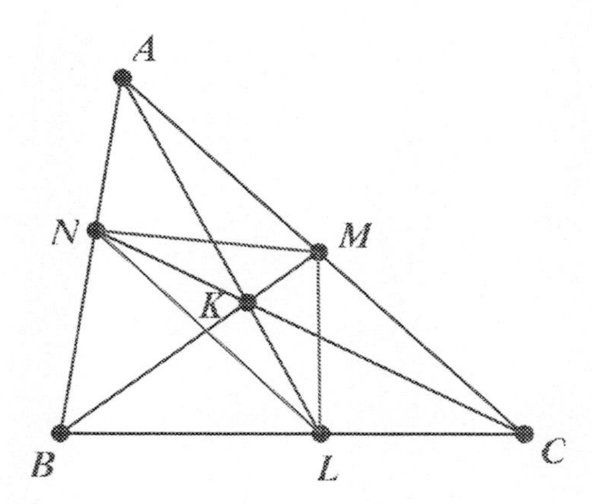

Figure 11.1: Example 11.3.1

An inequality arises for a fixed triangle ABC when the co-ordinates of one or more points depend on the values of certain parameters, typically the position of some variable point P in the configuration, and Δ has a maximum or minimum for certain positions of P. An alternative situation is when Δ depends on the shape of triangle ABC, which is allowed to vary subject to some constraint, such as having a constant circumradius.

Example 11.3.1

K is a point internal to triangle ABC and AKL, BKM, CKN are the Cevians through K. Prove that $[LMN] \leq \frac{1}{4}[ABC]$, with equality if, and only if, K is the centroid of ABC. See Fig. 11.1. This is probably the easiest example of its type.

Let K have areal co-ordinates (l, m, n), then the normalised co-ordinates of L, M, N are $L(0, m, n)/(m+n)$, $M(l, 0, n)/(n+l)$, $N(l, m, 0)/(l+m)$. It follows that

$$\frac{[LMN]}{[ABC]} = \frac{2lmn}{(m+n)(n+l)(l+m)}.$$

It is now sufficient to prove that $(m+n)(n+l)(l+m) \geq 8lmn$. Since K is internal we have $l, m, n > 0$, and so we may use the Arithmetic Mean/Geometric Mean inequality, which gives $m + n \geq 2\sqrt{mn}$, together with two similar inequalities for other the other pairs of variables. Multiplying up provides the required result.

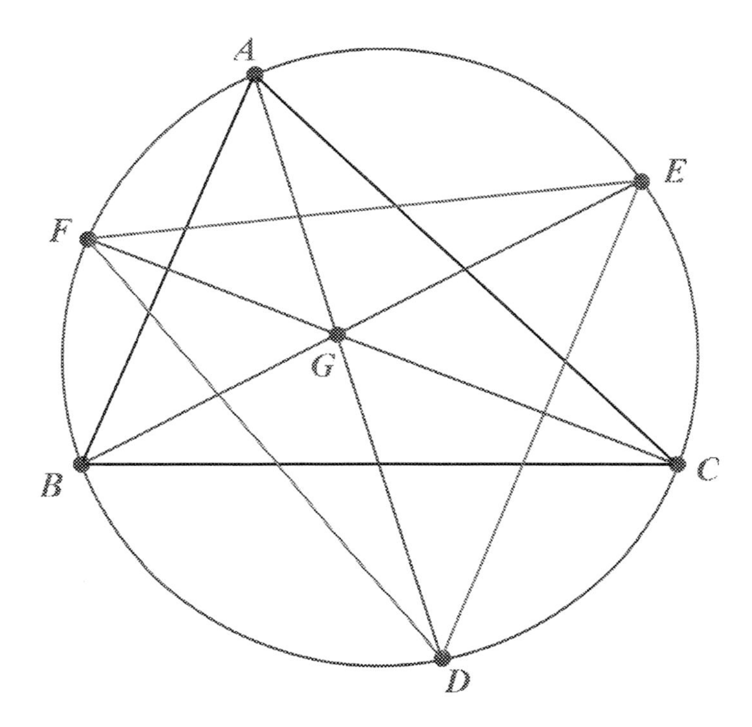

Figure 11.2: Example 11.3.2

Example 11.3.2

ABC is a triangle, centroid G. The Cevian lines AG, BG, CG meet the circumcircle of ABC at D, E, F respectively. Prove that $[DEF] \geq [ABC]$.

See Fig. 11.2. The equation of the circumcircle is

$$a^2yz + b^2zx + c^2xy = 0,$$

and the equation of AG is $y = z$. Solving these, with the side condition $x + y + z = 1$, we find the co-ordinates of D are $(-a^2, b^2 + c^2, b^2 + c^2)$, with similar expressions for the co-ordinates of E and F, by cyclic change of a, b, c and x, y, z. Now put $a^2 = u$, $b^2 = v$, $c^2 = w$, then we find

$$\frac{[DEF]}{[ABC]} = \frac{(u+v+w)^3}{(2v+2w-u)(2w+2u-v)(2u+2v-w)}.$$

Note that each bracketed term in the denominator is positive, and since their average is $(u + v + w)$, it follows by the Arithmetic Mean/Geometric Mean that $[DEF]/[ABC] \geq 1$.

Exercise 11.3.3

1. ABC is an acute-angled triangle. D is the reflection of A in the side BC, E is the reflection of B in the side CA, F is the reflection of C in the side AB. Prove that $[DEF] \geq 4[ABC]$.

2. ABC is an acute-angled triangle. The altitudes meet the circumcircle at D, E, F. Prove that $[DEF] \leq [ABC]$.

3. ABC is a triangle with centroid G. The line AG meets the circle BGC again at D, BG meets the circle CGA again at E, CG meets the circle AGB again at F. Prove that $[DEF] \geq 4[ABC]$.

4. ABC is an equilateral triangle and P is an internal point of ABC. AP meets circle BPC at D and E, F are similarly defined. L is the circumcentre of triangle BPC and M, N are similarly defined. Prove that $[LMN] \geq [ABC]$ and $[DEF] \leq 4[LMN]$.

5. ABC is an acute-angled triangle. L, M, N are the centres of circles BOC, COA, AOB respectively. AO meets circle BOC again at D, BO meets circle COA again at E, CO meets circle AOB again at F. Prove that

   (i) $[DEF] \geq 4[ABC]$;

   (ii) $[LMN] \geq [ABC]$;

   (iii) $[DEF] \leq 4[LMN]$.

6. ABC is a triangle, centroid G. L, M, N are the midpoints of BC, CA, AB respectively and U, V, W are the midpoints of AL, BM, CN respectively. AU meets circle BUC again at D, BV meets circle CVA again at E and CW meets circle AWB again at F. Prove that

$$[DEF] \geq 9[ABC]/4.$$

7. ABC is an acute-angled triangle and P is an internal point of ABC. L, M, N are the circumcentres of triangle BPC, CPA, APB respectively. Prove that $[ABC] \leq [LMN]$. When does equality hold? Is the result true if the triangle is obtuse? Is the result true if P is external to ABC?

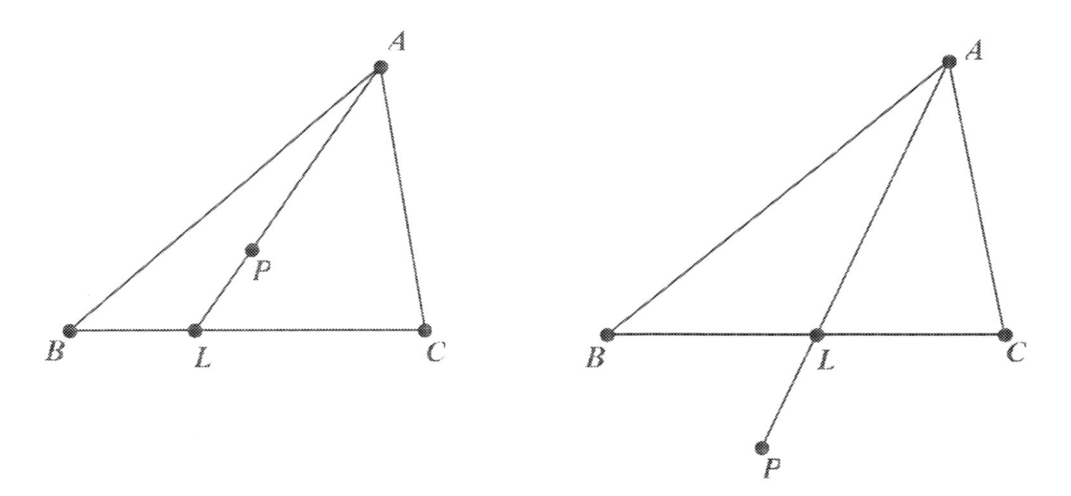

Figure 11.3: Configurations for Part 11.4

11.4 Inequalities involving distances

Inequalities involving ratios of distances sometimes just depend on the definition of areal co-ordinates. For example in both configurations in Fig. 11.3, $PL/AL = [BPC]/[ABC]$, provided we pay regard to the sign convention for the ratio of lengths of line segments and for the ratio of areas. It follows that if the areal co-ordinates of P are (l, m, n), then $AP/PL = (m + n)/l$ in both cases, provided signs are taken into account. Establishing such inequalities also tends to involve the use of the areal distance function (see Equation (2.5)). For example, since the displacement $AP = (-(m + n), m, n)$, we have

$$\begin{aligned} AP^2 &= -a^2mn + b^2n(m + n) + c^2m(m + n) \qquad (11.5) \\ &= c^2m^2 + b^2n^2 + (b^2 + c^2 - a^2)mn. \end{aligned}$$

Exercise 11.4.1

1. By considering an equilateral triangle, Equation (11.1) and Equations like (11.5) prove that, if $u + v + w = 1$ and $l + m + n = 1$, then

$$(m+n)u^2 + (n+l)v^2 + (l+m)w^2 + lvw + mwu + nuv \geq mn + nl + lm.$$

Example 11.4.2

Lines are drawn from the vertices A, B, C of a triangle through the variable point P within the triangle to meet the opposite sides in L, M, N respectively. If p, q, r are fixed positive numbers with $p > q > r$, between what limits does $S = pAP/AD + qBP/BE + rCP/CF$ lie?

Let P have areal co-ordinates (l, m, n). Then it is known that $AP/AD = (m+n)$, $BP/BE = (n+l)$ and $CP/CF = (l+m)$. We show $p+q > S > q+r$. For

$$p + q - S = (p + q)(l + m + n) - p(m + n) - q(n + l) - r(l + m)$$

$$= pl + qm - r(l + m) = (p - r)l + (q - r)m > 0,$$

since $p > r$ and $q > r$ and $l, m, n > 0$. Similarly

$$S - q - r = p(m + n) + q(n + l) + r(l + m) - (q + r)(l + m + n)$$

$$= p(m + n) - qm - rn = (p - q)m + (p - r)n > 0.$$

Example 11.4.3

ABC is an equilateral triangle of side a. P is any point in the plane of ABC. Find the minimum value of $AP^2 + BP^2 + CP^2$.

From the formula (11.5) we have

$$AP^2 + BP^2 + CP^2 = a^2(2l^2 + 2m^2 + 2n^2 + mn + nl + lm)$$

$$= a^2(2(l + m + n)^2 - 3(mn + nl + lm)) = a^2(2 - 3(mn + nl + lm)).$$

But for any l, m, n $1 = (l+m+n)^2 \geq 3(mn+nl+lm)$, so $AP^2+BP^2+CP^2 \geq a^2$, with equality if, and only if, $l = m = n = 1/3$, that is when P is at the centroid.

Example 11.4.4

ABC is a triangle with centroid G and symmedian point S. Prove that

$$AS/AG + BS/BG + CS/CG \leq 3.$$

Also prove that

$$AG/AS + BG/BS + CG/CS \geq 3.$$

First, by Apollonius's theorem (or by using the areal distance function) $AG^2 = (2b^2 + 2c^2 - a^2)/9$. The areal co-ordinates of S are $(a^2, b^2, c^2)/(a^2 + b^2 + c^2)$ so $\mathbf{AS} = (-(b^2 + c^2), b^2, c^2)$. It follows from Equation (2.5) that

$$AS^2 = \frac{-a^2b^2c^2 + 2b^2c^2(b^2 + c^2)}{(a^2 + b^2 + c^2)^2}$$

$$= \frac{9b^2c^2(AG)^2}{(a^2 + b^2 + c^2)^2}.$$

Hence

$$\frac{AS}{AG} + \frac{BS}{BG} + \frac{CS}{CG} = \frac{3(bc + ca + ab)}{(a^2 + b^2 + c^2)} \leq 3.$$

Secondly

$$\frac{AG}{AS} + \frac{BG}{BS} + \frac{CG}{CS} = \frac{a^2 + b^2 + c^2}{\frac{3}{bc} + \frac{3}{ca} + \frac{3}{ab}}$$

$$= \frac{(a^2 + b^2 + c^2)(a + b + c)}{3abc} \geq 3,$$

by the Arithmetic Mean/Geometric Mean inequality. (Alternatively, it is known for all $p, q, r > 0$ that $(p + q + r)(1/p + 1/q + 1/r) \geq 9$.)

Exercise 11.4.5

1. P is an interior point of triangle ABC. APX, BPY, CPZ are the Cevians through P. The images of A, B, C after a clockwise rotation of $90°$ about P are L, M, N respectively. Prove that $5/3 \leq (XL/XA)^2 + (YM/YB)^2 + (ZN/ZC)^2 < 3$. (BG)

2. P is an interior point of an equilateral triangle ABC with normalised areal co-ordinates (l, m, n). Q is the isotomic conjugate of P. Prove that $AP/AQ + BP/BQ + CP/CQ \geq 3$. Is it also true that $AQ/AP + BQ/BP + CQ/CP \geq 3$?

3. ABC is an acute-angled triangle and O is the circumcentre. AO, BO, CO meet the circumcircles of triangles BOC, COA, AOB respectively at D, E, F. Prove that $OD + OE + OF \geq 6R$, where R is the radius of the circumcircle of ABC.

4. ABC is a triangle with circumradius R. The circle through A touching BC at its midpoint has radius R_1. R_2 and R_3 are similarly defined. Prove that $R_1^2 + R_2^2 + R_3^2 \geq 27R^2/16$.

5. Using the notation of Problem 4, prove that

$$\frac{abc}{2[ABC]} < R_1 + R_2 + R_3 \leq \frac{3(a^3 + b^3 + c^3)}{16[ABC]}.$$

6. ABC is a triangle, centroid G. AG, BG, CG meet the circles BGC, CGA, AGB again at D, E, F respectively. Prove that $GD+GE+GF \geq 2(AG + BG + CG)$.

11.5 Trigonometrical inequalities

Many inequalities involve an application of trigonometry that makes use of the fact that the angles A, B, C of a triangle satisfy $A + B + C = 180°$.

Example 11.5.1

Prove that if A, B, C are the angles of a triangle then $\cos A+\cos B+\cos C \leq 3/2$. Hence show that if H is the orthocentre of ABC, then $AH+BH+CH \leq 3R$, where R is the circumradius of triangle ABC.

$$\cos A + \cos B + \cos C = 2\cos\frac{A+B}{2}\cos\frac{A-B}{2} + 1 - 2\sin^2 C/2$$

$$\leq 2\sin\frac{1}{2}C + 1 - 2\sin^2\frac{1}{2}C,$$

since $\cos\frac{A+B}{2} = \sin\frac{1}{2}C > 0$ and $\cos\frac{A-B}{2} \leq 1$. Equality holds if, and only if, $B = C$. Now $2s + 1 - 2s^2 \leq 3/2$ (by calculus) with equality if, and only if, $s = 1/2$, that is if, and only if, $C = 60°$. So the inequality holds, with equality if, and only if, the triangle is equilateral. Since $AH = 2R\cos A$ etc., the second inequality is immediate.

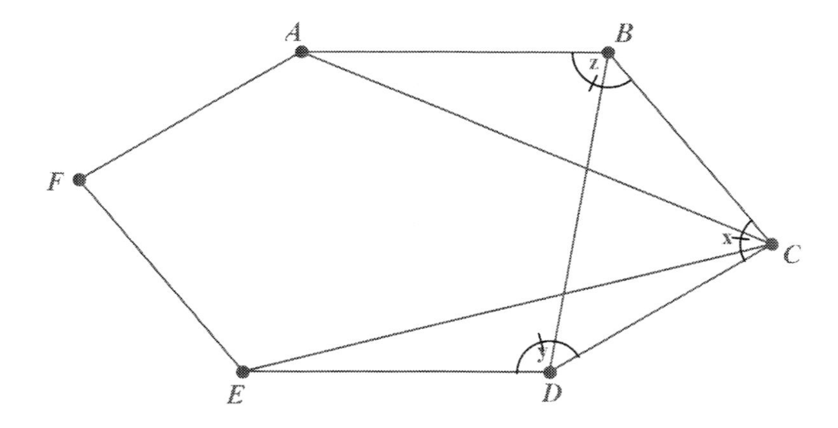

Figure 11.4: Example 11.5.2

Example 11.5.2

$ABCDEF$ is a hexagon (not necessarily regular) in which AB is equal and parallel to DE, BC is equal and parallel to EF and CD is equal and parallel to FA. Prove that

$$AC^2 + BD^2 + CE^2 \leq 3AB^2 + 3BC^2 + 3CD^2.$$

See Fig. 11.4. Let $AB = p$, $BC = q$, $CD = r$. Denote the angles $\angle ABC = z, \angle BCD = x, \angle CDE = y$. Use the cosine formula for triangles ABC, BCD, CDE and add to get

$$AC^2 + BD^2 + CE^2 = 2p^2 + 2q^2 + 2r^2 - 2qr\cos x - 2rp\cos y - 2pq\cos z.$$

To prove what is required it is sufficient to show that

$$p^2 + q^2 + r^2 + 2qr\cos x + 2rp\cos y + 2pq\cos z \geq 0.$$

But this is so, since the left-hand side is equal to $(p + r\cos y + q\cos z)^2 + (r\sin y - q\sin z)^2$, where we have used $\cos x = -\cos(y+z) = -\cos y\cos z + \sin y\sin z$. Equality holds when $p : q : r = \sin x : \sin y : \sin z$.

Exercise 11.5.3

1. If x, y, z are real numbers and A, B, C the angles of a triangle, prove that
$$x^2 + y^2 + z^2 \geq 2yz\cos A + 2zx\cos B + 2xy\cos C.$$

2. If A, B, C are the angles of a triangle, prove that

$$1 + \cos A \cos B \cos C \geq \sqrt{3} \sin A \sin B \sin C. \qquad \text{(JW)}$$

Hint: If a triangle is not equilateral, then either two angles are greater than 60° or two angles are less than 60°.

3. If A, B, C are the angles of a triangle, prove that

$$(1 - \cos A)(1 - \cos B)(1 - \cos C) \geq \cos A \cos B \cos C.$$

Hint: Express in terms of the sides of the triangle.

4. Prove that if x, y, z are real numbers such that $x+y+z = 0$ and A, B, C are the angles of a triangle, then $x^2 \cot A + y^2 \cot B + z^2 \cot C \geq 0$.

5. If A, B, C are the angles of a triangle, prove that

$$\sin 2A + \sin 2B + \sin 2C \leq \sin A + \sin B + \sin C.$$

6. If A, B, C are the angles of a triangle, prove that

$$\sin A + \sin B + \sin C \leq \cos \frac{1}{2}A + \cos \frac{1}{2}B + \cos \frac{1}{2}C.$$

7. ABC is a triangle with side lengths a, b, c. G is the centroid of ABC. AG meets the circle BGC again at D, and E, F are similarly defined. Prove that the perimeter p of the hexagon $AECDBF$ satisfies

$$p \geq 2(\sqrt{bc} + \sqrt{ca} + \sqrt{ab}).$$

8. If A, B, C are the angles of a triangle, prove that

$$\frac{\cos A}{\cos B \cos C} + \frac{\cos B}{\cos C \cos A} + \frac{\cos C}{\cos A \cos B} \geq 6.$$

9. If A, B, C are the angles of a triangle, prove that

   (i) $\sin A \sin B \sin C \leq 3\sqrt{3}/8$;

   (ii) $\cos^2 A + \cos^2 B + \cos^2 C \geq \frac{3}{4}$;

(iii) $6\sin A\sin B\sin C$

$$\le 8(\sin^3 A\cos B\cos C+\sin^3 B\cos C\cos A+\sin^3 C\cos A\cos B)$$

$$\le \frac{3\sqrt{3}}{8}(\cos^2 A+\cos^2 B+\cos^2 C).$$

10. If A, B, C are the angles of a triangle, prove that (i) $\tan\frac{1}{2}A+\tan\frac{1}{2}B+\tan\frac{1}{2}C\ge\sqrt{3}$; (ii) $\tan\frac{1}{2}A\tan\frac{1}{2}B\tan\frac{1}{2}C\le 1/3\sqrt{3}$.

11. If A, B, C are the angles of a triangle prove that

$$\sin 2A\sin^2 A+\sin 2B\sin^2 B+\sin 2C\sin^2 C-\sin 2A\sin 2B\sin 2C\le\frac{3\sqrt{3}}{4}.$$

12. If A, B, C are the angles of a triangle, prove that $\cos^2 A+\cos^2 B+\cos^2 C$ is greater than, equal to, or less than 1, according as to whether one of A, B, C is obtuse, a right angle or all are acute. *Hint: You first need to prove the useful equality*

$$\cos^2 A+\cos^2 B+\cos^2 C+2\cos A\cos B\cos C=1.$$

13. If A, B, C are the angles of a triangle, prove that

$$\sin A+\sin B+\sin C\ge 4\sin A\sin B\sin C.$$

14. Circles S_A, S_B, S_C, S are drawn with equal radius ρ, so that all four circles lie inside a triangle ABC that has no angle less than $30°$. S_A touches AB, AC and S, S_B touches BA, BC and S and SC touches CA, CB and S. Prove that $\rho\le R/4$, where R is the circumradius of ABC.

15. If A, B, C are the angles of a triangle, prove that

$$\frac{1}{2}(\sin A+\sin B+\sin C)\le\cos A\cos\frac{1}{2}A+\cos B\cos\frac{1}{2}B+\cos C\cos\frac{1}{2}C.$$

Hint: Let D, E be the feet if the altitudes from A and B. Prove that $CD+CE\le 4R\cos C\cos\frac{1}{2}C$.

16. Let x, y, z be angles each lying strictly between 0° and $45°$ with $x+y+z=90°$. Prove that $1\le\tan^2 x+\tan^2 y+\tan^2 z<2$.

17. If A, B, C are the angles of an acute-angled triangle, prove that

$$\sin A+\sin B+\sin C>\cos A+\cos B+\cos C.$$

Chapter 12

Triangles and quadrangles in perspective

12.1 Desargues's theorem

We refer the reader to Section 3.2, where a proof of Desargues's theorem is given. In the theorem ABC and DEF are two triangles and AD, BE, CF are concurrent at O (the vertex of perspective). In the projective proof of the theorem, which states that $L = BC \wedge EF$, $M = CA \wedge FD$, and $N = AB \wedge DE$ are collinear, we took ABC to be the triangle of reference, with $A(1, 0, 0)$, $B(0, 1, 0)$, $C(0, 0, 1)$ and O to be the unit point $O(1, 1, 1)$. Then, without loss of generality, we can take $D(d, 1, 1)$, $E(1, e, 1)$, $F(1, 1, f)$. The equation of the line LMN is

$$(1 - e)(1 - f)x + (1 - f)(1 - d)y + (1 - d)(1 - e)z = 0. \tag{12.1}$$

We observe that the converse of the theorem is its dual, which therefore does not require separate proof. The following additional result is a consequence.

12.1.1 Six intersection points on a conic

The other six intersections of the sides of two triangles in perspective lie on a conic The six points we identify from Fig. 12.1 are

$$L' = BC \wedge FD, L'' = BC \wedge DE, M' = CA \wedge DE,$$

$$M'' = CA \wedge EF, N' = AB \wedge EF, N'' = AB \wedge FD.$$

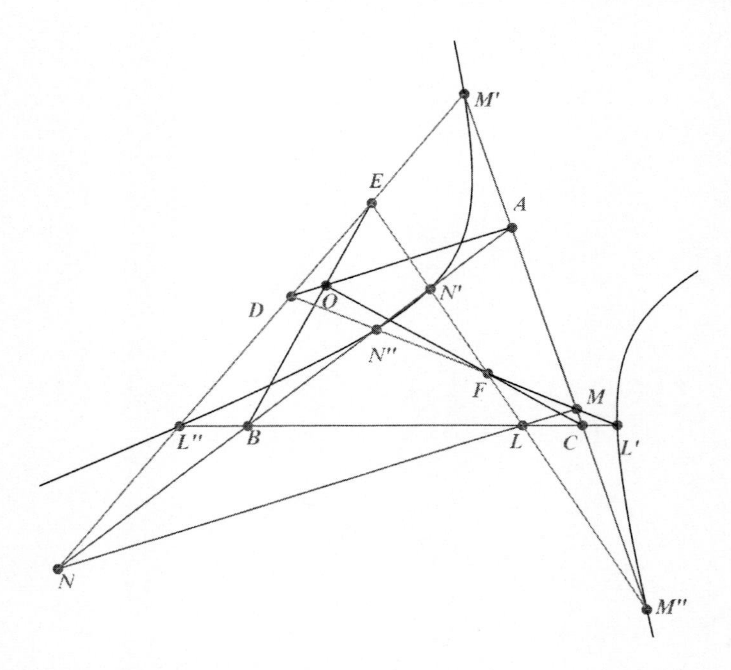

Figure 12.1: Triangles in perspective and the associated conic

This additional result shows the connection between Desargues's theorem and Pascal's theorem. This is because LMN is related to the hexagon $L'N''N'M''M'L''$ by

$$L = M''N' \wedge L'L'', M = N''L' \wedge M'M'', N = L''M' \wedge N'N''.$$

Hence the six points lie on a conic if, and only if, LMN is a straight line, which is so if, and only if, the two triangles are in perspective. It is interesting, however, to determine the equation of the conic. The equation of EF is

$$(ef - 1)x + (1 - f)y + (1 - e)z = 0.$$

This meets AB, $z = 0$, at $N'((1-f), (1-ef), 0)$ and CA, $y = 0$, at $M''((1-e), 0, (1-ef))$. It follows by cyclic change of letters that $L'(0, (1-d), (1-fd))$, $N''((1-fd), (1-f), 0)$, $M'((1-de), 0, (1-e))$, $L''(0, (1-de), (1-d))$. It can be verified that these six points lie on the conic with equation

$$(1 - ef)x^2 + (1 - fd)y^2 + (1 - de)z^2$$
$$-\left(\frac{(1-fd)(1-de)}{1-d} + (1-d)\right)yz - \left(\frac{(1-de)(1-ef)}{1-e} + (1-e)\right)zx$$

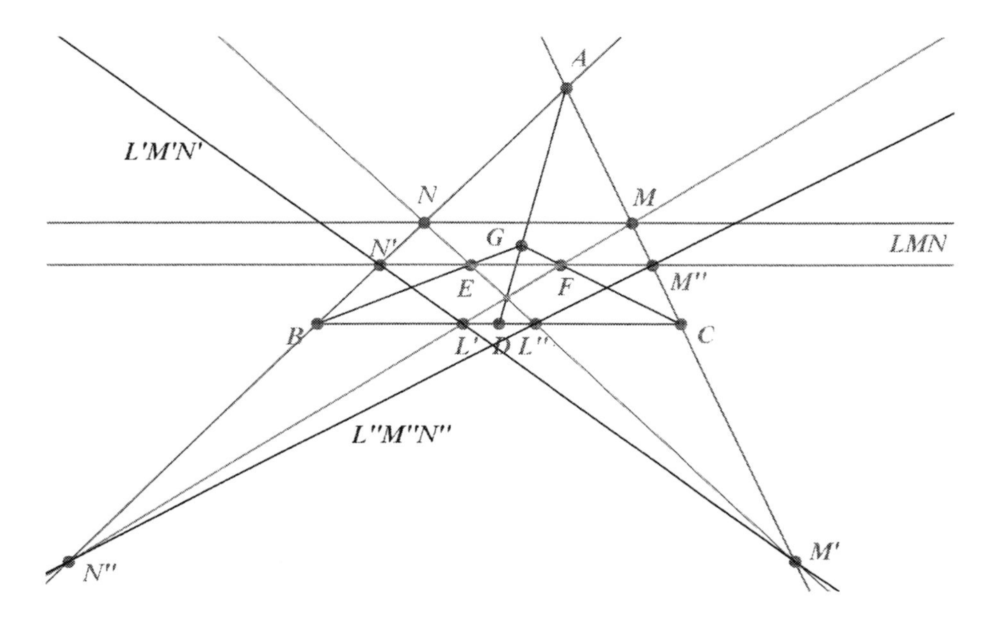

Figure 12.2: When the associated conic is degenerate

$$-\left(\frac{(1-ef)(1-fd)}{1-f}+(1-f)\right)xy=0. \tag{12.2}$$

We call this conic the *Desargues conic* of the perspective. The reader is invited to consider what happens if in addition the points A, B, C, D, E, F lie on a conic. It should be noted that this conic degenerates into two straight lines if $def = 1$ or $2 - d - e - f + def = 0$. An example of this is shown in Fig. 12.2. The vertex of perspective (perspector) is the centroid G, and the axis of perspective (perspectrix) is parallel to BC. Using areals, D is the point with co-ordinates $(1/9, 4/9, 4/9)$, E is the point with co-ordinates $(1/4, 1/2, 1/4)$ and F is the point with co-ordinates $(1/4, 1/4, 1/2)$.

There is another way of thinking about the Desargues axis of perspective and the Desargues conic. Consider the two triangles to be degenerate cubic curves. These meet in nine points. It can be shown that any cubic curve passing through eight of these points must pass through the ninth. If therefore three of the points lie on a line (of degree one in the variables), then the cubic concerned must degenerate into a line and a conic (of degree two in the variables), and the conic must therefore pass through all six remaining points. Similarly if six points lie on a conic the three remaining points must lie on a line. To make the argument rigorous it has to be shown that no more than three of the points can lie on a line, and no more than six of the points on a conic.

12.2 A configuration with many sets of triangles in perspective

ABC is a triangle with L, M, N the midpoints of the sides BC, CA, AB respectively. Let S be any conic through L, M, N that cuts BC, CA, AB again at points D, E, F respectively that are distinct from L, M, N, are such that AD, BE, CF are not parallel and are such that none of D, E, F coincides with A, B, C. Points are defined as intersections of lines as follows:

$$A' = DM \wedge FL, B' = EN \wedge DM, C' = FL \wedge EN;$$

$$A'' = EL \wedge DN, B'' = FM \wedge EL, C'' = DN \wedge FM,$$

$$H = EN \wedge FM, J = FL \wedge DN, K = DM \wedge EL;$$

$$P = EF \wedge MN, Q = FD \wedge NL, R = DE \wedge LM;$$

$$P' = DF \wedge LM, Q' = ED \wedge MN, R' = FE \wedge NL;$$

$$P'' = DE \wedge NL, Q'' = EF \wedge LM, R'' = FD \wedge MN.$$

See Fig. 12.3, in which lines, such as EF, are not drawn. Note that when S is a circle, it is the nine-point circle. The following results now hold:

12.2.1 A, H, Q, R and B, J, R, P and C, K, P, Q are sets of collinear points

12.2.2 H, J, K are collinear

12.2.3 Triangles ABC, $A'B'C'$, $A''B''C''$ are in perspective with vertex U.

12.2.4 Triangles ABC, $P'Q'R'$, $P''Q''R''$ are in perspective with vertex T.

12.2.5 Triangles $A'B'C'$, PQR, $P''Q''R''$ are in perspective with vertex V.

12.2.6 Triangles $A''B''C''$, PQR, $P'Q'R'$ are in perspective with vertex W.

12.2.7 T, U, V, W are collinear.

12.2.8 Triangles $A'B'C'$, $P'Q'R'$ are in perspective with vertex X.

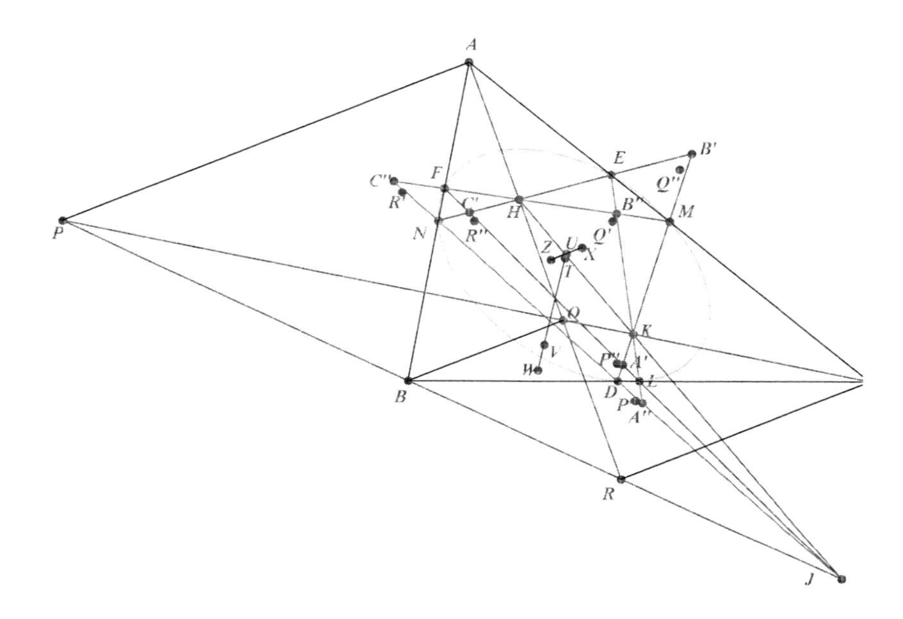

Figure 12.3: Configuration for Part 12.2

12.2.9 Triangles $A''B''C''$, $P''Q''R''$ are in perspective with vertex Z.

12.2.10 U, X, Z are collinear

12.2.11 Lines AP, BQ, CR, UXZ are parallel (so that triangles ABC, PQR are in perspective) and meet at a point Y on the line at infinity

Outline proof of results

Areal co-ordinates are used with ABC the triangle of reference. It is known that if a conic meets the sides of a triangle at six points L, M, N, D, E, F as described above and AL, BM, CN are concurrent (at the centroid in this case), then AD, BE, CF are also concurrent or parallel. We are told they are not parallel, so let the point of concurrence have co-ordinates (l, m, n), where since this point does not lie on any of the sides of ABC, we have $l, m, n \neq 0$. Define $u = 1/l$, $v = 1/m$, $w = 1/n$. Then the equation of S is

$$ux^2 + vy^2 + wz^2 - (v+w)yz - (w+u)zx - (u+v)xy = 0, \tag{12.3}$$

as it is easily verified that S passes through $L(0,1,1), M(1,0,1), N(1,1,0), D(0,m,n), E(l,0,n), F(l,m,0)$. When S is the nine-point circle u, v, w are proportional to $\cot A, \cot B, \cot C$ respectively.

The twelve lines we need have equations that can be checked to be:

$$MN: -x+y+z=0; NL: x-y+z=0; LM: x+y-z=0.$$

$$EF: -ux+vy+wz=0; FD: ux-vy+wz=0; DE: ux+vy-wz=0.$$

$$DM: wx+vy-wz=0; EN: -ux+uy+wz=0; FL: ux-vy+vz=0.$$

$$DN: vx-vy+wz=0; EL: ux+wy-wz=0; FM: -ux+vy+uz=0.$$

The eighteen required points of intersection can now be found and they have co-ordinates as follows:

$$A'(v(w-v), w(u+v), v(w+u)); B'(w(u+v), w(u-w), u(v+w));$$

$$C'(v(w+u), u(v+w), u(v-u)); A''(w(v-w), w(u+v), v(w+u));$$

$$B''(w(u+v), u(w-u), u(v+w)); C''(v(w+u), u(v+w), v(u-v));$$

$$H((u^2-vw), u(u-w), u(u-v)); J(v(v-w), (v^2-wu), v(v-u));$$

$$K(w(w-v), w(w-u), (w^2-uv));$$

$$P(w-v, w-u, u-v); Q(v-w, u-w, u-v); R(v-w, w-u, v-u);$$

$$P'(v-w, w+u, u+v); Q'(v+w, w-u, u+v); R'(v+w, w+u, u-v);$$

$$P''(w-v, w+u, u+v); Q''(v+w, u-w, u+v); R''(v+w, w+u, v-u).$$

12.2.1 Proof

A, H, Q, R are collinear on the line with equation $y(u-v)+z(w-u)=0$. Similarly B, J, R, P are collinear and C, K, P, Q are collinear.

12.2.2 Proof

H, J, K are collinear on the line with equation $u(v-w)x+v(w-u)y+w(u-v)z=0$.

This is a familiar result in connection with the nine-point circle configuration and HJK may be thought of as the axis of the figure, being the Desargues axis of perspective of triangles $A'B'C'$ and $A''B''C''$, and also of each of these triangles with triangle PQR.

12.2.3 Proof

Triangles ABC, $A'B'C'$, $A''B''C''$ are in perspective at the point

$$U\left(\frac{1}{u(v+w)}, \frac{1}{v(w+u)}, \frac{1}{w(u+v)}\right).$$

The point U is perhaps the most important point in the configuration.

12.2.4 Proof

Triangles ABC, $P'Q'R'$, $P''Q''R''$ are in perspective at the point T with co-ordinates $(v+w, w+u, u+v)$. When S is the nine-point circle then T is the symmedian point.

12.2.5 Proof

The equation of PP'' is $(u^2 - vw)x + v(w-v)y - u(w-v)z = 0$ and it may be verified that A' lies on this line. The equation of $B'QQ''$follows by cyclic change of letters and the two lines meet at V, the vertex of perspective of triangles $A'B'C'$, PQR and $P''Q''R''$. The three co-ordinates of V are $(w-v)(uvw + uv^2 - vw^2 - wu^2), (u-w)(uvw + vw^2 - wu^2 - uv^2)$, and $(v-u)(uvw + wu^2 - uv^2 - vw^2))$.

12.2.6 Proof

Similarly triangles $A''B''C''$, PQR, $P'Q'R'$ are in perspective at the point W. The three co-ordinates of W are $(v-w)(uvw + uw^2 - vu^2 - wv^2)$, $(w-u)(uvw + vu^2 - wv^2 - uw^2)$ and $(u-v)(uvw + wv^2 - uw^2 - vu^2)$.

12.2.7 Proof

T, U, V, W are collinear on the line with equation

$$u(u^2-vw)(w^2-v^2)x+v(v^2-wu)(u^2-w^2)y+w(w^2-uv)(v^2-u^2)z = 0. \quad (12.4)$$

12.2.8 Proof

The equation of $A'P'$ is

$$(w-v)(u^2-vw)x+v(v-w)(2u+v+w)y-(v-w)(uv+wu+2vw)z=0$$

and those of $B'Q'$ and $C'R'$ follow by cyclic change of letters. These lines are concurrent at a point X, which is the point from which triangles $A'B'C'$ and $P'Q'R'$ are in perspective. The co-ordinates of X are (x,y,z), where

$$x=3u^2vw+u^2w^2-uv^3+4uv^2w+5uvw^2+3v^2w^2+vw^3$$

and y,z follow by cyclic change of letters.

12.2.9 Proof

Similarly triangles $A''B''C''$, $P''Q''R''$ are in perspective at a point Z with co-ordinates (x,y,z), where

$$x=3u^2vw+u^2v^2-uw^3+4uvw^2+5uv^2w+3v^2w^2+wv^3$$

and y,z follow by cyclic change of letters.

12.2.10 Proof

U,X,Z are collinear on the line with equation

$$u(u^2-vw)(v+w)^2x+v(v^2-wu)(w+u)^2y+w(w^2-uv)(u+v)^2z=0. \quad (12.5)$$

12.2.11 Proof

The equation of AP is $(w-u)z=(u-v)y$, with similar equations by cyclic change of letters for BQ and CR. These three lines appear to meet at the point Y with co-ordinates $(v-w,w-u,u-v)$. But these three co-ordinates sum to zero and so the point lies on the line at infinity. It follows that AP,BQ,CR are parallel. Finally, the line UXZ also contains Y, so UXZ is parallel to AP,BQ,CR.

First generalization

The first question to ask is what happens if L, M, N are not the midpoints of BC, CA, AB respectively, but are the feet of another set of Cevians. The only change in the results is that Y no longer lies on the line at infinity. Triangles ABC and PQR are in perspective at the point Y and the line UXZ passes through Y, so that U, X, Y, Z are collinear. The proofs of all the results are more complicated, because in addition to the three variables u, v, w associated with D, E, F, one has in addition another three variables r, s, t associated with L, M, N.

Second generalization

If L, M, N and D, E, F are the feet of two sets of Cevians then a conic always passes through the six points. But a second question to ask is what happens if a conic S meets BC at L, D; CA at M, E; and AB at N, F but L, M, N are *not* the feet of a set of Cevians and consequently D, E, F are not the feet of a set of Cevians. This is the more general situation and there is a necessary and sufficient condition for it, called Carnot's theorem (see Part 12.3), which states that the six points lie on a conic if, and only if,

$$\left(\frac{BL}{LC}\right)\left(\frac{CM}{MA}\right)\left(\frac{AN}{NB}\right)\left(\frac{BD}{DC}\right)\left(\frac{CE}{EA}\right)\left(\frac{AF}{FB}\right) = 1. \tag{12.6}$$

Now the question arises as to which of the results still hold? In fact all the results true for the first generalization hold also for the second generalization! The reason for this is that if you consider the hexagon $FLENDM$ then its Pascal line is HJK, showing that the axis is still present. Also if you consider the hexagon $FLDNEM$ the Pascal line is $CC'C''$. These facts and similar ones show that the configuration is not dependent on the existence of two sets of Cevians, but on the relationship between Desargues's theorem and Pascal's theorem.

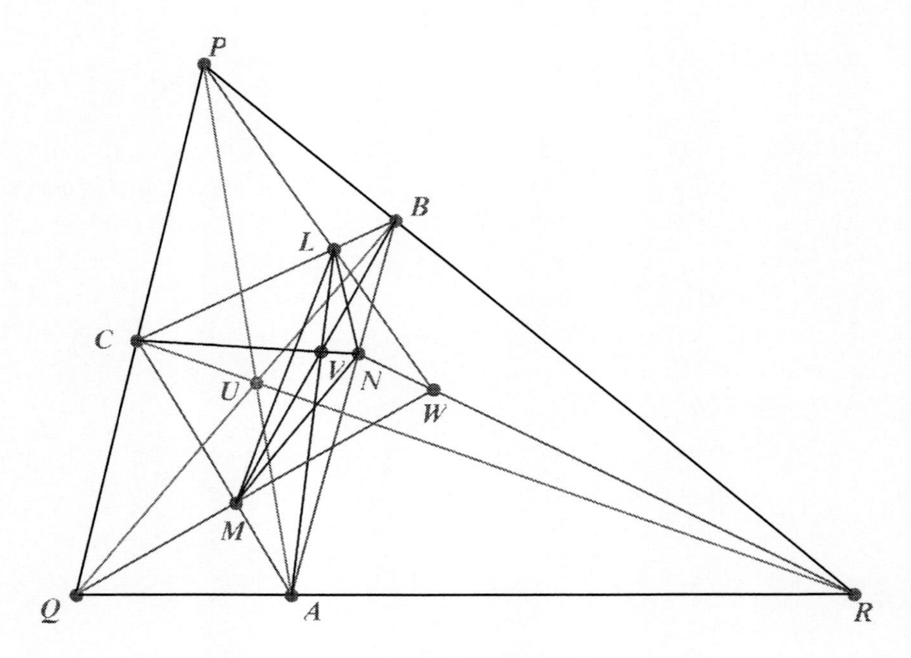

Figure 12.4: de Silva's theorem

12.3 Triangles in perspective arising from circles and a conic

12.3.1 de Silva's theorem [16]

PQR is a triangle and A lies on QR, B lies on RP and C lies on PQ. L lies on BC, M lies on CA, N lies on AB. Consider the following three statements:

(i) PQR and ABC are in perspective;

(ii) ABC and LMN are in perspective;

(iii) PQR and LMN are in perspective.

The theorem is that if any two of these statements are true, then the third statement is also true. See Fig. 12.4.

Let PQR be the triangle of reference. Since A, B, C lie on QR, RP, PQ respectively then (in an obvious notation) constants p, q, r exist (not 0 or 1) such that $A = pQ + (1-p)R$, $B = qR + (1-q)P$, $C = rP + (1-r)Q$ and statement (i) holds, by Ceva's theorem, if, and only if, $pqr = (1-p)(1-q)(1-r)$.

Similarly constants l, m, n exist (not 0 or 1) such that $L = lB + (1-l)C$, $M = mC + (1-m)A$, $N = nA + (1-n)B$ and statement (ii) holds, by Ceva's theorem, if, and only if, $lmn = (1-l)(1-m)(1-n)$.

Now the co-ordinates of L are $(l(1-q) + (1-l)r, (1-l)(1-r), lq)$, so the equation of PL is $ylq = z(1-l)(1-r)$. Similarly the equations of QM and RN are $zmr = x(1-m)(1-p)$ and $xnp = y(1-n)(1-q)$. Now these equations can subsist simultaneously if, and only if, $x = y = z = 0$ (a point which does not exist) or at a point with finite non-zero x, y, z if, and only if,

$$lmnpqr = (1-l)(1-m)(1-n)(1-p)(1-q)(1-r), \tag{12.7}$$

which is the necessary and sufficient condition for statement (iii). The result now follows.

12.3.2 Carnot's theorem

Let Σ be a conic that meets the side BC of a triangle at L, D, the side CA at M, E and the side AB at N, F. Then, supposing that none of these points coincides with a vertex,

$$\left(\frac{BL}{LC}\right)\left(\frac{BD}{DC}\right)\left(\frac{CM}{MA}\right)\left(\frac{CE}{EA}\right)\left(\frac{AN}{NB}\right)\left(\frac{AF}{FB}\right) = 1. \tag{12.8}$$

We use areal co-ordinates. Let the equation of Σ be

$$ux^2 + vy^2 + wz^2 + 2pyz + 2qzx + 2rxy = 0. \tag{12.9}$$

This meets BC, $x = 0$, where $vy^2 + 2pyz + wz^2 = 0$. Treating this as an equation in y and z, let the two solutions be y_1/z_1 and y_2/z_2, then we have $y_1y_2/z_1z_2 = w/v$. Now if $(0, y, z)$ are the co-ordinates of a point X on BC, then $BX/XC = z/y$, so in the case of our conic we have

$$\left(\frac{BL}{LC}\right)\left(\frac{BD}{DC}\right) = \frac{v}{w}.$$

Similarly

$$\left(\frac{CM}{MA}\right)\left(\frac{CE}{EA}\right) = \frac{w}{u}$$

and
$$\left(\frac{AN}{NE}\right)\left(\frac{AF}{FB}\right)=\frac{u}{v}$$
and the result follows.

The converse of the result in Section 12.3.2 also holds. For, suppose Equation (12.8) holds and Σ is the conic through the points L, M, N, D, E. Let Σ meets AB again at F'. Then, by Section 12.3.2, Equation (12.8) also holds with F' replacing F. Dividing out we get $AF/FB = AF'/F'B$ and so F and F' coincide, and Σ therefore passes through all six points.

Note that the analysis still holds when some of the points coincide. For example if L and D coincide and there are no other coincident points, then the conic touches BC at L. If in addition M and E coincide and N and F coincide, then either the conic touches all three sides or the conic is a degenerate coincident pair of straight lines and LMN is a transversal.

Note also constants l, m, n exist so that
$$(BL)(BD)=lv,\ (CL)(CD)=lw,\ (CM)(CE)=mw,\ (AM)(AE)=mu,$$
$$(AN)(AF)=nu \text{ and } (BN)(BF)=nv. \tag{12.10}$$

We now come to the main result of this section, which is as follows:

12.3.3 Applications involving de Silva's theorem

Let ABC be a triangle, with circumcircle S. And let Σ be a conic that cuts BC at L, D; CA at M, E; and AB at N, F (where L, D or M, E or N, F may coincide, but none of L, M, N, D, E, F lies at a vertex of ABC). Let S_1 be any circle passing through L and D (or touching BC at L if L and D coincide). Let S_2 be any circle passing through M and E (or touching CA at M if M and E coincide). And let S_3 be any circle passing through N and F (or touching AB at N if N and F coincide). Define PQR to be the triangle such that QR is the radical axis of S and S_1, RP is the radical axis of S and S_2 and PQ is the radical axis of S and S_3. Then triangles ABC and PQR are in perspective. Furthermore if S_1 passes through A, S_2 passes through B and S_3 passes through C, and triangles ABC and LMN are in perspective, then triangles ABC and DEF are in perspective, as are triangles LMN and PQR and triangles DEF and PQR. Let Equation (12.9) define the conic Σ. As established in the note to Section 12.3.2, Equations (12.10) hold. We now use the fact that the equation of any circle in areal co-ordinates has the form

$$a^2yz + b^2zx + c^2xy - (x+y+z)(\alpha x + \beta y + \gamma z) = 0, \qquad (12.11)$$

where a, b, c are the sides of ABC and α, β, γ are the powers of A, B, C respectively, with respect to the circle. See Equation (2.12). The earliest reference to this result we have been able to find is Problem 1317 of Wolstenholme [37]. Now consider the circle S_1. As proved in the note to Section 12.3.2 there exists a constant l such that $\beta = lv$ and $\gamma = lw$. Writing $\alpha = lu'$, where u' is an extra parameter that locates S_1, we see that the equation of S_1 is

$$a^2yz + b^2zx + c^2xy - l(x+y+z)(ux + vy + wz) = 0. \qquad (12.12)$$

Now the equation of the circumcircle S is

$$a^2yz + b^2zx + c^2xy = 0. \qquad (12.13)$$

Subtracting equation (12.3.6) from equation (12.3.7) we get

$$u'x + vy + wz = 0, \qquad (12.14)$$

which, being linear, must be the equation of the radical axis of S and S_1. Similarly, there exist constants v', w' defining the locations of S_2, S_3 respectively, such that the radical axes of S and S_2, and S and S_3 are respectively

$$ux + v'y + wz = 0 \qquad (12.15)$$

and

$$ux + vy + w'z = 0. \qquad (12.16)$$

Equations (12.14), (12.15), (12.16) are the equations of the lines QR, RP, PQ respectively, and solving these equations in pairs we find the co-ordinates of P to be $(-vw+v'w', u(w-w'), u(v-v'))$ and the equation of the line AP is thus $(v-v')y = (w-w')z$. Similarly BQ, CR have equations $(w-w')z = (u-u')x$, $(u-u')x = (v-v')y$ and we see that AP, BQ, CR are concurrent at the point X with co-ordinates $(1/(u-u'), 1/(v-v'), 1/(w-w'))$, provided $u' \neq u$, $v' \neq v$, $w' \neq w$. See Fig. 12.5. It is interesting to understand what happens if $u' = u$ (or $v' = v$ or $w' = w$). If $u' = u$, then Q lies on AB and R lies on CA, so that X coincides with A. The reader should check other possible

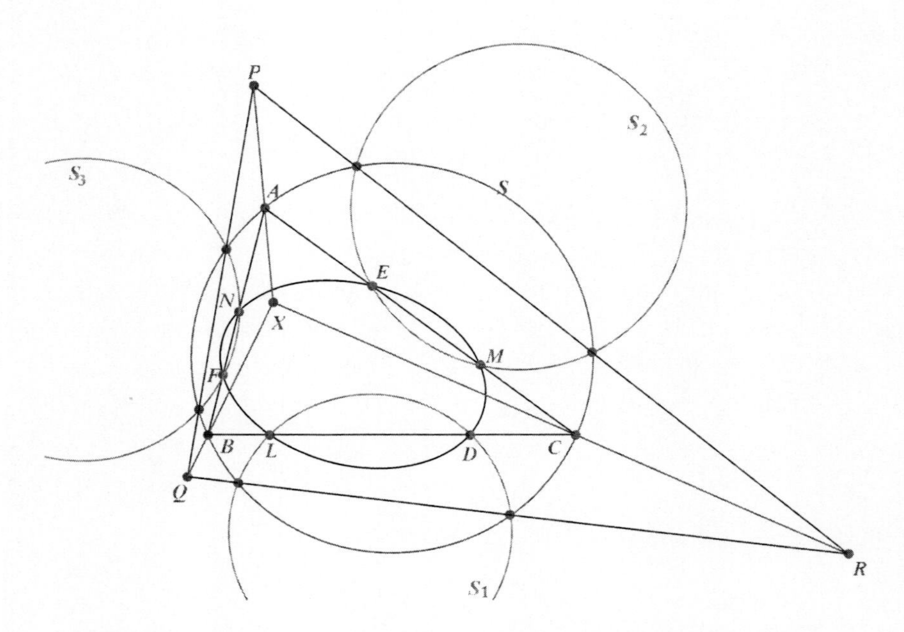

Figure 12.5: First configuration for Section 12.3.3

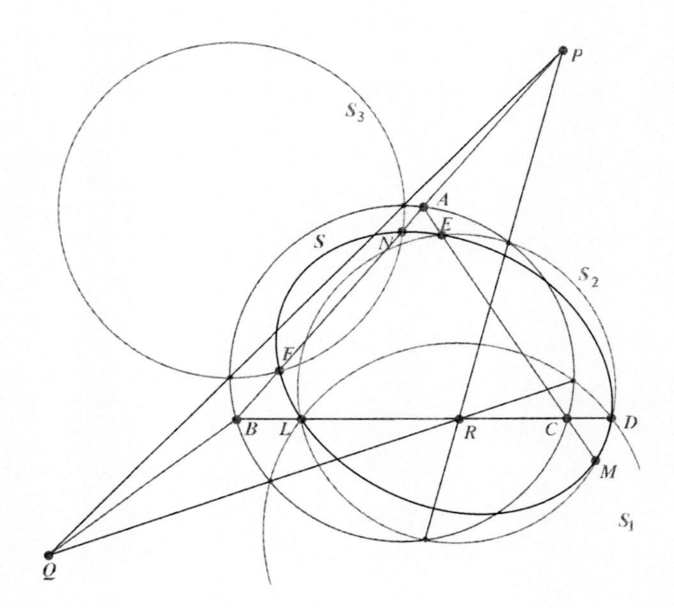

Figure 12.6: Second configuration for Section 12.3.3

singular cases. Fig. 12.6 shows a case with $v' = v$, in which P lies on AB, R lies on BC and X coincides with B.

Suppose now that AL, BM, CN are concurrent at a point K, so that AL, BM, CN are a set of Cevians for triangle ABC. Then, by Ceva's theorem we have $(BL/LC)(CM/MA)(AN/NB) = 1$. From Equation (12.6) it now follows that $(BD/DC)(CE/EA)(AF/FB) = 1$, so by the converse of Ceva's theorem AD, BE, CF are concurrent or parallel. In either case the two triangles ABC and DEF are in perspective. If in addition S_1, S_2, S_3 pass through A, B, C respectively, then QR passes through A, RP passes through B and PQ passes through C. Now the conditions of Section 12.3.1 apply and not only are ABC and PQR in perspective from a point X, but PL, QM, RN are in perspective from a point Y and PD, QE, RF are in perspective from a point Z.

In cases in which S_1 passes through A, S_2 passes through B and S_3 passes through C, it is evident that $u' = v' = w' = 0$, and hence P, Q, R have co-ordinates $P(-1/u, 1/v, 1/w)$, $Q(1/u, -1/v, 1/w)$, $R(1/u, 1/v, -1/w)$ and X has co-ordinates $X(1/u, 1/v, 1/w)$. Suppose further that L, M, N have co-ordinates $L(0, m, n)$, $M(l, 0, n)$, $N(l, m, 0)$ and that D, E, F have co-ordinates $D(0, e, f)$, $E(d, 0, f)$, $F(d, e, 0)$ (the condition for AD, BE, CF to be parallel is $d + e + f = 0$). Then the equation of the conic Σ is

$$efmnx^2 + fdnly^2 + delmz^2 \tag{12.17}$$

$$-dl(en + fm)yz - em(fl + dn)zx - fn(dm + el)xy = 0.$$

Hence we may take $u = 1/dl$, $v = 1/em$, $w = 1/fn$. Thus X has co-ordinates (dl, em, fn). The co-ordinates of Y are $(dl(e + f - d), em(f + d - e), fn(d + e - f))$ and those of Z are $(dl(m + n - l), em(n + l - m), fn(l + m - n))$. See Fig. 12.7.

Example 12.3.1

Take S_1, S_2, S_3 to be the circles touching BC, CA, AB at their midpoints and passing through A, B, C respectively. Then $d = e = f$ and $l = m = n$. And X, Y, Z coincide at the centroid of ABC. In this case LMN is the median triangle of ABC and ABC is the median triangle of PQR. See Fig. 12.8.

Exercise 12.3.2

1. Take S_1, S_2, S_3 to be the circles on AL, BM, CN as diameters respectively, where L, M, N are the midpoints of the sides. Find the co-ordinates of the points $D, E, F, P, Q, R, X, Y, Z$. (See Fig. 12.9)

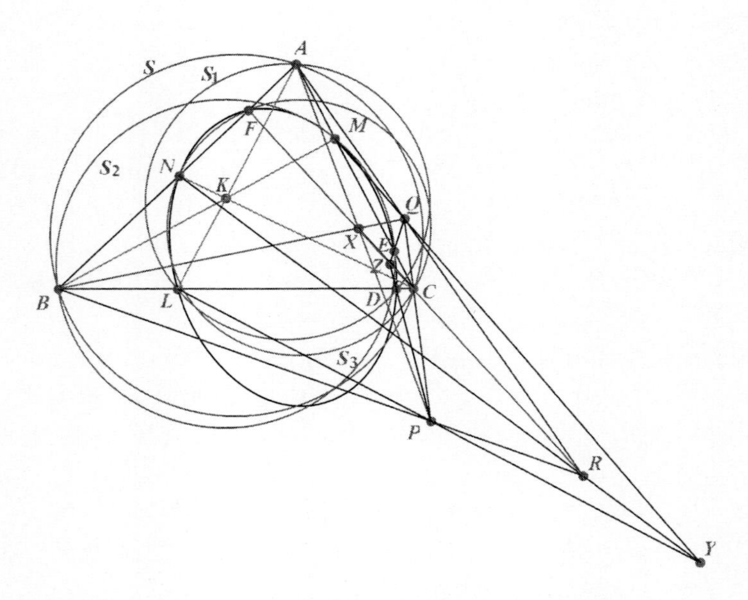

Figure 12.7: Third configuration for Section 12.3.3

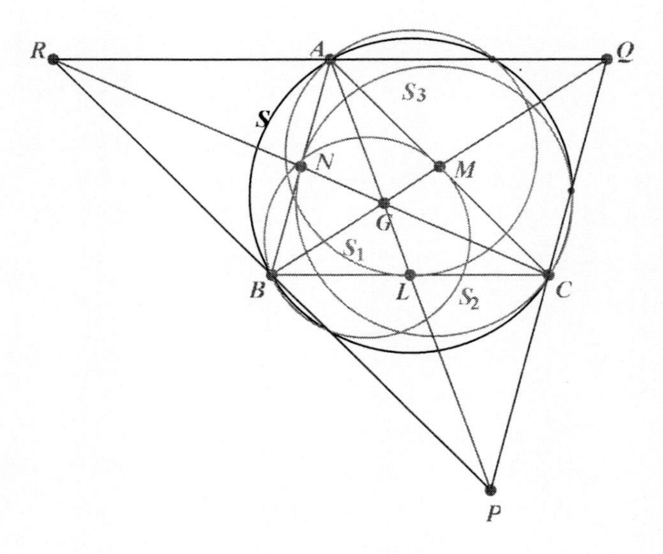

Figure 12.8: Example 12.3.1

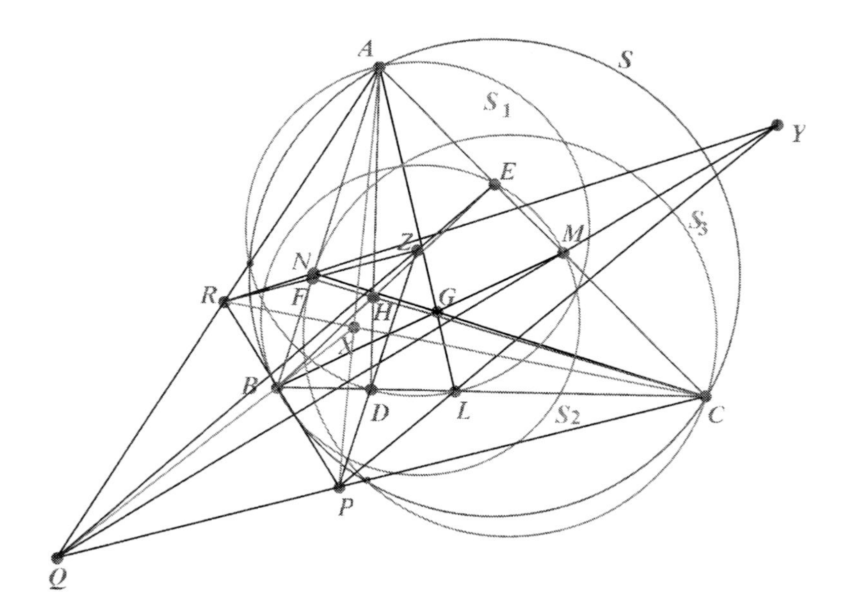

Figure 12.9: Example 12.3.2

Example 12.3.3

Let LMN be a transversal with $L(0, n, -m), M(-n, 0, l)$ and $N(m, -l, 0)$. Since the transversal is not parallel to a side $l, m, n \neq 0$. Let S_1 be the circle touching BC at L and passing through A, with S_2, S_3 similarly defined. Then it is easy to show that QR has equation $m^2y+n^2z = 0$, with similar equations for RP and PQ by cyclic change of letters. X is the point with co-ordinates $(1/l^2, 1/m^2, 1/n^2)$. L, M, N are not the feet of a set of Cevians, so the points Y, Z do not exist. See Fig. 12.10.

Exercise 12.3.4

1. Let L, M, N be the points where the incircle touches BC, CA, AB respectively. Let S_1 be the circle touching BC at L and passing through A, with S_2, S_3 similarly defined. Here AL, BM, CN are the Cevians through Gergonne's point Ge with co-ordinates $Ge(1/(s-a), 1/(s-b), 1/(s-c))$. Find the co-ordinates of P, Q, R and X. Find also the co-ordinates of Y. See Fig. 12.11.

2. Let AKL, BKM, CKN be a set of Cevians of triangle ABC. Suppose that S_1, S_2, S_3 are the circles AMN, BNL, CLM. It is known that these

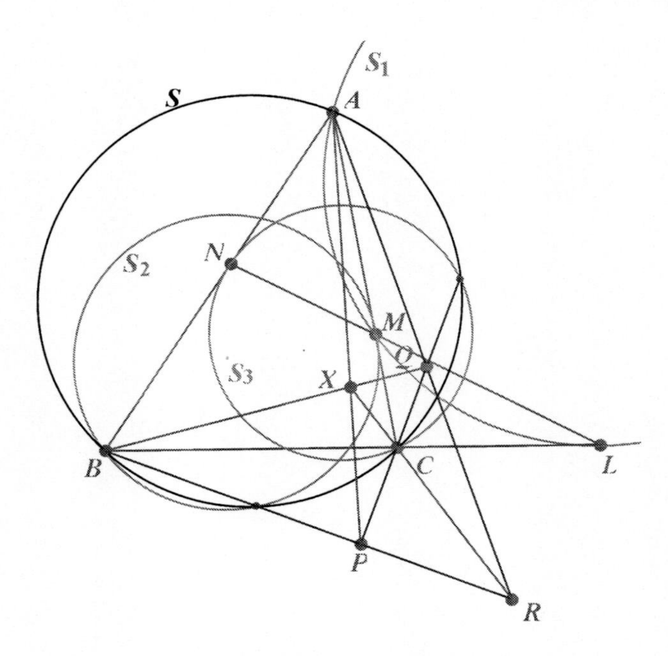

Figure 12.10: Example 12.3.3

circles have a common point T (the pivot point). Let S be the circumcircle and AQR the radical axis of S and S_1, with BRP, CPQ similarly defined. Prove that AP, BQ, CR are concurrent at a point X and that PL, QM, RN are concurrent at a point Y. Locate the position of Y relative to S_1, S_2, S_3.

3. What happens to the results of Problem 2 if instead of being the feet of a set of Cevians, L, M, N are points on a transversal, with L on BC, M on CA and N on AB?

12.4 Quadrangles in perspective

In this section we establish a sufficient condition for when a pair of quadrangles have what may be appropriately called a Desargues axis of perspective. It is also shown how the condition is automatically satisfied when corresponding pairs of vertices of the two quadrangles lie on a conic and are in

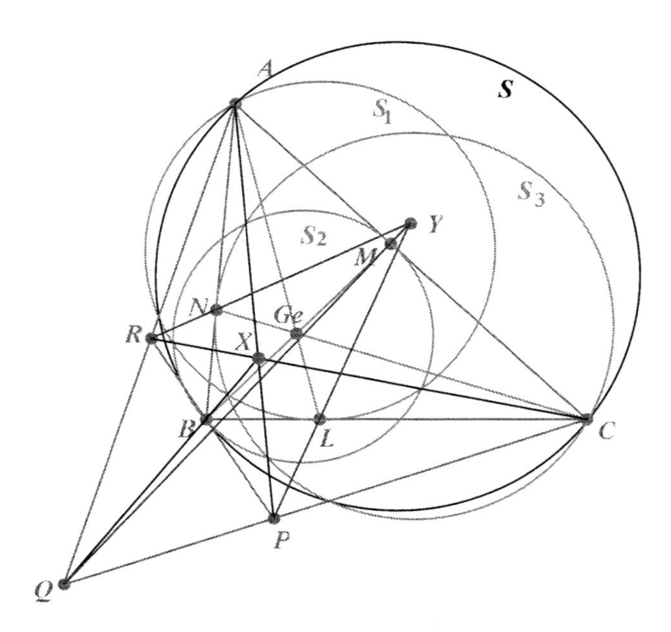

Figure 12.11: Exercise 12.3.4 Problem 1

involution by means of a perspectivity. The ideas are then extended to provide a guide as to what happens for polygons with an arbitrary number of sides. The analysis is carried out using projective co-ordinates. We define two types of perspective in dealing with the polygons $A_1A_2 \ldots An$ and $B_1B_2 \ldots B_n$. *Partial perspective* is defined to mean that $A_1B_1, A_2B_2, \ldots, A_nB_n$ are concurrent at a point X. *Complete perspective* is defined to mean that all the points $A_jA_k \wedge B_jB_k$, $j, k = 1$ to $n, j \neq k$ are collinear. So complete perspective is when an axis exists that may be termed a Desargues axis of perspective.

We have already seen in Part 12.1 that when $n = 3$ and we are dealing with triangles, then partial perspective $\Leftrightarrow$ complete perspective.

12.4.1 Quadrangles in complete perspective

We give a sufficient condition for quadrangles to be in complete perspective. Let $ABCD$ and $PQRS$ be in partial perspective with vertex X, so that P lies on AX, Q on BX, R on CX, and S on DX. Then, if $AC \wedge BD$, $PR \wedge QS$ are collinear with X, $ABCD$ and $PQRS$ are in complete perspective. What

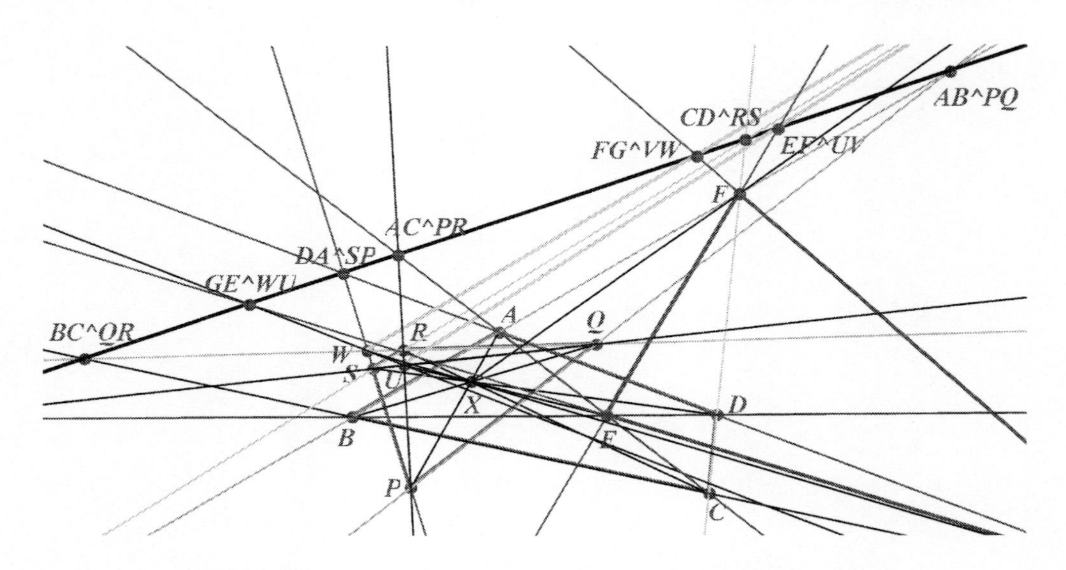

Figure 12.12: Quadrangles in complete perspective

the extra condition means is that if one corresponding pair of diagonal points $E = AC \wedge BD$ and $U = PR \wedge QS$ are such that EU also passes through X, and if $ABCD$ and $PQRS$ are also in partial perspective with vertex X, then they are in complete perspective. See Fig. 12.12.

In the real projective plane we take a co-ordinate system such that A is $(-1,1,1)$, B is $(1,-1,1)$, C is $(1,1,-1)$, D is $(1,1,1)$; the diagonal point triangle is the triangle of reference and in particular $E = AC \wedge BD$ is $(0,1,0)$. Let X be (f,g,h). Then P has co-ordinates $(f-\lambda, g+\lambda, h+\lambda)$ for some λ, Q has co-ordinates $(f+\mu, g-\mu, h+\mu)$ for some μ, R has co-ordinates $(f+\nu, g+\nu, h-\nu)$ for some ν, and S has co-ordinates $(f-\rho, g-\rho, h-\rho)$ for some value of ρ.

For the intersection of PR with EX we want a linear combination of P and R satisfying $hx = fz$. This is easily found to be

$$\nu P + \lambda R = (f(\lambda+\nu), g(\lambda+\nu) + 2\lambda\nu, h(\lambda+\nu)). \tag{12.18}$$

Similarly the intersection of QS with EX is

$$\rho Q + \mu S = (f(\mu+\rho), g(\mu+\rho) + 2\mu\rho, h(\mu+\rho)). \tag{12.19}$$

We require these to be the same point, the condition for which is

$$\frac{2\lambda\nu}{\lambda+\nu} = \frac{2\mu\rho}{\mu+\rho}$$

or

$$\lambda\mu\nu + \lambda\mu\rho + \lambda\nu\rho + \mu\nu\rho = 0. \tag{12.20}$$

Now consider the axis of perspective of triangles ABC and PQR. One point on it is the intersection of AB ($x + y = 0$) and PQ. This is $P - Q = (-\lambda - \mu, \lambda + \mu, \lambda - \mu)$. Another is the intersection of BC ($y + z = 0$) and QR, which is $Q - R = (\mu - \nu, -\mu - \nu, \mu + \nu)$. The line joining these is easily calculated to be

$$\lambda(\mu + \nu)x + \mu(\nu + \lambda)y + \nu(\lambda + \nu)z = 0. \tag{12.21}$$

From Desargues's theorem for the triangle, $CA \wedge RP$ must lie on this line. We must show that $AD \wedge PS$ lies on this line. AD has equation $y = z$ and the intersection is

$$P - S = (\rho - \lambda, \rho + \lambda, \rho + \lambda).$$

So the condition, from Equation (12.21), is

$$\lambda(\mu + \nu)(\rho - \lambda) + \mu(\nu + \lambda)(\rho + \lambda) + \nu(\lambda + \mu)(\rho + \lambda) = 0 \tag{12.22}$$

and this reduces to Equation (12.20). Similarly $BD \wedge QS$, $CD \wedge RS$ lie on the axis of perspective. The above proof is due to Monk [28].

12.4.2 Triangles in complete perspective

Under the same conditions as Section 12.4.1, the diagonal point triangles EFG of $ABCD$ and UVW of $PQRS$ are in perspective, with vertex of perspective X and with the same axis of perspective as the quadrangles themselves. The co-ordinates of F are $(0, 0, 1)$. By working analogous to that in Section 12.4.1, Equation (12.21) ensures that the intersection of PQ and FX is the same point as the intersection of RS with FX, and this shows that $AB \wedge CD$, X and $PQ \wedge RS$ are collinear. Similarly $AD \wedge BC$, X and $PS \wedge QR$ are collinear. This shows that triangles EFG and UVW are in perspective with vertex X. It is straightforward, though tedious, to show that the axis of perspective of the two diagonal point triangles is the same axis of perspective as the quadrangles themselves and a proof is omitted. Fig. 12.12 illustrates Sections 12.4.1 and 12.4.2.

An alternative theorem to that in Section 12.4.1 is possible. If one starts with triangles ABC and PQR in perspective, so that what will become the Desargues axis is located. Then D can be placed anywhere and the position of S may be fixed, not by the conditions of Section 12.4.1, but by requiring $AD \wedge PS$ and $BD \wedge QS$ to lie on the perspectrix. It then follows that $CD \wedge RS$ automatically lies on the perspectrix and the quadrangles are in complete perspective, as before. It follows, since D, S, X are then collinear, that for quadrangles complete perspective implies partial perspective.

It is now possible to see how to generalize the result for quadrangles to polygons with an arbitrary number of sides. As before partial perspective plus a number of other conditions ensure complete perspective. To get the idea let us consider pentagons $ABCDE$ and $PQRST$. Let AP, BQ, CR, DS, ET be concurrent at X. Fix D and S as in Section 12.4.1 so that quadrangles $ABCD$ and $PQRS$ are in complete perspective. Then E and T are selected in similar fashion so that $AC \wedge BE$, X, $PR \wedge QT$ are collinear. Then it follows that $ABCDE$ and $PQRST$ are in complete perspective. Essentially one builds up the pentagon from the two pairs of complete quadrangles $ABCD$, $PQRS$ and $ABCE$, $PQRT$. Admittedly this is merely a guide to what happens and does not constitute a proof.

12.5 Quadrangles inscribed in a conic

We now consider a special case when two quadrangles in partial perspective are automatically in complete perspective.

12.5.1 Quadrangles in perspective inscribed in a conic

Let $ABCD$ and $PQRS$ be two distinct quadrangles in partial perspective with vertex Y, and suppose all the vertices lie on a conic Γ not passing through Y, then the quadrangles are in complete perspective. First we give a co-ordinate proof. We show that $AC \wedge BD, Y, PR \wedge QS$ are collinear, so that Section 12.4.1 applies. Let the conic Γ have equation $y^2 = zx$, with points U on Γ having co-ordinates given parametrically as $(u^2, u, 1)$. Then the equation of UV is $x + uvz = (u + v)y$. Suppose now that Y has co-ordinates $(0, 1, 0)$, where Y does not lie on Γ, then points U, V at opposite ends of chords through Y are in involution by means of the perspectivity through Y, given analytically by $v = -u$. Let A, B, C, D have parameters a, b, c, d, then P, Q, R, S have parameters $-a, -b, -c, -d$. The equation of AC is $x - (a + c)y + acz = 0$ and that of BD is $x - (b + d)y + bdz = 0$ and

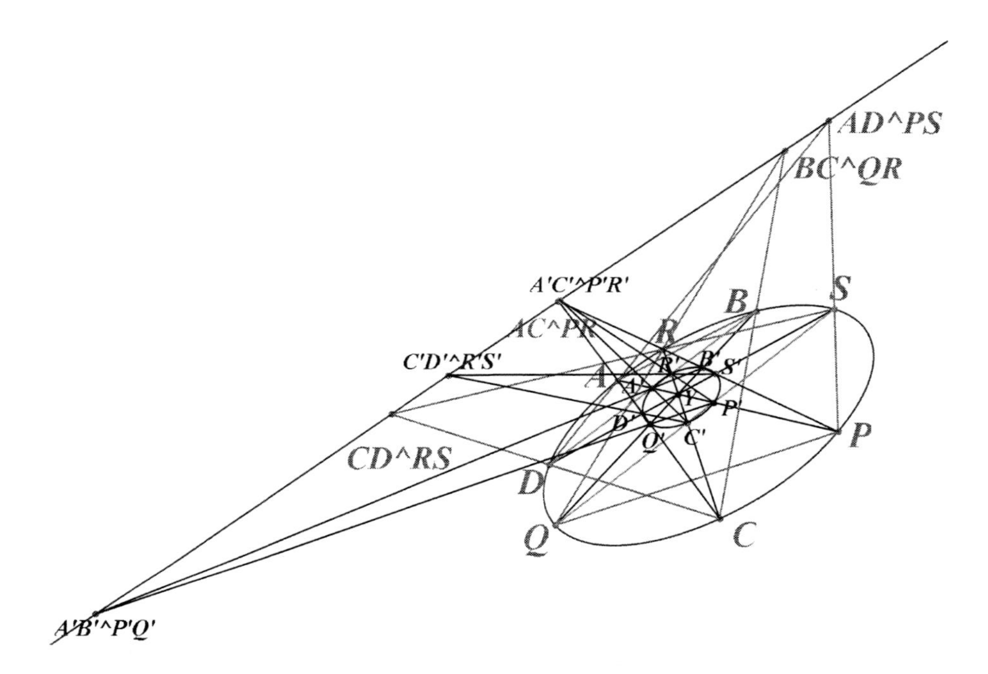

Figure 12.13: Configuration for Sections 12.5.1 and 12.5.2

$AC \wedge BD$ has co-ordinates

$$(ac(b+d) - bd(a+c), ac - bd, a + c - b - d).$$

Similarly $PR \wedge QS$ has co-ordinates

$$(-ac(b+d) + bd(a+c), ac - bd, -a - c + b + d).$$

Their collinearity with Y is now obvious.

An observation is that if $ac = bd$, then the two points above coincide. This is perfectly possible, but does not invalidate the theorem as we may then take two other corresponding diagonal points, and not all pairs of diagonal points can coincide unless the quadrangles coincide, which we have assumed not to be the case. The synthetic proof is trivial. From the inscribed quadrangle $ADPS$ the intersection $AD \wedge PS$ lies on the polar of Y and similarly for the other five intersections. Since A and P may be interchanged and also B and Q, C and R, and D and S it is also the case that such intersections as $AB \wedge PQ$, $AC \wedge PR$, $BC \wedge QR$, $BD \wedge QS$, $CD \wedge RS$ also lie on the polar of Y. For the purpose of clarity and because of available space only a few of the intersections are shown in Fig. 12.13.

12.5.2 Quadrangles induced in a second conic

In the same configuration as discussed in Section 12.5.1, let AYP meet BD and QS at A' and P', let CYR meet BD and QS at C' and R', let BYQ meet AC and PR at B' and Q' and let DYS meet AC and PR at D' and S'. Then $A', B', C', D', P', Q', R', S'$ lie on a conic and the polar of Y with respect to this conic is the same Desargues axis as for the quadrangles $ABCD$, $PQRS$.

The equation of BD is $x-(b+d)y+bdz=0$ and the equation of AYP is $x=a^2z$ and these meet at A', with co-ordinates $(a^2(b+d), a^2+bd, (b+d))$. Likewise P' has co-ordinates $(-a^2(b+d), a^2+bd, -(b+d))$. The co-ordinates of C' and R' follow by replacing a^2 by c^2. The co-ordinates of the other four points now follow by interchanging the roles of a, c and b, d. The equation of the conic that passes through all eight points may now be verified to be

$$\begin{aligned}((a+c)^2+(b+d)^2)x^2-(a+c)^2(b+d)^2y^2+(b^2c^2d^2+a^2c^2d^2+a^2b^2d^2+a^2b^2c^2 \\ +2abcd(ac+bd))z^2-((a^2+c^2)(b^2+d^2)-4abcd)zx=0.\end{aligned} \tag{12.23}$$

Furthermore the fact that xy and yz terms in Equation (12.23) are absent means that the polar of Y with respect to this conic has equation $y=0$, which is also the polar of Y with respect to the original conic $y^2=zx$.

It now follows that the quadrangles $A'B'C'D'$ and $P'Q'R'S'$ are in perspective with vertex Y and since they lie on a conic are in complete perspective. Thus intersections such as $A'B' \wedge P'Q'$ also lie on the same Desargues axis of perspective. See Fig. 12.13 as illustration of this.

12.6 Cross ratio

Consider the quadrangle $ABCD$ inscribed in the conic with equation $y^2=zx$, where A, B, C, D have parameters a, b, c, d then $(A, C; B, D) = (a, c; b, d)$. If the quadrangle with vertices P, Q, R, S is defined to be in perspective through Y, as in Section 12.5.1, then its parameters are $-a, -b, -c, -d$ respectively and since $(-a, -c; -b, -d) = (a, c; b, d)$, it follows that $(A, C; B, D) = (P, R; Q, S)$. Since the point Y is at our disposal we have proved the following theorem:

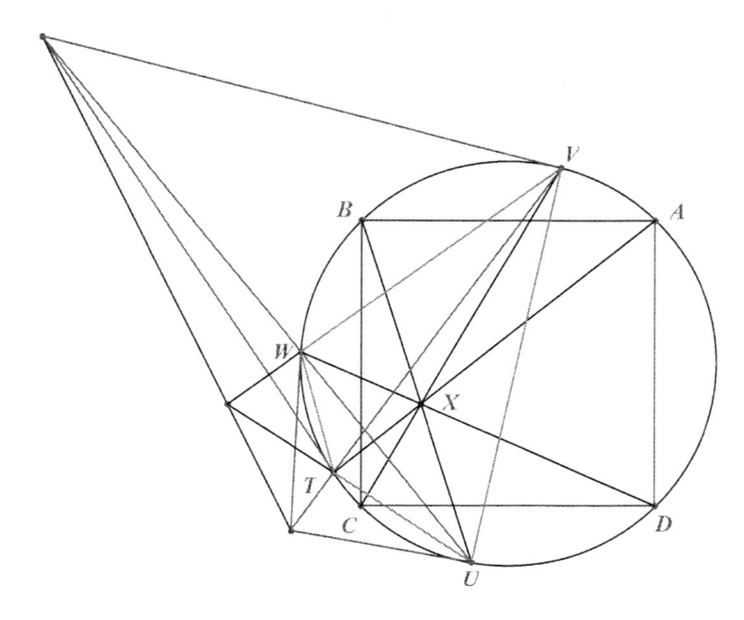

Figure 12.14: A harmonic quadrangle inscribed in a circle

12.6.1 Cross-ratio

Two quadrangles $ABCD$ and $PQRS$ in partial perspective on a conic are such that $(A, C; B, D) = (P, R; Q, S)$.

All one is saying is that cross-ratio on a conic is preserved by a perspectivity, but this result has interesting consequences. Consider a square inscribed in a circle. Using the $\tan(\frac{1}{2}\theta)$ parameter for its vertices, then its vertices have parameters $0, 1, \infty, -1$. Since $(0, \infty; 1, -1) = -1$ we have the result that any quadrangle inscribed in a circle that arises from a square by a partial perspective has vertices that form pairs of conjugate points, or if you prefer has vertices that separate one another harmonically.

We remind the reader of the key properties of a harmonic quadrangle $TUVW$. They are that $TU \wedge VW$, $TW \wedge UV$ define the polar line and the tangents at T, V and the line UW concur on the polar line as do the tangents at U, W and the line TV. See Fig. 12.14. In the case of the square these intersections lie on the line at infinity. It is the case for all quadrangles $TUVW$ that the tangents at pairs of opposite vertices meet on the polar line, but it is only true for harmonic quadrangles that UW and TV have the special property noted above.

A projective transformation of the real projective plane on to itself maps any four non-collinear points on to any other four non-collinear points. It has the property that it leaves invariant the incidence properties of a configuration consisting of points, lines and conics and leaves cross-ratios of points on conics invariant. If therefore we take a square $ABCD$ inscribed in a circle and map $A \to P$, $B \to Q$, $C \to R$, $D \to S$, where P, Q, R, S are any four non-collinear points, then the configuration resulting is a quadrangle $PQRS$ inscribed in the image of the circle, which is some conic. Since the cross-ratio is left invariant we know that $PQRS$ must be a harmonic quadrangle.

Chapter 13

Hexagons

13.1 The Pascal theorem configuration

Two proofs of Pascals theorem have been given previously in Sections 3.6.8 and 3.8.7, and we now give a more detailed review of this famous configuration, together with illustrations of some of its main features. We start by stating the theorem again.

13.1.1 Pascal's theorem

Let S be a conic and let D, E, F, L, M, N be six points lying on S. The notation (DEF, LMN) is taken to denote the set of points $\{11, 22, 33\}$ whose elements are the three cross-join intersections $11 = EN \wedge FM$, $22 = FL \wedge DN$, $33 = DM \wedge EL$. Pascal's theorem states that these three points are collinear. Note that if we say that $11, 22, 33$ is the Pascal line arising from (DEF, LMN) the order of the points is important. The first point 11 arises from the cross-join of the $2 \leftrightarrow 3$ and $3 \leftrightarrow 2$ letters and so on.

13.1.2 Pascal lines

This particular Pascal line may also be thought of as arising from the hexagon $DNELFM$, where DN, LF are opposite sides of the hexagon, as are NE, FM and EL, MD. We can always insist that D is chosen as the first vertex of the hexagon. Then, since there are $\binom{5}{2} = 10$ selections of letters that can be associated with D as third or fifth letters; and 6 permutations of the remaining three letters, it follows that there are at most 60 different Pascal lines. In fact in a general situation, without any special symmetry, such as the hexagon being regular, it can be shown that the 60 Pascal lines are distinct.

Fig. 13.1 shows six of the Pascal lines. These are, in addition to the line 11 22 33 specified above, 21 32 13 arising from (DEF, MNL); 31 12 23 arising from (DEF, NLM); A, A', A'' arising from (DEF, LNM); B'', B, B' arising from (DEF, NML); and C', C'', C arising from (DEF, MLN).

13.1.3 Steiner points

It is apparent that we can draw 6 significant triangles from these 18 points. They are triangles 11 32 23; 22 13 31; 33 21 12, $ABC, A'B'C', A''B''C''$. The significance lies in the fact that A, A', A'' are collinear, B, B', B'' are collinear and C, C', C'' are collinear and that these three lines are concurrent at the point labelled W in Fig. 13.1. This means that triangles ABC, $A'B'C'$ and $A''B''C''$ are in perspective with vertex W. Similarly 11 22 33 are collinear, 32 13 21 are collinear and 23 31 12 are collinear and these three lines are concurrent at the point labelled 4 in Fig. 13.1. Thus the 6 Pascal lines based on the choice $(DEF, \ldots)$ fall into two sets of three lines that are concurrent. W and 4 are called *Steiner points* and it can be shown they are conjugate points with respect to S. Since there are 60 Pascal lines altogether they fall into 20 sets of 3 concurrent lines. The points of concurrence define 20 Steiner points. To prove analytically that three suitably chosen Pascal lines are concurrent at a Steiner point, a projective proof is easiest. Choose S to have the equation $yz + zx + xy = 0$, with DEF the triangle of reference, so that $D(1, 0, 0)$, $E(0, 1, 0)$, $F(0, 0, 1)$ lie on S. A parameter θ may be chosen on S so that the points of S have co-ordinates $(\theta(\theta + 1), (\theta + 1), -\theta)$, so let L, M, N have co-ordinates with parameters $\theta = l, m, n$ respectively. As we assume all six points are distinct, no two of l, m, n are equal and $l, m, n \neq \infty, 0, -1$, these being the parameters of D, E, F respectively. Now EN has equation $x+(n+1)z = 0$ and FM has equation $x = my$ and hence 11 has co-ordinates $11(m(n + 1), (n + 1), -m)$. FL has equation $x = ly$ and DN has equation $ny + (n + 1)z = 0$ and hence 22 has co-ordinates $22(l(n + 1), (n + 1), -n)$. DM has equation $my + (m + 1)z = 0$ and EL has equation $x + (l + 1)z = 0$ and hence 33 has co-ordinates $33(m(l + 1), (m + 1), -m)$. It may be verified that 11 22 33 are collinear on the line with equation

$$(m - n)x + m(n - l)y + (n + 1)(m - l)z = 0. \tag{13.1}$$

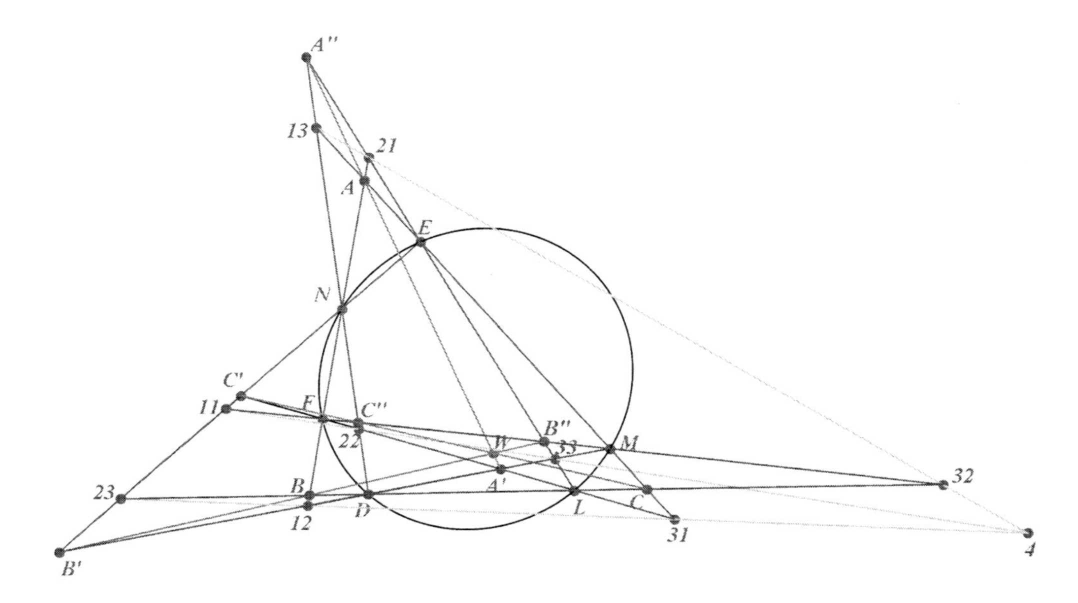

Figure 13.1: Six Pascal lines

This is yet another proof of Pascal's theorem.

The points $21, 32, 13$ have co-ordinates that may be deduced from $11, 22,$ 33 by cyclic change of l, m, n (but not x, y, z), so they are collinear on the line with equation

$$(n-l)x + n(l-m)y + (l+1)(n-m)z = 0. \tag{13.2}$$

Similarly $31, 12, 23$ lie on the line with equation

$$(l-m)x + l(m-n)y + (m+1)(l-n)z = 0. \tag{13.3}$$

Equations (13.1), (13.2), (13.3) are linearly dependent since their left hand sides obviously sum to zero, and hence these three lines are concurrent at the Steiner point labelled 4 in the Fig. 13.1. We do not obtain the co-ordinates of 4, as they involve cumbersome expressions. For an approach that embeds the configuration into three dimensions, the ellipse being the intersection of a quadric surface with a plane, see Pedoe [29].

Once a single Steiner point has been shown to exist, it is evident that there are 20 of them altogether. These 20 Steiner points have some interesting properties, which, in due course, we illustrate but do not prove.

Exercise 13.1.1

1. Obtain the co-ordinates of $A, A', A'', B, B', B'', C, C', C''$ and show they lie on three distinct lines that are concurrent at the Steiner point labelled W in Fig. 13.1.

13.1.4 Desargues axes

We also have 20 sets of 3 triangles that are mutually in perspective. A second observation we can make follows from the fact that triangles in perspective have an axis of perspective. Thus, for example, since triangles $A'B'C'$ and $A''B''C''$ are in perspective we have a *Desargues axis,* which we can conveniently express as $(A'B'C', A''B''C'')$ meaning that it consists of the cross-joins $B'C' \wedge B''C'', C'A' \wedge C''A'', A'B' \wedge A''B''$. Perusal of Fig. 13.1 shows that these three points are precisely the points 11 22 33. Similarly the Desargues axis of triangles ABC and $A'B'C'$ is 23 31 12 and that of triangles ABC and $A''B''C''$ is 32 13 21. In this situation the property is reciprocal. Thus the Desargues axis of triangles 22 13 31 and 33 21 12 is $AA'A''$, the Desargues axis of triangles 33 21 12 and 11 32 23 is $BB'B''$ and the Desargues axis of triangles 11 32 23 and 22 13 31 is $CC'C''$. In other words, each of the 60 Pascal lines is also a Desargues axis of perspective. This link between Pascal's theorem and Desargues's theorem has already been noted in Chapter 12.

13.1.5 Kirkman points and Cayley lines

We now consider another 18 Pascal lines associated with the Fig. 13.1. They derive from the hexagons stated (with care being taken about the order of the vertices) and are as follows:

(1) $P'P''A$ from $FDEMLN$; (2) $Q'Q''B$ from $DEFNML$;
(3) $R'R''C$ from $EFDLNM$; (4) $P''PA$ from $DEFLNM$;
(5) $Q''QB'$ from $EFDMLN$; (6) $R''RC'$ from $FDENML$;
(7) $PP'A''$ from $EFDNML$; (8) $QQ'B''$ from $FDELNM$;
(9) $RR'C''$ from $DEFMLN$; (10) $QR11$ from $FDENLM$;
(11) $Q'R'32$ from $DEFMNL$; (12) $Q''R''23$ from $EFDLMN$;
(13) $RP22$ from $DEFLMN$; (14) $R'P'13$ from $EFDNLM$;
(15) $R''P''31$ from $FDEMNL$; (16) $PQ33$ from $EFDMNL$;
(17) $P'Q'21$ from $FDELMN$; (18) $P''Q''12$ from $DEFNLM$.

With reference to Fig. 13.2 lines (1), (2), (3) are concurrent at a point labelled T, lines (4), (5), (6) are concurrent at a point labelled V and lines (7), (8), (9) are concurrent at a point labelled U. Furthermore T, V, U are collinear on a line passing through W.

Also lines (10), (11), (12) are concurrent at a point labelled 1, lines (13), (14), (15) are concurrent at a point labelled 2 and lines (16), (17), (18) are concurrent at a point labelled 3. Furthermore 1, 2, 3 are collinear on a line that passes through 4. Points such as T, V, U and 1, 2, 3 are called *Kirkman points.* Evidently there are 60 Kirkman points and they have the property that they lie 3 at a time on lines through the 20 Steiner points. These 20 lines, 1 for each Steiner point and each containing 4 important points of concurrence, are called *Cayley lines.*

13.1.6 More Desargues axes

From the previous paragraph we see that there are 6 sets of 3 triangles mutually in perspective corresponding to the above 6 Kirkman points. They are:

(1) $ABC, P'Q'R', P''Q''R''$ with the 3 Desargues axes
$1\ 2\ 3, 23\ 31\ 12, 32\ 13\ 21$;

(2) $PQR, A'B'C', P''Q''R''$ with the 3 Desargues axes
$23\ 31\ 12, 1\ 2\ 3, 11\ 22\ 33$;

(3) $PQR, P'Q'R', A''B''C''$ with the 3 Desargues axes
$32\ 13\ 21, 11\ 22\ 33, 1\ 2\ 3$;

(4) $11\ 32\ 23, QQ'Q'', RR'R''$ with the 3 Desargues axes
$TVU, BB'B'', CC'C''$;

(5) $22\ 13\ 31, RR'R'', PP'P''$ with the 3 Desargues axes
$TVU, AA'A'', CC'C''$;

(6) $33\ 21\ 12, PP'P'', QQ'Q''$ with the 3 Desargues axes
$TVU, BB'B'', AA'A''$.

13.1.7 Unexpected triangles in perspective

In addition to all this there are two sets of 3 pairs of triangles in perspective that seem to be interesting. These are (1) ABC, PQR with vertex of perspective labelled $G; A'B'C', P'Q'R'$ with vertex of perspective labelled $G'; A''B''C'', P''Q''R''$ with vertex of perspective labelled G'' and (2)11 32 23, $PP'P''$ with vertex of perspective labelled H; 22 13 31, $QQ'Q''$ with vertex of perspective labelled H'; 33 21 12, $RR'R''$ with vertex of perspective labelled H''. Moreover the points G, G', G'' are collinear and the line $GG'G''$ passes through the Steiner point W and H, H', H'' are collinear and the line $HH'H''$ passes through the Steiner point 4. See Fig. 13.2. These lines are very similar to Cayley lines, except that their points do not arise directly as the intersection of Pascal lines, but from triangles unexpectedly found to be in perspective. In the Pascal line configuration there are 20 of these lines, one for each Steiner point. These lines are new to this writer (and until someone supplies a proof, their existence must remain formally conjectural, though supported by overwhelming experimental evidence).

13.1.8 Plücker lines

There are 20 Steiner points of which 2 have been introduced and are labelled W and 4 in Figs. 13.1 and 13.2. These points and the other 18 Steiner points we now relabel to provide a consistent notation for them. W is now relabelled as EF' and 4 is relabelled as EF. They carry the letters E and F because they arise from the 6 Pascal lines $(DEF, \ldots)$. As we always put D first that letter may be omitted. If the three remaining letters L, M, N now appear cyclically in alphabetic order no prime is attached. Thus EF arises as the point of concurrence of the three Pascal lines (DEF, LMN), (DEF, MNL), (DEF, NLM). But if they appear in the reverse order to cyclic then a prime is attached. Thus EF' arises as the point of concurrence of the three Pascal lines (DEF, LNM), (DEF, MLN), (DEF, NML). And, as a second example, ML' arises as the point of concurrence of the three Pascal lines (DML, ENF), (DML, FEN), (DML, NFE), the prime occurring because the letters E, N, F appear out of their natural alphabetic order. Note that in this convention the point EF is the same as the point FE', so we may dispense with Steiner points with primes, leaving 20 such points, as has already been stated. Fig. 13.3 shows 16 Steiner points labelled in this consistent fashion, the other 4 being off the bottom of the figure as indicated.

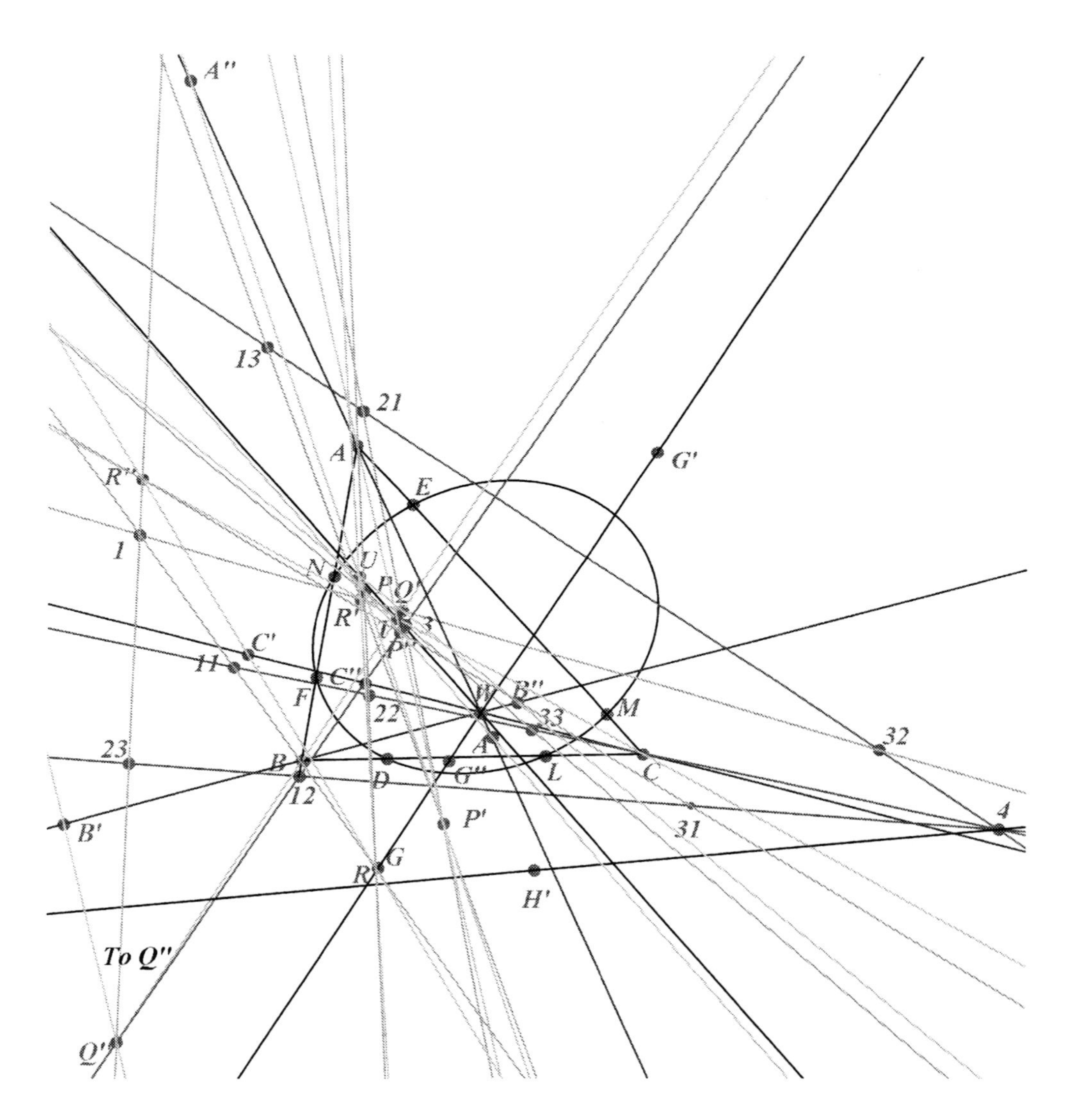

Figure 13.2: Twenty Pascal lines

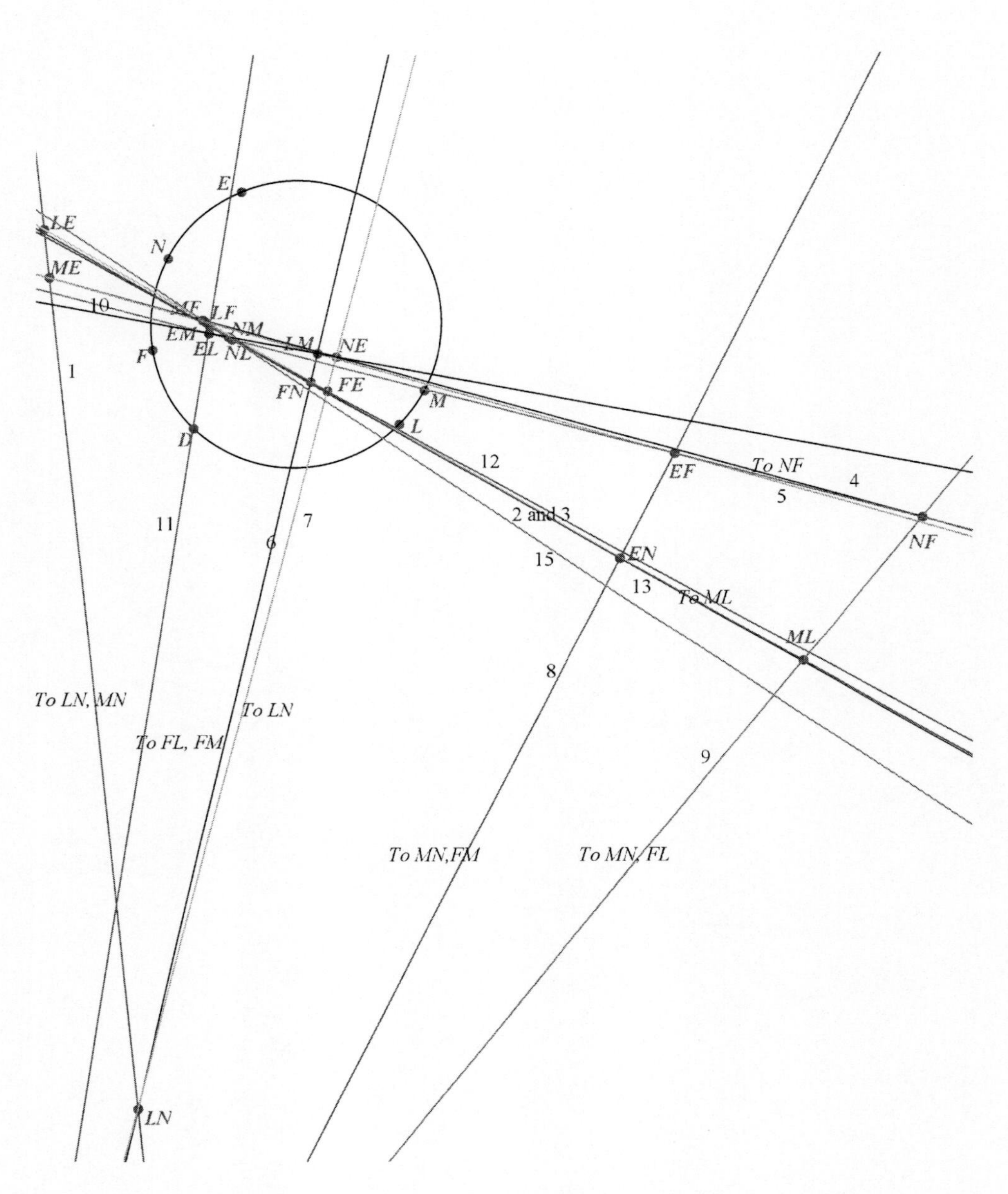

Figure 13.3: Twenty Steiner points

(I have tried many hexagons, but have been unable to find a setting in which the Steiner points are all reasonably close together.) The 20 Steiner points appear 4 at a time on 15 lines (so that each Steiner point appears on 3 such lines). These *Plücker* lines, as they are called, are shown in the figure. It also appears to be impossible to choose the points D, E, F, L, M, N on the conic (here chosen as a circle) to prevent some of the Steiner points being clustered together, and consequently some of the 15 lines appear virtually coincident. The points on each of the 15 lines are indicated as follows and in the diagram the lines concerned are marked with the appropriate number.

(1)LE, ME, LN, MN; (2)LE, EN, FN, LF; (3)LE, ML, FE, MF;
(4)ME, NF, NE, MF; (5)ME, EF, LM, LF; (6)LN, FM, FN, LM;
(7)FL, LN, FE, NE; (8)MN, FM, EN, EF; (9)MN, FL, ML, NF;
(10)EL, NL, EF, NF; (11)FL, FM, EL, EM; (12)NM, FN, FE, EM;
(13)EN, ML, EM, NL; (14)EL, NE, NM, LM; (15)LF, MF, NM, NL.

What have not been illustrated in this review are the 20 Cayley lines, which are attached as described above, one to each Steiner point. If such an illustration were to be drawn it would show that they meet 4 at a time at 15 *Salmon points* (so that each Cayley line passes through 3 Salmon points). There is thus a symmetric relationship between the 60 Pascal lines and the 60 Kirkman points, between the 20 Steiner points and the 20 Cayley lines and between the 15 Plücker lines and the 15 Salmon points.

13.1.9 Dual theorems

Since Pascal's theorem and its consequences are results in projective geometry, there is a collection of dual results that are equally true. The dual of Pascal's theorem is called Brianchon's theorem. The reader is invited to provide details with the aid of Fig. 13.4, in which the conic is shown as a circle with six tangents d, e, f, l, m, n and points of intersection such as ef.

Exercise 13.1.2

1. ABC is a triangle and a circle is drawn to touch the line segments AB and AC at S and T respectively and to cut BC at U and V with U nearer B than C. Let SU meet TV at Y. Prove that AY, BT, CS are concurrent at a point X. If the circle varies, find the locus of X.

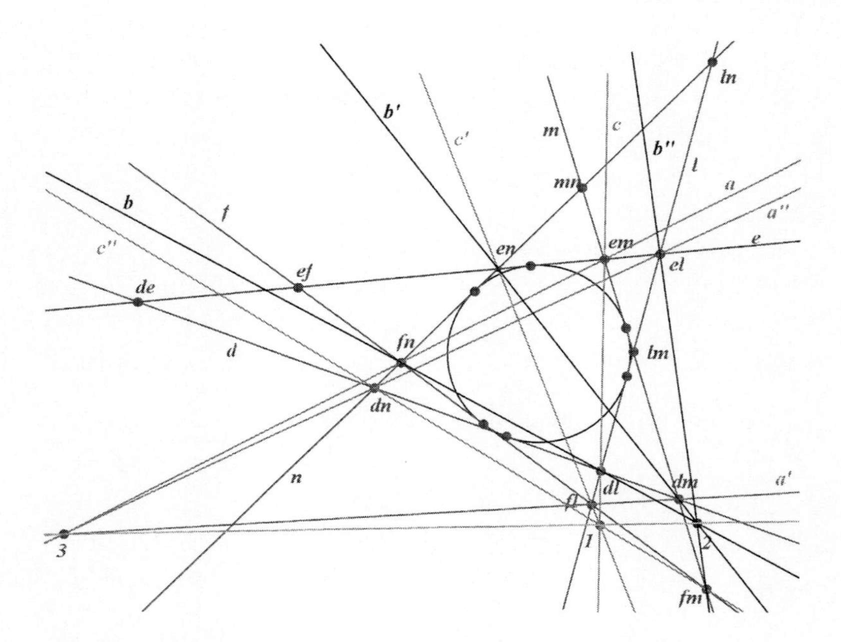

Figure 13.4: Brianchon's theorem

2. $ABCDEF$ is a cyclic hexagon in which AF is parallel to CD, CB meets EF at M and BA meets DE at L. Prove that $LMBE$ is cyclic and that LM is parallel to DC.

13.2 Hexagons with opposite sides parallel

An expanded version of this section appears in Bradley [4]. Figs. 13.5 and 13.6 show configurations in which a hexagon $PUQVRW$ has opposite sides parallel. The first shows a convex hexagon, the second shows one that is re-entrant. Fig. 13.5 and 13.6

Exercise 13.2.1

1. Let $PUQVRW$ be a hexagon with sides $PU, RV; UQ, WR; QV, PW$ parallel in pairs. Prove, by a synthetic argument, that the points P, U, Q, V, R, W lie on a conic.

2. Let the sides QV, RW, PU, extended where necessary, determine the sides BC, CA, AB of a triangle ABC, as shown in Figs. 13.5 and 13.6.

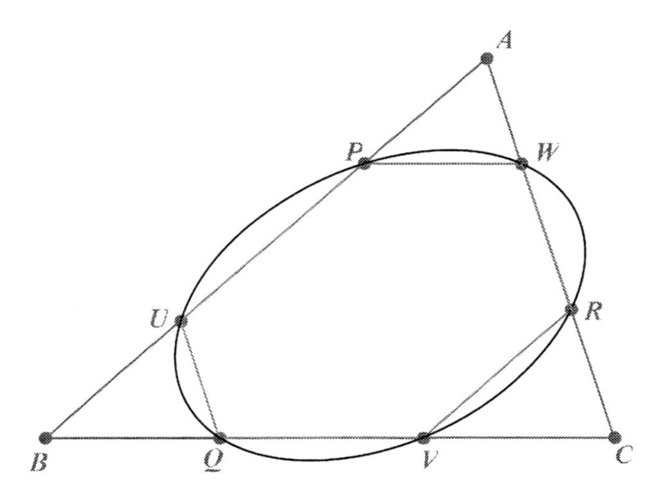

Figure 13.5: A convex hexagon with opposite sides parallel

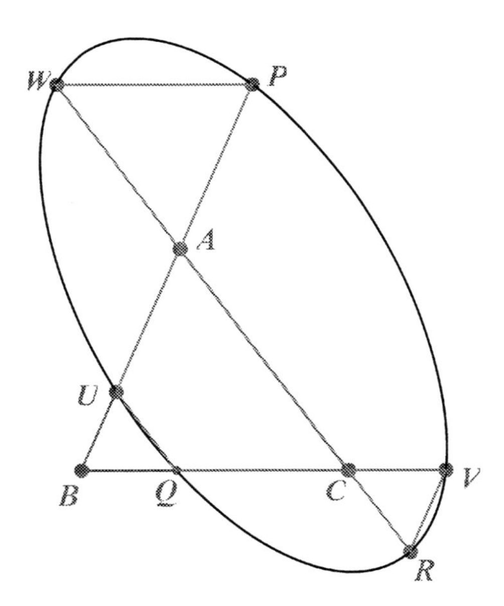

Figure 13.6: A re-entrant hexagon with opposite sides parallel

Take ABC to be the triangle of reference in an areal co-ordinate system. Then, because of the parallels, constants $p, q, r \neq 0, 1$ exist such that $P(p, 1-p, 0), Q(0, q, 1-q)$, $R(1-r, 0, r), U(1-q, q, 0), V(0, 1-r, r), W(p, 0, 1-p)$. For example, $\mathbf{PW} = (0, -(1-p), (1-p)) = (1-p)\mathbf{BC}$. Show, by direct substitution, that P, U, Q, V, R, W lie on the conic with equation

$$(1-p)qrx^2 + (1-q)rpy^2 + (1-r)pqz^2 - p(1-q-r+2qr)yz \quad (13.4)$$

$$-q(1-r-p+2rp)zx - r(1-p-q+2pq)xy = 0.$$

3. Given a hexagon inscribed in a conic having two pairs of opposite sides parallel, prove that the third pair of opposite sides are also parallel.

4. Prove that the conic with Equation (13.4) is a circle if, and only if, a constant $\mu \neq 0$ exists such that $p = \mu a^2, q = \mu b^2, r = \mu c^2$.

We now suppose that $UQ \wedge RV = A'$, $RV \wedge PW = B'$, $PW \wedge UQ = C'$. See Fig. 13.7.

Exercise 13.2.2

1. Prove, by a synthetic argument, provided $p, q, r, p+q+r \neq 0, 1$, that triangles ABC and $A'B'C'$ are in perspective.

2. Prove that, with co-ordinates as given in Problem 2 of Exercise 13.2.1, A' has co-ordinates $(1-r-q, q, r)$ and that AA', BB', CC' are concurrent at the point T with normalised co-ordinates $T(p, q, r)/(p+q+r)$.

For a finite T and real $A'B'C'$ we cannot have $p+q+r = 0$ or $p+q+r = 1$. The first exclusion covers cases in which AA', BB', CC' are parallel. See Fig. 13.8. The second exclusion covers cases in which A', B', C', T coincide. See Fig. 13.9 In these two cases we nonetheless have hexagons with the required properties, but either no finite point T or no triangle $A'B'C'$. Note that when the conic is a circle the co-ordinates of T, by Problem 4 of Exercise 13.2.1, are $T(a^2, b^2, c^2)$ and T is the symmedian point S of triangle ABC. Since triangles ABC and $A'B'C'$ are interchangeable, in this case T must also be the symmedian point $S'$of triangle $A'B'C'$.

We draw the reader's attention once more to Fig. 13.7. So far it has been used to demonstrate the properties of a given hexagon $PUQVRW$ with pairs

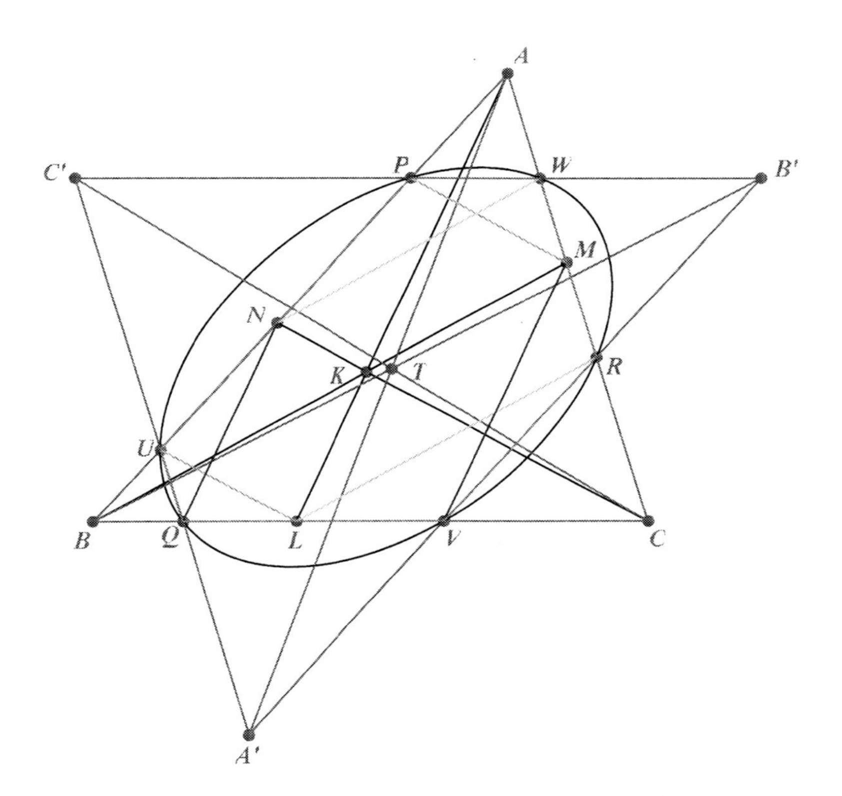

Figure 13.7: First configuration for Part 13.2

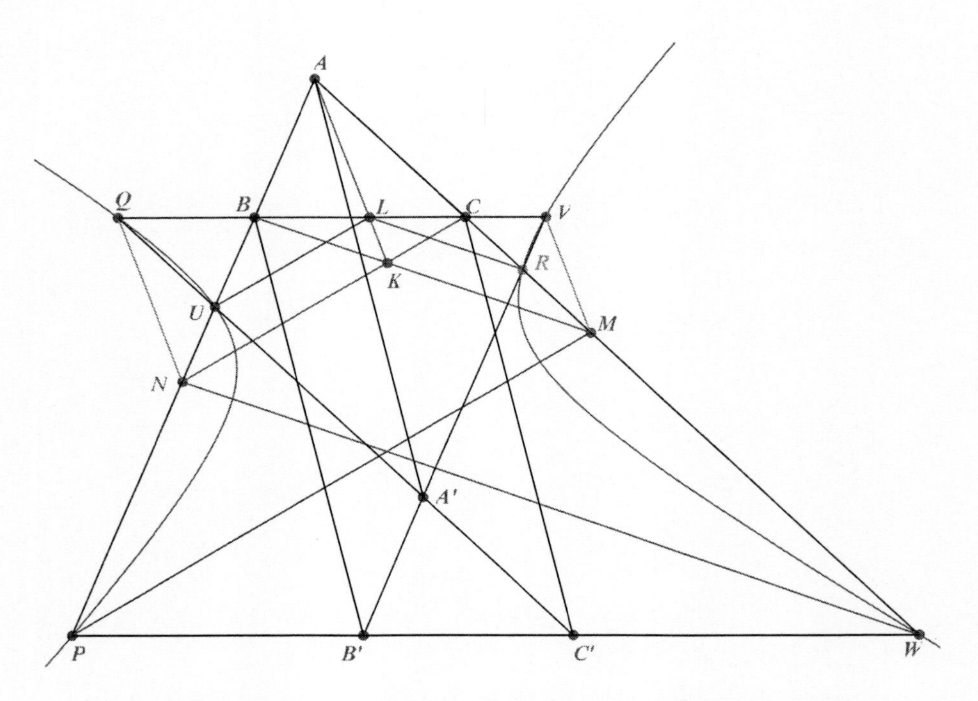

Figure 13.8: First singular configuration for Part 13.2

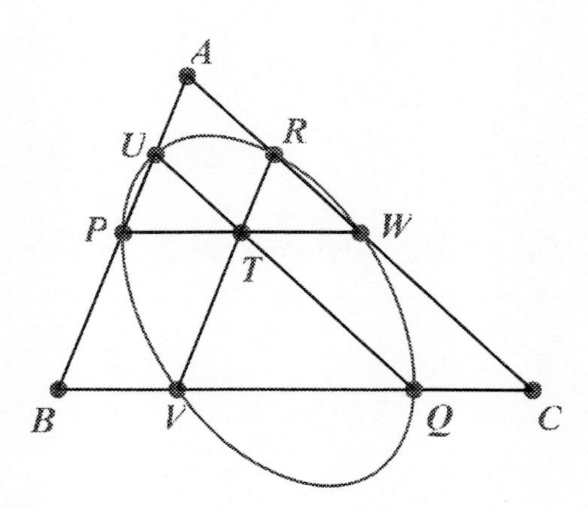

Figure 13.9: Second singular configuration for Part 13.2

of opposite sides parallel, leading to two triangles ABC and $A'B'C'$ that are in perspective. We now adopt another point of view. We start with triangle ABC and a point K in the plane of the triangle, not on a side or on a line through a vertex parallel to the opposite side. We draw the Cevians AKL, BKM, CKN through K with L on BC, M on CA and N on AB. Through L we draw lines parallel to BKM and CKN to meet CA and AB at R and U respectively. Through M we draw lines parallel to CKN and AKL to meet AB and BC at P and V respectively. Through N we draw lines parallel to AKL and BKM to meet BC and CA at Q and W respectively.

Exercise 13.2.3

1. With P, U, Q, V, R, W as defined in the last paragraph, prove by a synthetic argument that $PUQVRW$ is a hexagon with pairs of opposite sides parallel.

2. Let K have normalised co-ordinates $K(l, m, n)$. Obtain the co-ordinates of Q as $Q(0, m, nl)$, and those of V as $V(0, lm, n)$. Show, by cyclic change of l, m, n and x, y, z that the co-ordinates of R, P, W, U are $R(lm, 0, n), P(l, mn, 0), W(l, 0, mn), U(nl, m, 0)$. Show also that $\mathbf{PW} = (0, -mn, mn) = mn\mathbf{BC}$. And hence show that $PUQVRW$ is a hexagon with pairs of opposite sides parallel.

3. Identify the values of p, q, r for the conic circumscribing this hexagon as being given by the formulas $p = l/(l + mn), q = m/(m + nl), r = n/(n + lm)$. Show that these equations may be solved to give

   $$l^2 = (1 - q)(1 - r)/qr, m^2 = (1 - r)(1 - p)/rp, n^2 = (1 - p)(1 - q)/pq.$$

   From these equations we see that if we start with a set of non-zero values of l, m, n, that is, if we start with the point K, then we always get a hexagon with pairs of opposite sides parallel. But given values of p, q, r, that is, given one of our hexagons, there are not always valid values of l, m, n to be obtained from these formulas. Indeed, since we must have $l+m+n = 1$, it is clear that p, q, r have to be specially related for there to be a construction in which the feet of a set of Cevians lead to a hexagon under consideration.

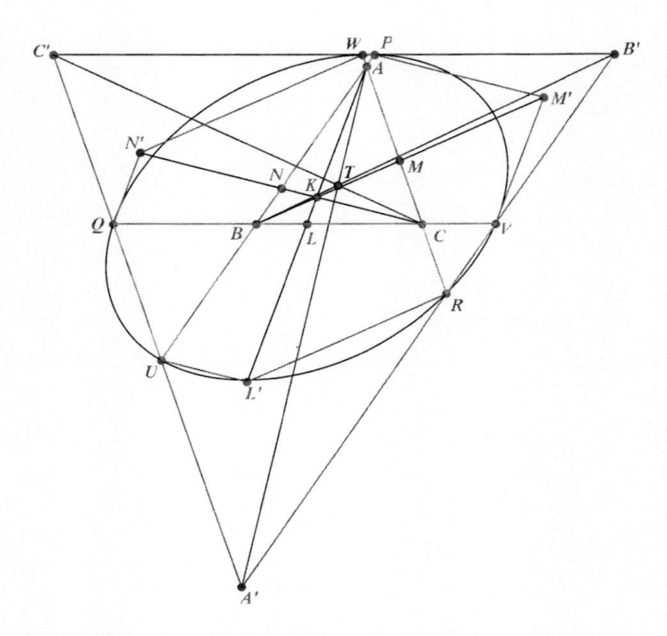

Figure 13.10: Configuration for Section 13.2.1

Exercise 13.2.4

1. Show that there is one fairly obvious solution of the equations in Problem 3 of Exercise 13.2.3, which is when $p = \sin^2 A$, $q = \sin^2 B$, $r = \sin^2 C$. Show that, in this case K is the orthocentre H and that P, U, Q, V, R, W lie on a circle.

We now enquire what the geometrical procedure is to obtain other possible hexagons whose pairs of opposite sides are parallel. In Fig. 13.10 the triangle ABC and the points K, L, M, N are defined as before, and L', M', N' are defined so that $AL'/AL = BM'/BM = CN'/CN = t$. In the figure $t = 2$, chosen only for the purpose of providing a clear illustration, and the number 2 can be replaced by any other non-zero constant. Lines through L', M', N' are drawn parallel to the Cevians through K to provide the points P, U, Q, V, R, W and these points also lie on a hexagon with pairs of opposite sides parallel.

13.2.1 How such hexagons arise

Let ABC be a triangle and K a point in the plane, not on a side of the triangle or on a line through a vertex parallel to the opposite side. Let AKL, BKM, CKN be the Cevians through K with L, M, N on BC, CA, AB respectively. Define L', M', N', by the equations $AL'/AL = BM'/BM = CN'/CN = t$. Through L' draw lines parallel to CKN and BKM to meet AB and CA at U and R respectively. Through M' draw lines parallel to AKL and CKN to meet BC and AB at V and P respectively. Through N'draw lines parallel to BKM and AKL to meet CA and BC at W and Q respectively. Then $PUQVRW$ is a hexagon with pairs of opposite sides parallel. Moreover, if $QU \wedge RV = A'$, $RV \wedge PW = B'$, $PW \wedge QU = C'$, then triangles ABC and $A'B'C'$ are in perspective at a point T, which is independent of t.

Exercise 13.2.5

1. Prove the truth of Section 13.2.1 by establishing the following results:

   (i) $P(lt, mn+l(1-t), 0), Q(0, mt, nl+m(1-t)), R(lm+n(1-t), 0, nt)$.

   (ii) $U(nl+m(1-t), mt, 0), V(0, lm+n(1-t), nt), W(lt, 0, mn+l(1-t))$.

   (iii) $\mathbf{PW} = (0, -mn - l(1-t), mn + l(1-t)) = (mn + l(1-t))\mathbf{BC}$, so that PW is parallel to BC. Similarly QU is parallel to CA and RV is parallel to AB.

   (iv) $p = lt/(mn + l), q = mt/(nl + m), r = nt/(lm + n)$.

   (v) ABC and $A'B'C'$ are in perspective and that the centre of the perspective has unnormalised co-ordinates $T(l/(mn + l), m/(nl + m), n/(lm + n))$, which is a point independent of t.

   (vi) $l^2 = (t-q)(t-r)/qr, m^2 = (t-r)(t-p)/rp, n^2 = (t-p)(t-q)/pq$.

It is not easy to see what is happening without a numerical example. Suppose that we take $K(1/2, 1/3, 1/6)$ and we produce the hexagon from lines through L, M, N, the feet of the Cevians. Then the formulas in result (iv) give $p = 9/10, q = 8/10, r = 5/10$ and conversely, given a hexagon with these values of p, q, r, it arises from K by means of lines through L, M, N. However, if we are given the hexagon with $p = 9/7, q = 8/7, r = 5/7$ we can see that since $p : q : r$ is the same as before, the points K and T are the same, but now the hexagon arises from points L', M', N' on the Cevians with $AL'/AL = BM'/BM = CN'/CN = t = 10/7$.

In particular $PUQVRW$ is a circle circumscribing one of our hexagons if, and only if, K is the orthocentre H and T is the symmedian point S of ABC. And L', M', N' are chosen on AH, BH, CH respectively so that $AL' = tAL, BM' = tBM, CN' = tCN$ for some constant t. Here L, M, N are now the feet of the altitudes. Then the six vertices of the hexagon are found by dropping perpendiculars from L' on to AB, CA, from M' on to BC, AB and from N' on to CA, BC. And p, q, r are given by $p = t\sin^2 A, q = t\sin^2 B, r = t\sin^2 C$.

Finally we investigate the position of the centres of these conics. See Equations (2.8) and (2.9). We recall that the polar of the centre of a conic is the line at infinity, with equation $x+y+z=0$. The result is that if (X, Y, Z) is the centre of a conic with Equation (2.7), then $X : Y : Z = vw - gv - hw - f^2 + fg + hf : wu - hw - fu - g^2 + gh + fg : uv - fu - gv - h^2 + hf + gh$.

13.2.2 Centres of Tucker circles and conics

The centre of the conic with Equation (13.4) is therefore given by

$$X : Y : Z = p(pq + rp - q^2 - r^2 - p + q + r) : \tag{13.5}$$

$$q(qr + pq - r^2 - p^2 - q + r + p) : r(rp + qr - p^2 - q^2 - r + p + q).$$

If, in this set of ratios, we replace p by pt, q by qt, r by rt we get the centres of all conics that result from one set of Cevians, and some straightforward algebra shows that all these centres lie on the line with equation

$$(q - r)x/p + (r - p)y/q + (p - q)z/r = 0. \tag{13.6}$$

Note that this line passes through $T(p, q, r)$, the centre of perspective of triangles ABC and $A'B'C'$. And interestingly it also passes through the point with areal co-ordinates (p^2, q^2, r^2). In particular the centres of all Tucker circles lie on the line passing through O, the centre of the circumcircle of triangle ABC (the circumcircle is a limiting case of a Tucker circle), S the symmedian point of triangle ABC and the fourth power point with areal co-ordinates (a^4, b^4, c^4). It does, of course, also pass through O', the centre of the circumcircle of triangle $A'B'C'$, but O' is a variable point, depending on t. The line of centres is the common Brocard diameter of the triangles ABC and $A'B'C'$. This result is illustrated in Fig. 13.11.

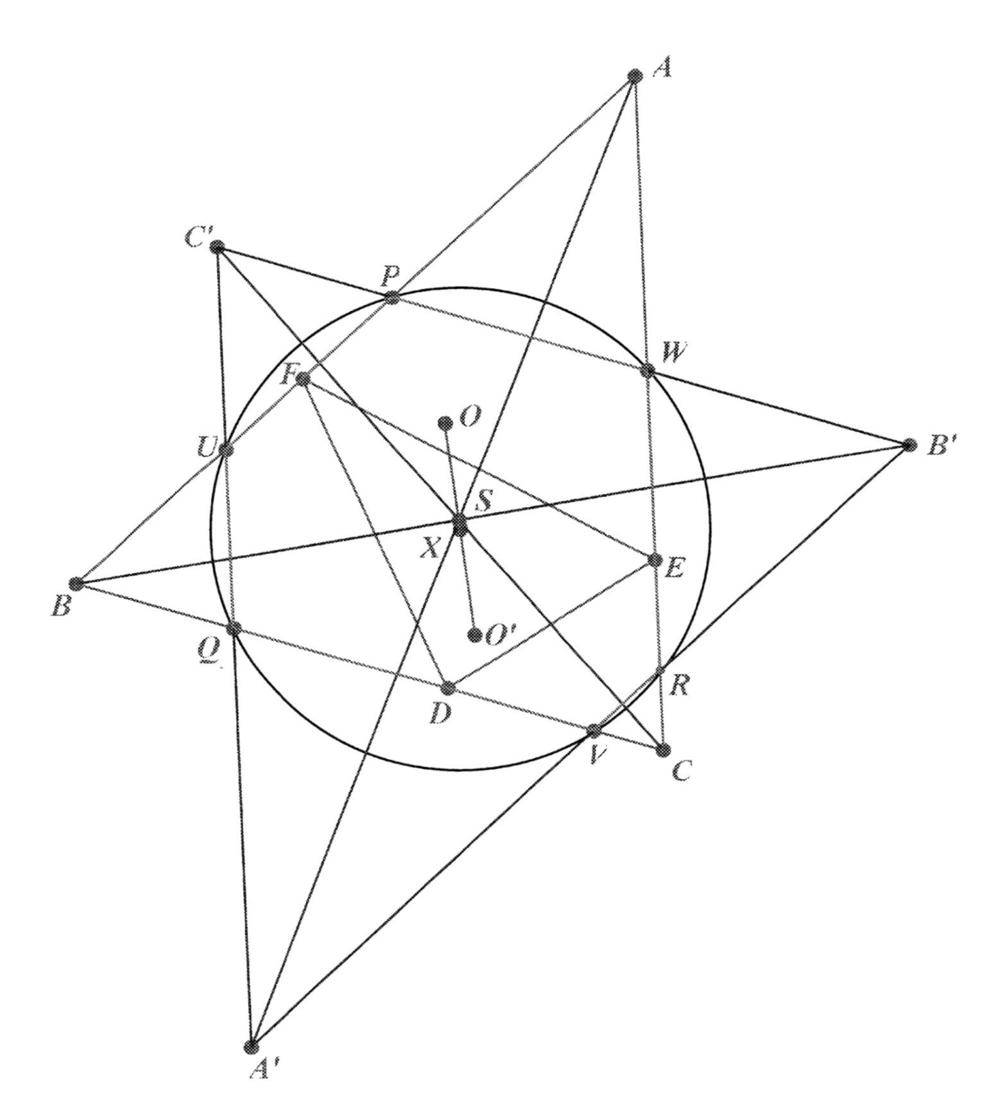

Figure 13.11: A Tucker circle

Exercise 13.2.6

1. Let D, E, F be the vertices of the pedal triangle (the feet of the Cevians through the orthocentre H). Show that the centre of the Tucker circle that arises from perpendiculars through D, E, F on to the sides is also the centre of the incircle of the triangle whose vertices are the midpoints of the sides of triangle DEF. This is Taylor's circle with its centre at the Spieker centre of triangle DEF.

2. Let D', E', F' be defined so that D, E, F are the midpoints of AD', BE', CF' respectively. Show that the centre of the Tucker circle that arises from perpendiculars through D', E', F' is the orthocentre of the pedal triangle DEF.

3. Show that the centre X of a Tucker circle is the midpoint of OO', where O is the circumcentre of ABC and O' the circumcentre of $A'B'C'$.

Chapter 14

Complex numbers and Inversion

14.1 Basic formulas

We use the notation that a point A in the Argand diagram is represented by the complex number $a = s + it$, where s and t are real numbers. Similarly for the point B we use the complex number $b = u + iv$, where u and v are real numbers. The variable point Z is represented by $z = x + iy$, where x and y are real numbers. Whereas the real Euclidean plane is enriched by a line at infinity, the complex Euclidean plane is enriched by a point at infinity.

The following basic results are assumed known. Readers should regard any result that is unfamiliar as an exercise to be completed before proceeding.

14.1.1 Addition and Subtraction

$$a \pm b = (s \pm u) + i(t \pm v). \tag{14.1}$$

14.1.2 Multiplication

$$ab = (su - tv) + i(sv + tu). \tag{14.2}$$

14.1.3 Complex Conjugate

This is the result obtained by reflecting a point in the x-axis of the Argand diagram and we denote the *complex conjugate* of z by $z^* = x - iy$. Note that

$(z^*)^* = z$.

14.1.4 Scalar product

$$\mathbf{OA}.\mathbf{OB} = su + tv = \frac{ab^* + a^*b}{2} = (a, b). \tag{14.3}$$

14.1.5 Exterior product

$$\mathbf{OA} \times \mathbf{OB} = sv - tu = \frac{ab^* - a^*b}{2i} = a \times b. \tag{14.4}$$

14.1.6 Modulus

$(a, a) = aa^* = |a|^2 = OA^2 = s^2 + t^2$. The quantity $|a|$ is called the *modulus* of a and denotes the distance of A from the origin O of the Argand diagram.

14.1.7 Distance

$$\text{Distance } AB = |b - a| = \sqrt{(u - s)^2 + (v - t)^2}. \tag{14.5}$$

14.1.8 Displacement

$$\mathbf{AB} \text{ is represented by } (b - a). \tag{14.6}$$

14.1.9 Argument

If $a = |a|(\cos\theta + i\sin\theta)$ and $\theta \in (-\pi, \pi]$ then θ is called the *principal argument* of a. Note that, in general, the value of θ can be altered by any multiple of 2π and when θ does not necessarily lie in the interval $(-\pi, \pi]$ it is called the *argument* of a and is denoted by $\arg a$; its value must be interpreted as being equal to $\theta (\text{mod } 2\pi)$. Note that the argument of -1 is π (mod 2π). Note also that radians are used for angles in the complex plane.

14.1.10 Angle

$\arg(b - a)$ represents the angle the directed line segment AB makes with the positive direction of the x-axis.

14.1.11 Formulas obeyed by modulus and argument

$|ab| = |a||b|$, $|a/b| = |a|/|b|$ and in particular $|1/a| = 1/|a|$. Also arg $(ab) =$ arg $a +$ arg b, arg $(a/b) =$ arg $a-$ arg b and in particular arg $(1/a) = -$ arg a. Finally $|a^*| = |a|$, arg $a^* = -$ arg a.

14.1.12 Division

$1/a = a^*/(aa^*) = a^*/|a|^2$. This formula is particularly important as it allows us to divide one complex number by another, resulting in a real denominator.

14.1.13 de Moivre's theorem

$$(\cos\theta + i\sin\theta)^n = (\cos n\theta + i\sin n\theta) \tag{14.7}$$

for all integers n.

If n is rational this formula provides one of the values of the rational power. For example

$$(\cos\theta + i\sin\theta)^{\frac{1}{3}} = \cos\frac{\theta + 2k}{3} + i\sin\frac{\theta + 2k}{3},$$

providing three distinct cube roots with $k = 0, 1, 2$.

14.1.14 $e^{i\theta}$

It turns out that

$$\cos\theta + i\sin\theta = \exp(i\theta) = e^{i\theta}. \tag{14.8}$$

This is, in fact, a sophisticated result, depending on Taylor's series for complex functions. It allows us, for example, to write $e^z = e^x(\cos y + i\sin y)$ and is responsible for remarkable formulas such as $e^{i\pi} = -1$.

14.1.15 $\ln z$

$\ln z = \ln|z| + i\text{arg } z$, and whereas, for example, $z^{1/3}$ has three values, $\ln z$ has an infinite number of values, since arg z has an infinite number of values. See Section 14.1.9. So, for example, $\ln i = i\text{arg } i = (\frac{1}{2}\pi + 2k\pi)i$, for any integer k. And thus, $\ln i^i = i\ln i = -(\frac{1}{2}\pi + 2k\pi)$, and $i^i = e^{-(\frac{1}{2}\pi + 2k\pi)}$, which is real, but infinite-valued.

Complex numbers appear to provide a synthesis of Cartesian co-ordinates and vectors, and at first sight appear to be the ideal framework for solving geometrical problems, and indeed much has been done along these lines, see, for example, Hahn [22]. The fact that such procedures are not entirely satisfactory is because the equations for lines and circles, using complex numbers, as we see below, are rather cumbersome. There are, however, certain problems for which complex numbers are extremely useful. This is because rotations are not easily handled using co-ordinates or vectors. Whereas in complex numbers, if $b = ae^{i\theta}$, then $|b| = |a|$ and arg $b =$ arg $a + \theta$, so the line segment represented by b has the same length as the line segment represented by a, but is rotated with respect to it by the angle θ. It is also the case that complex numbers form the natural framework for describing the process of inversion.

14.2 Further properties of complex numbers

14.2.1 The triangle inequality

This may be deduced, as in any scalar product space, from the Cauchy-Schwarz inequality $(a, a)(b, b) \geq (a, b)^2$. It is exactly what you would expect, namely that

$$|a \pm b| \leq |a| + |b|. \tag{14.9}$$

14.2.2 The section theorem

The point P dividing the line AB in the ratio $m : n$ is associated with the complex number

$$p = (na + mb)/(m + n). \tag{14.10}$$

14.2.3 Equation of a line

The equation of the line AB is

$$z(b^* - a^*) - z^*(b - a) = ab^* - a^*b. \tag{14.11}$$

In terms of the exterior product this may be written as $z \times a + a \times b + b \times z = 0$.

The equation of a line may also be expressed in the form $(z, u) = d$, where the line is perpendicular to the direction OU and the perpendicular distance from O on to the line is $|d|/|u|$. Since a scalar product is real, d is real, but may be positive or negative, with the usual interpretation given to its sign.

14.2.4 Area of a triangle

The area of triangle OAB is $\frac{1}{2}|a \times b|$ and the area of triangle ABC is $\frac{1}{2}|a \times b + b \times c + c \times a|$. Provided the exterior product is used, these have the same form as when using vectors. If the modulus is omitted there is a positive or negative sign, with the same interpretation as when using vectors.

14.2.5 Angle

The angle θ between the line segments represented by a and b is given by $\cos\theta = (a, b)/(|a||b|)$.

14.2.6 Equation of a circle

The equation of the circle, centre A and radius r may be expressed in three possible ways:

(i)
$$|z - a| = r; \tag{14.12}$$

(ii)
$$z - a = re^{i\theta}, \theta \in (-\pi, \pi]; \tag{14.13}$$

(iii)
$$zz^* - a^*z - az^* + aa^* - r^2 = 0. \tag{14.14}$$

Equations (14.11) and (14.14) are decidedly ugly and cumbersome to use.

The circle on AB as diameter has equation

$$(z - a, z - b) = 0. \tag{14.15}$$

14.2.7 Equation of a tangent and the polar of a point

If P lies on the circle, centre c, radius r, then the equation of the tangent at P is given by $(z - c, p - c) = r^2$. If P does not lie on the circle, then this equation is the equation of the polar of P.

14.2.8 Perpendicular bisector

The equation of the perpendicular bisector of the line segment AB is

$$z(a^* - b^*) + z^*(a - b) = aa^* - bb^*. \tag{14.16}$$

14.2.9 Orthogonal circles

The condition that two circles, centres C_1 and C_2, with equations $zz^* - c_1z^* - c_1^*z + k_1 = 0$ and $zz^* - c_2z^* - c_2^*z + k_2 = 0$, are orthogonal is

$$2(c_1, c_2) = k_1 + k_2. \tag{14.17}$$

14.2.10 Half line

The equation of the half line AP, excluding A, making an angle θ with the positive direction of the x-axis may be written in the form $\arg\,(z - a) = \theta$.

14.2.11 Angles in the same segment

If A and B are two fixed points then the locus of the point P such that the directed angle $\angle APB = \theta \pmod{\pi}$, is a circle through A and B, but excluding them, and it may be written in the form

$$\arg\,(z - b) - \arg\,(z - a) = \theta \pmod{\pi}. \tag{14.18}$$

Example 14.2.1 *(The circle of Apollonius)*

Let A and B be points associated with $a = -1$ and $b = 1$. Find the locus of points P such that $AP = kBP$, $k > 1$.

Using z for the point P, from $AP = kBP$ we obtain $|z + 1| = k|z - 1|$. Now put $z = x + iy$ and square. We then obtain the equation of the locus of P to be the circle, centre C and radius r, where $c = (k^2 + 1)/(k^2 - 1)$ and $r = \sqrt{c^2 - 1}$. Since P_1 and P_2 satisfy $AP_1 = kBP_1$ and $AP_2 = kBP_2$, it follows that PP_1 and PP_2 are the internal and external bisectors respectively of $\angle APB$ and $\angle P_1PP_2 = 90°$, providing a synthetic proof of the fact that the locus of P is a circle. Fig. 14.1 illustrates the case $k = 2, c = 5/3$ and $r = 4/3$.

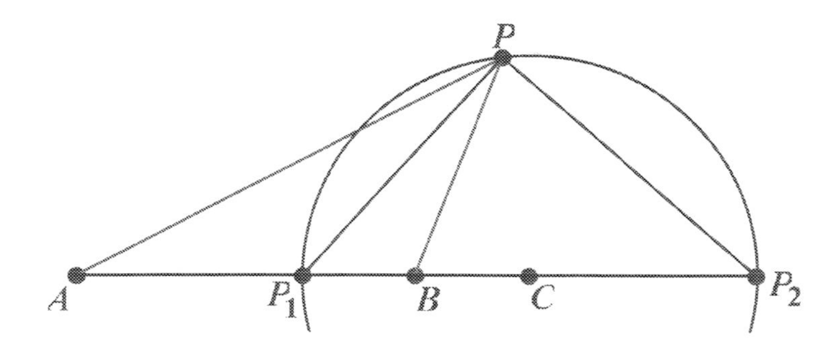

Figure 14.1: The circle of Apollonius

Example 14.2.2

Show that ABC is an equilateral triangle if, and only if, the complex numbers a, b, c representing their vertices satisfy

$$a^2 + b^2 + c^2 - bc - ca - ab = 0. \tag{14.19}$$

Equation (14.19) is equivalent to

$$\frac{a-b}{c-b} = \frac{b-c}{a-c}.$$

The modulus of this gives $AB/BC = BC/CA$ and its argument gives $\angle B = \angle C$. So $AC = AB = BC$ and the triangle is equilateral. The converse is true since the steps of this argument are reversible.

Example 14.2.3 *(Napoleon's Theorem)*

Equilateral triangles are erected externally on the sides of a triangle ABC. Prove that their centres are the vertices of an equilateral triangle.

Let A, B, C be represented by the complex numbers a, b, c. We claim that if $A'BC$ is the equilateral triangle erected on BC external to triangle ABC, then A' is represented by the complex number $a' = b - \omega(c - b)$, where $\omega = exp(2\pi i/3)$. (See the last paragraph of Part 14.1). Since $\omega^3 = 1$ and $\omega^2 + \omega + 1 = 0$, we have $a' = -\omega^2 b - \omega c$. The centroid of triangle $A'BC$ is therefore represented by the complex number $d = (1/3)(b(1 - \omega^2) + c(1 - \omega))$. The centroids e, f of triangles CAB', ABC' respectively follow by cyclic

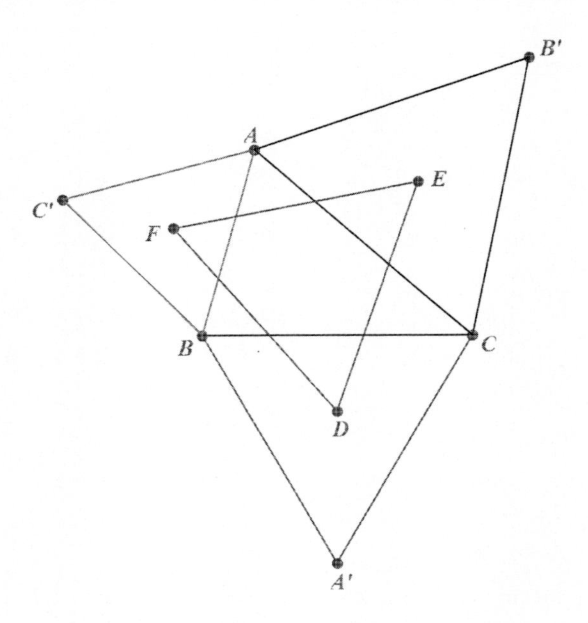

Figure 14.2: Napoleon's theorem

change of a, b, c. It may now be checked that $d^2 + e^2 + f^2 - ef - fd - de = 0$ and hence, by Example 14.2.2, triangle DEF is equilateral.

The reader is invited to modify this solution to cover the case when the equilateral triangles are constructed internally rather than externally.

Exercise 14.2.4

1. Let the altitude from A meet the side BC at D and the circumcircle again at P. Let the circumcentre be the origin. Prove that the complex numbers representing D and P are given respectively by $d = (a^2 + ab - bc + ca)/(2a)$ and $p = -bc/a$. Prove also, using complex numbers, that if T is the nine-point centre, represented by $t = \frac{1}{2}(a + b + c)$, then $TD = \frac{1}{2}R$, where R is the circumradius.

2. $ABCD$ is a parallelogram. On its sides squares are constructed externally. Prove that the centres P, Q, R, S of these squares themselves form a square.

3. If the parallelogram in Problem 2 is replaced by any convex quadrilateral, prove that $PR = QS$.

4. On the sides of a convex quadrilateral $ABCD$, equilateral triangles are constructed, alternately outwards and inwards. Prove that their vertices, other than A, B, C, D form a parallelogram.

5. For the configuration of Problem 2, show that the midpoints of AC, BD, PR, QS form a square.

6. $ABCD$ is a cyclic quadrilateral inscribed in a circle S. Assume that A, B, S are fixed and C, D variable so that the length CD is constant. Points X, Y on the rays AC, BC respectively are such that $AX = AD$ and $BY = BD$. Prove that XY is constant both in length and direction.

7. Prove that triangles ABC and DEF are similar if, and only if, either the 3×3 determinant formed by rows with entries $a, b, c; d, e, f; 1, 1, 1$ vanishes or the 3×3 determinant formed by rows with entries $a, b, c; d^*$, $e^*, f^*; 1, 1, 1$ vanishes.

8. Prove that A, B, C, D are concyclic if, and only if, the cross ratio $(a, c; b, d)$ of the complex numbers a, b, c, d representing those points is real.

9. Four circles S_1, S_2, S_3, S_4 are given. Suppose that S_1 and S_2 intersect at A and P, S_2 and S_3 intersect at B and Q, S_3 and S_4 intersect at C and R, S_4 and S_1 intersect at D and S. Prove that A, B, C, D are concyclic if, and only if, P, Q, R, S are concyclic.

10. Let ABC be a triangle, with L the midpoint of BC. Prove that the perpendicular from L on to the tangent to the circumcircle at A passes through T, the centre of the nine-point circle.

11. $ABCD$ is a parallelogram and on the sides AB, BC equilateral triangles ABP, BCQ are drawn. Prove that triangle PQD is equilateral.

14.3 Inversion

The *inverse* of a point P with respect to a circle S, centre A, radius r, is the point P' lying on AP or its extension such that $(AP)(AP') = r^2$ and where P' is on the same side of A as P. See Part 1.2 Section 1.3.7, where this definition first appears. In terms of complex numbers representing these points this becomes

$$(p-a)(p'^{*}-a^{*})=r^{2}. \tag{14.20}$$

If A is the origin, this takes on the form $pp'^{*}=r^{2}$. The circle S is called the *circle of inversion*, and very often in exercises its radius is immaterial and is taken to be of unit length. A is called the *centre of inversion.*

The idea behind inversion as a technique for solving problems is this. You are given a problem involving lines and circles in the z-plane. You then perform the transformation $z=r^{2}/w^{*}=r^{2}w/|w|^{2}$, which is the same as $w=r^{2}z/|z|^{2}$, mapping $z\to w$ for all points in the z-plane configuration, thereby producing a transformed configuration and hence a transformed problem in the w-plane. The skill lies in recognizing the sort of problem that is suitable for inversion, and in choosing the centre of inversion A in such a way that the transformed problem is significantly easier to analyze and solve. Note that a change in the value of r does nothing more than alter the scale of the configuration in the w-plane, so unless the original problem is specifically concerned with distances it is customary not to specify the value of r. Some points are worth mentioning at the outset. First, sometimes it is appropriate to draw the inverse figure on the same sheet of paper as the original figure, but sometimes to use a different sheet is desirable. This is not just a matter of experience, as there is a subjective element involved and also a question of expediency. It may depend on whether one wants to see the point A in the inverse figure. Note that $|w|>r$ corresponds to $|z|<r$ and $|w|<r$ corresponds to $|z|>r$. Note also that the inverse of A is the point at infinity. It is advisable to have a good working notation. We often label the point corresponding to P in the z-plane by P' or P^{*} in the w-plane. This helps to keep track of the relationship between the problem and the inverse problem. Note also that the inverse of the inverse figure is the original figure, when the same centre of inversion is used.

The following results hold:

14.3.1

If a straight line is inverted with respect to a point O on the line, the inverse is a straight line through O having the same orientation.

14.3.2

If a circle is inverted with respect to a point O on the circle, then the inverse is a straight line perpendicular to the diameter through O.

14.3.3

If a straight line is inverted with respect to a point O not on the line, the inverse is a circle through O with centre on the perpendicular from O on to the line.

14.3.4

If a circle is inverted with respect to a point O not on the circle, the inverse is a circle, in which O lies on the line of centres of the two circles.

Let the equation of the line or circle in the z-plane be

$$azz^* - c^*z - cz^* + d = 0. \tag{14.21}$$

Here c is a complex number and a, d are real numbers. We take the origin in the z-plane to be the centre of inversion and choose, without loss of generality, $r = 1$.

Section 14.3.1 corresponds to $a = d = 0$. Section 14.3.2 corresponds to $a \neq 0, d = 0$. Section 14.3.3 corresponds to $a = 0, d \neq 0$. Section 14.3.4 corresponds to $a \neq 0, d \neq 0$. Under the inversion $z = 1/w^*$ Equation (14.21) becomes

$$dww^* - c^*w - cw^* + a = 0. \tag{14.22}$$

If $a = d = 0$, then the line in the w-plane has the same equation as the line in the z-plane (though points are interchanged). This proves the result in Section 14.3.1.

If we take $c = -g - if$, write $z = x + iy, w = u + iv$, then Equations (14.21) and (14.22) become $a(x^2 + y^2) + 2gx + 2fy + d = 0$ and $d(u^2 + v^2) + 2gu + 2fv + a = 0$ respectively.

In Section 14.3.2 the circle through the origin has equation $a(x^2 + y^2) + 2gx + 2fy = 0$ and the inverse figure is the line in the w-plane with equation $2gu + 2fv + a = 0$. The slope of the line is $-g/f$ and the slope of the diameter of the circle through O is f/g. These lines are perpendicular, which proves the result in Section 14.3.2. The result in Section 14.3.3 is the same as that in Section 14.3.2 in reverse, and does not require separate proof.

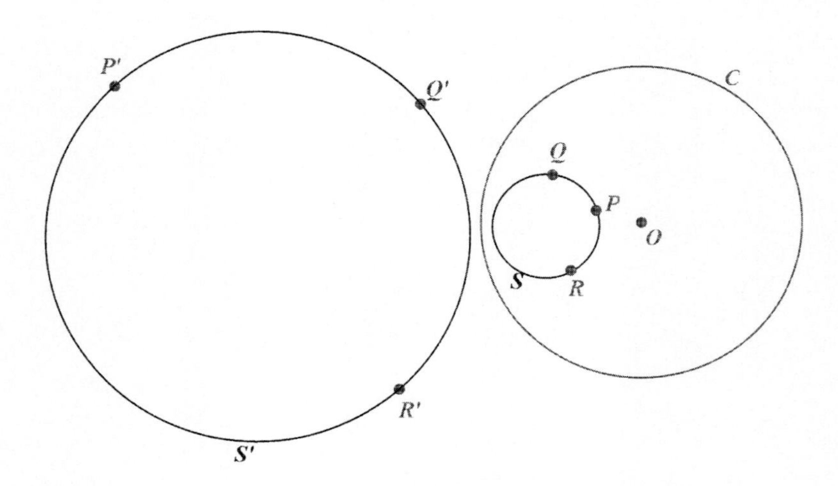

Figure 14.3: Configuration for Section 14.3.4

To prove the result in Section 14.3.4 note that the circle in the z-plane has centre $(-g/a, -f/a)$ and radius $\sqrt{g^2 + f^2 - ad}/a$, and the circle in the w-plane has centre $(-g/d, -f/d)$ and radius $\sqrt{g^2 + f^2 - ad}/d$. Their centres are on the same line through O and their radii are in the same proportion as the distances of their centres from O. It is important to realize that the centres of the two circles are not corresponding points in the inversion.

In Fig. 14.3 the circle S' is the inverse of the circle S, C being the circle of inversion and O the centre of inversion. Three pairs of corresponding points are shown.

Example 14.3.1

Prove that at the corresponding points of intersection, the angle at which two smooth curves intersect is equal to the angle at which their inverse curves intersect. (It is assumed that the point of intersection is not at the centre of inversion.)

Denote the curves by C_1 and C_2, and their images under inversion by C_1' and C_2'. Let Z be the intersection of C_1 and C_2. The intersection of C_1' and C_2', which is the image of Z, is denoted by W. Let Z_1 and Z_2 lie on C_1 and C_2 respectively, close to Z, and let W_1 and W_2 on C_1' and C_2' respectively be their images. Let the points be represented in the complex plane by

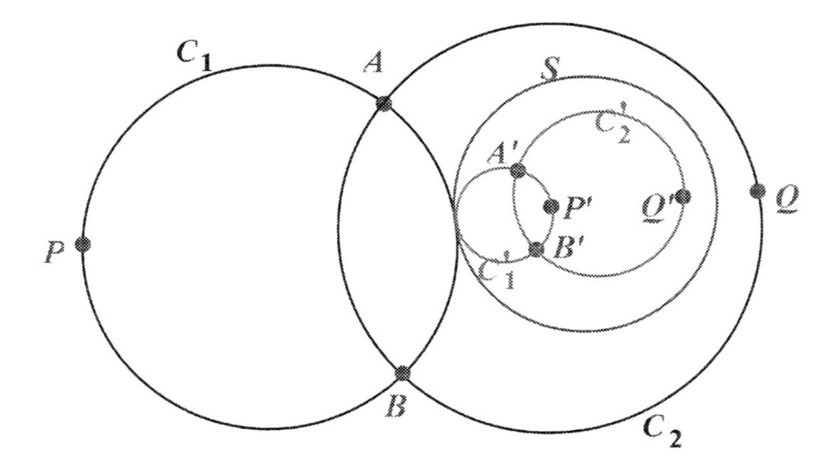

Figure 14.4: Configuration for Section 14.3.5

z, w, z_1, z_2, w_1, w_2 respectively. Making the transformation $w = 1/z^*$ we obtain

$$\frac{w - w_1}{w - w_2} = \frac{(z^* - z_1^*)z_2^*}{(z^* - z_2^*)z_1^*}. \tag{14.23}$$

It follows that arg $(w - w_1)-$ arg $(w - w_2) =$ arg z_1- arg z_2+ arg $(z - z_2)-$ arg $(z - z_1)$. Now in the limit as Z_1 and Z_2 tend to Z, arg z_1 tends to arg z_2, the expression on the left becomes the angle between C_1 and C_2 and the expression on the right becomes the angle between C_1' and C_2'. It may be significant in some applications to appreciate that these angles are actually measured in opposite directions.

Example 14.3.1 implies the following results.

14.3.5

A pair of orthogonal circles, in general (see Section 14.3.4), inverts into a pair of orthogonal circles. See Fig. 14.4.

14.3.6

Suppose a pair of orthogonal circles meet at A and B, and A is used as the centre of inversion. Then (see Section 14.3.2) their inverses are a pair of

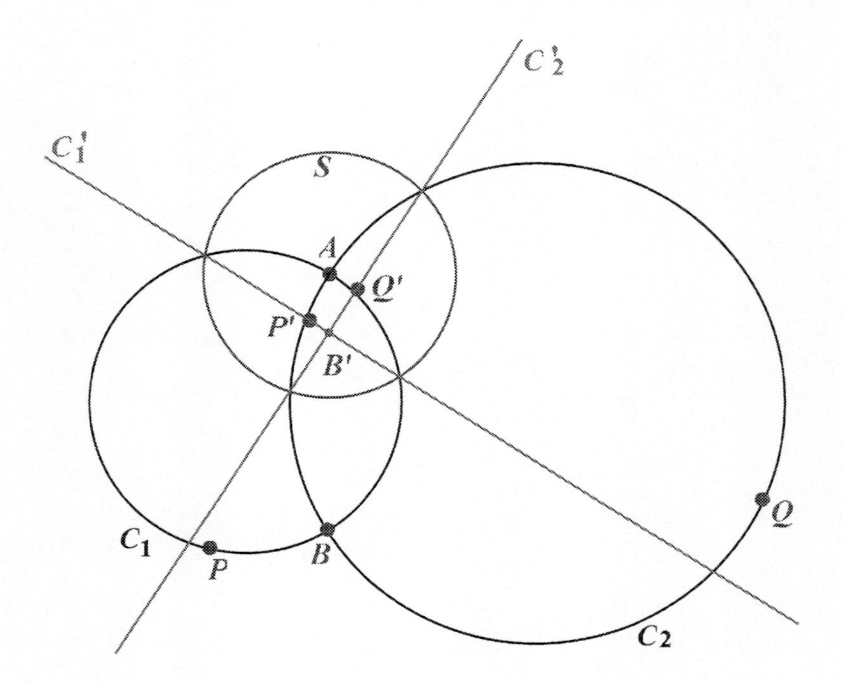

Figure 14.5: Configuration for Section 14.3.6

straight lines at right angles to each other intersecting at B', the inverse of B. See Fig. 14.5.

14.3.7

If two orthogonal circles are such that one passes through the centre of inversion and the other does not, then the inverse figure is a circle and its diameter. See Fig. 14.6.

14.3.8

If two curves touch, then their inverses touch, or if their inverses are lines, then those lines are parallel. See Fig. 14.7 and Fig. 14.8.

Example 14.3.2

(i) What is the inverse of an intersecting system of coaxal circles, with respect to one of their points of intersection?

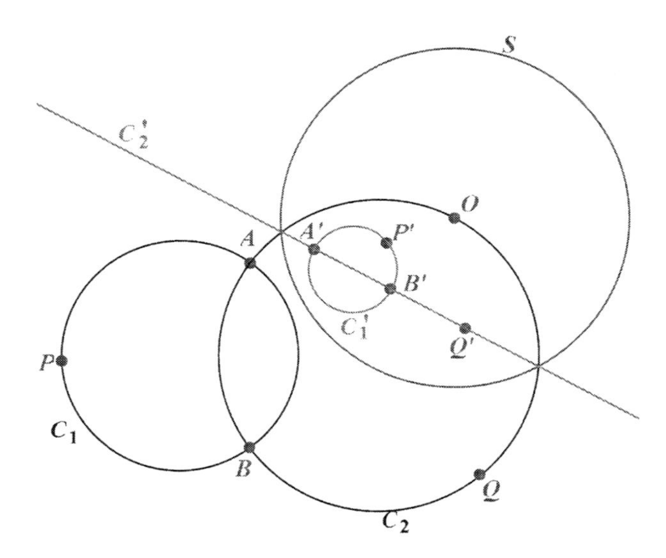

Figure 14.6: Configuration for Section 14.3.7

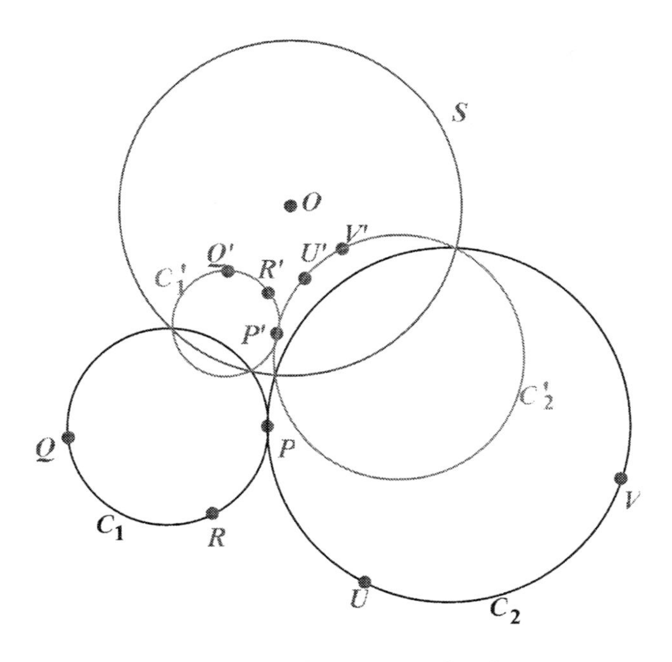

Figure 14.7: First configuration for Section 14.3.8

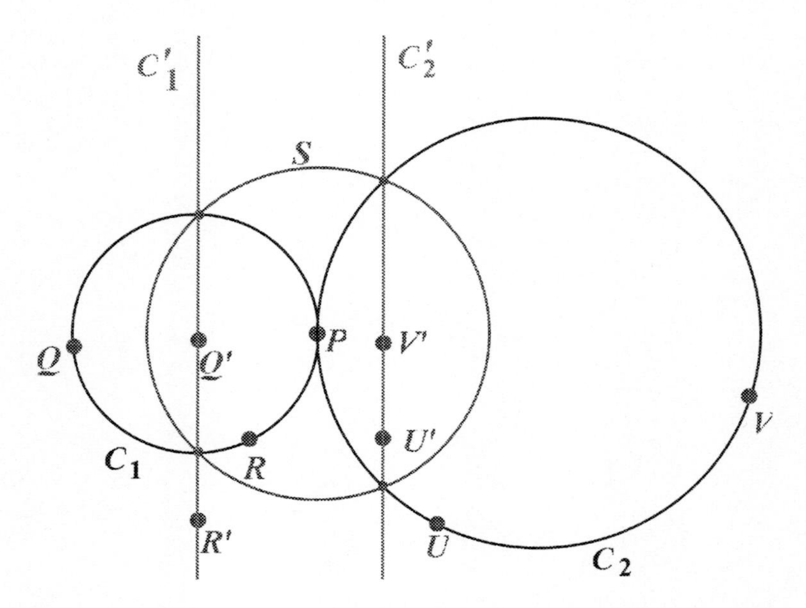

Figure 14.8: Second configuration for Section 14.3.8

(ii) What is the inverse of a non-intersecting system of coaxal circles, with respect to one of their limiting points?

(i) If the circles are an intersecting coaxal system through A and B and the centre of inversion is A, then the images under inversion with circle of inversion S, centre A, are lines through B', the inverse of B. If C_k is a member of the coaxal system, then its image C_k', being a line, is the line of intersection of S and C_k, since the common points of S and C_k map into themselves. See Fig. 14.9.

We have proved the following result: if S is a fixed circle centre A, and C is a variable circle passing through A and another fixed point B, then the common chord of C and S passes through a fixed point.

(ii) Every circle of a non-intersecting system of coaxal circles is orthogonal to every circle of the complementary intersecting system of coaxal circles. Moreover, the points of intersection of the latter are the limiting points of the former. If we form the inverse figure with respect to the limiting point A, then the intersecting system, by part (i) become lines through B', the other limiting point. The images of the non-intersecting coaxal system must cut

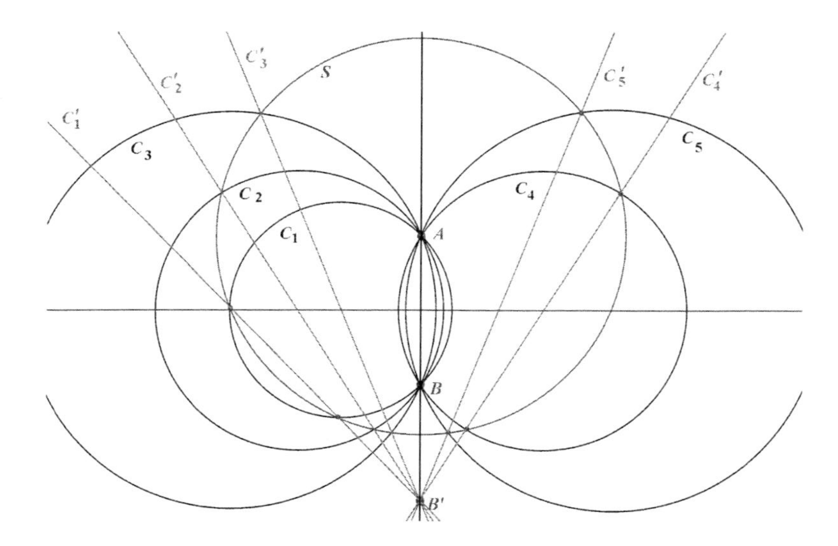

Figure 14.9: Example 14.3.2

these lines orthogonally, and therefore become a system of concentric circles centre B.

Exercise 14.3.3

1. What is the inverse of a circle with respect to its centre?

2. What is the inverse of a pair of parallel straight line?

3. What is the inverse of a rectangle with respect to any point?

4. Invert with respect to O the following theorem: from an external point P two lines can be drawn to touch a given circle OAB.

5. Show that the theorem, which states that angles in the same segment are equal, inverts into the alternate segment theorem.

6. Recall that if ABC is a triangle and P is a point on the circumcircle and L, M, N are the feet of the perpendiculars from P on to BC, CA, AB respectively, then LMN is a straight line. By inverting with respect to P, prove the following theorem: let A', B', C' lie on a line and let P be a point not on that line, then the circumcentres of triangles $PB'C', PC'A', PA'B'$ are concyclic with P.

7. If A', B', C', D' are the inverses of A, B, C, D and $ABCD$ forms a harmonic range, prove that $A'B'C'D'$ either forms a harmonic range on a line or on a circle.

8. Invert with respect to H the theorem that the altitudes of a triangle are concurrent at H.

9. O, A, B, C, H are five points such that the circles OAB, OCH are orthogonal and circles OBC, OAH are orthogonal. Prove that circles OCA, OBH are also orthogonal.

10. PQ, RS are common tangents to two circles PAR, QAS. Prove that the circles PAQ, RAS touch each other.

11. A circle OAB is given, with diameter AB. P is a point on the line segment AB. Prove that the circles APO and BPO are orthogonal.

12. ABC is a line segment, with B an internal point of AC. O is a point not on the line. Circles OAB, OCB are drawn, and the tangents at A and C to these circles meet at D. Prove that $OADC$ is a cyclic quadrilateral.

13. P is a fixed point, l is a fixed line. Q, R are variable points on l such that $\angle QPR$ is constant. Prove that the circumcircle of triangle PQR touches a fixed circle.

14. ABD, CAE, BCF are three circles touching each other at points A, B, C. The common tangent at C passes through D, and DAE, DBF are straight lines. Prove that EF touches the circles at E and F.

15. The lines l and m meet at right angles at the point T. Two circles touch l and each other externally at T. Two more circles touch m and each other externally at T. Prove that the four points of intersection of pairs of these circles (other than T) are concyclic.

16. Prove that if O is the intersection of the exterior common tangents of two circles, then it is possible with O as centre of inversion to invert either circle into the other. (This can cause a misinterpretation, as O is a centre of similitude of the two circles. However, it is best to put that thought aside, as the inversion and the similarity are different mappings. For if $OABCD$ is a straight line cutting the two circles, with

points in that order, then in the inversion we have $A \leftrightarrow D, B \leftrightarrow C$, whereas in the similarity we have $A \leftrightarrow C, B \leftrightarrow D$.)

17. C_1 and C_2 are two coplanar circles. A_0 is a general point in the plane and A_{2n+1} is the inverse of A_{2n} in C_1 $(n = 0, 1, 2, \ldots)$ and A_{2n} is the inverse of A_{2n-1} in C_2 $(n = 1, 2, \ldots)$. Prove that A_n $(n = 0, 1, 2, \ldots)$ all lie on a circle. Identify this circle.

Example 14.3.4

Prove there is at least one circle of an intersecting system of coaxal circles that is orthogonal to a given fixed circle, and at most two circles of an intersecting coaxal system that touch a given circle.

Invert the figure with respect to one of the points of intersection. As shown in Example 14.3.2, the coaxal system becomes a system of straight lines passing through the inverse of the other point of intersection. In general the inverse of the fixed circle is a circle in general position. In that case there is only one line that is a diameter of this circle and two lines that touch it. An exception is when the image of the fixed circle has its centre at the inverse of the other point of intersection, in which case every line is a diameter. In the original figure this corresponds to when the fixed circle is a member of the complementary coaxal system, when it is orthogonal to every member of that system. The other exceptional case is when the image of the fixed circle is a line, when there is only one line of the system at right angles to it and only one line of the system coincident with it or parallel to it. In the original figure this corresponds to when the fixed circle passes through one and only one of the two points of intersection of the coaxal system. The problems in Exercise 14.3.5 are about coaxal circles, and may be solved by means other than inversion.

Exercise 14.3.5

1. AB is a common tangent to two non-intersecting circles. Prove that the circle on AB as diameter cuts the two circles orthogonally.

2. A common tangent to two non-intersecting circles C, D touches them at P, R respectively. K and L are internal to C and D respectively, and are the limiting points of the coaxal system determined by C and D. PL meets C again at Q and RL meets D again at S. Prove that QS is a common tangent to C and D.

3. Prove that the three pairs of limiting points of three circles taken in pairs are concyclic.

4. An exterior common tangent to two non-intersecting circles cuts their radical axis at P. K, L are the limiting points. Prove that PK, PL are parallel to the internal common tangents.

5. AB is a common tangent to two circles. Prove that A, B are conjugate points with respect to any circle coaxal with them.

6. A quadrilateral is inscribed in one circle and circumscribed about another. Prove that the point of intersection of the diagonals is a limiting point of the two circles.

14.4 Problems involving distance

Problems in previous sections have been concerned with incidence properties of lines and circles and also with angles between curves, so that questions about parallel lines, tangency and orthogonal circles have featured. There is a further class of problems, in which distance is involved, that are capable of being treated by inversion. In some of these it is even necessary to designate the value of r, the radius of the circle of inversion. In some cases one is required to draw the inverted figure on the original figure in order to deduce properties of the diagram as a whole.

The first step is to provide a key result between distances in the original and inverted figures. See Fig. 14.10.

14.4.1 Ptolemy's theorem and its extension

Let O be the centre of inversion and r the radius of the circle of inversion S, and suppose that Z_1, Z_2 are two points in the z-plane and W_1, W_2 are their images under the inversion $wz^* = r^2$. Then

$$\frac{W_1W_2}{Z_1Z_2} = \frac{r^2}{(OZ_1)(OZ_2)} = \frac{(OW_1)(OW_2)}{r^2} \tag{14.24}$$

We have $w_1 = r^2/z_1^*$ and $w_2 = r^2/z_2^*$, so

$$\frac{W_1W_2}{Z_1Z_2} = \frac{|w_1 - w_2|}{|z_1 - z_2|} = r^2 \frac{\left|\frac{1}{z_1^*} - \frac{1}{z_2^*}\right|}{|z_1 - z_2|}$$

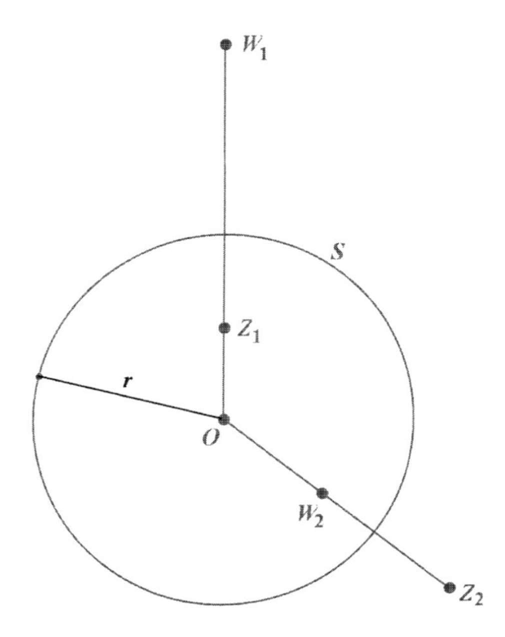

Figure 14.10: Configuration for Section 14.4.1

$$= \frac{r^2}{|z_1||z_2|} = \frac{r^2}{(OZ_1)(OZ_2)} = \frac{(OW_1)(OW_2)}{r^2},$$

the last step following since $(OW_1)(OW_2)(OZ_1)(OZ_2) = r^4$.

The first example provides another proof of Ptolemy's theorem and the proof leads naturally to its extension.

Example 14.4.1 *(Ptolemy's theorem and its extension)*

If A, B, C, D are any four points in a plane, then

$$(AB)(CD) + (BC)(DA) \geq (AC)(BD), \tag{14.25}$$

with equality if, and only if, A, B, C, D are concyclic (with vertices appearing in that order, either clockwise or anticlockwise).

Invert with respect to A. Points B, C, D become points B', C', D', which, in general form a triangle, for which the triangle inequality

$$B'C' + C'D' \geq B'D' \tag{14.26}$$

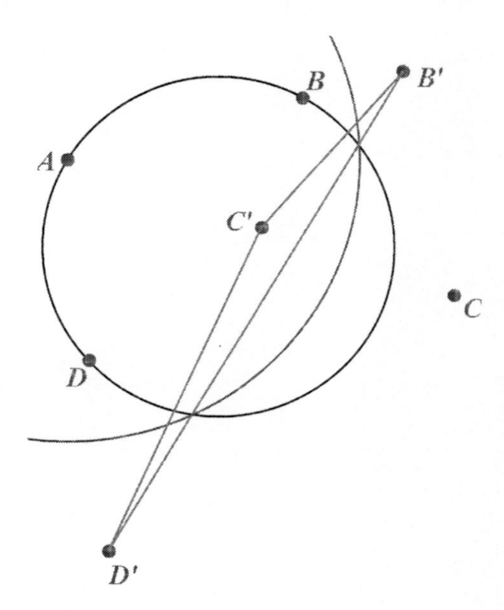

Figure 14.11: Ptolemy's theorem and its extension

holds. Also equality occurs if, and only if, B', C', D' lie on a straight line with C' lying between B' and D'. Now, from Example 14.4.1, Equation (14.26) becomes

$$\frac{BC}{(AB)(AC)} + \frac{CD}{(AC)(AD)} \geq \frac{BD}{(AB)(AD)}. \tag{14.27}$$

Clearing fractions, we obtain the Inequality (14.25). Moreover $B'C'D'$ is a straight line, with vertices in that order if, and only if, A, B, C, D are concyclic, with vertices appearing in that order. See Fig. 14.11.

Exercise 14.4.2

1. What theorem may be deduced from Menelaus's theorem by inverting with respect to a general point in the plane?

2. Suppose that circles LBC, MCA, NAB meet at a point P. Prove that if APL, BPM, CPN are straight lines, then

$$(BL/LC)(CM/NA)(AN/NB) = 1.$$

3. Generalize Ptolemy's theorem for n points lying on a circle.

4. Invert a rhombus $ABCD$ with respect to A and state the theorem that corresponds to the fact that the diagonals of a rhombus bisect each other at right angles.

5. A, B, C are three collinear points. O is a point such that $\angle AOB = \angle BOC$. Prove that $OA/OC = AB/BC$.

6. If in Problem 5 the two angles are equal to $60°$, prove that $1/OB = 1/OA + 1/OC$.

7. O is a point on the minor arc BC of the circumcircle of an acute-angled triangle ABC. OP, OQ, OR are the perpendiculars from O on to the sides BC, CA, AB respectively. Prove that $BC/OP = CA/OQ + AB/OR$.

8. $OABC, ODEF$ are two lines though O such that B, E are the mid-points of AC, DF respectively. Prove that the circles OAD, OBE, OCF have a common chord.

9. Two coplanar circles C_1 and C_2 have radii 9 and 2 units and their centres are 5 units apart. Prove that it is possible to draw six circles in the area between C_1 and C_2, each touching C_1 and C_2 and its two neighbours. (This is an example of a Steiner chain and it creates a porism. It is either impossible to draw such chains, or a chain is possible irrespective of where the first circle is situated.)

10. $ABCD$ is a cyclic quadrilateral. The sides DA and BC meet at P; the diagonals AC and BD meet at R. Z is the foot of the perpendicular from D to PR. M is the midpoint of CD. Prove, by inversion with respect to D, that circles $ABCD$ and DZM are orthogonal.

Chapter 15

Polar reciprocation

15.1 Introduction

Consider a configuration F in a plane consisting of points, with labels such as P, and lines with labels such as l. Let S be a non-degenerate conic in the plane of the configuration. For every point P in F, let p be the polar of P with respect to S. For every line l in F, let L be the pole of l with respect to S. Then the configuration f consisting of the lines, with labels such as p, and points, with labels such as L, is called the polar reciprocal of F with respect to S, or just the reciprocal of F with respect to S. It is clear from definition that the reciprocal of f with respect to S is F.

If C is a conic in F, then its polar reciprocal c in f may be interpreted as the locus of the poles of the tangents of C. Alternatively it may be interpreted as the envelope of the polars of the points of C. The two constructions provide the same curve. Later in the section we prove that c is, as you would expect, a conic.

If Part 15.2 we sometimes use Cartesian co-ordinates and in proofs we then often take S to be the circle with equation $x^2 + y^2 = 1$, centre O, radius 1. If S is a circle it is usually called the auxiliary circle. Generalisations in the projective plane are considered in Section 15.3. However, all the incidence properties chronicled in Part 15.2 are true whether or not S is a circle.

Example 15.1.1

Reciprocate with respect to a circle, centre O, the configuration in which the altitudes of a triangle are shown to be concurrent. Hence prove that if O is any point in the plane of a triangle abc, and lines perpendicular to $O\ bc, O\ ca, O\ ab$ through O meet the sides a, b, c at points L, M, N respectively, then L, M, N are collinear. Here the notation bc means the point of intersection of lines b and c and so on.

See Fig. 15.1. The configuration is a triangle ABC with altitudes AD, BE, CF concurrent at the orthocentre H. In the reciprocal figure a, b, c, d, e, f, h are the polars of A, B, C, D, E, F, H. Points such as ab in the reciprocal figure are the poles of lines such as AB in the original configuration.

cf is the pole of CF so $O\ cf$ is perpendicular to CF and hence parallel to AB, which in turn is perpendicular to $O\ ab$. It follows that $O\ ab$ and $O\ cf$ are at right angles to one another. Similarly $O\ bc$ is perpendicular to $O\ ad$ and $O\ ca$ is perpendicular to $O\ be$. Now cf lies on c, which is the line joining bc and ca. Hence cf is the point N and similarly ad is L and be is M.

Now H lies on AD so h passes through ad, that is L lies on the line h. Similarly M, N also lie on h, and thus L, M, N are collinear. l, m, n are, of course, the polars of L, M, N and are respectively the altitudes AD, BE, CF.

Note that $O\ ab$ is perpendicular to AB, $O\ ca$ is perpendicular to CA, and hence $\angle ab\ O\ ca = 180° - A$. When the circle of reciprocation, usually known as the auxiliary circle, is actually the circumcircle, then a is the tangent at A, b the tangent at B and c the tangent at C.

Exercise 15.1.2

1. By reciprocating the theorem that the internal bisectors of the angles of a triangle are concurrent at the incentre of the triangle, prove the following theorem:

   Let ABC be a triangle and O a point in the plane of the triangle not on its sides and let the external bisectors (or one external bisector and two internal bisectors) of angles $\angle BOC, \angle COA, \angle AOB$ meet the sides BC, CA, AB at L, M, N. Then L, M, N are concurrent.

15.2 Basic properties

The following results are immediate from definition.

15.2.1 Points

The reciprocal of a point is the polar of that point.

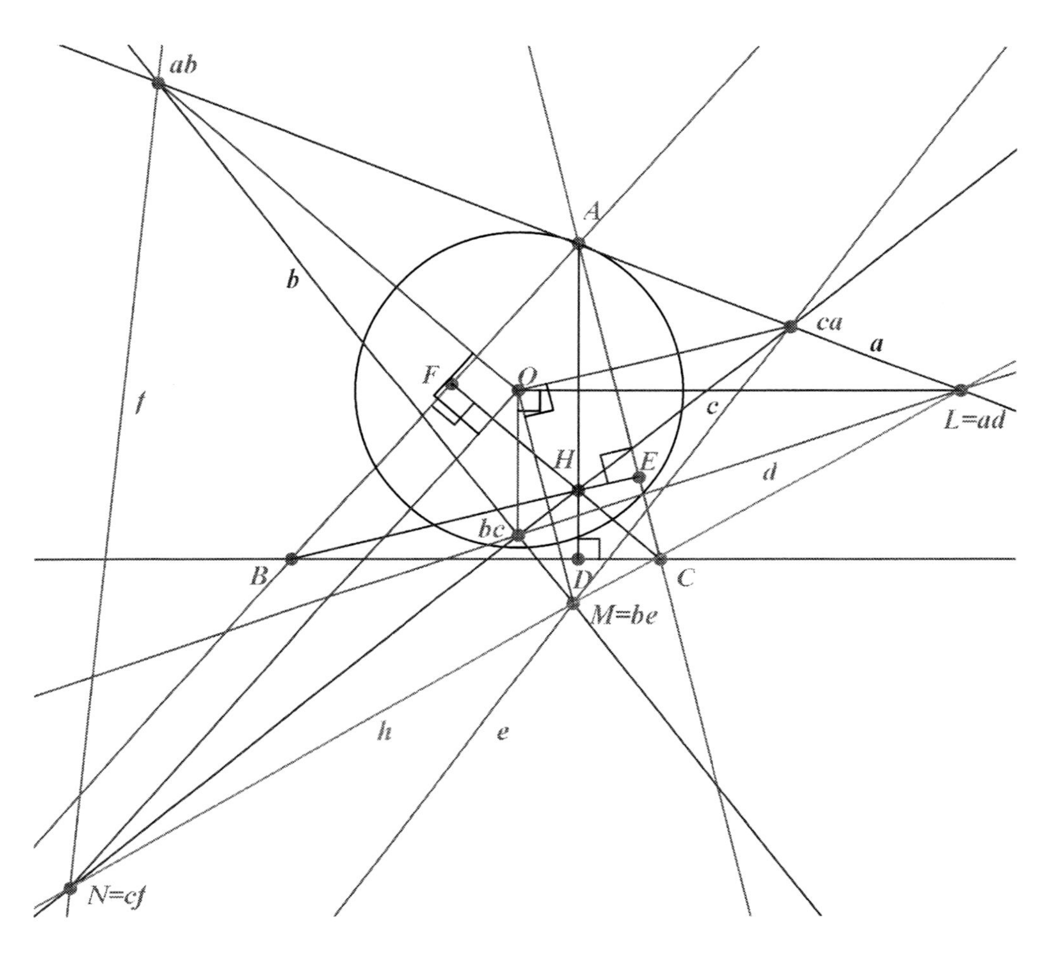

Figure 15.1: Example 15.1.1

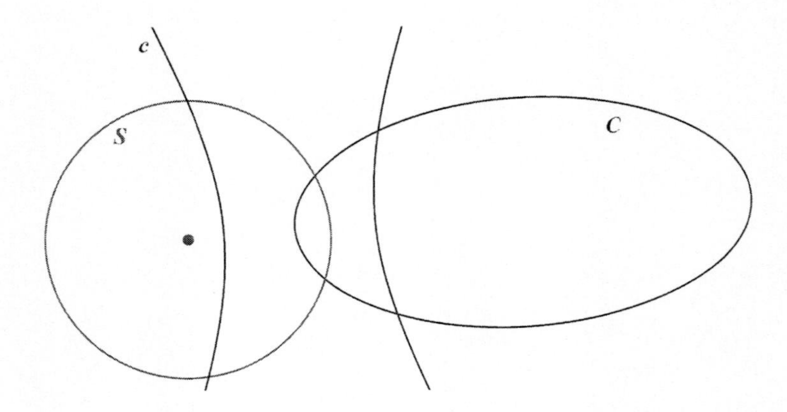

Figure 15.2: A polar reciprocal that is a hyperbola

15.2.2 Lines

The reciprocal of a line is the pole of that line.

15.2.3 Two intersecting lines

Their reciprocal is a line passing through two points, the poles of the two lines. The line is the polar of the point of intersection.

15.2.4 Line joining two points

Its reciprocal is a point through which two lines pass, the point being the pole of the line and the lines being the polars of the points.

The results in Section 15.2.1-15.2.4 underpin the principle of duality, first mentioned in Section 3.2.2. Thus four points and six lines (a quadrangle) reciprocate into four lines and six points (a quadrilateral) and conversely. We now prove

15.2.5 The reciprocal of a conic is a conic

It is called the *polar reciprocal conic.*

We reciprocate the conic C with equation

$$ax^2 + by^2 + cz^2 + 2fyz + 2gzx + 2hxy = 0 \quad (z = 1) \tag{15.1}$$

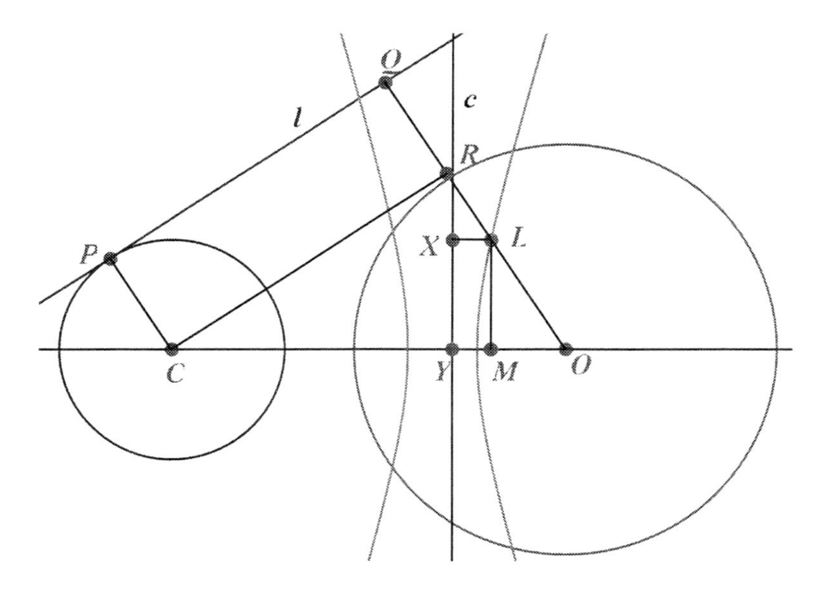

Figure 15.3: A configuration for Section 15.2.6

in the auxiliary circle S, with equation $x^2 + y^2 = 1$.

In Fig. 15.2, the polar reciprocal is a hyperbola.

The line equation of the conic with Equation (15.1) is

$$Al^2 + Bm^2 + Cn^2 + 2Fmn + 2Gnl + 2Hlm = 0, \tag{15.2}$$

where A, B, C, F, G, H are the minors $A = bc - f^2$, $H = fg - hc$ etc.. (See Equation 3.11.) Now the polar of (X, Y) with respect to S is $Xx + Yy = 1$, and if (X, Y) lies on the reciprocal curve, then this must be a tangent to the conic C. As its line co-ordinates $l : m : n = X : Y : -1$ it follows that the equation of c is

$$Ax^2 + By^2 + C - 2Fy - 2Gx + 2Hxy = 0. \tag{15.3}$$

First note that Equation (15.3) represents a conic, so that the polar reciprocal of a conic C is a conic c. (This is true if the conic of reciprocation is not a circle, as only the degree of the curve matters.) Further, since $(AB - H^2) = c\Delta$, where Δ is the discriminant defined by Equation 3.5 (and c is now the constant in Equation (15.1)), it follows that if $c = 0$, that is the conic C passes through the centre of the auxiliary circle, the polar reciprocal is a parabola. The polar reciprocal changes from being a hyperbola to an

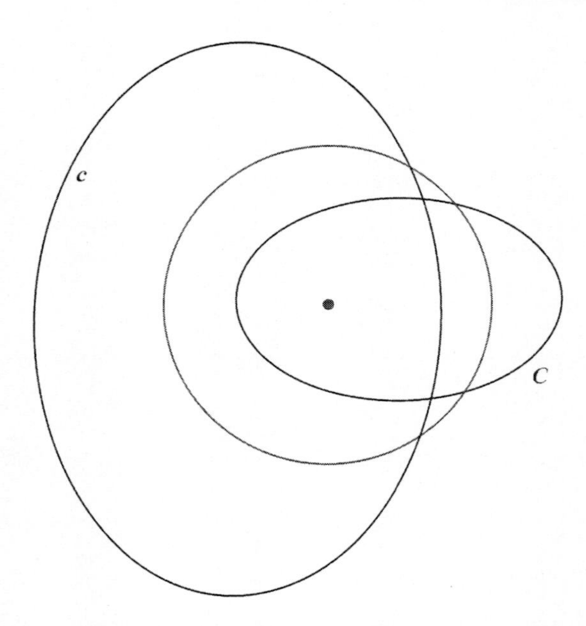

Figure 15.4: A polar reciprocal that is an ellipse

ellipse, or from being an ellipse to a hyperbola as the centre of the auxiliary circle crosses the boundary of C. See Fig. 15.4 and Fig. 15.5.

Fig. 15.3 illustrates the reciprocal polar of a circle centre C when the auxiliary circle has centre O. The tangent l reciprocates into the point L and the centre C reciprocates into the line c. The locus of L as l varies is a hyperbola, since O lies outside the circle centre C. Let the foot of the perpendicular from L on to c be denoted by X and let c meet OC at Y.

15.2.6 c is the directrix and O the focus of the reciprocal conic

This is an important result, as it allows focal properties of conics to be deduced from properties of a circle.

Use the notation in Fig. 15.3, so that $CPQR$ is a rectangle and $XLMY$ is a rectangle. If k is the radius of the auxiliary circle we have $(OC)(OY) = (OL)(OQ) = k^2$ and by similar triangles $(OC)(OM) = (OL)(OR)$. By subtraction $(OC)(MY) = (OL)(QR)$ or $(OC)(LX) = (OL)(PC)$. With PC and OC fixed we see that $(OL) = e(LX)$, where $e = (OC)/(PC)$. These are the

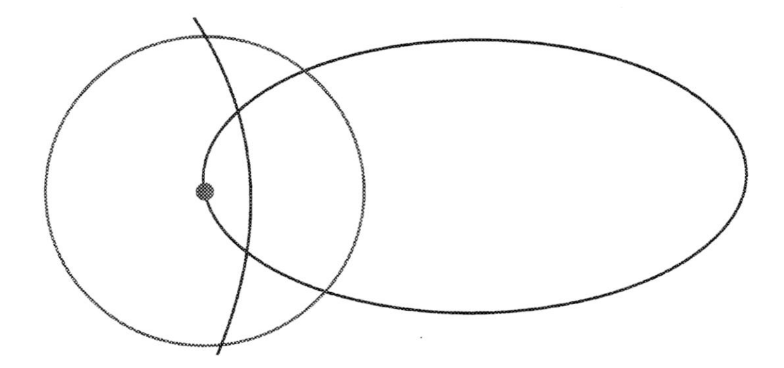

Figure 15.5: A polar reciprocal that is a parabola

defining equations for the focus and directrix of a conic.

The results in Sections 15.2.7 - 15.2.11 and Section 15.2.13 follow without too much work and are left as exercises for the reader. However, as Section 15.2.12 is so important, we provide a proof. Note that those properties that involve incidence are true when the conic of reciprocation is a general conic.

15.2.7 Chord of a conic

A fixed chord meets a conic in two points and these points reciprocate into a pair of tangents to the polar reciprocal conic through a fixed point. Likewise two tangents and the chord of contact to a conic reciprocate into two points on the polar reciprocal conic and the intersection of the tangents at those point.

15.2.8 Lines intersecting on a conic

These reciprocate into a pair of points and a line joining them that touches the polar reciprocal conic.

15.2.9 Pole and polar

If a point and a line are pole and polar with respect to a conic, then they reciprocate into polar and pole on the polar reciprocal conic.

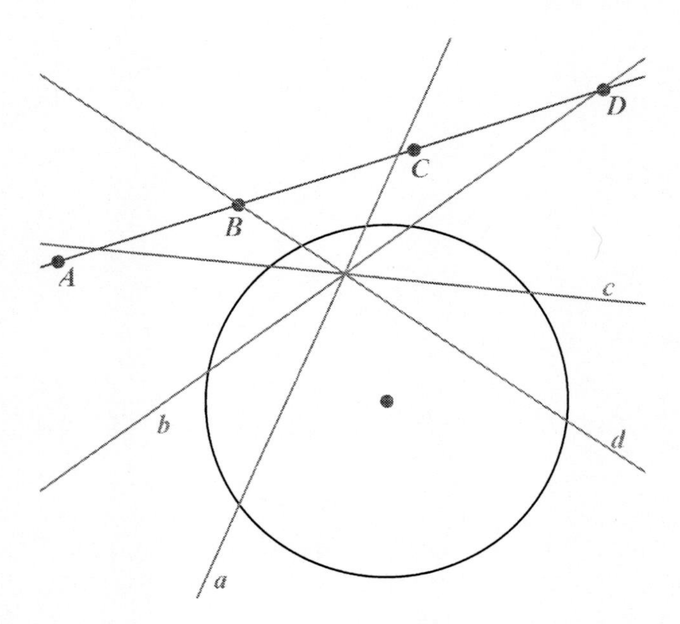

Figure 15.6: Configuration for Section 15.2.10

15.2.10 Range of points and pencil of lines

A range of points $A, B, C, D, \ldots$ reciprocates into a pencil of lines $a, b, c, d, \ldots$ and conversely. Also $(A, C; B, D) = (a, c; b, d)$. See Fig. 15.6.

15.2.11 Inscribed and circumscribed figures

A polygon inscribed in a conic reciprocates into a polygon circumscribing the polar reciprocal conic and conversely. Fig. 15.7 illustrates this for a triangle, and also illustrates Section 15.2.12, which is true if the conic of reciprocation is a circle.

15.2.12 The angle between two lines

This reciprocates into the equal angle subtended at the centre O of the auxiliary circle by the lines joining O to the points reciprocal to the two lines. This property was used in Example 15.1.1. The proof is given immediately after Section 15.2.13

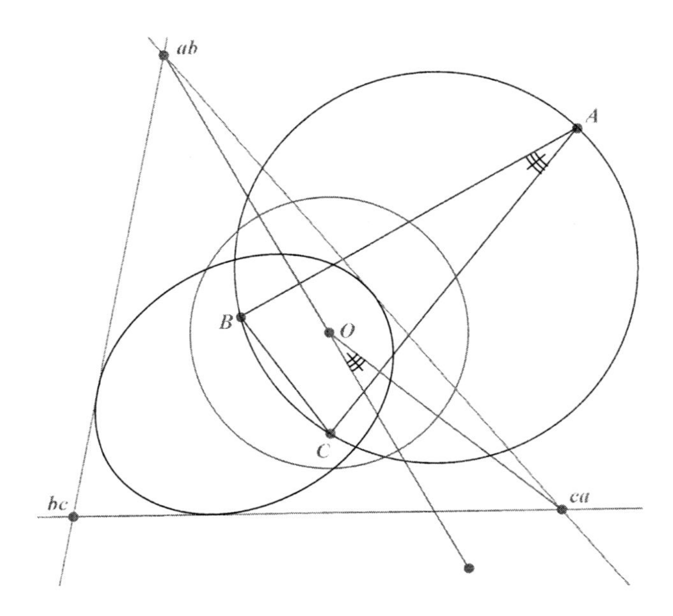

Figure 15.7: Configuration for Section 15.2.11

15.2.13 The centre O of the auxilary circle

This reciprocates into the line at infinity. Two lines through the origin O reciprocate into two points on the line at infinity. Two tangents through O to a conic, with O external to the conic become two points on the line at infinity on the reciprocal curve. The points of contact of those tangents become the asymptotes of the reciprocal polar conic. The angle between these tangents becomes the angle between the asymptotes.

The result stated in Section 15.2.12 is important, so we give a proof. Let the two lines be $lx + my = 1$ and $px + qy = 1$. These lines have gradients $-l/m$ and $-p/q$. The polar reciprocals of these lines with respect to the auxiliary circle $x^2 + y^2 = 1$ are the points L, P with conordinates (l, m) and (p, q) respectively. The gradients of the lines OL, OP are m/l and q/p respectively. Since OL is perpendicular to the line with gradient $-l/m$ and OP is perpendicular to a line with gradient $-p/q$, it follows that $\angle LOP$ is equal to the angle between the two lines. The reader should consider what happens when the conic of reciprocation is the rectangular hyperbola $x^2 - y^2 = 1$.

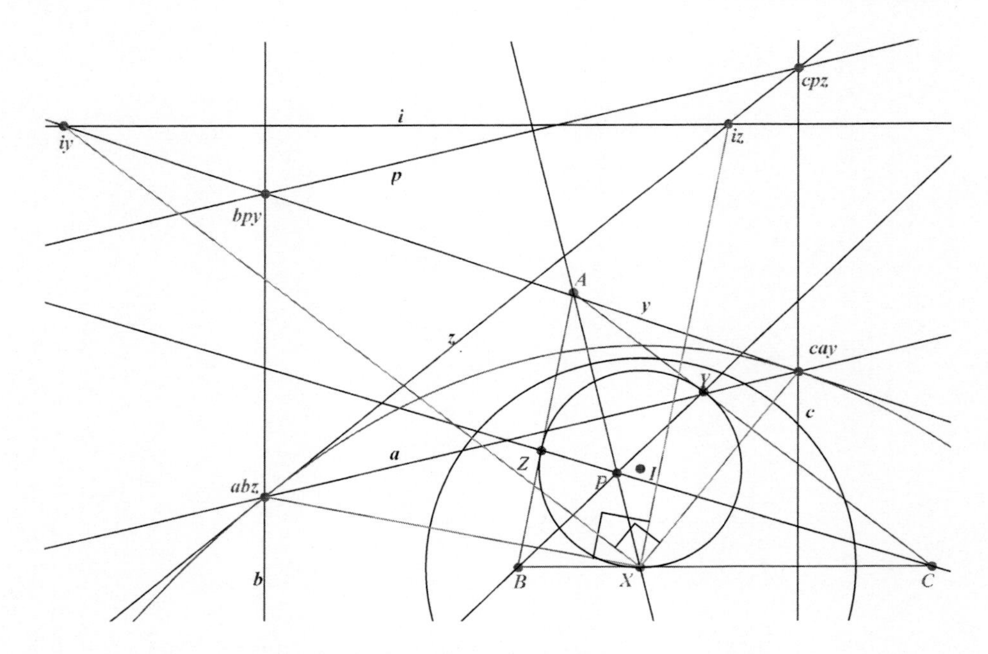

Figure 15.8: Example 15.2.1

Example 15.2.1

Lines a, b, c are given with b and c parallel. A point X and lines y, z, i are also given with the properties: (i) a parabola touches y and z and has focus X, (ii) a, b, z are concurrent, (iii) c, a, y are concurrent, (iv) $\angle iz\ X\ ab = \angle iy\ X\ ca = 90°$, (v) i is perpendicular to b and c. Prove that the line joining by and cz is parallel to a.

As Fig. 15.8 shows, this configuration is the reciprocal, with respect to a circle centre X, of the one in which ABC is a triangle, XYZ is the incircle, with X on BC, Y on CA and Z on AB. The theorem is the reciprocal of the known result that AX, BY, CZ are concurrent at P. The line i is, of course, the polar of the incentre and the angle conditions express the fact that IY and IZ are perpendicular to CA and AB respectively. The fact that i is perpendicular to b and c represents the fact that X lies on BC and IX is perpendicular to BC.

Many of the problems in Exercise 15.2.2 pose similar conundrums, which may be unravelled in the same sort of way. So the reader should study Fig. 15.8 carefully, in order to understand all the features represented.

Exercise 15.2.2

1. Prove that if a circle, centre C, is reciprocated with respect to a circle, whose centre O lies on the circle, then the polar reciprocal is a parabola, focus O and directrix the reciprocal of C.

2. Let a and b be two tangents that meet at a point ab on the directrix of a parabola. Let p be any third tangent that meets a and b at pa and pb respectively. Let O be the focus of the parabola. Prove that $\angle pa\ O\ pb = 90°$. To which circle theorem does this correspond?

3. This problem is concerned with 9 lines, which we denote by $a, b, k, d, e, f, l, m, n$. It is given that n, d, e are concurrent, as are a, k, d and e, k, b. Also a, b, f, n are concurrent. Further k, f are parallel, as are e, a, m and d, b, l. Prove that l, m, n are concurrent and that a lies midway between e and m, and b lies midway between d and l. *Hint: Concentrate on k and its relationship with the other lines.*

4. The lines a, b, c are three tangents to a parabola, focus P. These tangents form a triangle ABC. The line l is the perpendicular through A to PA, the line m is the perpendicular to B through PB, the line n is the perpendicular to PC through C. Prove that l, m, n are concurrent and locate the point of concurrence. *Hint: Reciprocate the Wallace-Simson line theorem or use Cartesian co-ordinates.*

5. An ellipse, focus O, is inscribed in a quadrilateral $ABCD$. Prove that $\angle AOB + \angle COD = 180°$.

6. An ellipse is inscribed in a triangle touching its sides at X, Y, Z. If the tangent at Y is parallel to ZX and the tangent at Z is parallel to XY, prove that the tangent at X is parallel to YZ.

7. What is the reciprocal figure when a set of confocal ellipses are reciprocated in a circle the centre of which is at one of the foci? What is the reciprocal of the other focus?

8. Describe how to construct a pair of intersecting confocal parabolas.

9. Lines a and b are tangents to a parabola at the points A and B respectively. l is the line parallel to b through A and m is the line parallel

to a through B. Lines a and b meet at C, and n is the line through C parallel to the axis of the parabola. Prove that l, m, n are concurrent.

10. a, b, c, l, m, n are six lines such that a, l are parallel as are b, m and c, n. Lines b, c, l are concurrent at P, lines c, a, m are concurrent at Q and a, b, n are concurrent at R. Lines m, n meet at U, lines n, l meet at V, lines l, m meet at W. Lines d, e, f have the following properties: line d passes through P, line e passes through Q, line f passes through R; d meets a at X, e meets b at Y, f meets c at Z in such a way that lines XU, YV, ZW are concurrent. Prove that d, e, f are concurrent.

11. ABC is a triangle and K is an internal point of ABC. Cevians AKL, BKM, CKN are drawn. The line through K parallel to MN meets BC at D, and E, F are similarly defined. Prove that D, E, F are collinear.

12. A conic is touched by six lines d, e, f, l, m, n. A further three lines a, b, c are such that b, l, d, c are concurrent, as are c, m, e, a and a, n, f, b. Prove that $a \wedge l$, $b \wedge m$, $c \wedge n$ are collinear if, and only if, $a \wedge d$, $b \wedge e$, $c \wedge f$ are collinear. Prove further that if the circle on $a \wedge d$, $b \wedge c$ as diameter meets the circle on $b \wedge e$, $c \wedge a$ as diameter at O, then the circle on $c \wedge f$, $a \wedge b$ as diameter passes through O.

13. ABC is a triangle. XYZ is another triangle such that A lies on YZ, B lies on ZX, C lies on XY. The line YZ meets BC at L, ZX meets CA at M and XY meets AB at N. Suppose L, M, N are collinear. Prove that a conic can be drawn to touch YZ at A, ZX at B and XY at C.

14. DEF is a triangle and ABC is its median triangle with A on EF, B on FD and C on DE. O is the circumcentre of DEF. Through O are drawn three lines parallel to the sides of DEF. The line parallel to EF meets DE at L and FD at L'. The line parallel to FD meets EF at M and DE at M'. The line parallel to DE meets FD at N and EF at N'. Prove that a conic can be drawn having $AL, AL', BM, BM', CN, CN'$ as tangents. *Hint: Reciprocate with respect to a circle with centre O.*

15. ABC is a triangle and L, M, N are the midpoints of BC, CA, AB respectively. Prove that the ellipse through L touching CA, AB at M, N respectively touches BC at L. *Hint: Reciprocate tangents to the circumcircle with respect to a circle with centre the symmedian point. Of course, the problem can easily be solved by areals.*

16. Given a fixed point and a fixed line, prove that there are an infinite number of parabolas having the point as focus and the line as a common tangent.

17. What do you get if you reciprocate the theorem that the circumcircles of four triangle formed by four lines in general position concur at a point?

18. State the dual of Pappus's theorem.

19. LMN is a triangle and ABC an inscribed triangle with A on MN, B on NL and C on LM. Prove that three confocal ellipses (1) touching BC, LM, LN; (2) touching CA, MN, ML; (3) touching AB, NL, NM have a common tangent.

20. ABC is a triangle, incentre I. The perpendicular bisectors of AI, BI, CI are denoted by p, q, r respectively. BC meets p at L, CA meets q at M, AB meets r at N. Prove that L, M, N are collinear.

21. Two conics have a common focus S and two common tangents A_1A_2 and B_1B_2. Prove that$\angle A_1SA_2$ and $\angle B_1SB_2$ are equal or supplementary. (JW)

22. A conic is given with focus S. A straight line is drawn from S to a point P on the conic and a line l at right angles to SP is drawn. Prove that l and the tangent at P meet on the directrix.

23. Explain how to construct a self-polar triangle of a conic.

24. ABC is a triangle, incentre I. DEF is homothetic to ABC through I and congruent to ABC. MN is the line through A perpendicular to IA, NL is the line through B perpendicular to IB and LM is the line through C perpendicular to IC. EF meets MN at U, FD meets NL at V and DE meets LM at W. Prove that U, V, W are collinear.

25. A point S is taken inside a triangle ABC such that each side subtends an equal angle at S. Prove that four conics may be drawn circumscribing triangle ABC having focus S, and that one of these conics touches the other three. (JW)

26. LMN is a triangle and ABC is its median triangle with A on MN, B on NL and C on LM. G is the centroid. The line p meets BC at U and UG meets MN at X. Similarly p meets CA at V and VG meets NL at Y and p meets AB at W and WG meets LM at Z. Prove that X, Y, Z are collinear.

27. a, b, c, x, y, z are six lines such that ay, bz, cx are collinear and az, bx, cy are collinear. Prove that ax, by, cz are collinear.

28. ABC is a triangle orthocentre H. An ellipse, focus H is inscribed in ABC. The line p is any other tangent to the ellipse. AC meets p at M, MH meets BC at U, and q is the line through U parallel to AB. AB meets p at N, NH meets BC at V, and r is the line through V parallel to AC. BC meets p at L, q and r intersect at W. Prove that L, W, H are collinear.

29. ABC is a triangle circumcentre O. A parabola is drawn with focus O to touch AB and AC. Prove that its directrix is parallel to BC.

30. ABC is a triangle and u is the line parallel to BC through A. s and t are distinct lines through C and B respectively and they are concurrent with u. The parabola is drawn with focus O, the circumcentre of ABC, that touches AB and AC. The line j is the tangent at the vertex of the parabola. t and j meet at P; s and j meet at Q. The line p passes through P and is parallel to s. The line q passes through Q and is parallel to t. Prove that p and q meet on BC.

31. AQR is an isosceles triangle with $AQ = AR$ and S is the excircle of triangle AQR opposite A. K is the centre of this excircle. The point X is chosen on the same side of QR as A and Y, Z are the points where the excircle touches AQ and AR respectively. XK meets YZ at U. Given that X and U are inverse points with respect to the excircle prove that XK bisects $\angle YXZ$.

32. The incircle of triangle ABC touches the sides BC, CA, AB at X, Y, Z respectively and I is the incentre. PQR is another triangle having the same incircle and AP, BQ, CR are concurrent at a point K, which is the incentre of triangle XYZ. Prove that a line k exists such that k, BC, QR are concurrent at L; k, CA, RP are concurrent at M; k, AB, PQ are concurrent at N and $\angle POL = \angle QOM = \angle RON = 90°$.

33. $TUVW$ (with vertices in that order) is a convex quadrilateral having an incircle, centre O. No pair of opposite sides of the quadrilateral are parallel. TU meets VW at Z and TW meets UV at Y. Prove that $\angle TOZ = \angle VOY$. What happens if the quadrilateral is a trapezium or a rhombus?

34. A circle S is given, centre I and Q is a point external to the circle. Tangents from Q to the circle are drawn and points P and R are variable points on these tangents such that PR is also a tangent to S (and I, therefore, is the incentre of triangle PQR). Prove that $\angle PIR$ is constant and is equal to the supplement of $\angle QIT$, where T is one of the points of tangency from Q to S.

35. Two conics have a common focus C. They each pass through three given points, and at one of these points they have a common tangent p and at another of these points the tangent to one conic is denoted by a and the tangent to the other conic by b. The lines a and p meet at U and the lines b and p meet at V. Prove that $\angle UCV = 90°$.

36. Two parabolas have a common focus A. They also have a common tangent. Lines b and b' are variable tangents to the first parabola and c and c' being parallel to b and b' respectively are both tangents to the second parabola. Tangents b and b' meet at B and tangents c and c' meet at C. Prove that $\angle BAC$ is constant.

37. P and Q are a pair of parabolas with a common focus B. They also have a common tangent a. The directrices of P and Q are denoted by c and d respectively. Q has a tangent q which is concurrent with c and a. Prove that a parabola exists with focus B, having c, d and q as tangents.

38. S is a fixed circle, centre O, and a and b are fixed lines. A variable conic Σ is drawn with focus O, having a and b as tangents. Lines c and d are the common tangents of S and Σ. Prove that c and d meet on a fixed line through the intersection of a and b.

39. Two parabolas are drawn with the following properties: they have a common focus A and a common tangent b. A tangent c is drawn to the first parabola and a tangent d is drawn to the second parabola parallel to c. The line c meets the directrix of the first parabola at S and d

meets the directrix of the second parabola at T. Prove that another parabola with focus A can be drawn to touch b, c, d and ST.

40. O is a fixed point and a is a fixed line. Lines b, c, p are variable lines that are concurrent. Two ellipses are drawn with common focus O, having a and p as common tangents. b is tangent to the first and c to the second. Lines a and b meet at C; a and c meet at B. Prove that $\angle CAB$ is constant.

41. ABC is a triangle, incentre I. Another tangent to the incircle meets BC at L, CA at M, AB at N. Points P, Q, R are located on LMN such that $\angle PIL = \angle QIM = \angle RIN = 90°$. Prove that AP, BQ, CR are parallel.

42. $PQRS$ is a quadrilateral with an incircle, centre I. Prove that the internal angle bisectors of $\angle PIR$ and $\angle QIS$ are at right angles.

43. Let $PQRS$ be a cyclic inscriptable quadrilateral in which PQ meets RS at F, and PS meets QR at G. Suppose that SP meets the incircle at A, PQ meets it at B, QR meets it at C and RS meets it at D. Prove that if the angle bisectors at F and G meet at O at right angles, then $ABCD$ is a cyclic quadrilateral with centre O and in which AC is perpendicular to BD.

44. Let $ABCD$ be a cyclic quadrilateral and let the tangents at A and B meet at P, with Q, R, S similarly defined. Prove that the lines AC, BD, PR, QS are concurrent.

45. With the notation of Problem 44, let O be the centre of $ABCD$. Let AB, CD meet at F and DA, BC meet at G. The circle on diameter FG has centre X. Prove that a hyperbola can be drawn with one focus O, centre X, touching PR and QS and touching the circle on FG as diameter.

46. ABC is a triangle and S is a circle centre A. The polar of C with respect to S meets AB at C' and the polar of B with respect to S meets AC at B'. Prove that $B'C'$ is parallel to BC.

47. $ABCD$ is a cyclic quadrilateral in which AC is perpendicular to BD. O is the centre of circle $ABCD$. a, b, c, d are the tangents at A, B, C, D

respectively. Line a meets OB at B_1. Line b meets OA at A_1. Line b meets OC at C_2. Line c meets OB at B_2. Line c meets OD at D_3. Line d meets OC at C_3. Line d meets OA at A_4. Line a meets OD at D_4. Prove that $A_1B_1, B_2C_2, C_3D_3, D_4A_4$ are concurrent at a point on the line joining $a \wedge c$ and $b \wedge d$.

48. $PQRS$ is a quadrilateral with an incircle, centre I. PS meets QR at F and PQ meets RS at G. Lines t, u, v, w are drawn with the following properties:

    (i) t and v pass through G and u and w pass through F;

    (ii) SP meets t at K, PQ meets u at L, QR meets v at M, RS meets w at N;

    (iii) $\angle KIG = \angle LIF = \angle MIG = \angle NIF = 90°$.

    Prove that an ellipse, focus I, can be drawn to touch t, u, v, w.

49. Let ABC be a triangle, incentre I, and S a circle centre A. Suppose that b, c, i are the polars of B, C, I with respect to S. Prove that $b \wedge c$ is the circumcentre of triangle $A\ b \wedge i\ c \wedge i$.

50. ABC is a triangle with incentre I. The line p is tangent to the incircle. Lines l, m, n are drawn with the following properties:

    (i) l passes through A, m passes through B, n passes through C;

    (ii) l meets p at L, m meets p at M, n meets p at N;

    (iii) $\angle LOA = \angle MOB = \angle NOC = 90°$.

    Prove that l, m, n are concurrent.

15.3 Conics with a common self-polar triangle

Problems involving a pair of conics are sometimes best solved using the method of reciprocation. It is a rewarding but vast area of study, and in this concluding section we merely give a hint of what is possible by returning to the topic raised in Part 7.1, that of the harmonic quadrangle. In this Part projective co-ordinates are used. It may be recalled that all conics passing through the four points A, B, C, D with co-ordinates $A(1,0,0)$, $B(0,1,0)$, $C(0,0,1)$ and $D(1,1,1)$ have equations of the form

$$axy + byz - (a+b)zx = 0, \tag{15.4}$$

for varying values of the parameters a and b. The cases $a = 0, b = 0$ and $a + b = 0$ are special as they lead respectively to the degenerate cases of line pairs AB, CD; BC, AD; and AC, BD. Also the diagonal point triangle EFG where $E(1,0,1)$, $F(1,1,0)$, $G(0,1,1)$ is self-polar with respect to all non-degenerate conics through A, B, C, D.

The first problem is to determine the reciprocal of the conic S with Equation (15.4) with respect to the conic R with equation

$$pxy + qyz - (p+q)zx = 0. \tag{15.5}$$

Given any two non-degenerate conics that intersect in four distinct points, we may take their points of intersection to be the vertices of the triangle of reference and the unit point, so this may be regarded as a representative case. Let $K(X,Y,Z)$ be a point of S, so that $aXY + bYZ - (a+b)ZX = 0$. The polar of K with respect to R has equation $(pY - (p+q)Z)x + (pX + qZ)y + (qY - (p+q)X)z = 0$, and is therefore the line k with line co-ordinates $l = pY - (p+q)Z$, $m = pX + qZ$, $n = qY - (p+q)X$. Solving for X, Y, Z we obtain

$$X = (m(p+q) - np + lq)/2p(p+q), Y = (m(p+q) + np + lq)/2pq,$$

$$Z = (m(p+q) + np - lq)/2q(p+q). \tag{15.6}$$

The envelope of k is therefore the conic with line equation

$$(2apq^3 + aq^4 - bp^2q^2)l^2 + (ap^2q^2 + 2apq^3 + aq^4 + bp^4 + 2bp^3q + bp^2q^2)m^2$$
$$+(bp^4 + 2bp^3q - ap^2q^2)n^2 + 2(bp^4 + 2bp^3q + bp^2q^2)mn$$
$$-2(ap^2q^2 - bp^2q^2)nl + 2(aq^4 + 2apq^3 + ap^2q^2)lm = 0. \tag{15.7}$$

In terms of x, y, z the conic Σ has equation

$$(aq-bp)^2(x^2+y^2+z^2)+2(a^2q^2+2abq(p+q)-b^2p^2)yz-2(a^2q^2+2ab(p^2+pq+q^2)$$
$$+b^2p^2)zx - 2(a^2q^2 - 2abp(p+q) - b^2p^2)xy = 0. \tag{15.8}$$

Σ does, of course, coincide with S, if R coincides with S.

Lemma The most general conic that has EFG as a self-polar triangle has equation

$$\lambda(x^2 + y^2 + z^2) + 2(\alpha - \lambda)yz + 2(\beta - \lambda)zx + 2(\gamma - \lambda)xy = 0, \qquad (15.9)$$

where $\alpha + \beta + \gamma = 2\lambda$ and we may take $\lambda \geq 0$.

The proof of this is left to the reader. It may be verified that the equation of Σ is of this form, so we have proved the interesting result that the reciprocal of a non-degenerate conic S with respect to a non-degenerate conic R has the same self-polar triangle as that shared by S and R.

A second problem is to discover the conics R that reciprocate two given conics S and S' into each other. We consider only a general case in which S has Equation (15.4) and S' has Equation (15.9). We then require p, q so that

$$\lambda = (aq - bp)^2, \alpha - \lambda = a^2q^2 + 2abq(p + q) - b^2p^2,$$

$$\beta - \lambda = -(a^2q^2 + 2ab(p^2 + pq + q^2) + b^2p^2), \quad \gamma - \lambda = -(a^2q^2 - 2abp(p+q) - b^2p^2),$$

where $\alpha + \beta + \gamma = 2\lambda \geq 0$. This leads to a pair of simultaneous quadratic equations and in general four solution pairs (p, q). We have proved that, in general, there are four conics R that can be used to reciprocate one conic into another.

Exercise 15.3.1

1. Show that the conics with equations $ax^2 + by^2 + cz^2 = 0$ for varying a, b, c share a common self-polar triangle whose vertices are those of the triangle of reference. Find the equations of the four conics that reciprocate the conic with equation $x^2 + y^2 = 9z^2$ into the conic with equation $x^2 + 4y^2 = z^2$.

2. Show that the conics with equations $a(y+z)^2 + b(z+x)^2 + c(x+y)^2 = 0$ for varying a, b, c share a common self-polar triangle whose sides have equations $y + z = 0$, $z + x = 0$, $x + y = 0$. What is the equation of the polar reciprocal of the conic with equation $5x^2 + 4y^2 + 3z^2 + 2yz + 4zx + 6xy = 0$ with respect to the conic with equation $x^2 + y^2 + z^2 + yz + zx + xy = 0$?

3. Find the equations of the four conics that reciprocate the conic with equation $2xy + yz - 3zx = 0$ into the conic with equation $x^2 + y^2 + z^2 + 22yz - 130zx + 106xy = 0$.

4. Find the equations of the conics that reciprocate the conic with equation $y^2 = zx$ into the conic with equation $y^2 = 4zx$. *Hint: These conics do not intersect in four distinct points so do not expect four conics for the solution. In fact there are two distinct cases to consider, one of which yields two solutions and the other an infinite number of solutions.*

5. Given two distinct triangles, is it possible to find a conic that reciprocates the vertices of one into the sides of the other and the sides of one into the vertices of the other? How does your result relate to Desargues theorem and its converse?

The theory of polar reciprocation may be used to prove general theorems on poristic systems of triangles. See, for example, Maxwell [25].

Chapter 16

Additional topics

16.1 Conics referred to tangents from a point

Consider a triangle ABC and a conic S, where $l(x, y, z) = 0$ is the equation of one tangent to S from B, touching it at A and $m(x, y, z) = 0$ is the equation of the other tangent to S from B, touching it at C. Let $c(x, y, z) = 0$ be the equation of CA, the chord of contact. In this context we speak of the tangents l and m and the chord of contact c. We have seen in Part 3.6 that if ABC is the triangle of reference then the equation of S, supposing that the unit point lies on it, may be taken as $y^2 = zx$. In the more general case we have the following result.

16.1.1 Equation of such a conic

The equation of S is $kc^2 = lm$, where k is some constant.

Consider a conic S circumscribing a quadrilateral $TUVW$ with sides TU denoted by l', WV by m', WT by c_1 and VU by c_2. The equation of the conic is $l'm' = kc_1c_2$ for some constant k. The reason for this is that it is of the second degree and clearly contains the points T, U, V, W. For example T is the intersection of $c_1 = 0$ and $l' = 0$. The constant k may be adjusted so that the conic passes through some distinct fifth point or satisfies one additional constraint. Now let c_1 and c_2 tend to c, so that T and U tend to A, and V and W tend to C. Then TU and VW become the tangents at A and C respectively, and the equation of the conic becomes $lm = kc^2$. See Fig. 16.1.

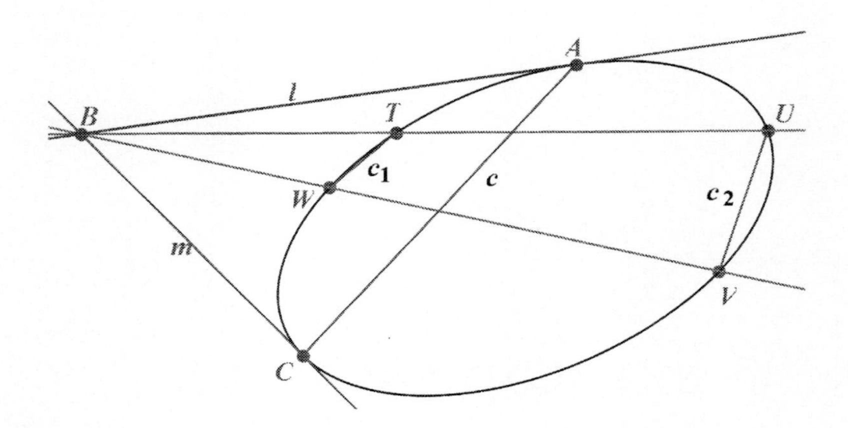

Figure 16.1: Configuration for Section 16.1.1

Example 16.1.1

Using Cartesian co-ordinates, we suppose l and m have equations $x+2y=4$ and $2x-y=3$ and the chord of contact has equation $x=0$. We find the equation of S, supposing it also passes through the point with co-ordinates $(-3,0)$.

The equation of S is of the form $(x+2y-4)(2x-y-3)=kx^2$. Since it passes through $(-3,0)$ we have $k=7$ and the equation of S is

$$5x^2+2y^2-3xy+11x+2y-12=0. \tag{16.1}$$

The points of contact of the tangents are $A(0,2), C(0,-3)$ and the tangents meet at the point $B(2,1)$.

16.1.2 A parameter for the conic

For every point P on the conic S there exists a number λ_P such that the equations of AP, CP, BP are respectively $\lambda_P l = kc, m = \lambda_P c$ and $km = \lambda_P^2 l$ respectively

Suppose that P is an arbitrary point of S. Consider the line with equation $\lambda l = kc$. This line passes through A and for varying values of λ it forms a pencil of lines through A. For one value of λ, say λ_P, the line passes through P. Since a line meets a conic in one point besides A, it follows that there is a $1-1$ correspondence between points P on S and values λ_P. The line with equation $\lambda_P l = kc$ therefore represents AP. Consider next the line with equation $m = \lambda_P c$. Since $m = 0, c = 0$ defines the point C, and since $lm = kc^2$ the line passes through P. It therefore represents the line CP. Finally consider the line with equation $km = \lambda_P^2 l$. Since $l = 0, m = 0$ represents the point B and again since $lm = kc^2$ on the conic, it also passes through P, so this line represents BP. It is worth mentioning that when $\lambda_P = 0$, P is at C and when $\lambda_P = \infty$, P is at A. Note also that since BP contains only λ_P^2, the lines through B meet S at points P and P' (defined to have $\lambda = -\lambda_P$). The lines through B do, of course, meet S at pairs of points in involution with A and C as double points. The points A, C, P and P' cut out a harmonic range on S.

Example 16.1.2

We continue Example 16.1.1 when λ_P has co-ordinates (i) $(-3, 0)$ and (ii) $(-1, 2)$. (i) AP has equation $3y - 2x = 6$, which is the same as $\lambda_P(x + 2y - 4) = kx = 7x$ when $\lambda_P = 3$. It may then be verified that CP has equation $x + y + 3 = 0$ and BP has equation $x - 5y + 3 = 0$. (ii) AP has equation $y = 2$, which is the same as $\lambda_P(x + 2y - 4) = 7x$ when $\lambda_P = 7$. It may then be verified that CP has equation $5x + y + 3 = 0$ and BP has equation $x + 3y = 5$.

16.2 Equation of a chord of the conic

The equation of the chord PQ is given by

$$\lambda_P \lambda_Q l - k(\lambda_P + \lambda_Q)c + km = 0 \tag{16.2}$$

Equation (16.2) can be written in the form $\lambda_Q(\lambda_P l - kc) - k(\lambda_P c - m) = 0$ showing that it represents a line through P. By symmetry it passes through Q. The equation of the tangent at P is therefore

$$\lambda_P^2 l - 2\lambda_P kc + km = 0. \tag{16.3}$$

As a corollary it follows that the tangents at P and P' (defined by $\lambda = -\lambda_P$) meet on the chord c.

Example 16.2.1

Continuing with the above example, we find the equation of the chord joining $P(-3,0)$ with $\lambda_P = 3$ and $Q(-1,2)$ with $\lambda_Q = 7$ is $21(x+2y-4) - 70x + 7(2x-y-3) = 0$, that is $x - y + 3 = 0$. The equation of the tangent at P is $9(x+2y-4) - 42x + 7(2x-y-3) = 0$, that is $19x - 11y + 57 = 0$.

16.2.1 Polar of a point with respect to the conic

Suppose that we are given a point $D(x', y', z')$ and we write $l' = l(x', y', z')$, $m' = m(x', y', z'), c' = c(x', y', z')$. Then the polar of D with respect to S is given by

$$m'l - 2kc'c + l'm = 0 \tag{16.4}$$

From Equation (16.3), the pair of tangents from D to the conic S have values of λ given by $\lambda^2 l' - 2\lambda kc' + km' = 0$. If we denote these values by λ_1 and λ_2, then $\lambda_1 + \lambda_2 = 2kc'/l'$ and $\lambda_1\lambda_2 = km'/l'$ from which we deduce from Equation (16.2) that the polar of D has equation $(km'/l')l - k(2kc'/l')c + km = 0$, that is $m'l - 2kc'c + l'm = 0$. A corollary is that the polars of a fixed point D with respect to all conics of the form $lm = kc^2$, as k varies, all pass through a fixed point given by the solution of the simultaneous pair $c = 0$ and $m'l + l'm = 0$. Notice that this fixed point lies on the chord of contact.

Example 16.2.2

Continuing with the example we find the polar of the point with co-ordinates $(1,2)$ with respect to the conic S. We have $l' = 1, m' = -3$ and $c' = 1$. Thus the polar has the equation $-3(x+2y-4) - 14x + (2x-y-3) = 0$, that is $15x + 7y = 9$. The fixed point, mentioned in the corollary, has co-ordinates $(0, 9/7)$.

16.2.2 Pairs of tangents to the conic

We suppose that the tangents at distinct points P and Q meet at R and we determine the equation of the line BR, showing it to be

$$\lambda_P\lambda_Q l = km \tag{16.5}$$

The equations of the tangents at P and Q are $\lambda_P^2 l - 2\lambda_P kc + km = 0$, $\lambda_Q^2 l - 2\lambda_Q kc + km = 0$. Any linear combination of these passes through R

and eliminating c we obtain $\lambda_P\lambda_Q l = km$. This passes through $l = 0, m = 0$, which is the point B. It therefore represents the line BR.

Example 16.2.3

Continuing with our example, let P be the point with co-ordinates $(-3, 0)$ and Q the point with co-ordinates $(-1, 2)$. As we have seen in Example 16.1.2, $\lambda_P = 3$ and $\lambda_Q = 7$. It follows in this case that the equation of BR is $3(x+2y-4) = (2x-y-3)$, that is $x+7y = 9$. Note that R has co-ordinates $(-25/12, 19/12)$.

16.2.3 Four concurrent lines

Let P, Q, P', Q' have parameter $\lambda_P, \lambda_Q, -\lambda_P, -\lambda_Q$ respectively. Then the line BR (defined in Section 16.2.2 and the chords PQ' and QP' are concurrent at a point on the chord of contact. The equation of the chord PQ' is $\lambda_P\lambda_Q l + k(\lambda_P - \lambda_Q)c - km = 0$ and the equation of the chord QP' is $\lambda_P\lambda_Q l - k(\lambda_P - \lambda_Q)c - km = 0$. These meet where $\lambda_P\lambda_Q l = km$ and $c = 0$. The first of these is the line BR and the second of these is the chord of contact. So the point of concurrence is the intersection of BR with the chord c.

Example 16.2.4

Continuing our example, we have already shown in Example 16.2.3 that the equation of BR is $x + 7y = 9$. The chord of contact is $x = 0$, so the point of concurrence we require has co-ordinates $(0, 9/7)$. Note that in this example P' and Q' have co-ordinates (3/4, 3/4) and (1/2, 3/2) respectively.

16.2.4 Three concurrent lines

The chord PQ, the chord of contact c and the line with equation $\lambda_P\lambda_Q l + km = 0$ are concurrent This result is immediate from Part 16.2.

Example 16.2.5

In our example the point of concurrence has co-ordinates $(0, 3)$.

In Exercises 16.2.6 the notation of Part 16.1 is used in all the problems. The aim of the exercise is to provide an opportunity to use the methods of Part 16.1, though the problems can all be solved by more elementary methods.

Exercise 16.2.6

1. Take the tangents to be $l : 3x + 4y - 15 = 0, m : x = 3$, and the chord to be $c : y + 2x - 5 = 0$. Find the co-ordinates of A, C and B and the value of k for which the conic S is a circle.

2. Take P to have co-ordinates (2, 2) and Q to have co-ordinates $(-2, 4)$. Find the values of λ_P and λ_Q and the equations of the lines AP, BP, CP, AQ, BQ, CQ. Also find the co-ordinates of the points $P'$and Q'.

3. Find the equation of the chord PQ and the equations of the tangents at P and Q.

4. Find the equation of the tangent at P' and show that it meets the tangent at P on the chord c.

5. Take D to be the origin (0, 0) and find the equation of the polar of D with respect to the circle.

6. If the tangents at P and Q meet at R, find the equation of the line BR.

7. Show that PQ', QP' and BR meet at a point on c whose co-ordinates should be determined.

8. Verify the result in Section 16.2.4 in this case, and find the point of concurrence.

16.3 A property of a self-polar triangle

Let S be a conic and A, B, C three distinct points that do not lie on S. We consider the possibility of the existence of points P, Q, R lying on S such that A lies on QR, B lies on RP and C lies on PQ. There are two questions to answer.

(i) If A, B, C are arbitrary fixed points is it possible to find points P, Q, R lying on S with the above property?

(ii) If A, B, C are specially selected points is it possible that for all points Q lying on S, then if QA meets S at R and RB meets S at P then PC passes through Q?

We provide an analysis of the problem when the conic is a circle S, though the results are true for all conics. Without loss of generality we may take S to have equation $x^2 + y^2 = 1$. We use the parameter t whereby points on the circle have co-ordinates $((1 - t^2)/(1 + t^2), 2t/(1 + t^2))$. See Part 1.3 for an account of these parameters.

(i) Let Q, R have parameters s, t and let A have co-ordinates (h, k). Then by Equation 1.6 we have $(1 - st)h + (s + t)k = (1 + st)$. Hence the parameter t of R in terms of s is given by

$$t = \frac{ks + (h - 1)}{(h + 1)s - k}. \tag{16.6}$$

Since A does not lie on S, $h^2 + k^2 \neq 1$, and so Equation 16.6 represents an algebraic 1-1 correspondence or projectivity between s and t. If we now let RB meet S at P with parameter u, then in the same way, there will be a projectivity between t and u, and hence between s and u. Likewise if PC meets S at Q' with parameter v, there will be a projectivity between s and v, that is $Q \leftrightarrow Q'$ is an algebraic 1-1 correspondence. In general, therefore, there will be two self-corresponding points in which Q and Q' coincide. Hence, in general, there will be two choices of the position of Q, for which the circuit $QRPQ$ is closed.

(ii) We now show that if ABC is a self-polar triangle of S, then, for all Q on S, the circuit $QRPQ$ is closed with QR passing through A, RP passing through B and PQ passing through C. See Fig. 16.2 below, which illustrates this situation.

We suppose, without loss of generality, that A has co-ordinates $(a, 0)$, B has co-ordinates $(1/a, b)$ and C has co-ordinates $(1/a, c)$. For ABC to be self-polar $bc = 1 - 1/a^2$. Start with Q, parameter s, then since QR passes through A we find the parameter t of R is given by $t = (a-1)/[(a+1)s]$. Then since RP passes through B we find the parameter u of P is given by

$$u = \frac{(a - 1)[(a + 1)s - ab]}{(a + 1)[abs - (a - 1)]}.$$

Then since PQ' passes through C we find the parameter v of Q' is given by $v = s$. That is Q' coincides with Q.

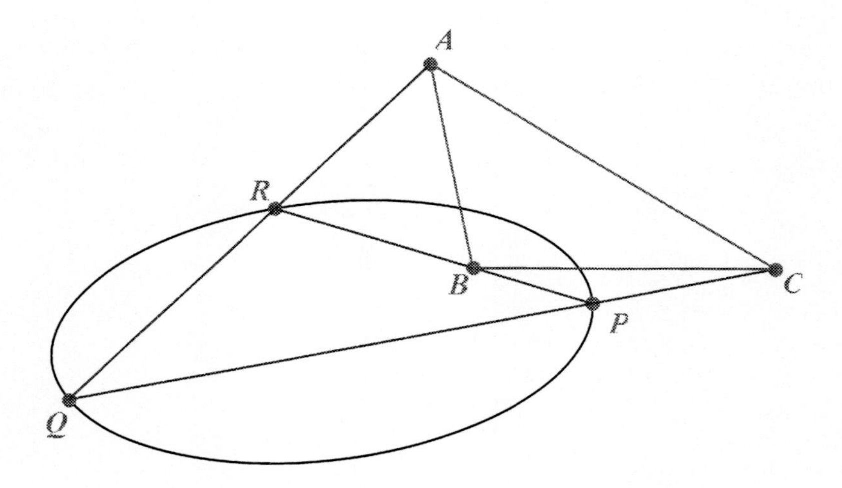

Figure 16.2: Configuration for Part 16.3

16.4 The quadrangle and the eleven-point conic

Let $ABCD$ be a quadrangle and EFG its diagonal point triangle, with $E = AC \wedge BD$, $F = AB \wedge CD$ and $G = AD \wedge BC$. For convenience we adopt a different co-ordinate system to that of Example 3.3.1, where the harmonic properties of the quadrangle are first deduced. We take projective co-ordinates with $A(-1, 1, 1)$, $B(1, -1, 1)$, $C(1, 1, -1)$, $D(1, 1, 1)$. The equation of AC is $z + x = 0$ and that of BD is $z = x$. It follows that E has co-ordinates $(0, 1, 0)$. Similarly F has co-ordinates $(0, 0, 1)$ and G has co-ordinates $(1, 0, 0)$. We take a fixed line L, which does not pass through any of A, B, C, D, and which has equation $lx + my + nz = 0$. We consider the conics passing through A, B, C, D in relation to the line L. Fig. 16.3 shows two conics of the system and the intersections of the line L with one of the conics and with four of the six sides of the quadrangle.

We suppose that L meets AB at Q and CD at Q', that it meets BC at P and AD at P' and that it meets a conic S of the system at R and R'.

16.4.1 Pairs of points in involution

The conics S of the system cut out on L pairs of points that are in involution, which using the above notation, contain pairs (P, P') and (Q, Q'). We remark

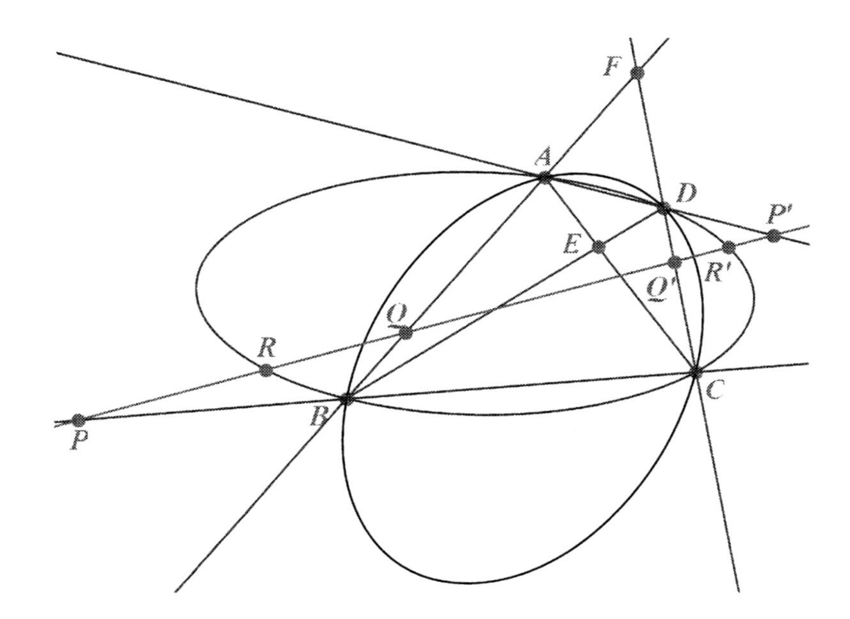

Figure 16.3: Configuration for Part 16.4

first that any conic passing through A, B, C, D has an equation of the form $x^2 - (1+k)y^2 + kz^2 = 0$, for some constant k. The equation of BC is $y + z = 0$ and this meets L at $P(n-m, l, -l)$. The equation of AD is $y = z$ and this meets L at $P'(-m-n, l, l)$. The equation of AB is $x + y = 0$ and this meets L at $Q(n, -n, m-l)$. The equation of CD is $x = y$ and this meets L at $Q'(n, n, -l-m)$. We use the ratio x/y for the parameters of these points so that P has parameter $s = (n-m)/l$, P' has parameter $t = (-m-n)/l$, Q has parameter $s = -1$, and Q' has parameter $t = 1$. We determine the values of a, b, c such that $ast + b(s+t) + c = 0$ is the involution determined by the pairs (P, P') and (Q, Q'). After some algebra we find $a = c = 2lm$ and $b = l^2 + m^2 - n^2$. Now L meets a conic of the system where $n^2x^2 - n^2(1+k)y^2 + k(lx+my)^2 = 0$, that is where

$$(n^2 + kl^2)x^2 + 2klmxy + [km^2 - n^2(1+k)]y^2 = 0. \tag{16.7}$$

Since the ratio x/y is being used as parameter we see that the parameters of the points $R, R'$are given by $s + t = -2klm/(n^2 + kl^2)$ and $st = [km^2 - n^2(1+k)]/(n^2 + kl^2)$ and it may now be checked that for all values of k we have $ast + b(s+t) + c = 0$, showing that the pairs (R, R') are in involution.

16.4.2 Key result on polars of a fixed point

The polars of a fixed point $K(\xi, \eta, \zeta)$ with respect to all members of the system of conics pass through a fixed point. The polar of K with respect to a member of the system of conics has equation $x\xi - (1+k)y\eta + kz\zeta = 0$ and for all values of k this passes through the point J with co-ordinates $(1/\xi, 1/\eta, 1/\zeta)$. The polars therefore form a pencil of lines through J.

Definition The *eleven-point conic* with respect to a quadrangle $ABCD$ and a line L, not passing through any of its vertices, is defined to be the conic (proved to exist in Section 16.4.3), which passes through the following points: the vertices E, F, G of the diagonal point triangle; the harmonic conjugate on AB of the point where L cuts AB with respect to A and B and the other five points when the line AB is replaced by AC, AD, BC, BD, CD; and the two points where L touches two (and only two) conics of the system of conics through A, B, C, D.

16.4.3 The eleven-point conic

The locus of the points U which have the property that the polar of U with respect to one member of the system of conics coincides with L, is the eleven-point conic. Let U be a variable point with co-ordinates (x', y', z'). The polar of U with respect to a member of the system of conics through A, B, C, D is $xx' - (1+k)yy' + kzz' = 0$. If this coincides with $lx + my + nz = 0$, then $x'/l = -(1+k)y'/m = kz'/n$. Solving these equations to find the value of k (and hence the conic for which the polar of U coincides with L), we find $k = -ny'/(ny' + mz'.)$. Since $nx' = lkz'$ we get $ly'z' + mz'x' + nx'y' = 0$, and hence the locus of U is the conic Σ_L with equation

$$lyz + mzx + nxy = 0. \tag{16.8}$$

This conic evidently passes through the vertices E, F, G of the triangle of reference, which is the diagonal point triangle of $ABCD$. Now AB has equation $x + y = 0$ and L meets AB at Q where Q has co-ordinates $Q(n, -n, m - l)$. The conic with Equation (16.8) meets AB at the point $Q_0(m - l, l - m, n)$. Taking z/y as parameter on this line we have the parameters of A, B, Q, Q_0 to be $1, -1, (l - m)/n, n/(l - m)$ respectively. Since $(1, -1; x, 1/x) = -1$, it follows that Q and Q_0 separate A and B harmonically. Similarly for the five pairs of points that are cut out by L and Σ_L on the remaining sides of the quadrangle. Finally L touches Σ_L at two points, which trivially have L as their polar.

Exercise 16.4.1

1. The circles with equation $x^2+y^2+2gx=1$, for varying g, form a coaxal system of circles. Show that the locus of the point U such that the polar of U with respect to one member of the coaxal system coincides with the fixed line $L: lx+my=n$ is a hyperbola H_L with asymptotes $x=0$ and $ly=mx$. Show that L and H_L intersect the line $x=0$ at a pair of points that separate the points $(0,1)$ and $(0,-1)$ harmonically.

2. What is the dual of the result of Section 16.4.3?

16.5 A property of two self-polar triangles

We prove the result that if ABC and DEF are two self-polar triangles of a conic S, then the six vertices A, B, C, D, E, F lie on another conic. First we give the proof of a particular case. Take S to have equation $x^2+y^2+z^2=0$, for which we have shown in Part 3.6 that the triangle of reference $A(1,0,0), B(0,1,0), C(0,0,1)$ is self-polar. To find another self-polar triangle we note that $E(a,b,c), F(b,c,a), G(c,a,b)$ is a self-polar triangle provided that $ab+bc+ca=0$. This is because the polar of E has equation $ax+by+cz=0$ and this contains both F and G provided that $ab+bc+ca=0$, and similarly for the other polars. Putting $a=-bc/(b+c)$ and multiplying up we obtain for the three points $E(-bc, b(b+c), c(b+c))$, $F(b(b+c), c(b+c), -bc)$, $G(c(b+c), -bc, b(b+c))$. It is now easily verified that all six vertices of the two triangles lie on the conic with equation $yz+zx+xy=0$.

The above is not a general proof, because the co-ordinates of E, F, G depend only on two parameters b, c. But it sets the scene for a general proof. We suppose E, F, G have co-ordinates (x_k, y_k, z_k), $k=1,2,3$. For E, F, G to be self-polar we must have

$$x_jx_k + y_jy_k + z_jz_k = 0, \; j>k. \tag{16.9}$$

There are three such equations with $(j,k)=(2,1),(3,1)$ and $(3,2)$. It is, of course, possible to find values of the co-ordinates in terms of parameters, as it is equivalent to finding three independent vectors in three-dimensional

Euclidean space that are at right angles to each other. However, this is not necessary, as we assert that E, F, G must lie (along with A, B, C) on a conic with an equation of the form $fyz + gzx + hxy = 0$, for some values of f, g, h. The condition for this is that the three equations

$$\frac{f}{x_k} + \frac{g}{y_k} + \frac{h}{z_k} = 0, \;\; k = 1, 2, 3. \tag{16.10}$$

should have a non-zero solution for f, g, h. This is so, if, and only if, the determinant, whose k-th row is $(1/x_k, 1/y_k, 1/z_k)$, $k = 1, 2, 3$, vanishes. Multiplying the columns by $x_1x_2x_3$, $y_1y_2y_3$, $z_1z_2z_3$ respectively, the vanishing occurs if, and only if, the determinant vanishes, whose first row is x_2x_3, y_2y_3, z_2z_3 with similar entries in the other rows. This determinant vanishes, since by adding elements in the rows we obtain a first column of three zeros.

Exercise 16.5.1

1. Suppose that in the above analysis S is a circle centre H, and J and K are on the line at infinity with HJ perpendicular to HK. HJK is a self-polar triangle. Suppose that ABC is a (finite) self-polar triangle. What can you say about the conic passing through A, B, C, H, J, K? What theorem about the triangle is thereby provided?

16.6 A problem on circles

The problem we consider is whether, in relation to a triangle ABC, a circle exists such that if D, D' are its intersections with BC; E, E' are its intersections with CA and $F, F'$are its intersections with AB, then AD, BE, CF are parallel and AD', BE', CF' are also parallel. If so how do we characterize such a circle? If there are many such circles, do they have anything in common apart from the property stated? See Fig. 16.4.

We use areal co-ordinates. Let $P(l, m, n)$ be a point on the line at infinity, so that $l + m + n = 0$ and the parallel lines PA, PB, PC meet BC, CA, AB respectively at D, E, F where $D(0, m, n), E(l, 0, n), F(l, m, 0)$ and let $P(l', m', n')$ be another point on the line at infinity, so that $l' + m' + n' = 0$ and $D'(0, m', n'), E'(l', 0, n'), F'(l', m', 0)$ are similarly defined so that AD', BE', CF' are parallel. None of l, m, n, l', m', n' may be zero.

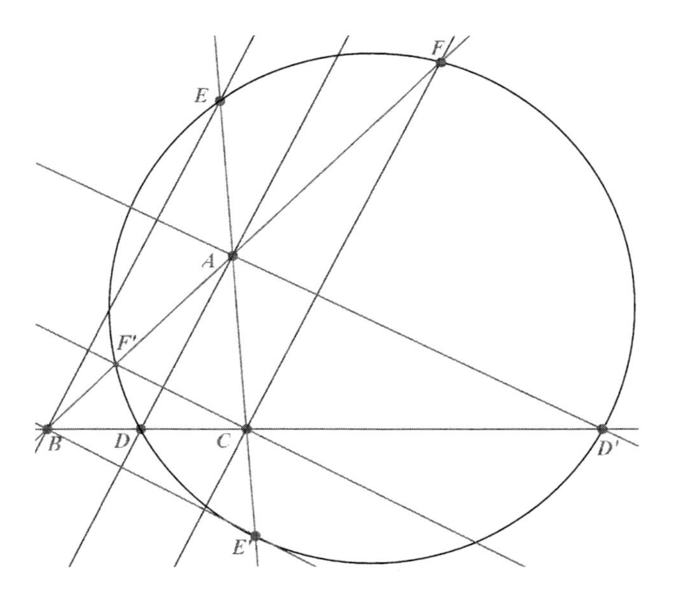

Figure 16.4: Configuration for Part 16.6

It is straightforward to show that the six points D, E, F, D', E', F' lie on the conic with equation

$$x^2/ll' + y^2/mm' + z^2/nn' \tag{16.11}$$
$$-(1/nm' + 1/mn')yz - (1/ln' + 1/nl')zx - (1/ml' + 1/lm')xy = 0.$$

We now put in the condition (2.13) that this should be a circle and find, after some algebra, that $a = \pm ll', b = \pm mm', c = \pm nn'$, where we have used $l + m + n = 0$ and $l' + m' + n' = 0$. Hence we have $\pm a/l \pm b/m \pm c/n = 0$ and $\pm a/l' \pm b/m' \pm c/n' = 0$. It follows that, if such a circle exists, then P, P' must be the points at infinity on one of the conics of the form

$$\pm ayz \pm bzx \pm cxy = 0. \tag{16.12}$$

Now if all the signs are the same in Equation (16.12) and you try to solve it subject to $x + y + z = 0$, you find that no such points exist. This is because if a, b, c are the sides of a triangle, then $a^2 + b^2 + c^2 - 2bc - 2ca - 2ab < 0$. In other words it is an ellipse. This means that for a triangle with fixed a, b, c there are only three such circles, corresponding to points P, P' that are the intersections of the three hyperbolas $-ayz+bzx+cxy = 0$, $ayz-bzx+cxy = 0$ and $ayz + bzx - cxy = 0$ with the line at infinity.

Exercise 16.6.1

1. Find the equation of the circle that corresponds to a triangle with $a = 2, b = 3, c = 4$ when P and P' are the intersections with the line at infinity of the conic with equation $2yz - 3zx + 4xy = 0$.

2. Let P be a point on the line at infinity and let PAD, PBE, PCE be three parallel lines meeting BC, CA, AB at D, E, F. Let the circle DEF have centre X and suppose it meets the sides BC, CA, AB again at L, M, N. It is known that AL, BM, CN are concurrent at a point Q. Is it possible to choose P so that XQ is parallel to PAD?

16.7 Hagge circles

The *Hagge circle* of a general point P in the plane of a triangle ABC is defined as follows. Extend AP to meet the circumcircle again at D. Let U denote the reflection of D in BC. Define E, V and F, W by cyclic change. The Hagge circle of P is the circumcircle of triangle UVW. Let Q be the centre of the Hagge circle of P. The following results hold. See Fig. 16.5.

16.7.1 Hagge circle passes through the orthocentre H

A co-ordinate proof of this result is indicated by the problems in Exercise 16.7.1.

16.7.2 The centre of a Hagge circle

For a general point P, the nine-point centre T is the midpoint of QP_g, where P_g is the isogonal conjugate of P. Though we cover only general points P, it is possible to extend the definition of Hagge circles to points lying on the sides of the triangle and even at the vertices, see Bradley and Smith [9]. The result in Section 16.7.2 is proved by complex number methods by Peiser [30]. Note that when P lies on the circumcircle, then the Hagge circle degenerates into a straight line through H. Exercise 16.7.1 covers an alternative approach to that employed in [9].

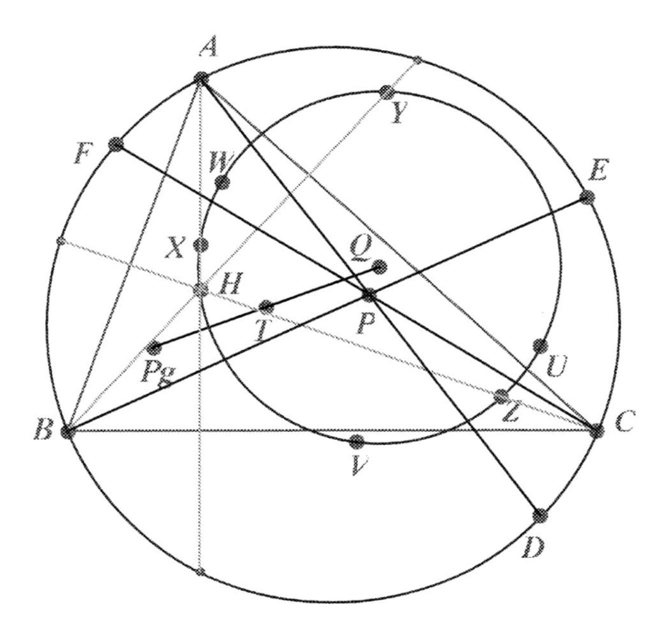

Figure 16.5: A Hagge circle

Exercise 16.7.1

In the problems P is a point in the plane of a triangle ABC not lying on its sides and not on the circumcircle. Notation for other points is defined above. See also Fig. 16.5. Areal co-ordinates are used throughout.

1. Show that if a, b, c are the side lengths of triangle ABC, then the first co-ordinate of H is given by

$$H_x = \frac{(c^2 + a^2 - b^2)(a^2 + b^2 - c^2)}{2b^2c^2 + 2c^2a^2 + 2a^2b^2 - a^4 - b^4 - c^4}.$$

2. Show that if a, b, c are the side lengths of triangle ABC, then the first co-ordinate of T is given by

$$T_x = \frac{2b^2c^2 - b^4 - c^4 + c^2a^2 + a^2b^2}{2(2b^2c^2 + 2c^2a^2 + 2a^2b^2 - a^4 - b^4 - c^4)}.$$

3. Given that the isogonal conjugate of P has co-oordinates

$$\frac{(a^2mn, b^2nl, c^2lm)}{a^2mn + b^2nl + c^2lm}.$$

Show that the point Q such that T is the midpoint of P_gQ has first co-ordinate Q_x given by

$$Q_x = a^4mn(a^2 - b^2 - c^2) + (b^2nl + c^2lm)(2b^2c^2 + c^2a^2 + a^2b^2 - b^4 - c^4)$$

all divided by $(a^2mn + b^2nl + c^2lm)(2b^2c^2 + 2c^2a^2 + 2a^2b^2 - a^4 - b^4 - c^4)$.

4. Prove that the co-ordinates of D are given by

$$\frac{(-a^2mn, m(nb^2 + mc^2), n(nb^2 + mc^2)}{n^2b^2 + m^2c^2 + mn(b^2 + c^2 - a^2)}.$$

5. Let K have co-ordinates (d, e, f). Let L be the foot of the perpendicular from K on to BC. Prove that the co-ordinates of L are $(0, 2a^2e + d(a^2 + b^2 - c^2), 2a^2f + d(c^2 + a^2 - b^2))/(2a^2)$.

6. Use the result of Problem 5 to show that the co-ordinates of the reflection of K in the line BC are $(-d, e + d(a^2 + b^2 - c^2)/a^2, f + d(c^2 + a^2 - b^2)/a^2)$.

7. Use the result of Problem 6 to show that the co-ordinates of U are $(a^2mn, -a^2mn + m(m + n)c^2, -a^2mn + n(m + n)b^2)/(n^2b^2 + m^2c^2 + mn(b^2 + c^2 - a^2))$.

8. Use a computer algebra package with the areal distance function (see Section (2.3.2)) to show that $QU = QH$.

Exercise 16.7.1 covers the general situation. Difficulties arise when

(a) P is the vertex of triangle ABC;

(b) P lies on a triangle side other than at a vertex;

(c) P is on the circumcircle, but not at a triangle vertex;

(d) P is a point at infinity.

In case (a) we make no attempt to cover the difficulty, though this can be done by an argument involving the removal of singularities. In case (b), if P is on the line BC but is neither B nor C, then V coincides with C and W coincides with B. The point U is well-defined and the Hagge circle is the circumcircle of BCH and is independent of P. In case (c) the Hagge

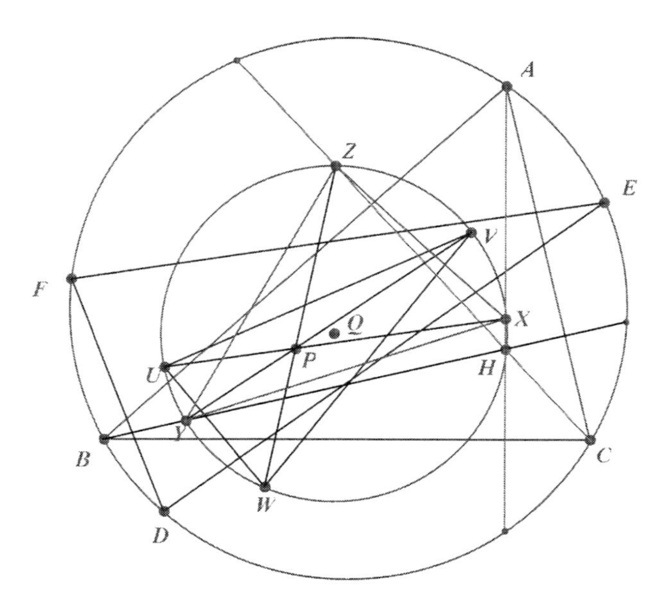

Figure 16.6: A Hagge circle and the two pairs of similar triangles

circle becomes the straight line through the reflections of P in the three triangle sides (which are known, by the Wallace-Simson line property, to be collinear on a line through H) and the Hagge centre becomes a point on the line at infinity. In case (d) a Hagge circle construction is possible in general. You draw lines parallel to each other through A, B, C to meet the circumcircle again at D, E, F. Then reflect to get the points U, V, W. The circle UVW passes through H and the centre Q lies on the circumcircle of triangle $A'B'C'$, which is the rotation through 180° of triangle ABC about its nine-point centre T.

Further properties of Hagge circles, including details of the great Hagge Theorem are covered in Bradley and Smith [10], and are generalized in Chapter 17. The Hagge construction is one of the most important in classical geometry, for reasons, which we now outline. If a Hagge circle meets the altitudes from A, B, C at X, Y, Z respectively, then triangles ABC and XYZ are similar triangles in perspective, but are labelled in the opposite sense. The same applies to triangles DEF and UVW. Hagge [21] himself proved that UPX, VPY, WPZ are straight lines. See Fig. 16.6. What he does not explicitly state, though perhaps he knew, is that the Hagge circle and its two triangles are connected to the circumcircle and its two triangles by an inverse spiral symmetry through P. Now two years earlier and quite independently Speckman [34], in what appears to be a little known paper (in Dutch), had made an extensive classification of the properties of indirectly similar triangles in perspective. If you combine the two papers and make the appropriate links you have a complete theory of indirectly similar triangles in perspective. A curious fact is that a completely satisfactory theory of directly similar triangles in perspective did not appear for another twenty-four years. See Wood [38]. We do not consider this topic in this book.

16.8 Triangles with vertices on more than one circle

We consider triangles constructed in a special way so that their vertices are located on two or three circles. It turns out that the loci of special points, such as triangle centres, and other well-defined points, are often curiously simple. We prove a number of general theorems and provide illustrative examples. The analysis on two circles is carried out using rectangular Cartesian co-ordinates and that on three circles using areal co-ordinates, though we also mention a possible synthetic proof. We set the scene by considering the simplest case, which involves two equal intersecting circles.

Let C_1 and C_2 be two equal circles, centres C and D respectively, intersecting at A and B and suppose that $\angle CAD = 120^o$. Since we are interested only in ratios of distances, we may as well take the radii of the circles to be 2 and the distance between their centres to be $2\sqrt{3}$. Then with origin the midpoint of CD and CD as x-axis the equations of the circles are

$$C_1: \quad x^2 + y^2 + 2\sqrt{3}x = 1; \tag{16.13}$$

$$C_2: \quad x^2 + y^2 - 2\sqrt{3}x = 1. \tag{16.14}$$

Let P be a variable point on C_1 (other than A or B), lying on a line of gradient m passing through A. Then its co-ordinates (x_P, y_P) are given by

$$x_P = -\frac{2(m + \sqrt{3})}{1 + m^2}, \tag{16.15}$$

$$y_P = -\frac{(m + \sqrt{3} - 2)(m + \sqrt{3} + 2)}{1 + m^2}. \tag{16.16}$$

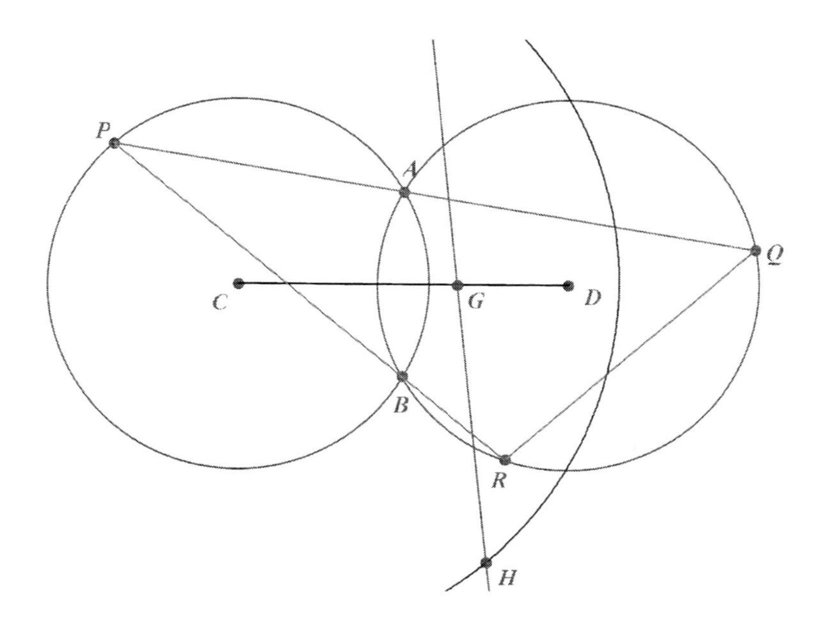

Figure 16.7: First configuration for Part 16.8

Suppose that PA meets C_2 again at Q, then from Equation (16.14) the co-ordinates of Q are (x_Q, y_Q), where

$$x_Q = \frac{2(\sqrt{3} - m)}{1 + m^2}, \tag{16.17}$$

$$y_Q = -\frac{(m - \sqrt{3} + 2)(m - \sqrt{3} - 2)}{1 + m^2}. \tag{16.18}$$

The line through P passing through B has gradient n given by $n = (m\sqrt{3} - 1)/(m + \sqrt{3})$ and from Equation (16.14) this line meets C_2 again at R, whose co-ordinates (x_R, y_R) are given by

$$x_R = \frac{(1 + m\sqrt{3})(m + \sqrt{3})}{1 + m^2}, \tag{16.19}$$

$$y_R = \frac{2(m - 1)(m + 1)}{1 + m^2}. \tag{16.20}$$

From Equations (16.15) - (16.20) the centroid G of triangle PQR has co-ordinates $(1/\sqrt{3}, 0)$. Note that this is a fixed point on CD independent of m, that is, independent of the point P initially chosen on C_1. Thus, in this case, the locus of G as P varies is a point. This point has the properties (i) that $CG/GD = 2$ and (ii) that $\angle AGB = 120^o$. See Fig. 16.7 for an illustration of the configuration. Furthermore we conclude that as P varies the Euler line of triangle PQR rotates about the point G on CD. It is of interest, therefore, to discover the loci of the circumcentre O and the orthocentre H of triangle PQR.

The configuration is in fact extremely interesting. The midpoint of PQ, from Equations 16.14 - 16.18, turns out to be

$$\left(\frac{-2m}{1+m^2}, \frac{1-m^2}{1+m^2}\right).$$

Note that this point lies on the unit circle with equation $x^2 + y^2 = 1$. Since the gradient of PQ is m, the perpendicular bisector has gradient $-1/m$ and so has equation $m(y+1) + x = 0$. This line meets C_2 at B and a point X with the same y-coordinate as P and such that $PX = 2\sqrt{3}$. Similarly the perpendicular bisector of PR passes through A and X and the midpoint of PR also lies on the unit circle.

It follows that X is the circumcentre O of triangle PQR and the radius of the circumcircle PQR is $2\sqrt{3}$. Note that this is independent of the position of P. Also since O lies on C_2, the locus of O as P varies is the circle C_2 itself. Since in any triangle, $GH = 2OG$ and $GH = 4GN$, we are able to deduce from Fig. 16.8 the loci of H and the nine-point centre N. Since the locus of O is a circle centre D, and G is the fixed point on CD such that $CG = 2GD$, it follows that the centres of the circles that are the loci of N and H are respectively the origin and the centre C of C_1. Also the radii of these circles are 1 and 4 respectively.

An interesting question is whether the locus of any other triangle centre is also a circle. CABRI indicates that this is the case (as one would expect) for other points on the Euler line, such as de Longchamps point, and also, interestingly, the Fermat points. There may be others, but not loci of familiar points such as the incentre or the symmedian point. Another interesting question is what happens when the radii of C_1 and C_2 are not equal, or even, if equal, when $\angle CAD$ differs from 120°. CABRI indicates that what happens is that the Euler line of triangle PQR still rotates about a fixed point T on CD. The loci of O, N, H are circles with centres at D, the midpoint of CD and C respectively, but of radii that vary with the dimensions of the circles

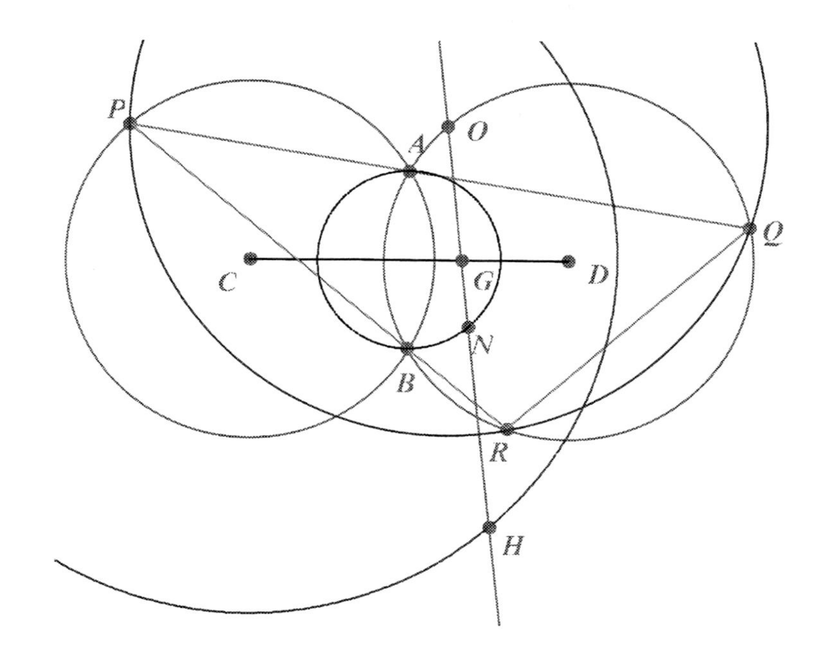

Figure 16.8: Second configuration for Part 16.8

C_1, C_2 and the distance CD. The locus of G is in general a circle, since except in particular cases, it does not coincide with the fixed point on CD. The centre of the locus of G is at the point of trisection of CD nearer D.

We now revise briefly some of the points mentioned in Chapter 2 on the use of areal co-ordinates. When areal co-ordinates are employed there has to be a sensibly chosen triangle of reference ABC. A, B, C are given homogeneous co-ordinates $(1, 0, 0), (0, 1, 0), (0, 0, 1)$. With the customary notation $a = BC, b = CA, c = AB$ it can then be shown that the equation of the circumcircle of ABC is

$$a^2yz + b^2zx + c^2xy = 0. \tag{16.21}$$

More generally, it can be shown that the equation of any circle has the form

$$\Sigma(u, v, w) = a^2yz + b^2zx + c^2xy - (x + y + z)(ux + vy + wz) = 0, \tag{16.22}$$

where u, v, w are the powers of A, B, C respectively with respect to the circle. The radical axis of the two circles $\Sigma(u, v, w)$ and $\Sigma(u', v', w')$ is found by subtracting their equations, and since $x + y + z \neq 0$ at any finite point, its equation is

$$(u-u')x+(v-v')y+(w-w')z=0. \tag{16.23}$$

We also need the result that three points are collinear if the determinant consisting of rows having the co-ordinates of those points vanishes. In many problems involving a given point it is sufficient to know the ratios of the co-ordinates of the point. That is, if (p,q,r) are the normalised co-ordinates with $p+q+r=1$, then one can use unnormalised co-ordinates (kp,kq,kr) instead, where k is any non-zero constant.

In what follows we refer to Fig. 16.9. Now C_1, C_2, C_3 are three circles with a common point A. C_1 and C_2 also meet at C; C_1 and C_3 also meet at B; and C_2, C_3 also meet at D. Following a similar procedure to that in dealing with two circles, we take a variable point P on C_1, let PC meet C_2 again at Q, and let PB meet C_3 again at R. In this way we construct a variable triangle PQR having one vertex on each of the three circles.

First we note that using the notation of Equations (16.21)-(16.23), with ABC as triangle of reference, the equations of the three circles are

$$C_1:\ \Sigma=0, C_2:\ \Sigma(0,v,0)=0, C_3:\ \Sigma(0,0,w)=0. \tag{16.24}$$

This is because the equation of the radical axis AC of C_1 and C_2 is $y=0$, and that of the radical axis AB of C_1 and C_3 is $z=0$. The constants v and w in the equations of C_2 and C_3 serve to distinguish one possible circle from another and enable us to say that the figure is quite general.

The next task is to find the co-ordinates of the points Q and R in terms of those of P. We start with a point P whose co-ordinates are in general parametric form for points on $\Sigma=0$. These are $(x_P,y_P,z_P)=(a^2t(1-t),-b^2(1-t),-c^2t)$ with a parameter t, which can take on all real values (and infinity). However we wish to exclude the points A,B,C so we exclude $t=\infty,0,1$. To obtain the co-ordinates of Q we need the equation of PC, which is $b^2x+a^2ty=0$ and find where this line meets C_2. This is tedious, but since one of the points is C itself, those of Q do not contain square roots and are (x_Q,y_Q,z_Q), where

$$x_Q=a^4t(1-t)-a^2tv, y_Q=-a^2b^2(1-t)+b^2v, z_Q=-a^2c^2t+v(a^2t-b^2) \tag{16.25}$$

Similarly the co-ordinates of R are

$$x_R=a^4t(1-t)-a^2w(1-t), y_R=-a^2b^2(1-t)+a^2w(1-t)-c^2w, \tag{16.26}$$
$$z_R=-a^2c^2t+c^2w.$$

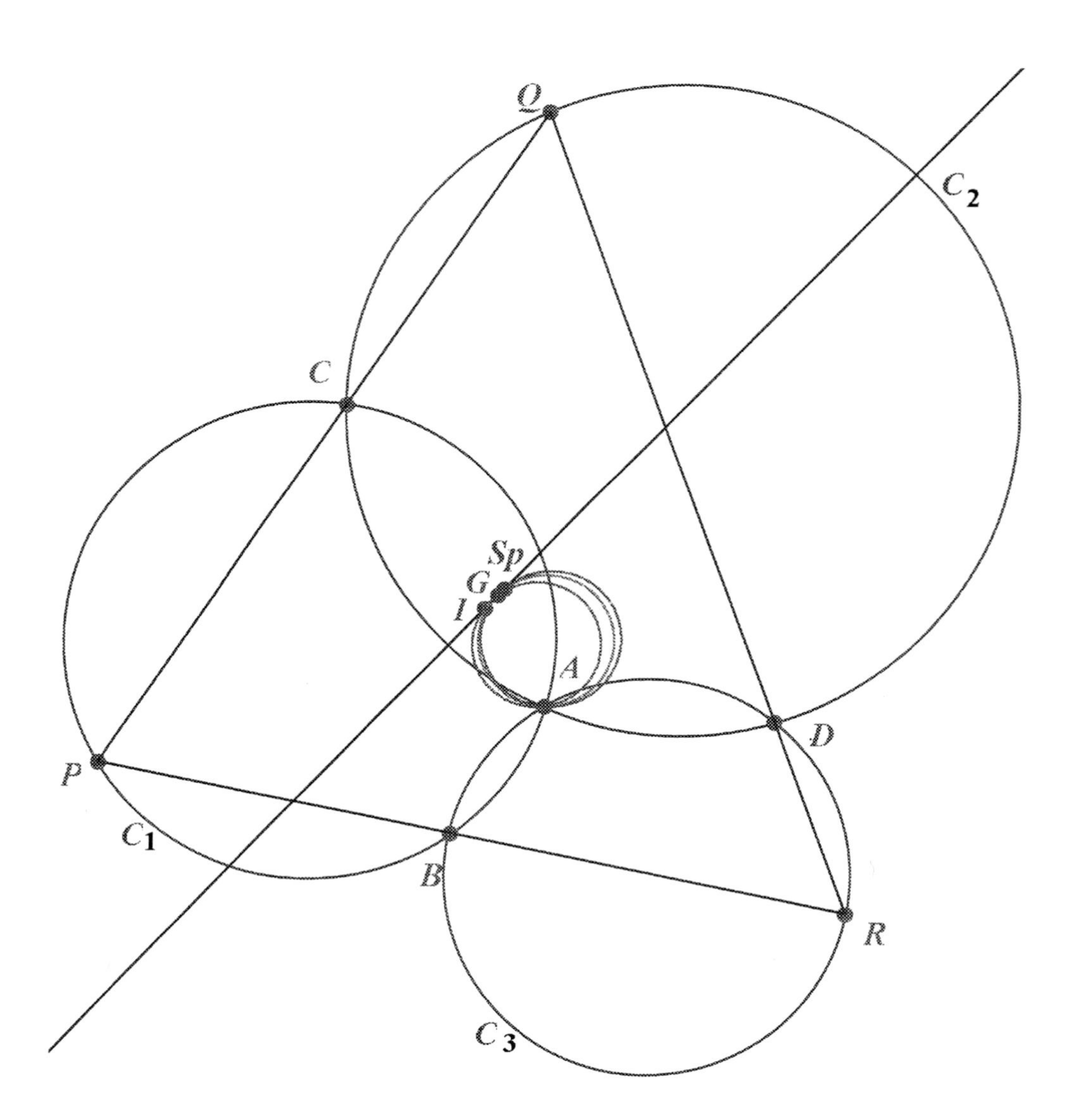

Figure 16.9: Configuration for Sections 16.8.3 and 16.8.4

16.8.1

Although the triangle PQR varies in position and size it remains the same shape. This is because the points A, B, C, D are fixed so the angles at P, Q, R are fixed.

16.8.2

QR always passes through D. To see this, observe that D lies on the radical axis of C_2 and C_3 whose equation is $vy = wz$. The actual co-ordinates of D are now easily obtained and are

$$x_D = vw(a^2-v-w), y_D = w(vw-b^2v-c^2w), z_D = v(vw-b^2v-c^2w). \quad (16.27)$$

The determinant whose entries are the co-ordinates of Q, R and D, given by Equations (16.25) - (16.27) may be shown to vanish.

16.8.3

As P varies, the locus of any point K that has fixed co-ordinates relative to triangle PQR is a circle through A. See Fig. 16.9, which shows, amongst others, the locus of the centroid G.

Suppose, using an obvious notation that $K = lP + mQ + nR$, where now we use normalised co-ordinates for P, Q, R and $l + m + n = 1$, so the co-ordinates of K are also normalised. Then, using the notation of Equation (16.22), K lies on the circle $\Sigma(0, mv, nw) = 0$. This appears at first sight to be independent of l, but remember that $l+m+n = 1$. The fact that $u = 0$ in the equation of this circle means that it always passes through $A(1, 0, 0)$. The working involved to check this result is not as formidable as might appear, since the denominators of the points of P, Q, R when the co-ordinates are normalised all have the factor $a^2(-a^2t^2 + (a^2 + b^2 - c^2)t - b^2)$, which then cancels. A reader checking the result, however, is advised to use computer algebra software.

16.8.4

The circles of all points K lying on a particular line relative to triangle PQR form a coaxal system. In Fig. 16.9 we show the line joining the incentre I and the centroid G. The loci of I, G, Sp are shown, where Sp is the Spieker centre.

The circles pass not only pass through A but pass through a second point X. We show this as follows: If the points K lie on a line, then l, m, n are related by a linear equation of the form $dl+em+fn = 0$. For example, for the line IG we have $d = (b-c)$, $e = (c-a)$, $f = (a-b)$. Now $l+m+n = 1$, so m and n are related by the equation $(d-e)m + (d-f)n = d$. Suppose now K_1 and K_2 lie on the line and correspond to the pairs (m_1, n_1) and (m_2, n_2). Then the circles that are their loci have equations $\Sigma(0, m_1v, n_1w) = 0$ and $\Sigma(0, m_2v, n_2w) = 0$. Since these circles both pass through A, they have a common chord, whose equation is obtained by subtraction and is $v(m_1 - m_2)y + w(n_1 - n_2)z = 0$. But $(d-e)(m_1 - m_2) + (d-f)(n_1 - n_2) = 0$. Hence the common chord has equation $(d-f)vy + (d-e)wz = 0$. This is independent of which points K_1 and K_2 are chosen on the line, so all pairs have the same common chord and so form a coaxal system.

It is interesting to outline a possible synthetic argument of the results in Sections 16.8.2 - 16.8.4. Consider Fig. 16.9, but with C_2 removed, leaving only C_1, C_3 and the line PR. Then take a point Q anywhere and draw lines QP and QR intersecting C_1 and C_3 at C and D respectively. Then we have a triangle PQR with points C, D, B on its sides and circles C_1, C_3 passing through A. It follows, by the converse of the pivot theorem that Q lies on circle DAC. It is now clear that triangle PQR is of fixed shape and that as P moves once round C_1, Q, R move once around circles C_2, C_3. We can now use K in this argument instead of Q, where K is any point and it too will move on a circle. This proof is due to Smith [33].

We now consider the consequences of the results for three circles, when we increase the number of circles, but retain the same construction. In Fig. 16.10 we show four circles C_1, C_2, C_3, C_4 all passing through a point A. Once the circles have been drawn they remain fixed, as do the other points of intersection. C_1 meets C_2, C_3, C_4 again at C, D, B respectively. C_2 meets C_3, C_4 again at E, F respectively and C_3, C_4 meet again at G. If we omit one of these circles, say C_2, then we have the same situation as above with three circles and a triangle PRS produced by the same construction. It follows that RS passes through G. Similarly by omitting other circles and concentrating on triangles PQR, PQS we deduce that QR passes through E and QS passes through F. Since all the points A to G are fixed, it follows that the angles of the quadrilateral $PQRS$ are fixed as the points move around the various circles on which they lie.

Furthermore since the component triangles QRS, PRS, PQS, PQR remain the same shape during the revolutions of the points, so does the quadri-

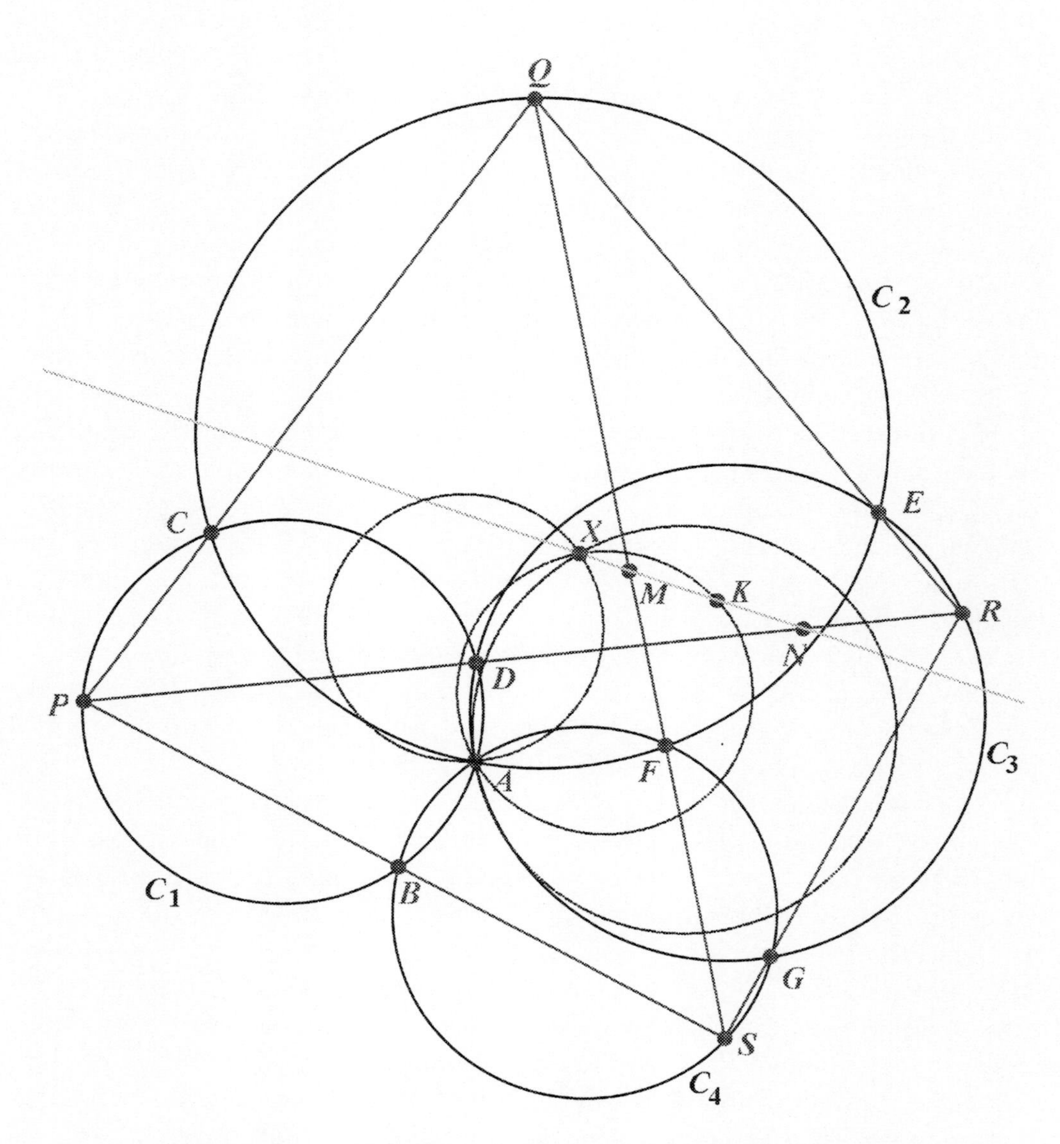

Figure 16.10: Second configuration for Section 16.8.4

lateral $PQRS$ as a whole. Thus the quadrilateral remains the same shape during any revolution. This means that relative to the vertices of triangle PQR the point S has fixed co-ordinates. Hence any point constructed from the vertices of the quadrilateral are in fact dependent only on the triangle PQR. The results in Sections 16.8.3 and 16.8.4 hold equally well for the quadrilateral. In Fig. 16.10 we show the loci of several points on the line MN, where M is the midpoint of QS and N is the midpoint of the segment from $PS \wedge QR$ and R. As anticipated they are members of a coaxal system with common chord AX.

Chapter 17

Affine and Projective generalizations

17.1 Introduction

In this chapter, we show how to generalize some of the famous results in Euclidean geometry about the triangle, into associated results that exist in the affine and projective planes. This work is foreshadowed in Chapter 7 and in a previous article by Bradley and Bradley [5], where the Wallace-Simson line theorem is generalized into the affine plane. We are also aware of similar ideas published quite independently by van Lamoen [36], who has pointed out the relevance of Desargues involution in defining perpendicularity. The generalisations are aesthetically very pleasing, but more than that they enable you to draw figures of the generalisations with geometrical software now available. It is also possible, as with generalisations in other areas of mathematics that they lead to the discovery of new results in the area from which the generalisation is made. Indeed we have discovered a new result in this way that has been made in connection with the orthopole construction. See Section 17.10.6. This chapter is a greatly expanded version of an article by Bradley and Smith [11].

It may be argued that one of the crucial differences between Euclidean, affine and projective geometry is that in Euclidean geometry there exist perpendiculars and parallels, but in affine geometry only parallels and in projective geometry neither. So the first step in the generalisation is to discover how to replace the notion of the perpendiculars to the sides of a triangle in affine geometry, and how to replace both parallels and the perpendiculars to the sides of a triangle, and indeed to other lines also, in projective geometry. In Euclidean geometry actual distances can be measured relative to some fixed unit, such as the radius of the circumcircle of a triangle. In affine geometry only ratios of distances along a line have meaning, and in projective geometry one has to make do with cross-ratios on lines and conics. In Euclidean geometry a line at infinity can be introduced, and in affine geometry this is a crucial fixed line. In projective geometry no such line exists and the best one can do is to designate some arbitrary line as special. In co-ordinate terms, in projective geometry one can specify any three points to form a triangle of reference and then one can select an arbitrary line as the unit line or one can select a fourth point as the unit point, but one cannot do both because they are related. These matters are explained in Section 3.2.5, where the term unit point is first defined. It is the point with co-ordinates $(1, 1, 1)$. Similarly the unit line is defined to be the line with equation x + y + z = 0. In Part 17.2 we describe the intimate relationship between the unit point and the unit line, which in this Chapter are called the P-centroid and the P-line at infinity respectively. See also Fig. 17.1, which illustrates the relationship.

17.2 The centroid of a triangle and the line at infinity

We take as read the definition of the centroid of a triangle ABC in the Euclidean plane. Since midpoints exist in affine geometry, then the centroid of a triangle in the affine plane is the same as in the Euclidean plane: it is the point of concurrence of the medians. If the midpoint of BC in the Euclidean or affine plane is denoted by L then the point at infinity on the line BC may be defined to be the point A_0 at infinity such that the cross-ratio $(B, C; L, A_0) = -1$. Points B_0 and C_0 on CA and AB respectively may be defined similarly and the points A_0, B_0, C_0 all lie on the line at infinity. Using areal co-ordinates its equation is $x + y + z = 0$ and the co-ordinates of A_0, B_0, C_0 are $(0, 1, -1)$, $(-1, 0, 1)$, $(1, -1, 0)$ respectively. It may be noted that we may also regard the line at infinity as being the polar of the centroid with respect to the conic with equation $yz + zx + xy = 0$ (the outer Steiner ellipse).

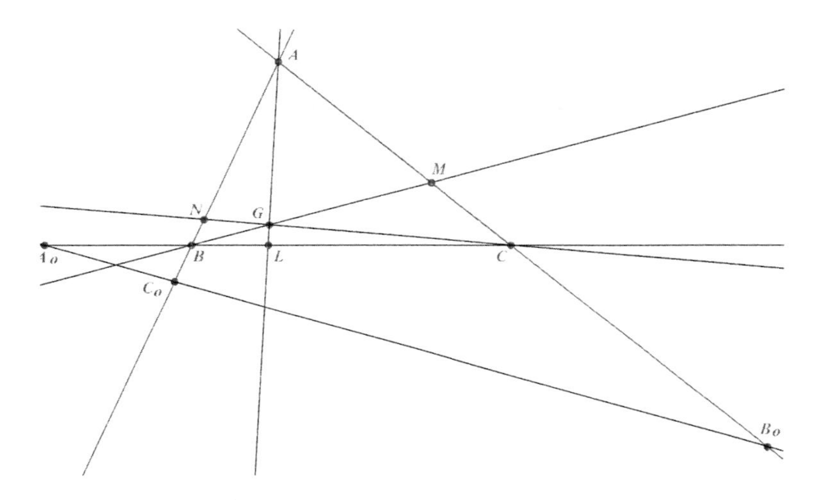

Figure 17.1: The P-centroid and the P-line at infinity of a triangle

These observations provide the clue as to how to define the P-centroid G and the P-line at infinity in the projective plane. Given triangle ABC we can define the P-centroid G to be anywhere (except on the sides of ABC or their extensions). Then we can draw AG, BG, CG to meet BC, CA, AB respectively at L, M, N. Finally we can define the points A_0, B_0, C_0 by the cross-ratios $(B, C; L, A_0) = (C, A; M, B_0) = (A, B; N, C_0) = -1$. Then it is a well-known theorem in projective geometry that $A_0B_0C_0$ is a straight line. This we take to be the P-line at infinity. If G is the unit point, then the P-line at infinity is the unit line. For convenience, in a co-ordinate presentation, we follow this practice. See Fig. 17.1 for an illustration of these concepts.

Thus, as is well known, two lines in the affine plane are parallel if, and only if, they meet on the line at infinity. In the projective plane, now that we have a designated P-line at infinity (unit line) we may define two lines to be P-parallel if, and only if, they intersect on the P-line at infinity. Thus, for example, any line (other than BC) that passes through A_0 may be said to be P-parallel to BC.

As for the position of G itself, it may be anywhere other than on the sides of the triangle or their extensions, and if G is moved, then, for a fixed triangle, so the P-line at infinity moves. In areal co-ordinates we are used to the notion that those points within the triangle of reference have all co-ordinates positive, but in the projective plane, it is the choice of the position

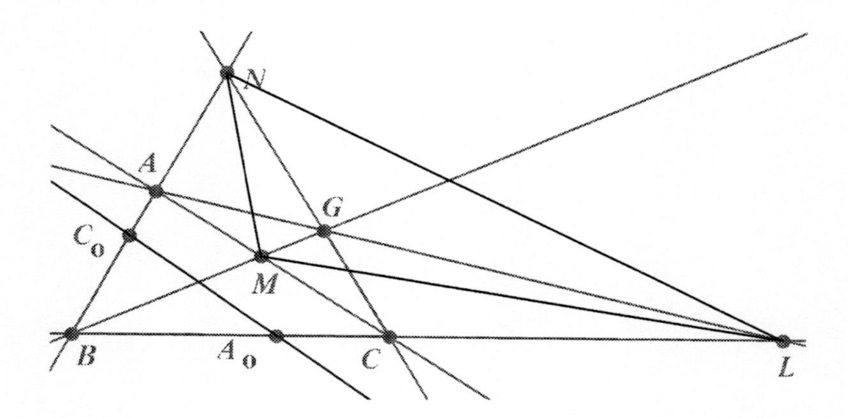

Figure 17.2: Suitable choice of P-centroid for a quadrangle

of G that determines which region in the real projective plane has all its co-ordinates positive. There are four regions in the projective plane, separated by the sides of the triangle of reference, those that have all co-ordinates positive and those that may be chosen to have just one co-ordinate negative. In all the diagrams we draw involving triangles, G will appear to be internal to the triangle ABC, but the reader must not presume this is for any reason other than visual convenience. As soon as one wants to draw a quadrangle $ABCG$, it is visually convenient to picture G as external to triangle ABC as in Fig. 17.2.

One sees immediately that the diagram exhibits the harmonic properties of the complete quadrangle, with LMN as the diagonal point triangle.

Finally, if one has a second triangle, then since the P-line at infinity is already determined by the centroid of the first triangle, it follows that the P-centroid of the second triangle is already predetermined by the harmonic properties of its sides relative to the P-line at infinity.

Exercise 17.2.1

1. Using Fig. 17.2, with G situated in the position indicated, locate the four regions of the real projective plane mentioned in the text and decide the sign of homogeneous projective co-ordinates in the four regions.

17.3 The orthocentre

In the Euclidean plane it is the orthocentre H that determines the perpendiculars from the vertices to the sides, and any line parallel to AH is also perpendicular to BC, and similarly for lines perpendicular to the other sides. Also the orthocentre has the property in the Euclidean plane that if AH meets BC at D and the circumcircle at D' then $HD = DD'$. It is these two concepts that allow us to make further generalisations.

First, when we move to the affine plane, it must be realised that in a generalisation H may be taken to be any Cevian point (that is any point not on a side or its extension, and not on a line through a vertex parallel to an opposite side). We may then regard all lines parallel to AH as being A-perpendicular (affine perpendicular) to BC and all lines parallel to BH, CH as being A-perpendicular respectively to CA, AB. The notion of lines P-perpendicular to the sides of a triangle in the real projective plane is now just a step away. We take any point H, not on the sides of the triangle and not on a line through a vertex P-parallel to an opposite side (that is, not on AA_0, BB_0, CC_0) and draw AH to meet the P-line at infinity at D_0 and then any line through D_0 is P-perpendicular to BC. Similarly lines P-perpendicular to CA and AB are defined by determining the points E_0 and F_0 on the P-line at infinity to be the points where it intersects the lines BH and CH. See Fig. 17.3. When we come to consider the properties of the P-circumcircle of triangle ABC we shall see there is a further mild restriction on the choice of the position of H, but otherwise it may be chosen arbitrarily, but subject to the conditions just stated.

We now give an analysis of the Desargues involution on the P-line at infinity. We take H to have projective co-ordinates (u, v, w). The line AH, with equation $wy = vz$, meets the P-line at infinity at the point D_0 with co-ordinates $(-(v+w), v, w)$ and E_0, F_0 are similarly defined as the intersection of BH, CH with the P-line at infinity. We take a parameter on the P-line at infinity to be equal to its y-co-ordinate divided by its z-co-ordinate. The points A_0, B_0, C_0 thus have parameters $s = -1, 0, \infty$ respectively and the points D_0, E_0, F_0 have parameters $t = v/w$, $-(w+u)/w$, $-v/(u+v)$ respectively. It may now be verified that the point pairs (A_0, D_0), (B_0, E_0), (C_0, F_0) are pairs in the involution defined by

$$w(u+v)st + vw(s+t) + v(w+u) = 0. \tag{17.1}$$

If S, T are points on the P-line at infinity with parameters (s, t) in this

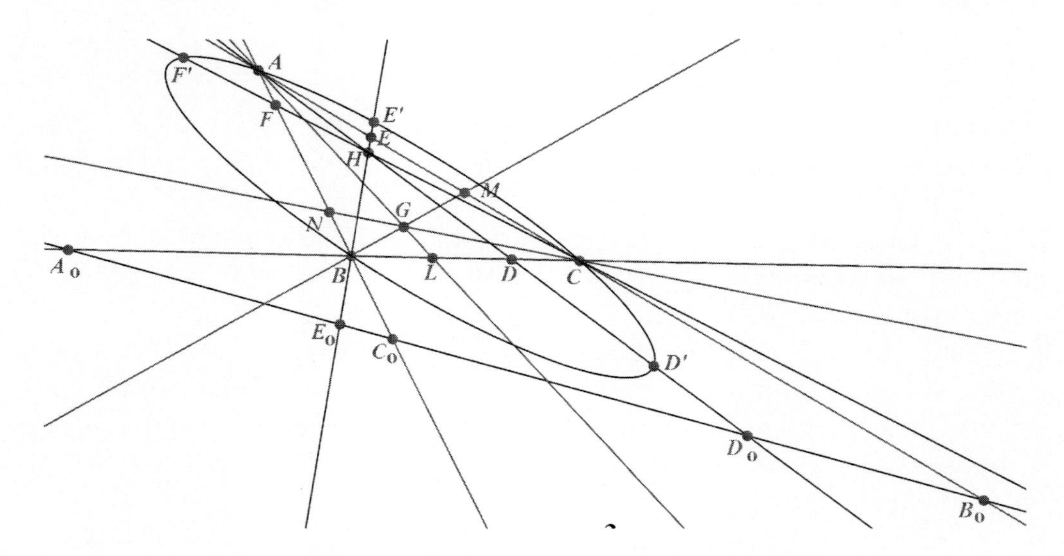

Figure 17.3: The P-orthocentre and P-circumcircle

involution and K is any point not on this line, then we define KS, KT to be P-perpendicular. If you have two points K_1 and K_2, then K_1S is P-parallel to K_2S and K_1T is P-parallel to K_2T, so it is consistent to say that if K_1S is P-perpendicular to K_1T, then K_2S is P-perpendicular to K_2T.

Double points of the involution are found by solving Equation (17.1) when $s = t$ and it turns out there are two real double points when $uvw(u+v+w) < 0$, there is one (repeated) real double point when this expression is zero and there are complex double points when $uvw(u + v + w) > 0$. Note that when $uvw(u+v+w) = 0$, since $uvw \neq 0$, then it must be the case that $u+v+w = 0$ and then H lies on the P-line at infinity. In this Chapter we exclude this possibility. Note that we have excluded the case that H lies on AA_0, BB_0, CC_0, so it is not possible for any of the pairs A_0, D_0 or B_0, E_0, or C_0, F_0 to coincide. However, in the real case, there can exist real lines passing through a double point, leading to the somewhat disquieting fact that such a line is both P-parallel and P-perpendicular to itself!

We now proceed to determine the P-circumcircle, using the following property, which ensures that the conic we obtain is a valid generalization of the circumcircle in the Euclidean plane. If the line AH meets BC at D, then D' is constructed so that $(H, D'; D, D_0) = -1$, reflecting the fact that in the Euclidean plane a key property of the orthocentre is that $HD = DD'$. Points

E' and F' are similarly defined.

17.3.1 The circumcircle

With points D', E', F' as defined in the text above, a conic passes through A, B, C, D', E' and F'. It is defined to be the P-circumcircle of triangle ABC with G as P-centroid and H as P-orthocentre. See Fig. 17.3.. From above we know that the co-ordinates of D_0, E_0, F_0 are respectively $(-(v+w), v, w)$, $(u, -(w+u), w)$, $(u, v, -(u+v))$. D is the foot of the P-altitude from A on to BC and has co-ordinates $(0, v, w)$. Similarly E, F have co-ordinates $(u, 0, w)$, $(u, v, 0)$ respectively. In order to work out the co-ordinates of D' we set up a parameter t on the line AH so that $t = x/(y+z)$. Then A, H, D, D_0 have parameters $\infty, u/(v+w), 0, -1$, respectively. If D' has parameter s, then to satisfy $(H, D'; D, D_0) = -1$, we require $(u/(v+w), s; 0, -1) = -1$. This gives $s = -u/(2u+v+w)$, so that D' has co-ordinates $(-u(v+w), v(2u+v+w), w(2u+v+w))$. Suppose now E', F' are similarly defined by cyclic change, then by inspection the six points A, B, C, D', E', F' lie on the conic with equation

$$u(v+w)yz + v(w+u)zx + w(u+v)zx = 0. \tag{17.2}$$

Equation (17.2) defines the P-circumcircle of triangle ABC with P-centroid G and P-orthocentre H. Whether it deserves the title depends on whether it satisfies the sort of properties one expects, such as the Wallace-Simson line property. The answer is emphatically yes, as we see in due course. However it turns out that for developments that include a P-incircle and P-excircles, then one must restrict H so that the signs of all of $u(v+w), v(w+u), w(u+v)$ are the same. The criterion for this is as follows: u, v, w must be proportional to $1/(g+h-f), 1/(h+f-g), 1/(f+g-h)$, where f, g, h are all positive (or negative). Reasons for this mild restriction will emerge.

17.4 The circumcentre, the nine-point centre, the nine-point circle and the Euler line

The property that defines the circumcentre of a triangle in the Euclidean plane is that the perpendicular bisectors of the three sides are concurrent at

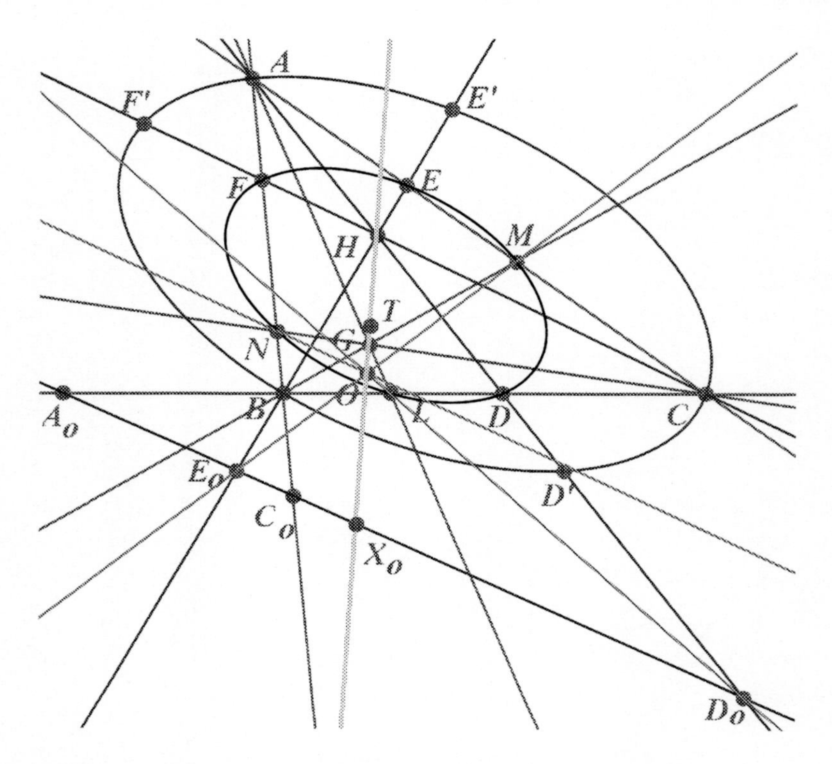

Figure 17.4: The Euler line and the nine-point circle

that point. For a generalization we require the lines LD_0, ME_0, NF_0 to be concurrent and then the point of concurrence O serves as the P-circumcentre of ABC.

The equation of LD_0 is $(w-v)x+(v+w)(z-y) = 0$. This passes through the point $O((v + w), (w + u), (u + v))$, and, by symmetry, this lies on E_0M and F_0N. It is therefore the required P-circumcentre of the P-circumcircle Σ with Equation (17.2). It is easily verified that this is the pole of the P-line at infinity with respect to the P-circumcircle.

17.4.1 The equation of the nine-point circle

We discuss the equation of the P-nine-point circle and the co-ordinates of the P-nine-point centre T. See Fig. 17.4. We use the fact that T is the P-midpoint of OH.

The line OGH, the P-Euler line has equation

$$(v-w)x+(w-u)y+(u-v)z=0. \tag{17.3}$$

This meets the P-line at infinity at the point $X_0(v+w-2u, w+u-2v, u+v-2w)$. The fact that $(O, H; T, X_0) = -1$ may now be used to deduce the co-ordinates of T which are $(2u+v+w, 2v+w+u, 2w+u+v)$. Since L, M, N, and D, E, F are the feet of the Cevians of G and H respectively, we are assured, by Carnot's theorem, that they lie on a conic. Its equation is

$$vwx^2+wuy^2+uvz^2-u(v+w)yz-v(w+u)zx-w(u+v)xy=0, \tag{17.4}$$

as may be verified directly by substitution. This is the equation of the P-nine-point circle.

Exercise 17.4.1

1. Show that the P-midpoints of AH, BH, CH lie on the P-nine-point circle.

2. Prove that the polar of the point O with respect to the P-circumcircle is the P-line at infinity. (This is an important result and has the implication that O is the P-circumcentre of all triangles inscribed in the P-circumcircle of ABC.)

17.4.2 Property of circumcentre

Let ABC be a triangle with P-circumcentre O and P-circumcircle Σ. Choose another triangle TUV inscribed in Σ (with the same P-circumcentre). Then the P-circumcircle of TUV is also Σ (though, in general, the P-centroid and P-orthocentre of TUV are different from those of ABC).

17.5 Isotomic conjugate, isogonal conjugate

We refer to Chapter 10, where these concepts are first introduced in the Euclidean plane. Their generalizations in the affine plane are fairly easy. We first consider the generalization of isotomic conjugate. Since midpoints exist in the affine plane and also ratios of segments on the same line, the definition of isotomic conjugate remains unchanged.

Thus if P is any point, not on the sides or their extensions, and not on a line through a vertex parallel to an opposite side, then we may draw Cevians AP, BP, CP to meet the sides BC, CA, AB respectively at D, E, F. Then, if L, M, N are the midpoints of the sides, we may define points D', E', F' on BC, CA, AB respectively such that L, M, N are the midpoints of DD', EE', FF'. Finally it is a well-known theorem that AD', BE', CF' are concurrent at a point P_t, which is the isotomic conjugate of P. Straightforward analysis shows that the if P has co-ordinates (l, m, n), then P_t has co-ordinates $(1/l, 1/m, 1/n)$. This result, in terms of co-ordinates, allows us to extend the definition of P_t when P lies on a line through a vertex parallel to an opposite side. The property that L is the midpoint of DD' extends into the projective plane without any trouble. The position of L follows from the position of the P-centroid. A_0 is the point on BC where it meets the P-line at infinity, so we may define D' by means of the relation $(D, D'; L, A_0) = -1$. If homogeneous projective co-ordinates replace areal co-ordinates the working is line for line the same as for the affine case, so if P has co-ordinates (l, m, n) then the P-isotomic conjugate P_t has co-ordinates $(1/l, 1/m, 1/n)$. The situation is illustrated in Fig. 17.5. Note that the P-centroid and the points $(-1, 1, 1), (1, -1, 1), (1, 1, -1)$ are their own conjugates in the involution $P \leftrightarrow P_t$.

In Chapter 10 we gave a construction for the isogonal conjugate that involved pedal triangles, but for the purpose of generalization the following procedure in the Euclidean plane is the best line of approach.

17.5.1 Incentre and excentres

A construction for the isogonal conjugate is as follows. Let ABC be a triangle and Σ the circumcircle. Let P be any point of the plane except on the sides of ABC or on their extensions or on Σ. Draw AP, BP, CP to meet the circumcircle at D, E, F respectively. Draw parallels to BC, CA, AB through D, E, F respectively to meet Σ again at D', E', F' respectively. Then AD', BE', CF' are concurrent at P_g, the isogonal conjugate of P. The proof is immediate since it is clear that AD, AD' make equal angles with AI, where I is the incentre of ABC, and similarly for the pairs BE, BE' and CF, CF', results given in Chapter 10.

This generalises to the affine case immediately, since parallels exist in the affine plane, so all one has to do is to make sure that the appropriate A-circumcentre has been chosen for the conic Σ being used as A-circumcircle.

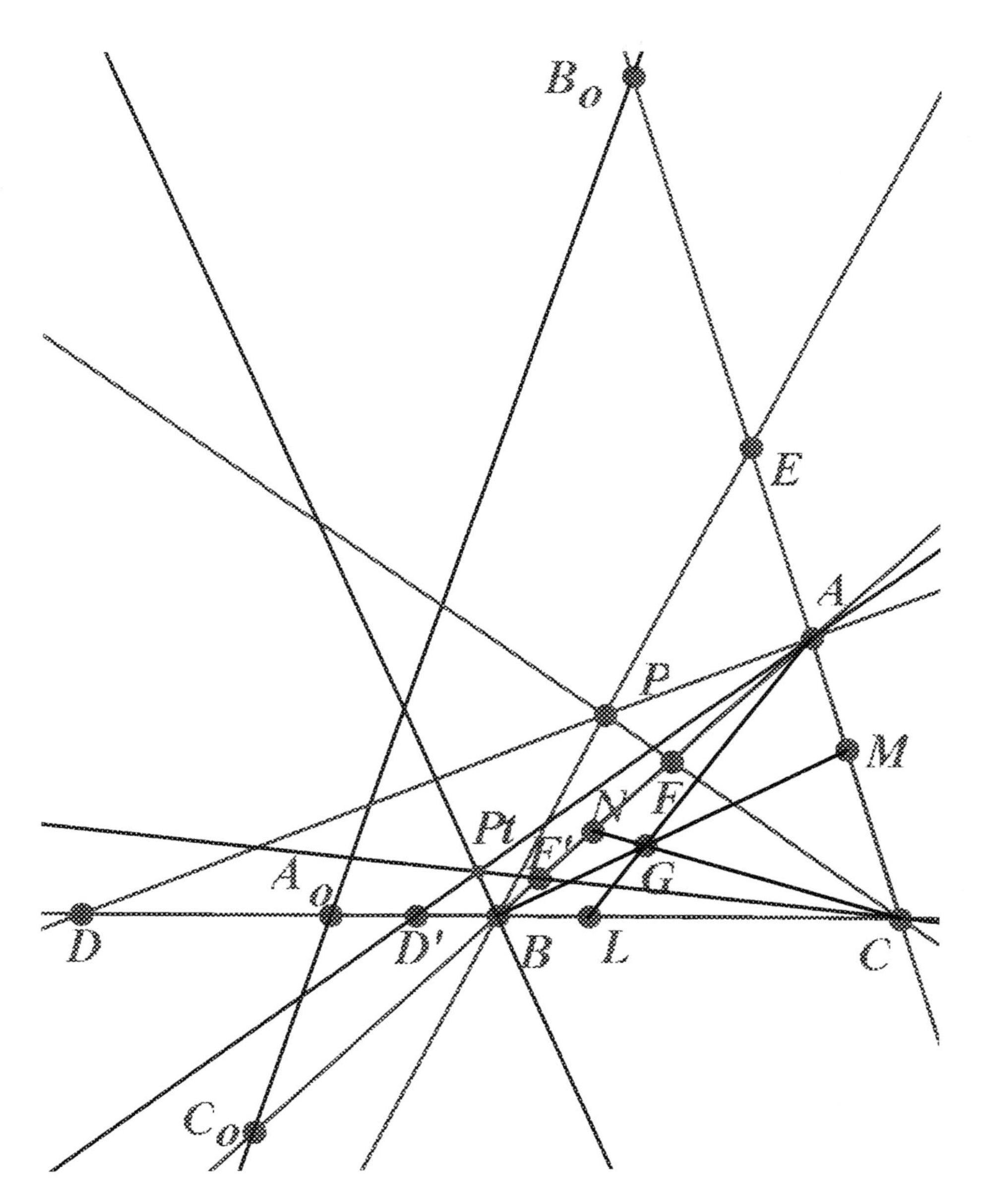

Figure 17.5: Isotomic conjugate

The generalization to the projective plane is now clear. A line P-parallel to BC is now one that intersects it at A_0. The result is shown in Fig. 17.6.

In general if H is the P-orthocentre and has co-ordinates (u, v, w), then Σ has Equation (17.2) and the P-isogonal conjugate has co-ordinates $P_g(u(v+w)/l, v(w+u)/m, w(u+v)/n)$, where (l, m, n) are the co-ordinates of P. These equations are valid whether one is in the affine plane using areal co-ordinates or in the projective plane using homogeneous projective co-ordinates. A first warning note is that in affine space, if one chooses H to be the same point as G, then $u = v = w$ and the isogonal conjugate and the isotomic conjugate become the same point. It may be thought this selection is slightly perverse, but it is actually quite an important construct, as the A-circumcircle is then the outer Steiner ellipse, and there is a famous porism (the double Steiner porism) based on this as circumconic. A second warning note is about the selection of the co-ordinates of H. In Euclidean space P_g has co-ordinates $(a^2/l, b^2/m, c^2/n)$, where a, b, c are the side lengths. This leads to the fact that in Euclidean space the four points which are their own isogonal conjugates are the incentre I and the four excentres I_1, I_2, I_3 with co-ordinates $(a, b, c), (-a, b, c), (a, -b, c), (a, b, -c)$.

This means that in affine space triangle ABC only has an A-incentre and A-excentres if H is chosen so that $ku(v + w), kv(w + u), kw(u + v)$ are all positive for some constant k. The implication is that triangle ABC with $H(u, v, w)$ has an A-incentre if, and only if, $u(v + w), v(w + u), w(u + v)$ are either all positive when we choose $k = 1$, or all negative when we choose $k = -1$. In either case the A-incentre and A-excentres then have co-ordinates $(a, b, c), (-a, b, c), (a, -b, c), (a, b, -c)$, where $a^2 = ku(v + w), b^2 = kv(w + u), c^2 = kw(u + v)$.

Exactly the same expressions hold, for the same reasons, in projective space. Of course, if the three expressions cannot all be made positive an isogonal conjugate of a point still exists, but in such cases, there are no real self-conjugate points and the P-incentre and P-excentres are not real points. A sign difficulty actually occurs with obtuse angled-triangles in Euclidean space. If you take $a = 2, b = 3, c = 4$ and put $u = 1/(b^2 + c^2 - a^2)$, $v = 1/(c^2 + a^2 - b^2)$, $w = 1/(a^2 + b^2 - c^2)$, you will find that $u(v+w), v(w+u), w(u+v)$ are all negative, being nonetheless proportional to a^2, b^2, c^2. See Fig. 17.6 for an illustration of the P-isogonal conjugate.

The P-isogonal conjugate of O is H, and that of the P-centroid G is the P-symmedian S. This is illustrated in Fig. 17.7. There it is shown that, if the tangent at A to the P-circumcircle meets BC at X, and X' is the harmonic

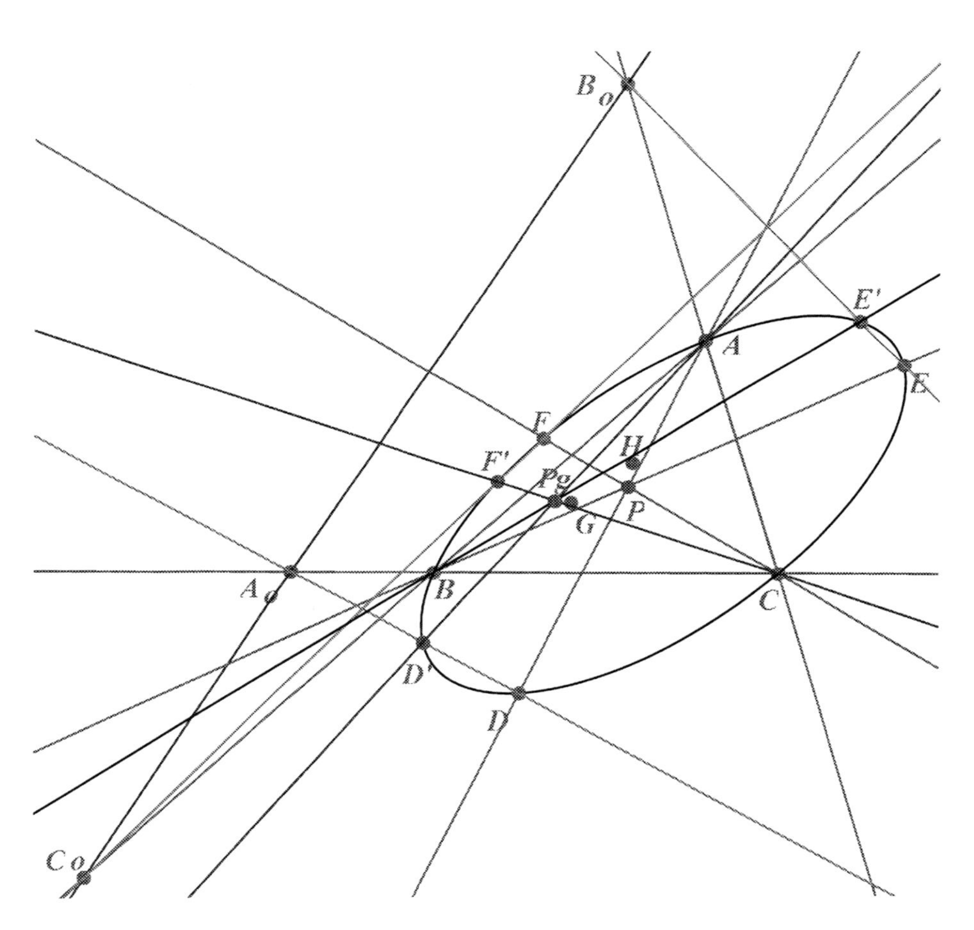

Figure 17.6: Isogonal conjugate

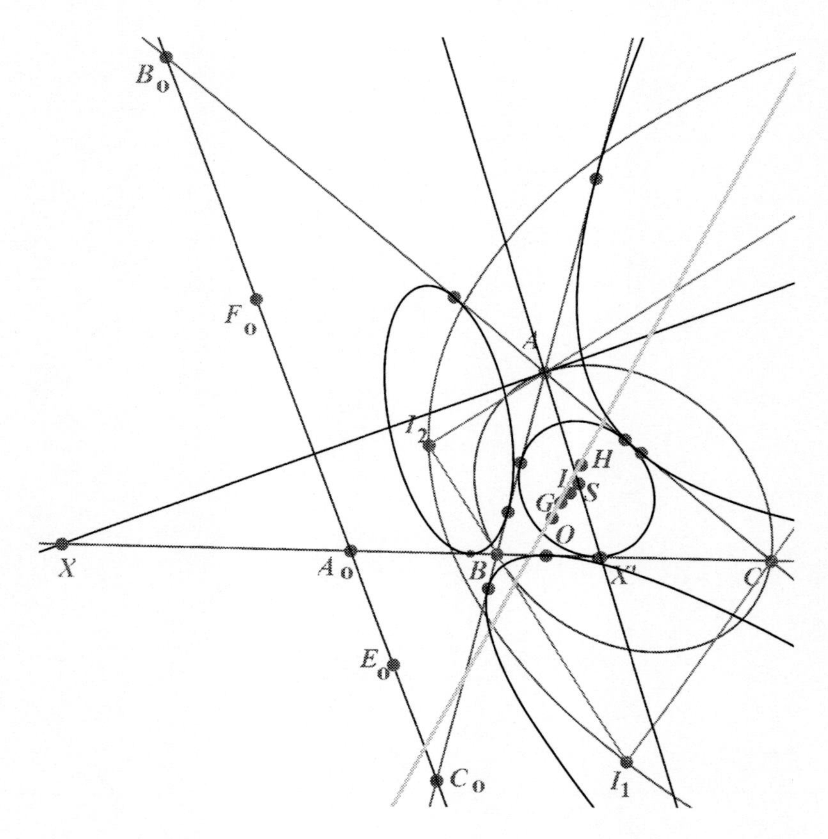

Figure 17.7: Incircle and excircles

conjugate of X with respect to B and C, then AX' passes through S. This shows that S is indeed the generalized symmedian point, significant in the theory of porisms.

Also shown in Fig. 17.7 are the P-incentre and excentres, the P-incircle, the triangle of excentres and its P-circumcircle, which is easy to construct, since its P-orthocentre is I. Parts of the P-excircles are also shown. The figure is very sensitive to the position of H and the relative situation of triangle ABC and the P-line at infinity.

Needless to say it is possible to continue to construct the P-versions of other triangle centres, but we have probably taken matters far enough. Some triangle centres depend on being able to draw a P-perpendicular to any line, and we show how this can be done in Section 17.12.

Exercise 17.5.1

1. Let P be a point and P_g its P-isogonal conjugate. Draw P-altitudes through P and P_g to the sides of triangle ABC and prove that the six intersections lie on a conic.

17.6 The Wallace-Simson line

In this Section ABC is a triangle with λ the P-line at infinity and H any point not on the sides of the triangle.

17.6.1 The projective version

We give a projective version of the Wallace-Simson line property. Suppose that HA meets λ at D_0, that HB meets λ at E_0 and that HC meets λ at F_0. Let P be a point that does not lie on λ. Suppose that PD_0 meets BC at L, PE_0 meets CA at M, PF_0 meets AB at N. Then L, M, N are collinear if, and only if, P lies on the P-circumcircle of ABC. Take P to have co-ordinates (p, q, r). D_0 has co-ordinates $(-(v+w), v, w)$, so the equation of PD_0 is

$$(qw - vr)x - (r(v+w) + pw)y + (pv + q(v+w))z = 0.$$

This meets BC at $L(0, pv+qv+qw, pw+rv+rw)$. Points M, N on CA, AB are similarly defined as intersections with PE_0, PF_0 respectively. Their co-ordinates are

$$M(qu + pw + pu, 0, qw + rw + ru), N(ru + pu + pv, rv + qu + qv, 0).$$

The points L, M, N are collinear if, and only if, the determinant of the co-ordinates of these points as rows vanishes. Since $u+v+w \neq 0$ and $p+q+r \neq 0$, we can factor out these terms leaving

$$u(v+w)qr + v(w+u)rp + w(u+v)pq = 0.$$

The locus of P with the property that L, M, N are collinear is therefore the P-circumcircle of ABC, having P-orthocentre H. See Fig. 17.8.

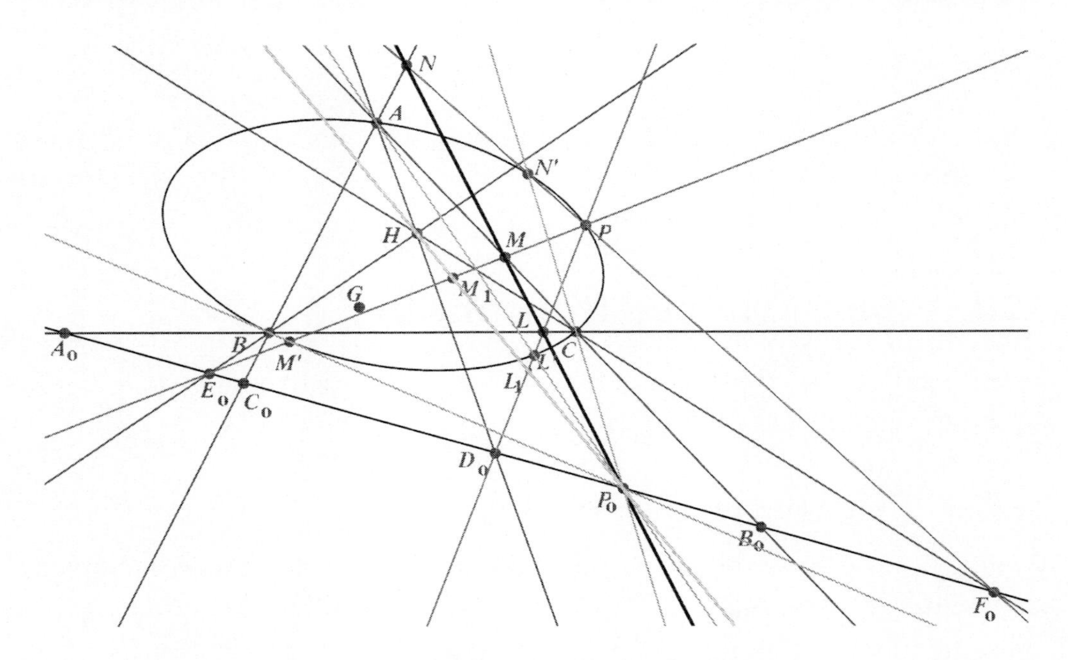

Figure 17.8: The Wallace-Simson line

Exercise 17.6.1

1. Let the P-Wallace-Simson line of P meet the P-line at infinity at P_0. Suppose that PD_0 meets the P-circumcircle again at L', with M', N' similarly defined. Prove that AL', BM', CN' are concurrent at P_0.

2. State and prove the dual theorem to the Wallace-Simson line theorem in the projective plane, thereby defining the P-Wallace-Simson point.

3. Suppose that PD_0 is extended beyond L to a point L_1 such that $(P, L_1; D_0, L) = -1$, with M_1, N_1 similarly defined. Prove that L_1, M_1, N_1 are collinear on a line that passes through H. (This line is sometimes known as the Double Wallace-Simson line.)

Fig. 17.8 illustrates the P-Wallace-Simson line and the results that can be obtained from Exercises 17.6.1.1 and 17.6.1.3.

17.7 Justification of the procedure

We now consider the justification of the method we use of creating and drawing projective generalisations. So far we have been intent on describing how to carry out the procedure. However, we now have a first significant result, showing that the P-circumcircle is indeed the conic that supports the generalisation of the Wallace-Simson line result.

When proving the Wallace-Simson theorem above it produces a proof that is valid whatever the values of u, v, w, provided H is placed to avoid forbidden positions, as previously detailed. Normally in the Euclidean plane the proof of the existence of the Wallace-Simson line uses angle considerations involving cyclic quadrilaterals. However, if you want, you can set up an analytic proof, and this turns out to be identical to the above analytic proof with $u = \tan A, v = \tan B, w = \tan C$, except that you never need to use these expressions, nor the fact that when u, v, w have these values the equation of the circumcircle is $a^2yz + b^2zx + c^2xy = 0$. The above proof is valid, but with these values hidden in the background. What is actually going on when you use a given set of values of u, v, w is to invoke the Desargues involution on the line at infinity, given by Equation (17.1) and use the perpendicularity implied. If you change the values of u, v, w to $\tan A, \tan B, \tan C$ respectively then it involves a different involution. A dual interpretation may now be given to this change. You can either change the shape of the triangle so that it becomes triangle $A'B'C'$ and the perpendicularity is then the standard perpendicularity for that triangle. Alternatively, you can stay with triangle ABC and then the change is equivalent to moving H to a different Cevian point and having a different perpendicularity, the P-perpendicularity. Now no-one would suggest that a different proof of the Wallace-Simson line result is required if you just change the shape of the triangle. Likewise no different proof is required if you move to a P-orthocentre, rather than using the real orthocentre.

In what follows in the final sections of this chapter the same considerations apply, as the next section makes clear.

17.7.1 Its consequence

If a theorem is known to be true in the Euclidean plane, and it does not depend for its validity on distances, then because its analytic proof (whether that analytic proof is actually written out or not) would just contain the symbols u, v, w and would not rely on their values in terms of the side lengths of the triangle, then the generalisation of that theorem in the Euclidean or projective planes does not require separate proof.

17.8 P-Hagge circles and the Great Hagge theorem

17.8.1 Statement of results

Given a triangle ABC, its P-orthocentre H, its P-circumcircle Σ_0 and a general point Y, let AY, BY, CY meet the P-circumcircle at U_0, V_0, W_0 respectively. Let the P-perpendiculars to BC, CA, AB respectively through U_0, V_0, W_0 meet the sides BC, CA, AB at U_1, V_1, W_1 respectively. Define U_2 so that U_1 is the P-midpoint of U_0U_2, with V_2, W_2 similarly defined. Now let U_1Y and U_2Y meet AH at P_1 and P_2 respectively, with Q_1, Q_2 on BH and R_1, R_2 on CH be similarly defined. Then $U_1, V_1, W_1, P_1, Q_1, R_1$ lie on a conic Σ_1 and $U_2, V_2, W_2, P_2, Q_2, R_2$ lie on a conic Σ_2 and furthermore Σ_2 passes through H.

An illustration of this result is shown in Fig. 17.9. The proof of the Great Hagge theorem in the Euclidean plane is difficult. A proof appears in Bradley and Smith [10]. In the Euclidean plane the conics Σ_0 and Σ_2 are, of course circles, but Σ_1 (called the midpoint conic) is not a circle. The points P_1, Q_1, R_1 are not labelled to prevent the diagram becoming too crowded.

17.9 The Miquel point

The point known as the Miquel point has a generalization in the projective plane as shown in Fig. 17.10.

17.9.1 The Miquel point

Let ABC be a triangle and let L, M, N points on BC, CA, AB respectively (but not at the vertices). Take a point H to act as P-orthocentre of ABC. Let H_1, H_2, H_3 be the P-orthocentres of triangles AMN, BNL, CLM respectively and let S_1, S_2, S_3 be the P-circumcircles of the triangles AMN,

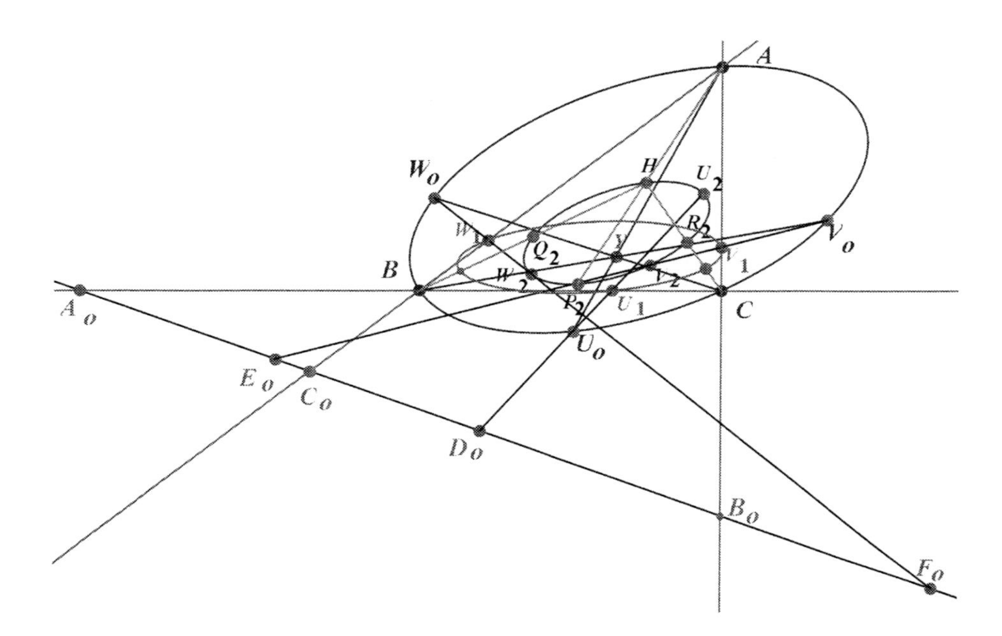

Figure 17.9: A Hagge circle and the midpoint conic

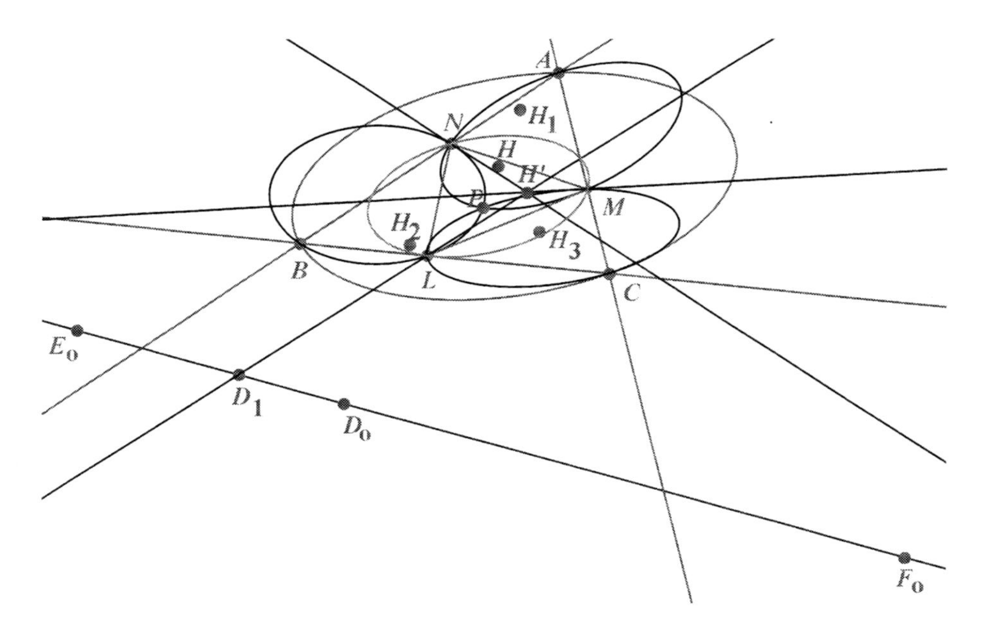

Figure 17.10: The Miquel point

BNL, CLM. Then S_1, S_2, S_3 have a point P in common, which we may call the P-Miquel point.

This follows from the Euclidean case, since the points H_1, H_2, H_3 are fixed by the condition that NH_1 is P-perpendicular to AC and MH_1 is P-perpendicular to AB and similarly for H_2 and H_3.

Exercise 17.9.1

1. Let ABC be a triangle and l a line, not passing through any of the vertices of ABC. Let L, M, N be points on the sides of ABC (but not at the vertices and not on l). Choose a point H (not on the sides of ABC and not on l). Let AH, BH, CH meet l at D, E, F. Suppose that FM and EN meet at P, DN and FL meet at Q, EL and DM meet at R. Let AP meet l at X, BQ meet l at Y, CR meet l at Z. Prove that LX, MY, NZ are concurrent.

17.10 The complete quadrilateral and coaxal circles

This is featured in Fig. 17.11, which shows the four lines BCL, CAM, ABN, LMN forming the quadrilateral.

The P-centroids are not shown. The first step in the construction is to choose an arbitrary point H, not on any of the four lines, as P-orthocentre of triangle ABC. Once this choice is made everything else follows. For, as the directions of P-perpendiculars to the sides of ABC are known, the P-orthocentres H_1, H_2, H_3 of triangles AMN, BNL, CLM follow automatically. Points O, O_1, O_2, O_3 of the four triangles are now automatically fixed. The following results now hold:

17.10.1 P-Orthocentres of the four triangles

The points H, H_1, H_2, H_3 are collinear.

17.10.2 P-Circumcircles of the four triangles

The P-circumcircles of the four triangles have a common point P. (The P-Miquel point of AMN, BNL, CLM and there is a common P-Wallace-Simson line of the triangles.)

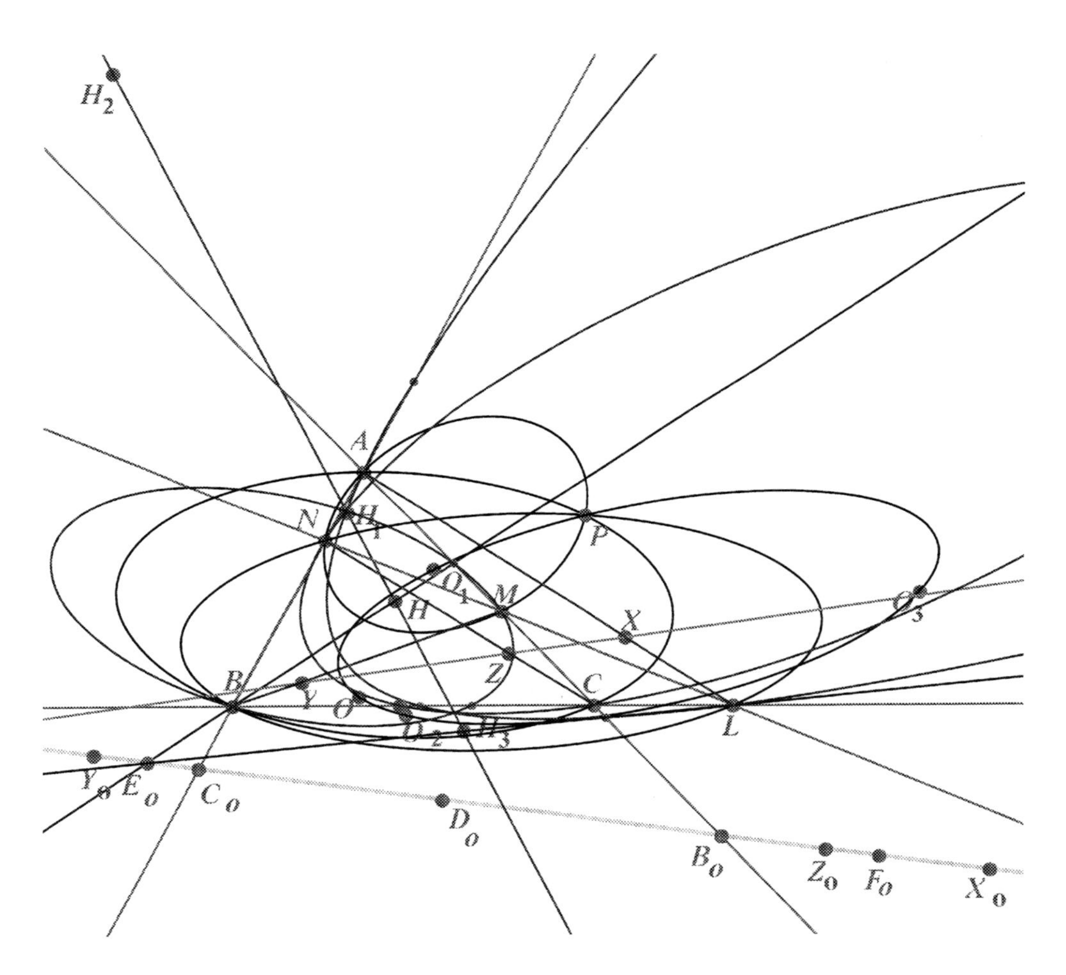

Figure 17.11: The complete quadrilateral

17.10.3 Property of the four P-circumcentres

The P-circumcircle $OO_1O_2O_3$ passes through P.(There is, of course a conic through these five points, but it has the special property that being derived from a circle in the Euclidean plane, it shares with the conics of Section 17.10.2 two complex points on the P-line at infinity)

17.10.4 Collinearity of P-midpoints of diagonals

The P-midpoints of AL, BM, CN are collinear

17.10.5 A P-coaxal system of circles

The P-circle on diameter AL consists of the following five points: A, L, the point where LH_3 meets CA, the point where LH_2 meets AB and the point where AH meets BC. The P-circles on diameters BM, CN are similarly defined. These three P-circles form a P-coaxal system with line of centres XYZ (the P-midpoints of AL, BM, CN) and P-radical axis the line of orthocentres.

17.10.6 More P-coaxal systems of circles

The complementary coaxal system are the P-polar circles of the four triangles, centres H, H_1, H_2, H_3 (These are not drawn in the figure).

Fig. 17.12 shows another instance of a system of P-coaxal circles. It consists of a triangle and the following P-circles, the P-circumcircle, the P-nine-points circle, the P-orthocentroidal circle and the P-polar circle, with centres O, N, J, H respectively. Note that in this particular representation the intersecting system has four real points of intersection. Two of them are labelled R and S and the other two lie on the P-line at infinity. Only one of these, with label Q, can be seen in the figure. The other is off the page to the left of the configuration.

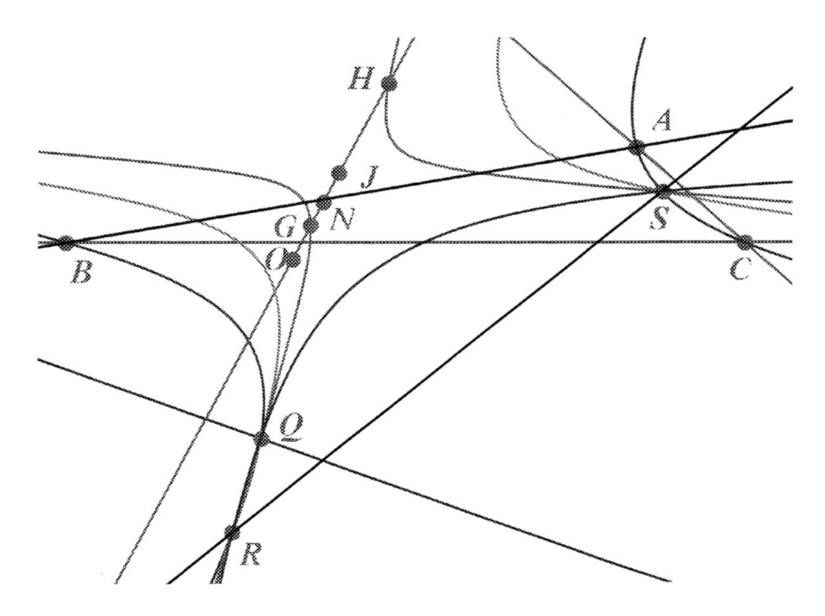

Figure 17.12: Four P-circles in a coaxal system

17.11 Triplicate ratio and Brocard circles, and the Brocard porism

In Chapter 10 we gave an account of the construction of the triplicate ratio circle and the Brocard or 7-point circle. In Fig. 17.13 we show how this looks in the projective plane. Some readers with the appropriate geometrical software might accept the challenge to reproduce the figure, though as it depends on the choices made for the P-centroid and P-orthocentre, no two figures will look the same. For the sake of clarity all construction lines have been deleted. The construction for obtaining the P-Brocard points as the intersection of conics (rather than as a by-product of the triplicate ratio circle construction) is possible, but rather tiresome and therefore not described. The P-triplicate ratio circle and the P-Brocard circle are shown in Fig. 17.13. Note that both these P-circles have P-centre at X, which is the P-midpoint of OS, where S is the P-symmedian point.

Once one can draw a triangle and a P-circumcircle, and obtain its P-symmedian point S, then one can draw the polar of S with respect to the P-circumcircle. Bradley and Smith [8] prove that triangles TUV that are

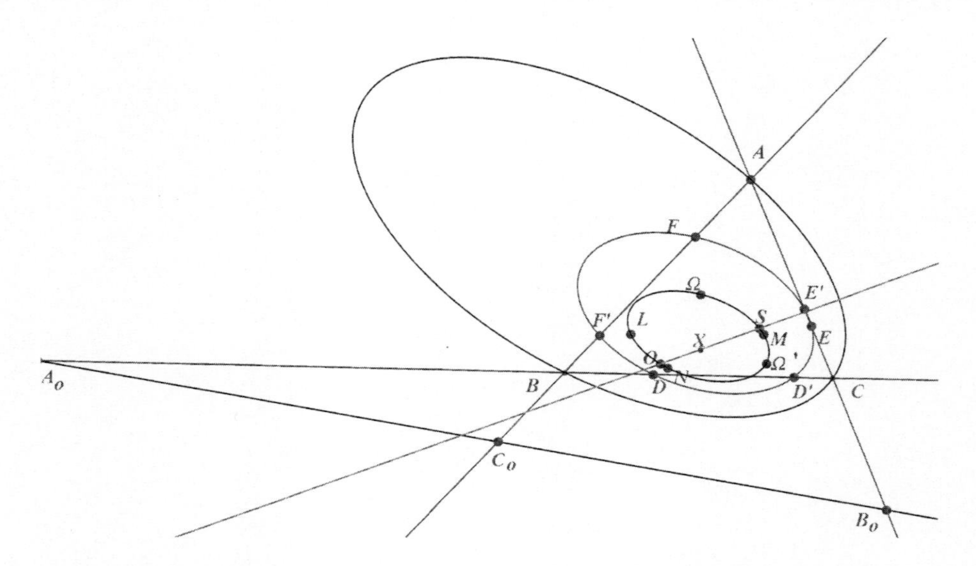

Figure 17.13: The P-triplicate ratio and P-Brocard circles

in triple reverse perspective with ABC (ABC with TVU, VUT, UTV) with vertices of perspective all lying on the polar of S form a porism, which we may call the P-Brocard porism. See also Part 3.9, where an independent coordinate proof is given of the projective generalization of the Brocard porism. In Fig. 17.14 we illustrate the P-Brocard porism with the P-circumcircle, the P-Brocard circle, two triangles ABC and TUV in the porism, along with the P-Brocard inellipse. S is the P-symmedian point.

17.12 The perpendicular from a point to a line

In this section we show how to draw a P-perpendicular to a line LMN through a given point X. Note that a line and the line P-perpendicular to it meet the P-line at infinity in pairs of points in the Desargues involution described in Part 17.3.

First let us consider a transversal cutting the triangle at points L, M, N on sides BC, CA, AB respectively (and not passing through a vertex). Let us see how to draw the P-perpendicular to it through A. We have already described how this may be done in Section 17.9.1 and Part 17.10 in dealing with

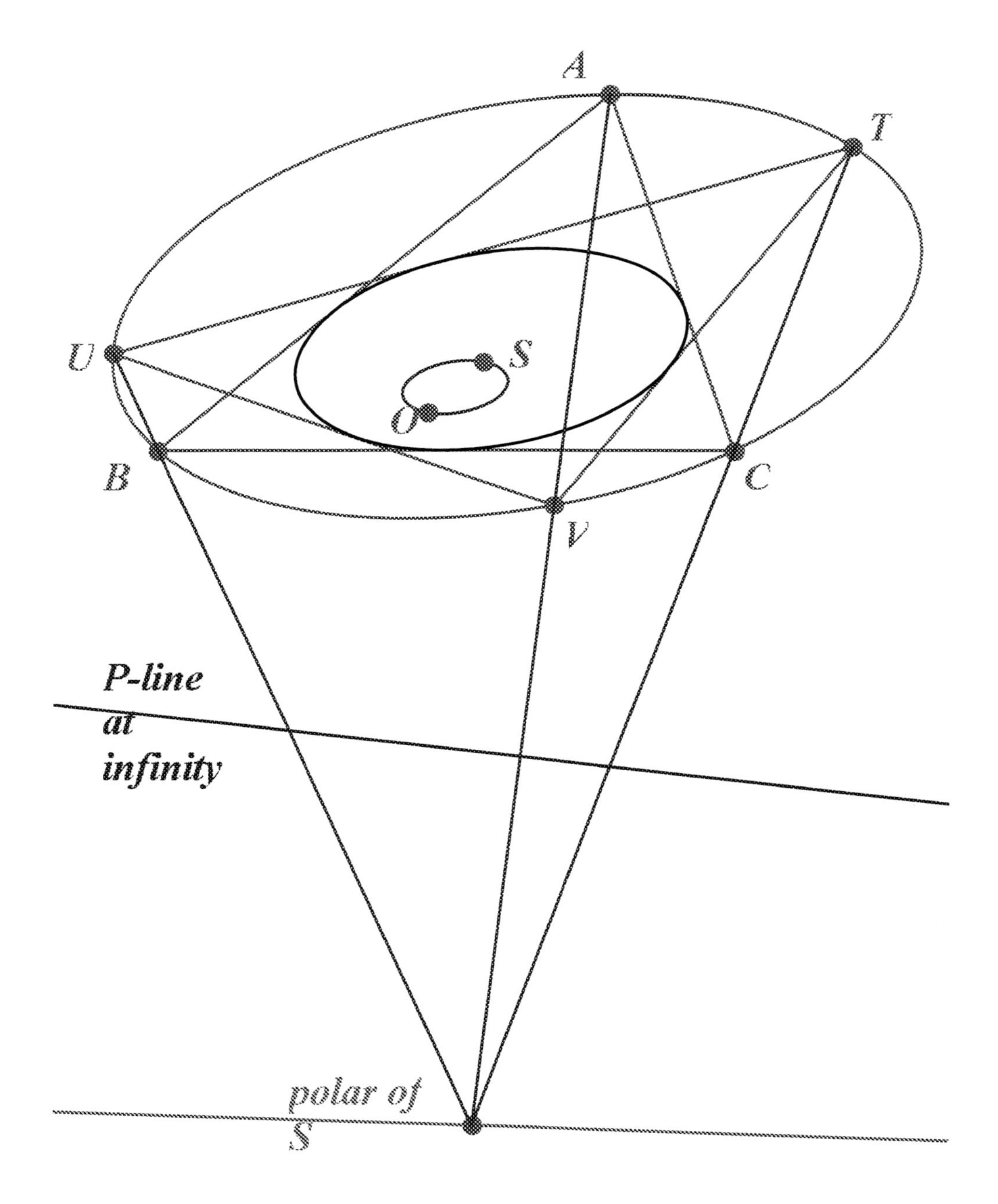

Figure 17.14: The P-Brocard porism

the Miquel point and the complete quadrilateral. But we now provide a justification of the procedure in terms of the Desargues involution. Consider the triangle AMN. Draw the line through N that is P-parallel to the P-altitude through B and draw the line through M that is P-parallel to the P-altitude through C. If these lines meet at H_1, then AH_1 is the P-perpendicular to the transversal LMN. See Fig. 17.15. Let AH_1 meet the P-line at infinity at Q.

The construction can now be modified to overcome the difficulty that the line LMN may pass through A. All one has to do is to choose another transversal that is P-parallel to it and so passes through the same point on the P-line at infinity. Secondly if the P-perpendicular to the transversal is required through a point X that is not the vertex A, then first find the P-perpendicular through A and then draw the P-parallel to it through X. This is the line XQ in Fig. 17.15.

In the above the point A may be regarded as a transition point before moving to X. The points B or C can be used equally well. Now it is clear that this construction is valid, as it produces point pairs in involution on the Q-line at infinity. Furthermore when L is fixed and the line moved, so that N coincides with B and M coincides with C, we obtain as a point pair in this involution the pair (A_0, D_0). Similarly the point pairs (B_0, E_0) and (C_0, F_0) are in the involution. The involution is therefore the same as the Desargues involution analysed in Part 17.3.

17.13 The orthopole

Section 17.12 points the way to the generalization of the orthopole properly. This property is more correctly thought of as a property of a complete quadrilateral, rather than as a property of a triangle, because the initial configuration consists of a triangle and a transversal. The collinear points H_1, H_2, H_3 are the P-orthocentres of the triangles with vertices A, B, C that are cut off by the transversal, but now L, M, N are the feet of the P-perpendiculars from A, B, C on to the transversal. The P-orthopole property is now that the lines LD_0, ME_0, NE_0 are concurrent at the P-orthopole P. The generalisation is illustrated in Fig. 17.16.

Once P-perpendiculars can be drawn there are many other problems in Euclidean geometry that can be put into the projective plane, provided distances and angles are not involved. For example, the theory of the Droz-Farny line can be generalized, and moreover the envelope of the P-Droz-Farny line

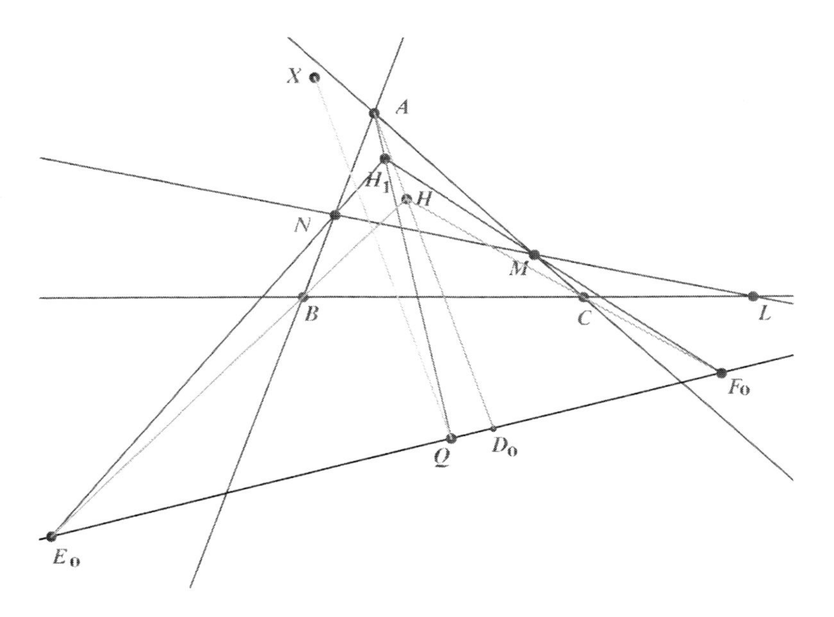

Figure 17.15: The P-perpendicular from a point to a line

is a conic. It also becomes possible to construct P-equilateral triangles on the sides of a triangle, and thereby construct the P-Fermat points of a triangle and their isogonal conjugates, the P-isodynamic points. The reader with appropriate software may like to attempt the construction, which uses the fact that in a P-equilateral triangle the lines from the vertices to the P-midpoints of the opposite sides coincide with the P-altitudes.

Exercise 17.13.1

(for those with appropriate geometry software)

1. Verify that the angle in a P-semi-circle is a P-right angle.
2. Verify that in a P-circle with P-centre O then, if Q is a point on the P-circle, then the tangent at Q is P-perpendicular to the line OQ.

17.14 Extension of orthopole property

Q is no longer defined to be the point where the P-perpendiculars from A, B, C to the line LMN meet, but is allowed to roam, as is the P-orthocentre

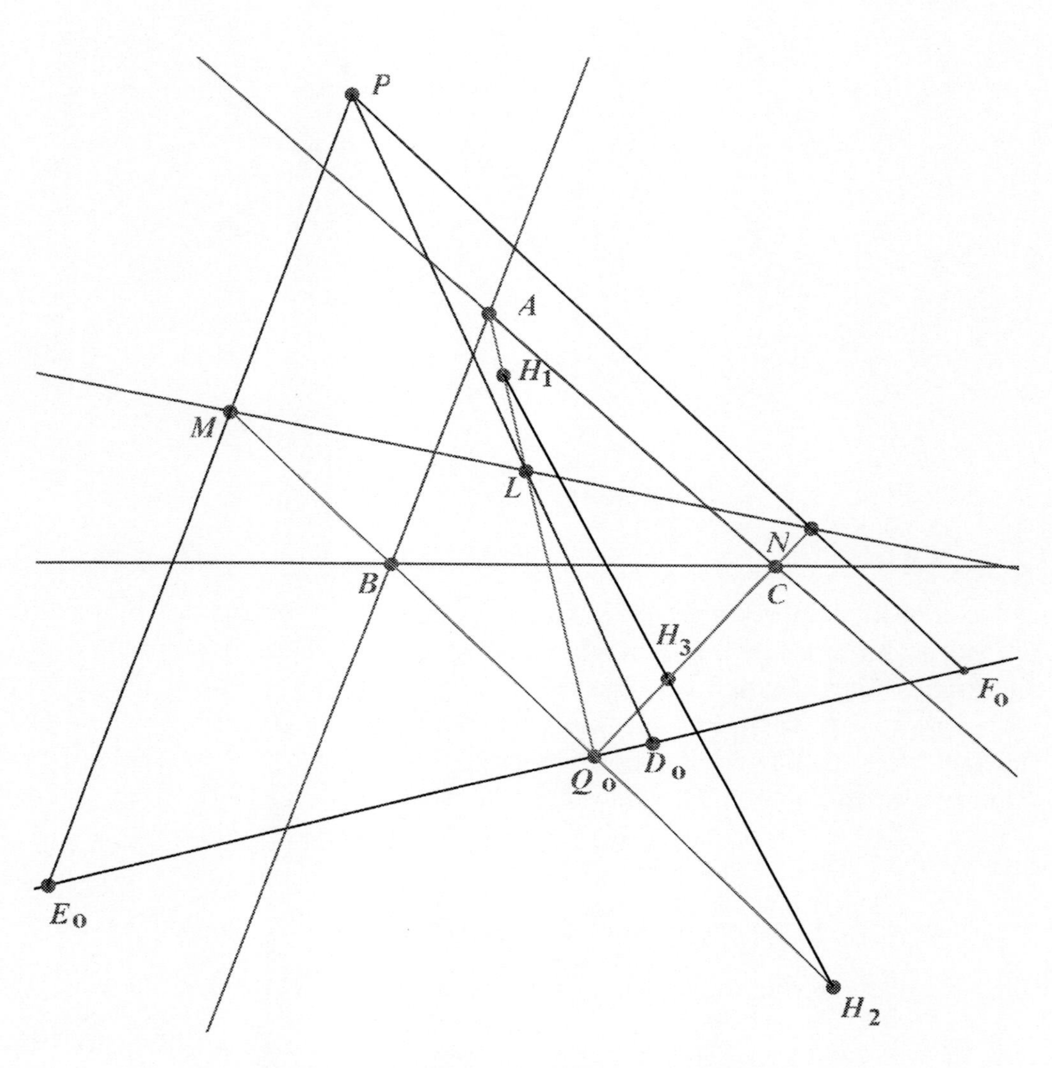

Figure 17.16: The P-orthopole property

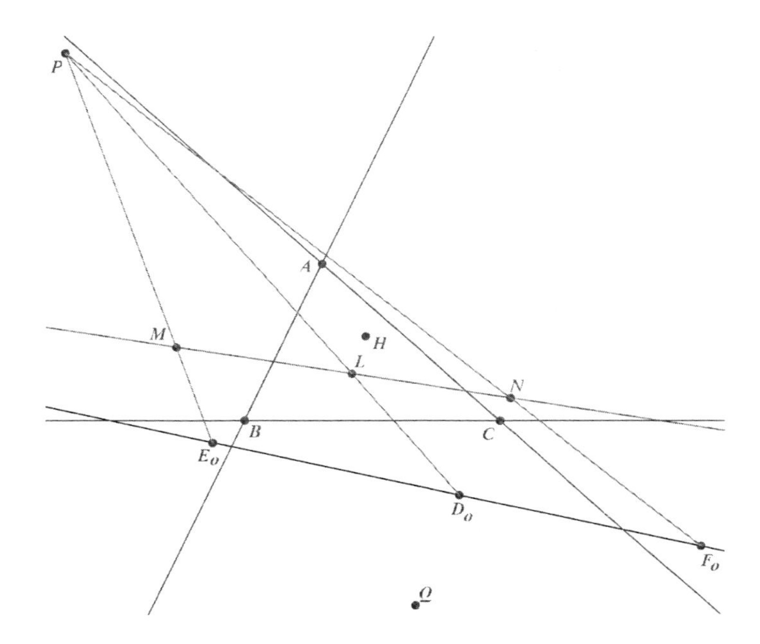

Figure 17.17: Extension of the P-orthopole property

of ABC and the transversal LMN itself. These freedoms of movement lead to the following satisfying results.

17.14.1

The lines D_0L, E_0M, F_0N are concurrent at a point P, when H and the transversal are fixed if, and only if, Q lies on a conic $S(H)$.

17.14.2

The lines are concurrent at a point P, when Q and the transversal are fixed if, and only if, H lies on a conic $S(Q)$.

17.14.3

The lines are concurrent, when H and Q are fixed if, and only if, the transversal envelopes a conic.

Take ABC as triangle of reference and let the P-line at infinity have, as usual, the equation $x+y+z=0$. Suppose that H has co-ordinates (u,v,w) and that Q has co-ordinates (p,q,r). Let the transversal have equation $lx+my+nz=0$.

AQ meets the transversal at L

The equation of AQ is $qz=ry$ and this meets $lx+my+nz=0$ where $lrx+mqz+nrz=0$, so that L has co-ordinates $(mq+nr,-lq,-lr)$. Similarly M,N have co-ordinates $(-mp,nr+lp,-mr)$, $(-np,-nq,lp+mq)$ respectively.

AH meets the P-line at infinity at D_0

We already know from previous work that $D_0(v+w,-v,-w)$. Similarly E_0 and F_0 have co-ordinates $(-u,w+u,-w),(-u,-v,u+v)$ respectively.

The equation of the line D_0L

Using the co-ordinates of L, D_0 just obtained we find the equation of the line D_0L to be

$$(lwq-lvr)x+(lr(v+w)-w(mq+nr))y+(v(mp+nr)-lq(v+w))z=0.$$

The equations of E_0M and F_0N are obtained from this by cyclic change of letters. The condition that these three lines meet is that the determinant of coefficients of x,y,z should vanish.

Now the construction is invalid if H lies on the P-line at infinity or if Q lies on the transversal, so $u+v+w\neq 0$ and $lp+mq+nr\neq 0$, so in working out this determinant the factors $(u+v+w)(lp+mq+nr)$ can be cancelled out. This leaves an equation which is homogeneous of the sixth degree, being quadratic in u,v,w and in p,q,r and in l,m,n. This equation (from DERIVE) is

$$\begin{gathered}-l^2pruv+lmpruv-lmqruv+lnpruv+lnqruv+m^2qruv\\ -mnpruv-mnqruv-lmqruw+lnqruw+mnqruw-n^2qruw\\ +l^2pquw-lmpquw-lnpquw+mnpquw+lmprvw-lnprvw\\ -mnprvw+n^2prvw+lmpqvw-lnpqvw-m^2pqvw+mnpqvw=0.\quad(17.5)\end{gathered}$$

We deduce that if H and the transversal are fixed then the condition for a pole to exist is that Q lies on a conic $S(H)$ and if Q and the transversal are fixed the condition for a pole to exist is that H lies on a conic $S(Q)$ and if H and Q are fixed the condition for a pole to exist is that the transversal envelopes a conic.

Reciprocal relationship

Furthermore, since Equation (17.5) is invariant under the exchange $p \leftrightarrow u$, $q \leftrightarrow v, r \leftrightarrow w$ it follows that if Q lies on $S(H)$, then H lies on $S(Q)$.

Chapter 18

Supplementary Exercises

1. Let C_1 and C_2 be two circles, centres X and Y, which intersect at A and B. Let AX meet C_1 at E and C_2 at C and let AY meet C_1 at D and C_2 at F. Prove that AB, DE and CF are concurrent at a point T. Let S be the midpoint of AT. Prove that S, C, Y, B, X, D are concyclic.

2. Let $ABCD$ be a convex quadrilateral. The extensions of BA beyond A and CD beyond D meet at F and the extensions of CB beyond B and DA beyond A meet at G. The incentres of triangles FBC and GCD are denoted by X and Y respectively. The excentre of triangle FAD opposite F is denoted by I and the excentre of triangle GAB opposite G is denoted by J. Assuming none of X, Y, I, J coincide, prove that they are concyclic.

3. ABC is a triangle with centroid G, circumcentre O, incentre I and $\angle BAC = 120^o$. Prove that AI is parallel to OG.

4. ABC is a triangle and AD is the altitude through A. L is the midpoint of BC. The point E is chosen on BC so that L is the midpoint of DE. Prove that the perpendicular at B to AB, the perpendicular at C to AC and the perpendicular at E to BC are concurrent.

5. Let ABC be a non-isosceles triangle and P a point in the plane of ABC, not on the sides of ABC. Let p be the polar of P with respect to the circumcircle of ABC. Suppose that the perpendiculars to AP, BP, CP at P meet the sides BC, CA, AB at L, M, N respectively. Prove that LMN is a straight line and that it is parallel to p if, and only if, P lies

on the orthocubic curve passing through the incentre and excentres of ABC.

Suppose now that P lies on this cubic curve, that OP meets p at P' and the line LMN at Q. Prove that if $PQ = QP'$, then P lies at either the incentre or an excentre of ABC.

6. Let ABC be a triangle and LMN the median triangle. Let P, Q, R be the excentres of triangle LMN with P the excentre opposite L and so on. Show that BQ and CR meet at a point U on MN, with V, W similarly defined. Show also that AU, BV, CW are parallel and that LU, MV, NW are concurrent. Are PU, QV, RW also parallel or concurrent?

7. ABC is a triangle all of whose angles lie between 45° and 90° inclusive. Its side lengths are a, b, c and the altitudes from A, B, C have lengths d, e, f respectively. Prove that

$$12 \leq (a^2 + b^2 + c^2)(1/d^2 + 1/e^2 + 1/f^2) \leq 16.$$

8. Let ABC be an acute-angled triangle, which is not equilateral. Prove that there are two points P such that

$$PA : PB : PC = \sin A : \sin B : \sin C$$

and that these points are collinear with any point Q such that

$$QA : QB : QC = \cos A : \cos B : \cos C.$$

How many such points Q exist and where is their precise location?

9. ABC is a triangle, circumcentre O, orthocentre H, and AOD, BOE, COF are diameters of the circumcircle of ABC. U, V, W are the reflections of D, E, F in the sides BC, CA, AB respectively. Prove that the circle UVW passes through H.

10. ABC is a triangle and D, E, F are the midpoints of BC, CA, AB respectively. P, Q, R are points on the sides BC, CA, AB respectively and P', Q', R' are points on BC, CA, AB respectively such that $PD = DP', QE = EQ', RF = FR'$. Let G_1 be the centroid of triangle AQR and G'_1 the centroid of triangle $AQ'R'$, with G_2, G'_2, G_3, G'_3 similarly defined. Prove that the areas of triangles $G_1G_2G_3$ and $G'_1G'_2G'_3$ are equal.

11. ABC is a triangle and D, E, F are the midpoints of BC, CA, AB respectively. P, Q, R are points on the sides BC, CA, AB respectively and P', Q', R' are points on BC, CA, AB respectively such that $PD = DP'$, $QE = EQ'$, $RF = FR'$. Let O_1 be the circumcentre of triangle AQR and O_1' the circumcentre of triangle $AQ'R'$, with O_2, O_2', O_3, O_3' similarly defined. Prove that triangles $O_1O_2O_3$ and $O_1'O_2'O_3'$are congruent.

12. ABC is a triangle and D, E, F are the midpoints of BC, CA, AB respectively. P, Q, R are points on the sides BC, CA, AB respectively and P', Q',R' are points on BC, CA, AB respectively such that $PD = DP', QE = EQ', RF = FR'$. Let H_1 be the orthocentre of triangle AQR and $H_1'$the orthocentre of triangle $AQ'R'$, with H_2, H_2', H_3, H_3' similarly defined. Prove that the areas of triangles PQR, $P'Q'R'$, $H_1H_2H_3$ and $H_1'H_2'H_3'$ are equal.

13. ABC is a triangle and D, E, F are the midpoints of BC, CA, AB respectively. P, Q, R are points on the sides BC, CA, AB respectively and P', Q', R' are points on BC, CA, AB respectively such that $PD = DP', QE = EQ', RF = FR'$. Let T_1 be the nine-point centre of triangle AQR and T_1' the nine-point centre of triangle $AQ'R'$, with T_2, T_2', T_3, T_3' similarly defined. Prove that the areas of triangles $T_1T_2T_3$ and $T_1'T_2'T_3'$ are equal.

14. Prove or disprove that the result of Problem 10 is true if 'centroid' is replaced by any point on the Euler line (such as deLongchamps point).

15. Let ABC be a triangle and P any point in the plane of ABC, but not on its sides or at its vertices. Let H, J, K be the orthocentres of triangles PBC, PCA, PAB respectively. Prove that $[HJK] = [ABC]$.

16. Prove that the result of Problem 15 is true if 'orthocentre' is replaced by 'centroid'.

In problems 17 - 22 $ABCD$ is a cyclic quadrilateral and $AC \wedge BD = E$, $AB \wedge DC = F$ and $AD \wedge BC = G$, so that EFG is the diagonal point triangle. We refer to triangles AFG, BFG, CFG, DFG as Set 1, and the triangles ACF, BDF, ACG, BDG as Set 2.

17. Let A', B', C', D' be the respective centroids of the triangles in Set 1. Prove that $A'B'C'D'$ is homothetic with $ABCD$ and one-third the size.

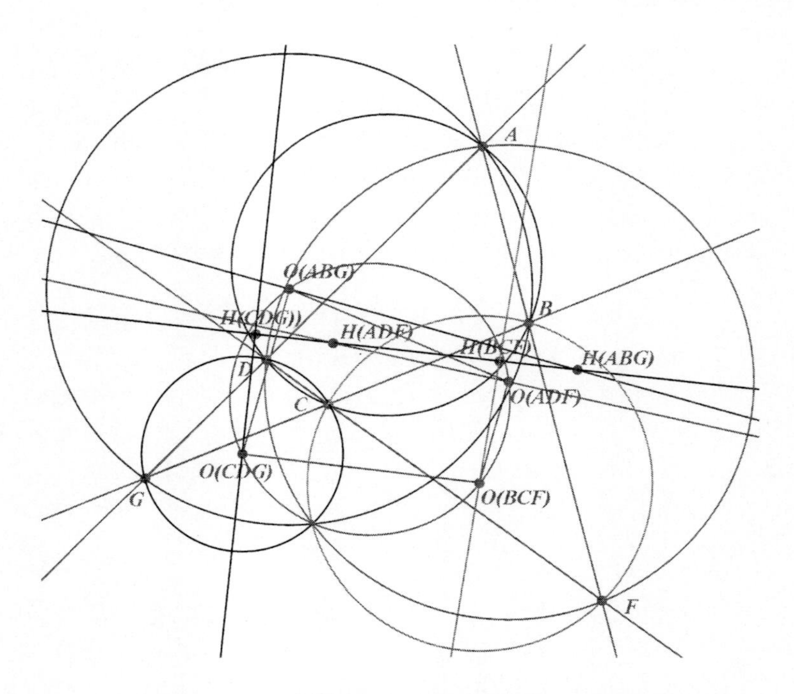

Figure 18.1: Configuration for Problem 23

18. Let $A'B'C'D'$ be the respective orthocentres of the triangles in Set 1. Prove that $A'B'C'D'$ is a cyclic quadrilateral with angles the same as those of $CDAB$.

19. Let A', B', C', D' be the respective nine-point centres of the triangles in Set 1, then $A'D'B'C'$ is an isosceles trapezium whose circumcircle passes through the midpoint of FG. (This is a conjecture.)

20. Let A', B', C', D' be the respective incentres of the triangles in Set 1. Prove that $A'B'C'D'$ is a quadrilateral with opposite angles summing to an odd multiple of $90°$.

21. Let A', B', C', D' be the respective centroids of the triangles in Set 2, then $A'B'C'D'$ is a parallelogram.

22. Let A', B', C', D' be the respective nine-point centres of the triangles in Set 2. Prove that if AC is perpendicular to BC, then A', B' lie on AC and C', D' lie on BD.

23. See Fig. S.1. $ABCD$ is a cyclic quadrilateral and $AB \wedge CD = F$, $AD \wedge BC = G$. Prove that orthocentres of the four triangles CDG, ADF, BCF, ABG are collinear and that their circumcentres form an isosceles trapezium. Prove also that the circumcircles of the four triangles and the circumcircle of the isosceles trapezium meet at a point.

24. A parabola is given with focus S and directrix d. a and b are tangents that meet on d. c is another tangent and $ac = X$, $bc = Y$ and $ab = Z$. Prove that X, Y, Z, S are concyclic.

25. Two intersecting circles have centres A, B. A common tangent touches them at C, D respectively and E is one of their points of intersection. G is the midpoint of EF. The lines AG and BD meet at P; BG and AC meet at Q. Prove that PQ is parallel to CD.

26. Three circles C_1, C_2, C_3 are drawn with a common point P. Let Q, R, S be the other points of intersection of C_2, C_3 and C_3, C_1 and C_1, C_2 respectively. A is a variable point on C_1. AR meets C_3 again at C and AS meets C_2 again at B. Prove that BC passes through Q. Prove also that as A varies on C_1, the Euler line of triangle ABC passes through a fixed point X and that the locus of the circumcentre of triangle ABC is a circle passing through P and X.

27. Let ABC be an isosceles triangle with $AC = BC$, whose incentre is I. Let P be a point on the circumcircle of triangle AIB lying inside the triangle ABC. The lines through P parallel to CA and CB meet AB at D, E respectively. The line through P parallel to AB meets CA and CB at F and G respectively. DF and EG meet at X and DG and EF meet at Y. Prove that D, E, X, Y are concyclic.

28. Let S be a circle and X an external point with tangents XC, XD to S. Let E be a point on S nearer to X than C or D. Let A, B be the reflections of E in XC, XD respectively. EF is a diameter of S. Let G be the point on EF produced such that $EF = FG$. The following circles are drawn: AEC, AEB, BED, GEA, GEB. Circle GEA meets circle BED again at P and circle GEB meets circle AEC again at Q. Prove that circles ACE and BED touch at E, that XE is a diameter of circle AEB and that circle PEQ touches S at E.

29. ABC is a triangle and l a transversal. Parallel lines are drawn through A, B, C to meet l at A', B', C' respectively. Lines are drawn through A', B', C' parallel to BC, CA, AB respectively. Show that these three lines form a triangle congruent to ABC?

30. ABC is a triangle and F_1, F_2 are its two Fermat points. Prove that the Euler lines of the six triangles $F_1BC, F_1CA, F_1AB, F_2BC, F_2CA, F_2AB$ are concurrent at a point X and that the circumcentres of these triangles fall three and three at points equidistant from X. Prove also that AF_1 is parallel to the Euler line of triangle F_1BC, with similar properties for the other five triangles.

31. Let ABC be a triangle and P a point not on its sides. Let l_1, l_2, l_3 be the Euler lines of triangles BPC, CPA, APB respectively. Cabri indicates that when P lies at certain points then l_1, l_2, l_3 are concurrent. It is conjectured that this is the case when P is

    (i) the circumcentre (trivial),

    (ii) the orthocentre (when the point of concurrence is the nine-point centre of ABC),

    (iii) the Fermat points (see Problem 30),

    (iv) the isodynamic or Hessian points,

    (v) the incentre

    (vi) the excentres.

    Supply proofs where possible.

32. Let ABC be a triangle, with vertices labelled in an anticlockwise direction, and l a transversal. Draw parallel lines through A, B, C making an angle $\theta$with l and meeting l at points A', B', C' respectively. Draw lines through A', B', C' making angles $180° - \theta$ with the increasing directions of BC, CA, AB respectively (that is the directions from B to C etc.). Show that these lines are concurrent.

33. This refers to the same configuration as Problem 32, but with $\theta = 90°$. Then the point L of concurrency is called the orthopole of the transversal L with respect to triangle ABC. Let l be a variable transversal passing through the circumcentre O of triangle ABC and rotating about

O. Prove that the locus of L is the nine-point circle of ABC. Prove also that if O is replaced by H the orthocentre, then the locus of L is the ellipse through the feet of the altitudes of ABC and having centre H.

34. In triangle ABC, the orthocentre and the incentre both lie on the circle OBC, where O is the circumcentre. Calculate $\angle BAC$.

35. The configuration is the same as in Problem 31, but now l_1, l_2, l_3 are the OS lines joining the circumcentre and the symmedian point. Prove or disprove that these lines are concurrent when P is

    (i) the circumcentre,
    (ii) the Fermat point(s),
    (iii) the incentre,
    (iv) the isodynamic or Hessian points,
    (v) the orthocentre,
    (vi) the excentres.

36. Let ABC be a triangle with circumcentre O. Prove that the circumcentres of triangles OBC, OCA, OAB lie on a circle whose centre lies on the Euler line of ABC.

37. Let ABC be an equilateral triangle and P a point in the plane not lying on the sides. Prove that the Euler lines of triangles BPC, CPA, APB are concurrent, and that the OS lines of these triangles are also concurrent. Prove that the points of concurrency and the centroid of ABC are collinear.

38. Let ABC be an acute-angled triangle and let the line through A parallel to the Euler line of ABC meet BC (possibly extended) at P. Prove that the Euler lines of triangle ABC, CPA and APB are concurrent.

39. Let $ABCD$ be a cyclic quadrilateral with diagonals AC and BD meeting at E. Prove that the Euler lines of triangles AEB, BEC, CED, DEA are concurrent.

40. Let $ABCD$ be a cyclic quadrilateral with diagonal point triangle EFG. Prove that the circumcentres of the four triangles FGA, FGB, FGC, FGD are collinear and that their four orthocentres are concyclic. Prove

that the same is true for the set of triangles GEA, GEB, GEC, GED and for the set of triangles EFA, EFB, EFC, EFD. Prove also that the three lines of collinear circumcentres are concurrent at K, the circumcentre of the diagonal point triangle EFG, and prove or disprove that K is the radical centre of the three circles of concyclic orthocentres.

41. $ABCD$ and $C'B'A'D'$ are two parallelograms with corresponding sides (such as AB and $C'B'$) parallel. Prove that AA', BB', CC', DD' are concurrent if, and only if, AC', BD', CA', DB' are concurrent.

42. Triangles ABC and $A'B'C'$ are such that AB is parallel to $B'A'$, BC is parallel to $C'B'$ and CA is parallel to $A'C'$. It is given that BC', CB', AA' are concurrent. Prove that BB', CC', AA' are also concurrent. If, furthermore AC', BB', CA' are concurrent, prove that AB', BA', CC' are concurrent. When there is a 4-fold perspective, show that one of the vertices of perspective is the centroid of the other three.

43. Let ABC be a non-isosceles triangle. The internal bisector of angle A meets the line through C parallel to the internal bisector of angle B at the point L. Points M and N are similarly defined by cyclic change of letters A, B, C. Prove that a conic exists that passes through the points A, B, C, L, M, N.

44. Prove that the result of Problem 43 is also true if 'internal bisector' is replaced by 'median'.

45. Prove that the result of the previous two problems is true if 'internal bisectors' or 'medians' are replaced by any set of Cevians through a point P. Determine the unique point P for which the conic is the circumcircle.

46. Two isosceles trapezia $ABCD$ and $A'B'C'D'$ with their parallel sides all parallel, and which if inscribed in a circle have their vertices in the order $AA'BB'C'CD'D$. Prove that if AB', BA', CD', DC' are concurrent, then AC', CA', BD', DB' are also concurrent.

47. $ABCD$ is a cyclic quadrilateral in which no pair of opposite sides are parallel. AB meets DC at F and AD meets BC at G. A line through G meets circle $ABCD$ at D' and A'. FD' meets the circle again at C' and

FA' meets the circle again at B'. Prove that G, C' and B' are collinear. Prove further that $AC, BD, A'C', B'D'$ are concurrent.

48. A circle centre I, the incentre of triangle ABC meets BC at A_1, A_2 with A_1 nearer B. Points B_1, B_2 on CA and C_1, C_2 on AB are similarly defined. Lines B_1C_2, C_1A_2, A_1B_2 are denoted by l, m, n. Define $A' = m \wedge n, B' = n \wedge l, C' = l \wedge m$. Prove that

    (i) AA', BB', CC' are concurrent at Ge, the Gergonne point of ABC;

    (ii) This point is the symmedian point of triangle $A'B'C'$;

    (iii) If l, m, n are concurrent, then the point of concurrence is Ge;

    (iv) IGe is the same line as the line joining the circumcentre and the symmedian point of triangle $A'B'C'$.

49. State and prove the converse of the result of Problem 48.

50. With the same configuration as in Problem 48, but with circle centre any point P in the plane of triangle ABC. Prove that AA', BB', CC' are concurrent.

51. Let ABC be a triangle with incentre I. A circle with centre I cuts the (possibly extended) sides BC at P and Q, CA at S and T and AB at U and V. Prove that the triangle whose vertices are the circumcentres of triangles IPQ, IST, IUV has itself got circumcentre I.

52. Let ABC be a triangle and O any point in the plane of ABC other than on the sides. A circle with centre O cuts the (possibly extended) sides BC at P and Q, CA at S and T, AB at U and V with PQ in the same direction as BC, ST in the same direction as CA and UV in the same direction as AB. Consider the triangle XYZ formed by the lines TU, VP, QS with $X = VP \wedge QS$ etc.. Prove that triangles ABC and XYZ are in perspective.

53. Prove the result of Problem 52 is true if the circle is replaced by a conic.

54. ABC and XYZ are two triangles in perspective. They meet in nine points, three of which lie on the Desargues axis of perspective. Prove that the other six points lie on a conic.

55. Let ABC be a triangle. Points P, Q lie on BC with PQ in the same direction as BC. S and T lie on CA with ST in the same direction as CA. U, V lie on AB with UV in the same direction as AB. The lines PT, QU, SV form a triangle $A'B'C'$ with $A = PT \wedge QU$, $B = PT \wedge SV$ and $C = QU \wedge SV$. Suppose now that AA', BB', CC' are concurrent. Prove that P, Q, S, T, U, V lie on a conic.

56. A fixed rectangular hyperbola H is given and B and C are two fixed points lying on it. A is a variable point on H. Show that the locus of the nine-point circle of triangle ABC as A varies is an intersecting coaxal system of circles. What two points lie on all these circles?

57. Suppose now we have the same configuration as in Problem 56, but that the hyperbola is not rectangular. Show that the locus of the nine-point centre is a hyperbola (and not a line).

58. Suppose now we have the same configuration as in Problem 56. Prove that the locus of the orthocentre of triangle ABC is the rectangular hyperbola H.

59. Suppose now we have the same configuration as in Problem 56. Prove that the locus of the centroid of triangle ABC is a rectangular hyperbola H'. Let O, O' be the centres of these hyperbolas and let C, C' be parallel chords of H, H' through O, O' respectively. Prove that the length of C is three times the length of C'.

60. BCD is a fixed triangle with circumcentre O. A is a variable point on the circle. The lines joining the midpoints of opposite sides meet at X. Prove that the locus of X as A varies is a circle.

61. Suppose now we have the same configuration as in Problem 60, but that X is now the point where the four lines through the midpoints of the sides perpendicular to the opposite sides meet. Show that the locus of X as A varies is also a circle.

62. A is a fixed point on a fixed circle with centre O. B is a variable point on the circle. Prove that the locus of the centroid of triangle ABO is a circle.

63. Suppose now we have the same configuration as in Problem 62. Prove that the locus of the symmedian point of triangle ABO is an ellipse.

64. Suppose we now have the same configuration as in Problem 62. Prove that loci of the isodynamic points are lines passing through A.

65. B and C are fixed points on a fixed circle and A is a variable point on the circumference of the circle. Prove that the locus of the symmedian point of triangle ABC is an ellipse.

66. Suppose now that we have the same configuration as in Problem 65. Prove that the loci of the isodynamic points are circles passing through B and C.

67. ABC is a right-angled isosceles triangle with $AB = AC$. Points L, M, N lie on BC, CA, AB respectively and are such that $BL = LC, AN = 3NB, MC = 3AM$. Prove that the circumcircle of triangle LMN passes through A and the midpoint of BL.

68. Four mutually intersecting circles are given. Prove that if their centres are concyclic, then the four radical centres define a similar cyclic quadrilateral. What theorem holds if the four centres just define some convex quadrilateral?

69. Let $ABCD$ be a fixed convex quadrilateral and X a fixed point not lying on any of its sides or their extensions. Prove that for every circle centre A, there exist three circles, centres B, C, D such that the four circles have a common radical centre X.

70. Let ABC be a triangle and LMN the first Brocard triangle (as defined in Example 10.2.1). Prove that lines through A, B, C parallel to MN, NL, LM respectively meet at the Steiner point X of ABC (the intersection other than the vertices of the circumcircle and the outer Steiner ellipse).

71. With the same configuration as in Problem 70, prove that the lines through A, B, C perpendicular to the sides MN, NL, LM respectively are concurrent at a point (called the Tarry point) diametrically opposite to X on the circumcircle of ABC.

72. Prove that the outer Steiner ellipse has four times the area of the inner Steiner ellipse.

73. Let ABC be a triangle with circumcentre O and let l, m, n be the perpendicular bisectors of the sides BC, CA, AB respectively. Let X be any point in the plane of the triangle, distinct from O. Draw the circle centre X through O to meet l, m, n at P, Q, R respectively. Prove that the lines through A, B, C parallel to QR, RP, PQ respectively are concurrent at a point U on the circumcircle of ABC. Prove that the Wallace-Simson line of U is parallel to OX.

74. Let ABC be a triangle with orthocentre H and let l, m, n be the perpendicular from A, B, C to the sides BC, CA, AB respectively. Let X be any point in the plane of the triangle, distinct from H. Draw the circle centre X through H to meet l, m, n at P, Q, R respectively. Prove that the lines through A, B, C parallel to QR, RP, PQ respectively are concurrent at a point U on the circumcircle of ABC. Prove that the Wallace-Simson line of U is parallel to OX.

75. Let ABC be a triangle with orthocentre H. Let D, E, F denote the centres of the circles HBC, HCA, HAB respectively. Prove that A, B, C, D, E, F lie on an ellipse and locate its centre.

76. This problem refers to Part 16.7. Let $UVWH$ be a Hagge circle and suppose it meets the altitudes through A, B, C at X, Y, Z respectively. Prove that UX, VY, WZ are concurrent. Locate the point of concurrence P.

77. With the same configuration as in Problem 76, suppose VW meets AH at L, WU meets BH at M and UV meets CH at N. Prove that L, M, N, P are collinear.

78. ABC is a triangle (which is not equilateral) with circumcentre O and orthocentre H. The point K lies on OH so that O is the midpoint of HK. AK meets BC at X, and Y, Z are the perpendiculars from X on to the sides AC, AB respectively. Prove that AX, BY, CZ are concurrent or parallel.

79. Let C be a circle through points P, Q, R and let l_1, l_2, l_3 be three chords PJ, QJ, RJ all passing through a point J lying on C. Suppose further that Y is any other point in the plane of C, but not lying on C and let PY, QY, RY meet C again at U, V, W respectively. Now let VW meet

l_1 at L, WU meet l_2 at M, UV meet l_3 at N. Then LMN is a straight line passing through Y.

80. Let H, K be two points in the plane of triangle ABC. AH, BH, CH meet BC, CA, AB at D, E, F respectively. DK, EK, FK meet EF, FD, DE at P, Q, R respectively. Prove that if QR passes through A, then RP passes through B and PQ passes through C. *Hint: Use projective co-ordinates.*

81. Taking Problem 80 into the Euclidean plane, let H be the orthocentre of ABC and K be the orthocentre of DEF. Prove or disprove that if QR passes through A, then AP passes through the nine-point centre of ABC.

82. $ABCD$ is a cyclic quadrilateral. A point X is chosen not on circle $ABCD$ and lines AX, BX, CX, DX meet circle $ABCD$ again at P, Q, R, S respectively. The diagonals of $ABCD$ meet at E and the diagonals of $PQRS$ meet at F. Prove that E, F, X are collinear. Prove that other pairs of diagonal points are also collinear with X.

83. Let P be a point in the plane of a triangle ABC, but not on its sides or its circumcircle and let P_g be its isogonal conjugate. Let AP, BP, CP meet the circumcircle at D, E, F respectively and let AP_g, BP_g, CP_g meet the circumcircle at L, M, N respectively. Prove that DL is parallel to BC and that the three lines DL, EM, FN form a triangle XYZ homothetic with ABC.

84. With the same configuration as in Problem 83 let the centre of similarity between the two triangles be denoted by Q. Suppose P and P_g have areal co-ordinates (u, v, w) and $(a^2/u, b^2/v, c^2/w)$, prove that the co-ordinates of Q are
$$\left(\frac{a^2}{u(AP)^2}, \frac{b^2}{v(BP)^2}, \frac{c^2}{w(CP)^2}\right).$$

85. Let P and P_g be a point and its isogonal conjugate in triangle ABC. Prove that
$$\frac{[CPA](BP)}{[APB](CP)} = \frac{b(BP_g)}{c(CP_g)}.$$

86. Let ABC be a triangle with circumcentre O and orthocentre H. P, Q, R are the reflections of O in the sides BC, CA, AB respectively. Locate the circumcentre of triangle PQR.

87. Let $ABCDEF$ and $ABCPQR$ be two conics circumscribing triangle ABC. Suppose further that all the following conics are well defined. Let conics $ABCER$, $ABCFQ$ meet at X, let conics $ABCDR$, $ABCFP$ meet at Y, and let conics $ABCDQ$, $ABCEP$ meet at Z. Prove that points A, B, C, X, Y, Z lie on a conic.

88. Let ABC be a triangle, K a point lying inside ABC and k a line in the same plane as ABC and K, but not cutting any of the segments BC, CA, AB. Let AK, BK, CK meet k at X, Y, Z respectively. Denote $CY \wedge BZ$ by D, $AZ \wedge CX$ by E, $BX \wedge AY$ by F. Prove that A, B, C, D, E, F lie on a conic. Is it possible to move K, keeping ABC and k fixed, so that D lies on AX, E lies on BY and F lies on CZ?

89. Let D be a point on the circumcircle of a triangle ABC and let A^*, B^*, C^* be the points of the circle diametrically opposite to A, B, C respectively. Then, if AD, BD, CD meet B^*C^*, C^*A^*, A^*B^* at L, M, N respectively, prove these points lie on a line through the circumcentre O.

90. If $ABCD$ is a cyclic quadrilateral, centre O, prove that there is a line through O meeting the sides AD, BD, CD, BC, CA, AB (none of which passes through O) in points L, M, N, L', M', N' such that O is the midpoint of LL', MM', NN'.

91. Triangle ABC is rotated by a given angle about its circumcentre to provide a second triangle TUV. The six points $AT \wedge BV$, $AT \wedge CU$, $BU \wedge CT$, $BU \wedge AV$, $CV \wedge BT$, $CV \wedge AU$ are constructed. Prove that these six points lie three and three on two parallel lines.

92. Let ABC be a triangle with $\angle BAC = 60°$. Denote $AB = c$ and $AC = b$ and let O be the circumcentre and H the orthocentre. Let OH meet the sides AB and AC at P and Q respectively. Prove that $AP = AQ = (1/3)(b + c)$. Prove also that if AD is the altitude from A, then $BD/DC = (c(2c - b))/(b(2b - c))$.

93. ABC is a triangle with circumcentre O. AOA^*, BOB^*, COC^* are diameters of the circumcircle. The tangents at A^*, B^*, C^* meet BC, CA, AB respectively at L, M, N. Prove that L, M, N are collinear.

94. Let ABC be a triangle and He, He^* the two Hessian (isodynamic points). Let the inner Steiner ellipse have foci S and S^*. Prove that the circles $AHeHe^*$, $BHeHe^*$, $CHeHe^*$, SS^*HeHe^* form part of a coaxal system.

95. Z^*AY, XY^*C, BZX^* are three parallel lines such that BXZ^*, ZY^*A, X^*CY are also parallel. Prove that XX^*, YY^*, ZZ^* are concurrent.

96. $ABCD$ is a cyclic quadrilateral in which BA produced and CD produced meet at F. Prove that the Euler lines of triangles FAD, FBD, FAC, FBC are concurrent.

97. Two orthogonal circles, centre C and D intersect at A and B. P is a variable point, other than A or B, on the circle centre C; PA, PB meet the circle with centre D at Q, R respectively. Prove that the triangle PQR has fixed nine-point centre N and calculate CN/ND. Prove further that:

    (i) the circumradius of triangle PQR is equal to CD;

    (ii) the locus, as P varies, of the circumcentre, centroid, nine-point centre and orthocentre of triangle PQR are circles, and locate the position of these circles.

98. A triangle ABC is given inscribed in a circle and He is the first isodynamic point of ABC. AHe, BHe, CHe meet the circle again at X, Y, Z respectively. Prove that triangle XYZ is equilateral.

99. Let a, b, c be the sides of a triangle. Prove that the sides x, y, z of any triangle with the same Brocard angle are given by $x^2 = a^2t^2 - (a^2 + b^2 - c^2)t + b^2$, with y, z given by cyclic change and where t is a real parameter (which may be infinite).

100. Let ABC be a triangle with P-centroid G. Let P be a fixed point and suppose the P-pedal triangle is LMN. Let Q be a variable point with P-pedal triangle DEF. Find the locus of Q such that $LMNDEF$ is a conic.

101. ABC is a triangle, circumcentre O, isodynamic point He and symmedian point S. PQ is a chord through S. PHe, QHe meet the circumcircle again at U, V. Prove that UV passes through O.

102. Let ABC be a triangle and S a rectangular hyperbola passing through its incentre and excentres. Prove that ABC is self-polar with respect to S. Now let DEF be another triangle that is self-polar with respect to S. Prove that its incentre and excentres lie on S.

103. $ABCD$ is a quadrangle and O is the intersection of AC and BD. P is a point on CD. PA meets BC at Q and OP meets DQ at R. Prove that the locus of R as P varies on CD is a conic.

104. ABC is a triangle with centroid G. P is a general point in the plane of ABC. Lines through P parallel to the medians are drawn to meet BC at D, CA at E and AB at F.

    (i) If K is the centroid of triangle DEF prove that K is the midpoint of PG;

    (ii) Prove that DEF is a straight line if, and only if, P lies on the outer Steiner ellipse of triangle ABC;

    (iii) Prove that $BD^2/a^2 + CE^2/b^2 + AF^2/c^2 = DC^2/a^2 + EA^2/b^2 + FB^2/c^2$.

105. Prove that the ellipse with centre the Spieker centre, which passes through the midpoints of the sides intersects the altitudes at points X, Y, Z where $AX = BY = CZ = r$, the radius of the incircle.

106. Let ABC be a triangle inscribed in a circle Γ and let V be a point on the tangent at C. VA, VB meet Γ at D, E respectively and AB, DE meet at N. Let VD meet BC at L and VE meet CA at M. Prove that L, M, N are collinear.

Chapter 19

Appendix

19.1 Projective transformations of the plane

19.1.1 Change of triangle of reference

Let triangle $A_1A_2A_3$ with $A_1(1,0,0), A_2(0,1,0), A_3(0,0,1)$ be a triangle of reference and P be a point with co-ordinates $P(p_1,p_2,p_3)$ with respect to $A_1A_2A_3$. We now select points B_1, B_2, B_3 with co-ordinates with respect to $A_1A_2A_3$ as $B_1(b_{11},b_{21},b_{31}), B_2(b_{12},b_{22},b_{32}), B_3(b_{13},b_{23},b_{33})$. We enquire what the co-ordinates of P are with respect to the triangle $B_1B_2B_3$. Suppose they are (q_1,q_2,q_3).

Using an obvious notation we have

$$P = p_1A_1 + p_2A_2 + p_3A_3 = q_1B_1 + q_2B_2 + q_3B_3$$
$$= q_1b_{k1}A_k + q_2b_{k2}A_k + q_3b_{k3}A_k = q_jb_{kj}A_k = p_kA_k,$$

using the summation convention. Since the A_k are independent, it is permissible to equate coefficients showing that $p^{\mathrm{T}} = Bq^{\mathrm{T}}$ or $q^{\mathrm{T}} = B^{-1}p^{\mathrm{T}}$.

If we define A_1, A_2, A_3 to have co-ordinates

$$A_1(a_{11},a_{21},a_{31}), A_2(a_{12},a_{22},a_{32}), A_3(a_{13},a_{23},a_{33})$$

with respect to triangle $B_1B_2B_3$, then similar working gives $P = p_kA_k = p_ka_{lk}B_l = q_lB_l$ so that $q^{\mathrm{T}} = Ap^{\mathrm{T}}$. Note that $A = B^{-1}$ as one would expect.

Implicit in these statements, since we can move either way, is that A and B are non-singular matrices. Note also, since we are dealing with homogeneous co-ordinates, that we may use the matrix kA rather than A, where k is any non-zero scalar.

19.1.2 Projective transformations

The projective transformations of the real projective plane consist of the elements of the group of 3×3 real non-singular matrices. Since the co-ordinates of a point are undetermined up to a non-zero constant multiplier, this means that a projective transformation exists that map any four points into any four points (no three being collinear in either case). If the points concerned have complicated expressions for their projective co-ordinates then the matrix concerned contains correspondingly difficult entries. In practice three of the points are often the vertices of the triangle of reference.

An example is sufficient to illustrate the construction of one of these matrices. Suppose we consider a general element of the group of automorphisms that preserve the Brocard porism. Thus we want to map the points A, B, C, S to T, U, V, S where ABC and TUV are two triangles in the Brocard porism and S the symmedian point. Then we may take

$$A(1,0,0), B(0,1,0), C(0,0,1),$$

$$T(-a^2t(1-t), b^2(1-t), c^2t), U(a^2t, -b^2t(1-t), c^2(1-t)),$$
$$V(a^2(1-t), b^2t, -c^2t(1-t)),$$

and

$$S(a^2, b^2, c^2).$$

Then it may be verified that the matrix required has its rows as follows:

$$-t(1-t), (1-t)a^2/b^2, ta^2/c^2,$$

$$tb^2/a^2, -t(1-t), (1-t)b^2/c^2$$

and

$$(1-t)c^2/a^2, tc^2/b^2, -t(1-t).$$

In the complex projective plane the only difference is that complex 3×3 non-singular matrices may be used.

19.1.3 The use of the complex projective plane

A few instances occur in which the complex projective plane is used without this being evident from the context. When a conic with equation $x^2+y^2+z^2 = 0$ is mentioned, it is surely evident that points lying on it must be regarded as having complex co-ordinates. However, it is less obvious that when one uses other special equations that this might also be the case. Consider the conic with equation $y^2 = zx$ also introduced in Part 3.7. In the real projective plane it is evident that the point $Y(0,1,0)$ is external to this conic, being the intersection of the two real tangent lines at $X(1,0,0)$ and $Z(0,0,1)$. Points exist through which no real tangents can be drawn. When dealing with conics in the real plane these points may be regarded as internal. One may pose a problem, which begins 'A chord PQ of a conic always passes through a fixed point B'. Then one may ask whether it is sufficiently general to take the conic as having equation $y^2 = zx$ and to take B to be the point Y. At first sight this does not seem to be general enough. However, if you work in the complex plane and choose B to be any point, then there are always (possibly complex) tangents through B that meet the curve at (possibly complex) points C and A. Then you can set up a complex projective transformation that takes A, B, C to X, Y, Z respectively, leaving the unit point invariant. The equation of the conic remains unchanged but the problem is transformed so that Y replaces B. It follows that by working over the complex field, it is the case that Y is a sufficiently general point to deal with the problem at hand.

Bibliography

[1] C. J. Bradley, *Challenges in Geometry*, Oxford, 2005.

[2] C. J. Bradley, "A theorem on concurrent Euler lines", *Math. Gaz.*, 90 (2006) 412-416.

[3] C. J. Bradley, "Cyclic quadrilaterals", *Math. Gaz.*, 88 (2004) 417-431.

[4] C. J. Bradley, "Hexagons with opposite sides parallel", *Math. Gaz.*, 90 (2006) 57-67.

[5] C. J. Bradley and J. T. Bradley, "Countless Simson Line configurations", *Math. Gaz.*, 80 (1996) 314-321.

[6] C. J. Bradley and G. C. Smith, "The Locations of Triangle Centers", *Forum Geom.*, 6 (2006) 57-70.

[7] C. J. Bradley and G. C. Smith, "The Locations of the Brocard Points", *Forum Geom.*, 6 (2006) 71-77.

[8] C. J. Bradley and G. C. Smith, "Stationary Triangle Porisms and Brocard Geometry", submitted for publication.

[9] C. J. Bradley and G. C. Smith, "Hagge circles and isogonal conjugation", to be published in *Math. Gaz.* (2007).

[10] C. J. Bradley and G. C. Smith, "On a construction of Hagge", submitted for publication.

[11] C. J. Bradley and G. C. Smith, "Projective generalisations of Euclidean theorems", submitted for publication.

[12] D. A. Brannan, M. F. Esplen, J. J. Gray, *Geometry,* Cambridge University Press, 1999.

[13] G. S. Carr, *A Synopsis of Elementary Results of Pure Mathematics,* Francis Hodgson, Cambridge, Macmillan & Bowes, 1886.

[14] H. S. M. Coxeter, *Introduction to Geometry,* John Wiley & Sons, 1969.

[15] L. Cremona, *Elements of Projective Geometry,* Dover 2005.

[16] V. de Silva, private communication.

[17] C. V. Durell, *Modern Geometry,* Macmillan, 1946.

[18] L. Euler, "Solutio facili problematum quorundam geometricorum difficillimorum", *Novi. Comm. Acad. Scie Petropolitanae* 11 (1765) reprinted in *Opera omnia, serie prima,* Vol 26 (ed. by A Speiser),139-157. (Latin)

[19] A. D. Gardiner and C. J. Bradley, *Plane Euclidean Geometry: Theory and Problems,* UKMT, 2005.

[20] A. P. Guinand, "Tritangent centers and their triangles", *Amer. Math. Monthly,* 91 (1984) 290-300.

[21] K. Hagge, "Der Fuhrmannsche Kreis und der Brocardsche Kreis als Sonderfälle eines allgemeineren Kreises", *Zeitschrift für Math. Unterricht,* 38 (1907) 257-269. (German)

[22] L. Hahn, *Complex Numbers & Geometry,* The Mathematical Association of America, 1994.

[23] R. Honsberger, *Episodes in Nineteenth and Twentieth Century Euclidean Geometry,* The Mathematical Association of America, 1995.

[24] C. Kimberling, "Triangle Centers and Central Triangles, *Congr. Num.,* (1998) 1-285.

[25] E. A. Maxwell, *The methods of Plane Projective Geometry based on the use of General Homogeneous Co-ordinates,* Cambridge, 1957.

[26] J. J. Milne, *Weekly Problem Papers,* Paper LVI no. 5, Macmillan (1886).

[27] J. J. Milne, *Weekly Problem Papers,* Paper XIX no. 5, Macmillan (1886).

[28] D. Monk, private communication.

[29] D. Pedoe, Geometry, *A Comprehensive Course*, Cambridge, 1970.

[30] A. M. Peiser, "The Hagge circle of a triangle", *Amer. Math. Monthly*, 49 (1942) 524-527.

[31] J. R. Silvester, *Geometry Ancient & Modern*, Oxford, 2001.

[32] G. C. Smith, "Statics and the moduli space of triangles", *Forum Geom.*, 5 (2005) 181-190.

[33] G. C. Smith, private communication.

[34] H. A. W. Speckman, "Over Omgekeerd Gelijkvormige Driehoeken", *Perspectief Gelegen, Nieuw Archief*, (2) 6 (1905) 179-188. (Dutch)

[35] A. Várilly, "Location of incenters and Fermat points in variable triangles", *Math. Mag.*, 7 (2001) 12-129.

[36] F. van Lamoen, "Pl-perpendicularity", *Forum Geom.*, 1 (2001) 151-160.

[37] J. Wolstenholme, *Mathematical Problems*,[1] Macmillan 1891.

[38] F. E. Wood, "Similar-Perspective Triangles", *Amer. Math. Monthly*, 36:2 (1929) 67-73.

[1] Problems labelled (JW) appear in this book.

Index